Stochastic Processes in Polymeric Fluids

Springer
Berlin
Heidelberg
New York
Barcelona
Budapest
Hong Kong
London
Milan
Paris
Santa Clara
Singapore
Tokyo

Hans Christian Öttinger

Stochastic Processes in Polymeric Fluids

Tools and Examples for Developing Simulation Algorithms

With 34 Figures and 3 Tables

Springer

Prof. Dr. Hans Christian Öttinger
Swiss Federal Institute of Technology Zurich
Institute of Polymers
CH - 8092 Zurich
Switzerland

The Fortran programs are available via ftp
from ftp.springer.de/pub/chemistry/polysim/

ISBN 3-540-58353-X Springer-Verlag Berlin Heidelberg New York

CIP-data applied for
Die Deutsche Bibliothek - CIP-Einheitsaufnahme
Öttinger, Hans Christian: Stochastic processes in polymeric fluids : tools and examples for developing simulation algorithms / Hans Christian Öttinger. -
Berlin ; Heidelberg ; New York ; Barcelona ; Budapest ; Hong Kong ; London ; Milan ; Paris ; Santa Clara ; Singapore ; Tokyo : Springer 1996
ISBN 3-540-58353-X (Berlin ...)
ISBN 0-387-58353-X (New York ...)

Typesetting: Camera-ready by author
SPIN:10114990 02/3020-5 4 3 2 1 0 - Printed on acid-free paper

To Ricarda

Preface

> A SPECTER is haunting the scientific world—the specter of computers. All the powers of traditional science have entered into a holy alliance to exorcise this specter: puristic theoreticians and traditionalistic experimentalists, editors and referees of prestigious journals, philosophers of science and mathematicians. Where is a pioneering computer simulation that has not been decried as unreliable by its opponents in power?
>
> **The Computer Manifesto**

As a result of the enormous progress in computer technology made during the last few decades, computer simulations have become a very powerful and widely applicable tool in science and engineering. The main purpose of this book is a comprehensive description of the *background and possibilities* for the application of *computer simulation techniques in polymer fluid dynamics*. Modeling and understanding the flow behavior of polymeric liquids on the kinetic theory level is not merely a great intellectual challenge but rather a matter of immense practical importance, for example, in connection with plastics manufacture, processing of foods, and movement of biological fluids.

The classical computer simulation technique for static problems in statistical mechanics is the *Monte Carlo method* developed in the early 1950s. The name of this method underlines how unusual and strange the idea of using random numbers in the exact sciences is at first glance. However, the Monte Carlo method is a rigorous and efficient means for evaluating moments and static spatial correlation functions for given probability distributions. It is based on the theory of Markov chains. An ensemble of properly distributed configurations is created by Monte Carlo moves, where the transition probability from one configuration to another one involves only the ratio of the probabilities of these two configurations. In statistical mechanics, this method has been most successfully applied to calculating static equilibrium averages with the Boltzmann distribution. In particular, many static properties of polymer solutions and melts are most accurately and most reliably known from Monte Carlo simulations. Since, in this book, we are interested in the dynamics of polymeric liquids, the Monte Carlo method is here mentioned only for comparison with the stochastic simulation techniques suited for obtaining dynamic properties.

What Monte Carlo simulations are to the theory of static polymer properties, *Brownian dynamics simulations* and other simulations based on the *numerical integration of stochastic differential equations* are to the theory of polymer dynamics. Both simulation methods are unique tools to obtain exact results. By the term "exact" it is meant that the only error in the result of a computer simulation is a statistical error which, in principle, can be made arbitrarily small by running the corresponding computer program sufficiently long (of course, for a simulation technique to be useful for practical purposes, "acceptably" small statistical error bars should be obtained with a "reasonable" amount of computer time; the meaning of the rather vague terms in quotation marks clearly depends on the particular problem and on the computers available).

Why do *stochastic differential equations of motion*, which are the starting point for all the simulations considered in this book, occur very naturally in the theory of polymer dynamics? In view of the immense number of degrees of freedom and the wide range of time scales involved in polymer problems, the derivation of tractable kinetic theory models requires some coarse-grained or trace description of such problems (e.g., by mechanical bead-rod-spring rather than atomistic models). The effects of the rapidly fluctuating degrees of freedom associated with very short length scales are usually taken into account through random forces which perturb the time evolution of the slower degrees of freedom. For that reason, the basic equations for most kinetic theory models are stochastic in nature. From a more intuitive point of view, thermal noise turns the equations of motion into stochastic differential equations.

In computer simulations based on the numerical integration of stochastic differential equations, we construct stochastic trajectories. It is therefore crucial to give a precise meaning to these random objects, the trajectories of stochastic processes, and not only to introduce their probability distribution, as it is usually done in the applied sciences and engineering. This is particularly important for constructing sophisticated integration schemes, that is, for developing efficient simulation algorithms. The *art of designing efficient algorithms*, and the required mathematical background, are the main subjects of this book. Moreover, the investigation of stochastic differential equations of motion for the polymer configurations has additional advantages compared to the usual consideration of partial differential equations for the time-evolution of probability densities in configuration space: we obtain a more direct understanding of the polymer dynamics and a better feeling for the degree of complexity of various types of models.

Brownian dynamics simulations have been applied in the field of polymer dynamics since the late 1970s. However, the great progress made with numerical methods for stochastic differential equations has not really been exploited yet in simulating polymer dynamics. Brownian dynamics simulations are not only a nice toy, and they should not only be used in their most rudimentary form. If the state-of-the-art techniques discussed in this book are used (such as higher-order integration schemes, predictor-corrector schemes, Runge-Kutta ideas, implicit and semi-implicit methods, time-step extrapolation, non-

Gaussian random numbers, shifted random numbers, and variance reduction), one is now in a position to attack the most important and exciting problems in polymer dynamics, and some progress in this direction has already been made. On the one hand, one can work out the *universal dynamic properties* of long chain molecules. This expected universality, which is related to the famous self-similarity and scaling laws for high-molecular-weight polymers, may also serve as a justification of the simple mechanical models used in polymer kinetic theory. On the other hand, one can simulate extremely large ensembles of molecules and, in combination with the finite element method, one can then directly *solve complex flow problems* with kinetic theory models. The stresses required in the finite element calculation of the flow field are read off from the molecular configurations. These exciting perspectives are the motivation for attempting a systematic approach to stochastic simulation techniques in this book.

In order to motivate the development of the theory of stochastic processes we first sketch the background and framework for our later applications. The basic tasks and goals of polymer kinetic theory and its relation to polymer fluid mechanics are hence described in a brief general introduction (Chapter 1). Our stochastic approach to kinetic theory models is contrasted with the classical formulation. For developing efficient simulation techniques based on the numerical integration of the stochastic differential equations resulting from polymer kinetic theory, it is indispensable to have or acquire a sound understanding of mathematical stochastics, in particular of the subtleties of stochastic calculus. *Stochastic calculus is fundamentally, unavoidably and undeniably different from deterministic calculus*. The most important concepts of stochastics are explained in Part I of this book (Chapters 2 and 3). Although the development of stochastics is strictly confined to what is actually needed in this book, a great variety of concepts and methods of modern stochastics is introduced because it is found to be very helpful. Therefore, this book not only provides the background for designing simulation algorithms in polymer kinetic theory but also gives a multitude of concrete applications of abstract stochastic concepts, and it may hence also be used as a collection of nontrivial practical examples and as a challenge for mathematicians. Since the required stochastic concepts are rather natural and intuitive, an attempt is made to give simple definitions which are sufficiently precise for the purposes of polymer scientists, hopefully without too much provoking the mathematicians who are accustomed to even more rigor. As a stepping-stone to stochastic processes, Part I should be helpful in many branches of science and engineering. Part II of the book contains the application of the theory of stochastic processes to polymer kinetic theory. *The power of stochastic simulation techniques is illustrated* through many examples. By applying the theory of stochastic differential equations to dilute solution models, one is immediately led to Brownian dynamics simulations for the polymer dynamics (Chapter 4). We then discuss the treatment of models with constraints by Brownian dynamics simulations (Chapter 5). In the last chapter, the stochastic concepts are applied to develop stochastic simulation algorithms for various reptation models for polymers in concentrated solutions and melts (Chapter 6).

For optimal benefitting from this book, some *mathematical background* and, perhaps even more important, the willingness for a certain mathematical abstraction are required. Indispensable is some familiarity with the most elementary ideas of set theory. Moreover, knowledge of the basic concepts of linear algebra (vector spaces, matrices, diagonalization, positive-definiteness) and of analysis (functions, convergence, calculus in $\mathbb{R}^d$, Fourier series and transforms, differential equations) is assumed. A background in differential geometry might simplify the reading and enhance the understanding of Sect. 5.1. Finally, some knowledge of measure theory, topology, symbolic computation, numerical mathematics, and computer science is not assumed, but it would be helpful for understanding certain details and examples. There are many exercises in which the reader is supposed to develop computer programs, where all the solutions are given in FORTRAN.

This book is based on several courses on polymer physics, kinetic theory and stochastic dynamical systems given over the last seven years to students of physics and material science. I owe an immeasurable debt of thanks to my teachers who introduced me in a very enlightening and fascinating manner to these and related fields: Josef Honerkamp, R. Byron Bird, and Joachim Meißner. A number of people have provided me with critical comments and suggestions for improving the manuscript: Jay D. Schieber, R. Byron Bird, Kathleen Feigl, Ludger Rüschendorf, Marshall Fixman, Manuel Laso, Martin Melchior, Ravi Prakash, Peter E. Kloeden, and many students. I wish to thank them all for their help. The assistance of Alain-Sol Sznitman and Hermann Rost in finding the mathematical background for processes with mean field interactions (Sect. 3.3.4) and Wesley P. Petersen's remarks on random number generators have been very helpful.

Zürich and Rafz, Switzerland Hans Christian Öttinger
September 1995

Table of Contents

Symbols and Notation

Latin Symbols

A, B, C, A_j, B_j, A_ω	Events
$(A_t)_{t\in\mathbb{T}}$, $(B_t)_{t\in\mathbb{T}}$	Coefficients in the stochastic differential of a real-valued Itô-process [(3.37)]
$(\boldsymbol{A}_t)_{t\in\mathbb{T}}$, $(\mathbf{B}_t)_{t\in\mathbb{T}}$	Coefficients in the stochastic differential of a vector-valued Itô-process [(3.42)]
A, $\boldsymbol{A}$, $\boldsymbol{A}_\mu$	Drift term or drift vector in a stochastic differential equation (in general, a function of time and configuration)
$\overline{\boldsymbol{A}}$	Drift term including a spurious drift [(3.132)]
A^{R}_{jk}	Elements of the Rouse matrix [(4.6)]
A^{Z}_{jk}	Elements of the Zimm matrix [(4.74)]
$\mathcal{A}$, $\mathcal{A}'$, $\mathcal{A}''$, $\mathcal{A}_j$	σ-algebras
$\mathcal{A}(\mathcal{E})$	σ-algebra generated by a set of events $\mathcal{E}$ [Definition 2.5]
$\mathcal{A}^X$	σ-algebra induced by a measurable function X [p. 36]
$(\mathcal{A}_t)_{t\in\mathbb{T}}$	Increasing family of σ-algebras (e.g., induced by a stochastic process)
$\bar{\mathcal{A}}_t$	σ-algebra induced by the future of the Wiener process
$\mathcal{A}^*$	Completion of a σ-algebra $\mathcal{A}$
a, b	Limits of intervals, real constants (b also used for finite extensibility parameter)
a_{b}	Bead radius
a_{ev}	Range of the excluded-volume potential [(4.116)]
$\boldsymbol{a}(t)$, $\mathbf{A}(t)$	Contributions to the drift vector $\boldsymbol{A}$ for linear equations [(3.53)]
a^{R}_j	Eigenvalues of the Rouse matrix [(4.8)]
a^{Z}_j	Eigenvalues of the Zimm matrix [(4.75)]
B, $B_{\mu\nu}$, $\mathbf{B}$, $\mathbf{B}_{\mu\nu}$	Noise prefactor in the diffusion term of a stochastic differential equation (in general, a function of time and configuration)
$\mathbf{B}_j(t)$	Contribution to $\mathbf{B}$ for linear equations [(3.53)]
$\overline{\mathbf{B}}$	$\mathbf{B}$ in a predictor-corrector scheme [(4.111)]
$\mathcal{B}$ (or $\mathcal{B}^d$)	Borel σ-algebra on $\mathbb{R}$ (or on $\mathbb{R}^d$) [Example 2.7]
$\mathcal{B}^{\mathbb{T}}$	Borel σ-algebra on the set of functions $\mathbb{R}^{\mathbb{T}}$
b	Finite extensibility parameter [(4.118)]
$\bar{b}$	Shifted finite extensibility parameter [Exercise 4.32]
$\boldsymbol{b}_j$	Column vector of $\mathbf{B}$ (or of a contribution to $\mathbf{B}$ for linear equations)
C^{R}_{jk}	Elements of the Kramers matrix [(4.26)]
C^{Z}_{jk}	Elements of the modified Kramers matrix [Exercise 4.21]
$C^{\mathbb{T}}$	Set of continuous real functions on $\mathbb{T}$
$\mathbf{C}(t,t')$	Cauchy strain tensor [(4.17)]
c, c', c_g, c_j, c'_j	Real constants

$\boldsymbol{c}$	Column vector
D, D_c, D_h, D_r	Diffusion coefficients
$\mathbf{D}$	Diffusion matrix
d, d', d''	Dimensions
d_p	Distance between plates
d	Differential
$\mathbf{E}(t, t')$	Displacement gradient tensor [(4.14)]
$E(X\|\mathcal{A}')$	Conditional expectation of X given $\mathcal{A}'$ [Definition 2.51]
$E(X\|Y = y)$	Conditional expectation of X for given value y of Y [(2.56)]
$\mathcal{E}$, $\mathcal{E}_j$, $\mathcal{E}_t$	Generating systems for σ-algebras [Definition 2.5]
e	2.7182818...
F_t^{B}	Brownian force, random noise
$F_{\mathrm{rel}}^{(\mathrm{m})}$	Effect of metric forces on the relative motion of two beads [Exercise 5.15]
$\boldsymbol{F}(\boldsymbol{r})$	Force field
$\boldsymbol{F}^{\mathrm{c}}$, $\boldsymbol{F}_k^{\mathrm{c}}$	Connector force (in spring k)
$\boldsymbol{F}_\mu$	Potential force on bead μ [(4.2)]
$\boldsymbol{F}_\mu^{(\mathrm{e})}$	External force on bead μ
$\boldsymbol{F}_\mu^{(\mathrm{m})}$	Metric force on bead μ [(5.62)]
$\overline{\boldsymbol{F}}_\mu^{\mathrm{B}}$	Smoothed Brownian force on bead μ [(4.66)]
$\mathcal{F}_k$	Generalized intramolecular forces [(5.24)]
$\mathcal{F}_k^{(\mathrm{e})}$	Generalized external forces [(5.25)]
f	Relative frequency of the occurrence of an event
f_c, f_c'	Correction factors for the bias in variance reduced simulations [(4.36), (4.41)]
$G(t)$	Relaxation modulus [(1.15)]
g, $\tilde{g}$, g_j, g_t, $g_t^{(n)}$	Generic names for functions with values in $\mathbb{R}$
$\boldsymbol{g}$, $\boldsymbol{g}_A$, $\boldsymbol{g}_B$, $\boldsymbol{g}_X$, $\boldsymbol{g}_Y$	Generic names for functions with values in $\mathbb{R}^d$ ($\boldsymbol{g}$ also used for gravitational field)
$\tilde{\mathrm{g}}_{jk}$, g_{jk}, $\bar{\mathrm{g}}_{jk}$, $\hat{\mathrm{g}}_{jk}$	Elements of various modified metric matrices [(5.7), (5.9), (5.48), (5.49)]; the corresponding symbols without subscripts indicate determinants; the elements of the inverse matrices are denoted by capital letters, e.g. $\tilde{\mathrm{G}}_{jk}$
H	Hookean spring constant
H_n	Hermite polynomials [(3.68)]
$H_{\mu\nu}$	Equilibrium-averaged hydrodynamic interactions between beads μ and ν [(4.69)]
$\mathbf{H}_{\mu\nu}$	Tensors associated with hydrodynamic interactions between beads μ and ν [(5.37)]
h^*, h	Hydrodynamic-interaction parameters [(4.70), p. 251]
$I(A)$	Set of indices of the elements contained in an event A [Example 2.13]
$I_t^{(n)}$, $\mathbf{I}_{t't}^{(2)}$	Iterated stochastic integrals [(3.65), (3.115)]

i	Imaginary unit
$\boldsymbol{J}$	Probability current [(3.83)]
j, k, l	Summation indices, integers
$\mathbf{K}$	Averaged structure tensor [Example 5.18]
k_{B}	Boltzmann's constant
L	Length of a rod
L_{C}	Contour length of a polymer chain [(6.20)]
$\mathcal{L}$, $\mathcal{L}_t$	Infinitesimal generator of a Markov process [(2.88)]
l_μ	Weight of bead μ in the center of resistance [(4.79)]
M, M_μ	Mass of a Brownian particle or bead μ
M_{e}	Molecular weight between entanglements [(6.15)]
M_{p}	Mass of a polymer chain
$\mathbf{M}$	Matrix
$\mathbf{M}_k$	Tensor describing the effect of a flow field on the generalized coordinate Q_k [(5.26)]
M_{jk}	Elements of a matrix with determinant M [Exercise 5.6]
m	Ratio of extensional rates [(1.17)]
N	Number of beads in bead-spring chains
N_{C}	Number of carbon atoms in the polymer backbone
N_{p}	Number of polymer chains in a system
N_{T}	Number of trajectories
N_t, N_y	Null sets labelled by t, y
n, n', n_j	Integers
n_{p}	Number density of polymers
n_A, n_B, n_{AB}	Number of occurrence of the events A, B, and $A \cap B$
$\boldsymbol{n}$	Unit normal vector
$o(\cdot)$	Term of lower order
P, P'	Probability measures
P_{u}	Probability for unobserved reflections [(6.18)]
P^X	Distribution of the random variable X [(2.27)]; $P(X \in \cdot) := P^X(\cdot)$
P^*	Completion of a probability measure
P_ω	Measure concentrated at a single point ω [Example 2.15]
$P(\cdot\|A)$	Conditional probability measure given the event A [(2.15)]
$P(X \in A\|Y = y)$	Conditional distribution of X for a given value y of Y [(2.54)]
$\overline{P}_{t_1 \ldots t_n}$	Family of finite-dimensional marginal distributions [(2.72)]
$\mathbf{P}$, $\mathbf{P}_{\mu\nu}$	Projection operators [Exercise 5.1, (5.41)]
$\mathcal{P}(\Omega)$	Set of all subsets of Ω
p, p_t	Probability densities
p^X	Probability density for P^X
p_ω	Probability density for P_ω (δ-function)

$p_{\alpha\Theta}$	Gaussian probability density with mean $\boldsymbol{\alpha}$ and covariance matrix $\boldsymbol{\Theta}$ [(2.21)]
$p_{\rm red}$	Probability density in variance reduced simulations [Exercise 4.14]
$p(\cdot\|\cdot)$	Conditional probability density [(2.59)]
$p_{tt'}(x\|x')$	Transition probability density [(2.85)]
$\overline{p}_{t_1\ldots t_n}$	Family of finite-dimensional marginal probability densities
$\widetilde{p}$	Characteristic function [(2.13)]
p_j	Probabilities in discrete spaces
$p_{\rm h}$	Hydrostatic pressure
Q, Q'	Length of $\boldsymbol{Q}$ [Example 4.13]
Q_j	Generalized coordinates
$\overline{\boldsymbol{Q}}$	Predicted value for $\boldsymbol{Q}$ in a predictor-corrector scheme [(4.33)]
$\widetilde{\boldsymbol{Q}}$	Biased random variables in variance reduced simulations
$\boldsymbol{Q}_j$, $\boldsymbol{Q}$	Connector vectors (or dummy variables) [Fig. 4.1]
$\boldsymbol{Q}'_k$	Normal modes associated with connector vectors
q, $\boldsymbol{q}$, $\boldsymbol{q}'$	Arguments of characteristic functions
R	Gas constant
$\overline{R}_{\rm g}$	Root-mean-square radius of gyration
$\overline{R}_{\rm h}$	Hydrodynamic radius [(4.91)]
R_F	Carlson's standard elliptic integral of the first kind [Example 4.17]
$\boldsymbol{R}_\mu$	Bead position vectors with respect to the center of mass [(5.1)]
r	Length of $\boldsymbol{r}$
$\boldsymbol{r}$, $\boldsymbol{r}'$	Position vectors
$\boldsymbol{r}_\mu$	Bead position vectors
$\boldsymbol{r}_{\rm c}$	Center of mass position vector [Fig. 4.1, (5.1)]
$\boldsymbol{r}_{\rm h}$	Center of resistance position vector [(4.78)]
$\widetilde{\boldsymbol{r}}_\mu$, $\overline{\boldsymbol{r}}_\mu$	Auxiliary bead positions in simulation algorithms for models with constraints
$S=(S_t)_{t\in\mathbb{T}}$, S', $\widehat{S}$	Wiener processes in the interval $[0,1]$
$\widetilde{S}_j$, S', S''	Approximations for the process S at discrete times
S_V	Area of an oriented plane
s, s', s_0	Dummy variable in the probability density of S, S_0
T	Absolute temperature
$T^{\rm FP}$	First passage time [(3.82)]
$T^{\rm LR}_t$	Last reflection time [(3.93)]
$\mathbf{T}(\boldsymbol{x})$	Matrix of partial derivatives [Example 3.43]
$\mathcal{T}$	Time ordering [(3.56)]
t, t', t'', t_j, $t_j^{(n)}$	Time variables, labels
$t_{\rm block}$	Time corresponding to a block of time steps
$t_{\rm max}$	Maximum time argument

U_{RD}, $U_{\eta\lambda}$, $U_{\eta\mathrm{R}}$, $U_{\Psi\eta}$, $U_{\Psi\Psi}$	Universal ratios for long polymer chains [(4.92), (4.95), (4.97), (4.98), (4.99)]
$\boldsymbol{U} = (\boldsymbol{U}_t)_{t\in\mathbb{T}}$, $\boldsymbol{U}'$, $\widehat{\boldsymbol{U}}$	Stochastic process on the unit sphere
$\widetilde{\boldsymbol{U}}_j$	Approximations for the process $\boldsymbol{U}$ at discrete times
$\boldsymbol{u}$	Dummy variable in the probability density of $\boldsymbol{U}$
$\hat{\boldsymbol{u}}$	Result of the deterministic time evolution of $\boldsymbol{u}$ [(6.6)]
V	Volume
$V^{(\mathrm{e})}$	External potential
V_{ev}	Excluded-volume potential [(4.116)]
$(V_t)_{t\in\mathbb{T}}$	Ornstein-Uhlenbeck velocity process
$\boldsymbol{V}$	Velocity
$\Delta\mathbf{V}$	Random matrix representing the antisymmetric part of the stochastic integral $\mathbf{I}^{(2)}_{t't}$ over a short time interval [(3.121)]
v_{ev}, U_{ev}	Strength of the excluded-volume potential [(4.116), (5.83)]
$\boldsymbol{v}(\boldsymbol{r}, t)$	Velocity field
$\boldsymbol{v}_0(t)$	Time dependent velocity vector
$\boldsymbol{v}_{\mathrm{rot}}(\boldsymbol{r})$	Rotational velocity field [Example 4.23]
$\Delta\boldsymbol{v}(\boldsymbol{r})$	Velocity perturbation
$W = (W_t)_{t\in\mathbb{T}}$	Wiener process [Example 2.79]
W^z, W^ϕ	Wiener processes
$(W^{(n)})_{n\in\mathbb{N}}$	Sequence of discretizations of the Wiener process
$\boldsymbol{W} = (\boldsymbol{W}_t)_{t\in\mathbb{T}}$	Wiener process in d dimensions [(2.83)]
$\boldsymbol{W}_\mu$	Wiener processes associated with bead positions
$\boldsymbol{W}'_\mu$	Wiener processes associated with normal modes [(4.10), (4.11)]
$\boldsymbol{W}_j$, $\boldsymbol{W}^{\mathrm{i}}$, W^{f}_j	Standard Gaussian random variables [Example 4.17]
X, X', Y, Y'	Real-valued random variables or stochastic processes, e.g. $X = (X_t)_{t\in\mathbb{T}}$ (whenever it is convenient, the time argument of a stochastic process is given in parentheses rather than as a subscript)
$\boldsymbol{X}$, $\boldsymbol{Y}$, $\boldsymbol{Y}^{\Delta t}$	Vector-valued random variables or stochastic processes
$\widetilde{X}_j$, $\widetilde{\boldsymbol{Y}}_j$, $\widetilde{W}_j$, $\widetilde{W}^{(n)}_j$	Random variables, stochastic processes at discrete times
$(X_n)_{n\in\mathbb{N}}$	Sequence of random variables
$(X^{(n)})_{n\in\mathbb{N}}$	Sequence of stochastic processes
x, x', x'', $\boldsymbol{x}$, y, $\boldsymbol{y}$	Dummy variables in functions or densities
$\overline{\boldsymbol{Y}}_j$	Predicted values in a predictor-corrector scheme [(3.133)]
$\Delta\widetilde{\boldsymbol{Y}}_j$	Auxiliary random variables [(4.115)]
$\mathbf{Z}$	Total effective friction tensor [(5.5)]
z	$\cos\theta$

Greek Symbols

α, $\boldsymbol{\alpha}$, α_t, $\boldsymbol{\alpha}_t$	Mean of a distribution
α_{B}	Strength of Brownian forces [(3.5)]
α_{s}	Exponent characterizing an α-stable distribution
β	Reduced shear rate [Fig. 4.15]
$\dot{\gamma}$, $\dot{\gamma}_0$	Shear rate [(1.4)]
γ_j	Lagrange multipliers
Δ	Correlation length parameter [(6.24)]
$\Delta\cdots$	Differences of variables
δ_{jk}	Kronecker's δ-symbol
$\delta(t)$	Dirac's δ-function
$\boldsymbol{\delta}$	Unit tensor or unit matrix (with components δ_{jk})
ϵ, $\bar{\epsilon}$, $\bar{\bar{\epsilon}}$	Dimensionless parameters in reptation models [(6.5), (6.30), (6.31)] (ϵ also used for small real numbers)
$\boldsymbol{\varepsilon}(t_j, t_{j+1}, t_n)$	Auxiliary tensor in variance reduced simulations [(4.52)]
$\dot{\epsilon}$	Extensional strain rate [(1.17)]
ζ	Friction coefficient
$\boldsymbol{\zeta}_\mu$	Bead friction tensors
$\boldsymbol{\zeta}_{\mu\nu}$	Effective friction tensors [(5.3)]
$\tilde{\boldsymbol{\zeta}}_{\mu\nu}$	Modified effective friction tensors [(5.4)]; $\tilde{\boldsymbol{\zeta}}_{\mu\nu} = \tilde{\zeta}_{\mu\nu}\,\boldsymbol{\delta}$ for isotropic tensors
η	Viscosity [(1.6)]
η_{p}	Polymer contribution to the viscosity
η_{s}	Solvent viscosity
$\bar{\eta}$	Trouton viscosity
$[\eta]_0$	Intrinsic viscosity [(4.96)]
Θ	Variance (special case of $\boldsymbol{\Theta}$)
Θ_{red}	Variance in variance reduced simulations
$\Theta_{\mathrm{f}}^{(j)}$	Eigenvalues of $\boldsymbol{\Theta}_{\mathrm{f}}$
$\boldsymbol{\Theta}$, $\boldsymbol{\Theta}'$, $\boldsymbol{\Theta}_{\mathrm{i}}$, $\boldsymbol{\Theta}_{\mathrm{f}}$, $\boldsymbol{\Theta}_t$	Covariance matrices
$\Theta_{tt'}$, $\boldsymbol{\Theta}_{tt'}$	Covariances of a Gaussian process
θ	Polar angle
κ	Elastic modulus of a rope
$\boldsymbol{\kappa}$	Transposed velocity gradient tensor
$\bar{\boldsymbol{\kappa}}$	Perturbed velocity gradient tensor
λ, λ_H, λ_h, λ_j, λ^{rd}	Time constants
λ_η	Characteristic time scale associated with the intrinsic viscosity [(4.94)]
$\boldsymbol{\lambda}_\mu$	Weight tensors [(5.6)]
μ_1, μ_2	Extensional viscosities [(1.19), (1.20)]
μ, μ', ν	Summation indices
$\overline{\mu}$	First passage time distribution for the Wiener process in $[0,1]$ [(3.91)]
$\overline{\nu}$	Memory function in reptation models [(6.33)]

ν_{ev}	Exponent characterizing the chain length dependence of the polymer size in the presence of excluded volume
π	3.1415926...
$\overline{\pi}$	Conditional distribution of first passage times for the Wiener process in $[0,1]$ [(3.92)]
ρ	Mass density
ρ_{p}	Polymer mass density
σ	Width of a one-dimensional distribution [(2.12)]
$\overline{\sigma}$	Width of a one-dimensional distribution with infinite variance
σ_{B}	Contribution to σ_{F} caused by Brownian forces
σ_{F}	Rope tension in the reptating-rope model [(6.20)]
σ_{sf}	Empirical steric factor [(4.119)]
$\boldsymbol{\sigma}$	Generalization of σ to d dimensions (a matrix with elements σ_{jk})
$\boldsymbol{\tau}$	Stress tensor [(1.2)]
$\boldsymbol{\tau}^{\mathrm{p}}$	Polymer contribution to the stress tensor [(4.18)]
$\boldsymbol{\tau}^{\mathrm{rv}}$	Random variable associated with the polymer contribution to the stress tensor [(4.29)]
$\widehat{\boldsymbol{\tau}}$, $\widehat{\boldsymbol{\tau}}^{n}$	Auxiliary tensors in a reptation model [(6.38)]
υ	Order of convergence [(3.111), (3.118)]
$\boldsymbol{\Upsilon}_j^n$, $\overline{\boldsymbol{\Upsilon}}_j^n$	Supporting values [(3.127), (3.129)]
$\boldsymbol{\Phi}_t$	Fundamental matrix of a system of homogeneous linear differential equations [(3.56), (3.62)]
$\boldsymbol{\Phi}_{\mathrm{app}}(t_{j+1}, t_n)$	Auxiliary tensor in variance reduced simulations [(4.52)]
ϕ	Polar angle
$(\phi_t)_{t\in\mathbb{T}}$	Increasing process [(3.88)]
χ_A	Indicator function of a set A
Ψ_1, Ψ_2	Normal stress coefficients [(1.7), (1.8)]
Ω, Ω', Ω'', Ω_j	Sure event, set of all possible outcomes
Ω_t	Time-dependent domain of a stochastic process
Ω_{jk}^{R}, Ω_{jk}^{Z}	Elements of orthogonal matrices [(4.7), (4.75)]
$(\Omega, \mathcal{A})$	Measurable space
$(\Omega, \mathcal{A}, P)$	Probability space
$\boldsymbol{\Omega}_{\Theta_{\mathrm{f}}}$	Orthogonal matrix diagonalizing $\boldsymbol{\Theta}_{\mathrm{f}}$
$\boldsymbol{\Omega}(\boldsymbol{r})$	Hydrodynamic-interaction tensor [(4.62), (4.67)]; $\boldsymbol{\Omega}_{\mu\nu} := \boldsymbol{\Omega}(\boldsymbol{r}_\mu - \boldsymbol{r}_\nu)$
ω_0	Frequency
ω, ω_j	Possible outcome in a random experiment, $\omega \in \Omega$

Subscripts

app	Approximate quantity (probability density, transition probability density, or expectation evaluated with an approximate probability density)

Ω, Υ	Different contributions to the viscometric functions for reptation models [(6.8)–(6.10)]

Superscripts

GA	Quantities associated with the Gaussian approximation
+	Quantity associated with stress growth at inception of steady flow
−	Quantity associated with stress decay after steady flow

Mathematical Symbols and Notation

$:=$	Equal by definition
$\sim$	Proportional
∂	Partial derivative
$\nabla := \frac{\partial}{\partial \boldsymbol{r}}$	Nabla operator
$\in$ ($\notin$)	Element of a set (not element of a set)
$\emptyset$	Empty set, impossible event
$\subset$	Subset of a set
$\cap$	Intersection of sets
$\bigcap_{j=1}^{\infty} A_j$	Intersection of a sequence of sets A_1, A_2, ...
$\cup$	Union of sets
$\bigcup_{j=1}^{\infty} A_j$	Union of a sequence of sets A_1, A_2, ...
$(\ldots)^C$	Complement of a set
$\mathbb{B}$	Boundary of a domain
$\mathbb{D}$	Continuous range of a label, domain of a function
$\mathbb{N}$	Natural numbers
$\mathbb{R}$	Real numbers
$\mathbb{R}^d$	Column vectors with d real components
$\mathbb{R}^{\mathbb{T}}$	Set of real functions on $\mathbb{T}$
$\mathbb{T}$	Range of time for a stochastic process, set of labels
$]a,b[$, $[a,b]$	Open and closed intervals ($a,b \in \mathbb{R}$), that is, $]a,b[:= \{x \mid a < x < b\}$ and $[a,b] := \{x \mid a \leq x \leq b\}$
$[a,b[$, $]a,b]$	Semi-open intervals ($a,b \in \mathbb{R}$), that is, $[a,b[:= \{x \mid a \leq x < b\}$ and $]a,b] := \{x \mid a < x \leq b\}$
$\langle \ldots \rangle$	Expectation (or average) of a random variable
$\langle \ldots \rangle_{\alpha\Theta}$	Expectation evaluated with $p_{\alpha\Theta}$
$\langle \ldots \rangle_{t'}^{\mathrm{rd}}$	Expectation for rigid dumbbells which were at equilibrium at the time t' and have been exposed to a flow after t'
$\lvert \ldots \rvert$	Absolute value, length (or norm) of a vector, (Eucledian) norm of a matrix

$(\ldots)^{-1}$	Inverse of a matrix, inverse of a function, or inverse image of a set
$(\ldots)^{T}$	Transpose of a matrix or a tensor
$\mathcal{L}^{\dagger}$	Adjoint of an operator $\mathcal{L}$
Cov	Covariance of two random variables [(2.45)]
inf	Infimum
lim	Limit; for random variables also almost sure limits (as-lim), mean-square limits (ms-lim), stochastic limits (st-lim), and limits in distribution (db-lim) are considered [Definitions 2.61 and 2.69]
max	Maximum
min	Minimum
sgn	Sign function
sup	Supremum
Tr	Trace of a tensor or matrix
Var	Variance of a random variable [(2.43)]
x_j, $j = 1, \ldots d$	Components of a d-dimensional column vector $\boldsymbol{x}$
B_{jk}, $j, k = 1, \ldots d$	Elements of a $d \times d$-matrix $\mathbf{B}$
$\cdot$	Contraction of tensors or matrix product
:	Double contraction or double dot product of tensors or matrices (e.g., $\mathbf{A} : \mathbf{B} = \mathrm{Tr}(\mathbf{A} \cdot \mathbf{B}) = \sum_{j,k} A_{jk} B_{kj}$)
$\times$	Product of sets
$\otimes$	Product of σ-algebras
$\circ$	Itô's circle (indicating Stratonovich integrals)
$\odot$	Itô's circle with a matrix multiplication implied

Computer Programs

`EULER`	Euler algorithm for single time step in `HIEUL` [Exercise 4.28]
`FENESI`	Semi-implicit scheme for FENE dumbbells [Exercise 4.34]
`HIEUL`	Euler scheme for Hookean dumbbells with hydrodynamic interaction [Exercise 4.28]
`HOOKE1`	Euler scheme for Hookean dumbbells [Exercise 4.11]
`HOOKE2`	Predictor-corrector scheme for Hookean dumbbells [Exercise 4.12]
`RANGLS`	Generates a random number with Gaussian distribution [Exercise 4.9]
`RANILS`	Initializes random number generators [Exercise 4.8]
`RANULS`	Generates a random number with uniform distribution [Exercise 4.8]
`RANUVE`	Generates a random unit vector [Exercise 4.10]
`REPT`	Simulation of the reptation process [Exercise 6.4]
`REPTIM`	Improved simulation algorithm for time step in `REPT` [Exercise 6.6]

REPTNA	Naive simulation algorithm for time step in REPT [Exercise 6.4]
RIGID2	Second-order scheme for rigid dumbbells [Exercise 5.22]
SECRES	Single time step in RIGID2 [Exercise 5.22]
SEMIMP	Single time step in FENESI [Exercise 4.34]
TEXTRA	Time step extrapolation [Exercise 4.11]

1. Stochastic Processes, Polymer Dynamics, and Fluid Mechanics

One can attempt to achieve a theoretical understanding of polymer fluid dynamics on two different levels: continuum mechanics and kinetic theory. *Continuum mechanics* deals with the formulation and solution of a system of macroscopic equations for the density, velocity, temperature, and possibly other fields describing the fluid structure, which are related to conservation laws for mass, momentum, energy, and maybe other quantities associated with the additional fields. In order to obtain a closed system of macroscopic equations one needs to supplement the fundamental conservation laws by certain empirically or microscopically founded equations of state for the fluxes of the conserved quantities. Such equations of state, which are characteristic for a given material, are often referred to as constitutive equations. The formulation of suitable constitutive equations and of general admissibility criteria for such constitutive equations is a central part of continuum mechanics. In *polymer kinetic theory*, one attempts to understand the polymer dynamics, the constitutive equation for the momentum flux or stress tensor, and eventually polymer fluid dynamics by starting from coarse-grained molecular models. Excellent reviews of the state of the art in both continuum mechanics and kinetic theory are given in the two volumes of the comprehensive introductory textbook *Dynamics of Polymeric Liquids* by R. B. Bird, C. F. Curtiss, R. C. Armstrong and O. Hassager [1, 2]. The more recent literature in continuum mechanics and in kinetic theory has been reviewed in [3] and [4], respectively.

This book deals almost exclusively with polymer kinetic theory and the mathematical background for its stochastic interpretation. The systematic stochastic approach to kinetic theory, which is clearly distinct from the conventional approach (see, e.g., [2]), is the main theme of this book. Rather than studying partial differential equations[1] for the time-evolution of probability densities in polymer configuration space we consider directly the stochastic differential equations of motion for the polymer molecules. As a motivation for developing some mathematical background, we describe in this short introductory chapter the role of kinetic theory in polymer fluid dynamics and the benefits of the stochastic approach. In the first section of this chapter, we set forth the

[1]These partial differential equations, which are often referred to as diffusion equations or Fokker-Planck equations, are of the parabolic type.

general approach to kinetic theory pursued in this book. We point out where our approach deviates from what is described in other textbooks, and what the goals and benefits of such an alternative approach are. In the second section of this chapter, we briefly discuss the relationship to the framework of continuum mechanics which yields the velocity fields for given initial and boundary conditions and hence is the key to the solution of the problems occurring in practical applications, for instance in polymer processing. In particular, we describe an idea that allows us to go directly from kinetic theory to complex flow calculations by using stochastic simulation techniques. In short, the purpose of this chapter is to motivate the subsequent introduction to stochastic processes by showing (i) how the stochastic approach can give additional insight into polymer dynamics, (ii) how stochastic simulation techniques can provide material information relevant to the flow behavior of polymeric liquids, and (iii) how stochastic simulations can be employed for calculating the complex flows occurring in polymer processing. The connections between stochastic differential equations, polymer motion, stress tensors, and polymer fluid dynamics are summarized in Fig. 1.1.

1.1 Approach to Kinetic Theory Models

Since polymeric liquids are extraordinarily complex systems, which necessarily involve a large number of degrees of freedom and a wide range of time scales, one of the most important and challenging problems in kinetic theory is the development of sufficiently simple models capturing as many large-scale characteristics of real polymer systems as possible. Beads, springs and rods are the typical building blocks in the construction kit of polymer kinetic theory models. For the coarse-grained molecules composed of these simple building blocks, one needs to derive suitable equations of motion which, as a consequence of eliminating rapid local details, are typically stochastic differential equations. In the traditional approach to kinetic theory, deterministic partial differential equations for the time evolution of probability densities in polymer configuration space (diffusion equations) are formulated rather than stochastic ordinary differential equations for the polymer configurations. In addition to the people developing coarse-grained molecular models, referred to as *model-builders* in the following, there are the *model-users* who try to predict material functions in given flows in order to compare the results to experiments, or who go further and calculate complex flows for established kinetic theory models. In other words, model-users solve the stochastic differential equations of motion for the polymer configurations or the diffusion equations previously derived by model-builders, and they then calculate material properties as suitable ensemble averages or integrals in polymer configuration space. The most important task of model-users is the development of analytical and numerical methods and of approximation schemes for solving time-evolution equations and for evaluating various material properties. Since model-builders would naturally like to see

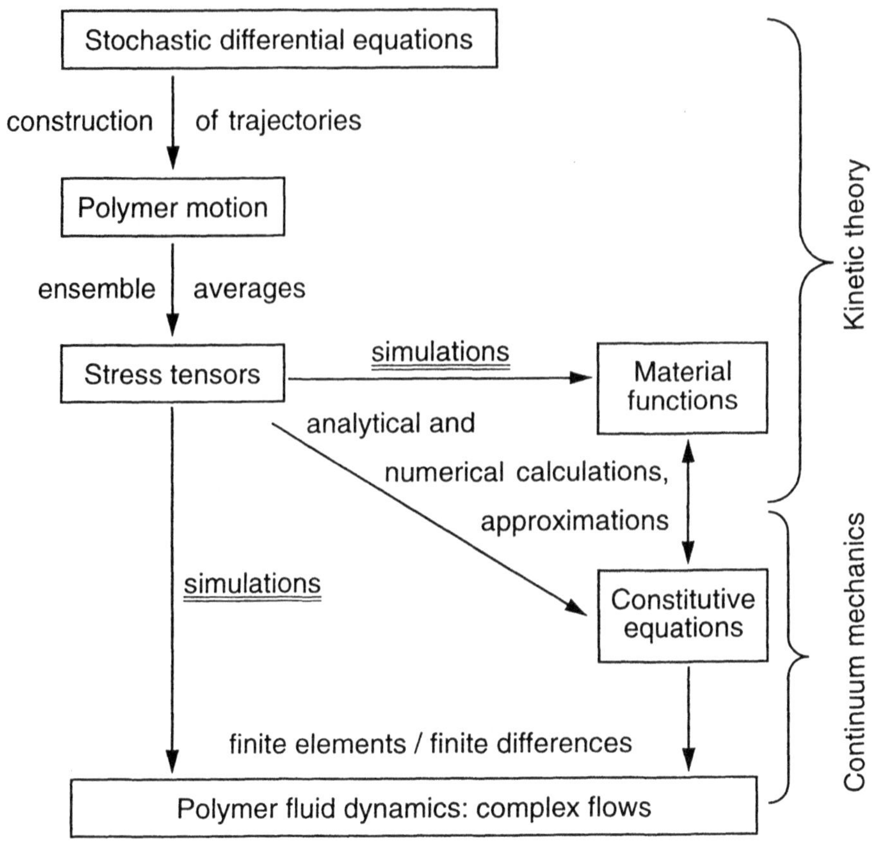

Fig. 1.1. Connections between stochastic differential equations, polymer motion, stress tensors, and polymer fluid dynamics.

the success—or equally naturally hate to see the failure—of their models they usually also work to some extent as model-users in the above sense.

The model-builders and model-users will presumably give very different answers to the question "What is a model?" When, on the one hand, a model-builder thinks about a particular model, he or she will probably have the complex physical situation and the more or less plausible, but unavoidable mathematical approximations and assumptions in mind. A model-builder's perception of what a model is may be represented by the following diagram:

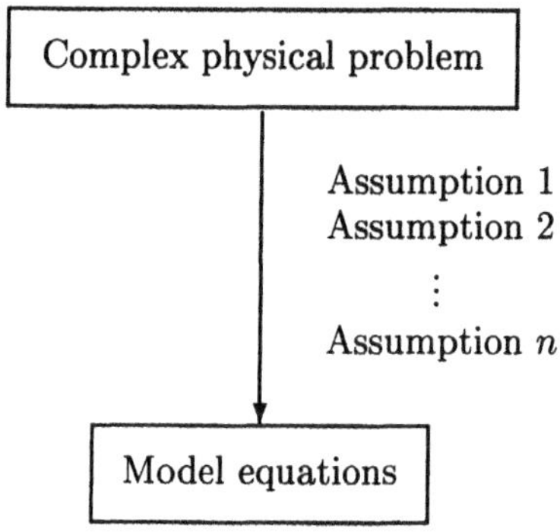

On the other hand, the model-user is interested only in the final model equations which he or she needs to solve in order to obtain concrete predictions.

The main goals of the stochastic approach pursued in this book are the development of computer simulation techniques for solving the final time-evolution equations of kinetic theory models and the evaluation of material properties as suitable averages. The computer simulations developed in Part II of this book are a model-user's method, and they are based on the translation of the final model equations for the polymer dynamics (diffusion equations) into the language of stochastic processes (stochastic differential equations). For a model-user following the stochastic approach, a kinetic theory model consists of a set of configurational variables, stochastic differential equations of motion for these variables, and functions of the configurational variables which, after averaging, give all the material properties of interest. Stochastic simulations are no more and no less than a rigorous, powerful and convenient *tool for solving given kinetic theory models*.[2] Since the development of simulation algorithms requires a detailed understanding of the stochastic content of the model equations, which is a consequence of eliminating rapidly fluctuating motions in favor of random

[2]We **do not** consider "first principle" computer simulations questioning the reliability of certain assumptions made in developing kinetic theory models, such as the numerous and controversial simulations performed to decide whether or not "reptation" exists in polymer melts. We **do not** rederive the diffusion equations of polymer kinetic theory. We merely reformulate diffusion equations as stochastic differential equations, we consider the plausibility of the resulting stochastic equations of motion, and we use them for developing numerical integration schemes or, in other words, simulation algorithms.

forces on larger building blocks such as beads, there is also a good chance of gaining more insight into the nature of a model by designing simulation algorithms. The more complicated a physical problem is, the more, and the more far-reaching, assumptions will be necessary in developing tractable model equations, and the more useful will such an additional insight into the stochastic dynamics implied by the final model equations be. In this sense, the reformulation of the model equations in a stochastic language suitable for computer simulations can be very useful not only as a tool for model-users but also as a framework for model-builders. By putting emphasis on the physical content of the final equations, simulations help us to recognize what remains in a model after simplifying a very complicated problem through a series of assumptions which may interfere and build up in an uncontrollable manner.

As a concrete example for our model-user's point of view, we mention the Doi-Edwards and Curtiss-Bird models for concentrated solutions and melts which are often described by (i) an illustration of the very complex entangled many-chain system and (ii) a list of the basic assumptions (e.g., "reptational motion" or "anisotropic friction," "independent alignment" or "mild curvature," etc.), that is in a model-builder's fashion (see the diagram on p. 4). For the non-experts in the field of modeling, these models then appear to be extremely complicated. By looking merely at the model-user's stochastic differential equations of motion and at the simulation algorithms described in Chap. 6, the non-experts can easily understand the final model equations, and the professional model-builders may even find clues for improving these models (for example, in such a manner that the more complicated, improved models can still be investigated by stochastic simulation techniques).

1.2 Flow Calculation and Material Functions

In order to understand the role of stochastic simulations of polymer dynamics in obtaining material properties related to flow behavior and in calculating complex flows, we need to consider the basic equations of fluid dynamics. Most kinetic theory models describe the dynamics of polymers in solutions or melts undergoing homogeneous flows, that is for velocity fields of the special form $\boldsymbol{v}(\boldsymbol{r},t) = \boldsymbol{v}_0(t) + \boldsymbol{\kappa}(t) \cdot \boldsymbol{r}$ where the vector $\boldsymbol{v}_0(t)$ and the tensor $\boldsymbol{\kappa}(t)$ are independent of the position $\boldsymbol{r}$ but may depend on the time t.[3] At least, the linear dependence of velocity on position is generally assumed on the polymer length scale. Moreover, it is usually assumed that the polymeric liquid to be modeled is incompressible so that the density $\rho(\boldsymbol{r},t)$ is constant in space and time. The general *continuity equation*,

$$\frac{\partial \rho}{\partial t} + \boldsymbol{\nabla} \cdot (\rho \boldsymbol{v}) = 0, \tag{1.1}$$

[3] We use boldface symbols for vectors and tensors; a formal distinction is made possible by using slanted symbols (italic style) for vectors and unslanted symbols (roman style) for tensors.

which expresses the fact that due to conservation of mass a change of the density in a given volume element can occur only if there is a mass flux through the surface of the volume element, implies the incompressibility condition $\boldsymbol{\nabla} \cdot \boldsymbol{v}(\boldsymbol{r}, t) = 0$. Hence, for homogeneous flows of incompressible liquids, the transposed velocity gradient tensor $\boldsymbol{\kappa}(t) = [\boldsymbol{\nabla}\boldsymbol{v}(\boldsymbol{r}, t)]^T$ must be traceless. Incompressibility is assumed throughout this book.

In nonhomogeneous flow situations it is natural that the composition of a liquid consisting of several components, such as a polymer solution, becomes a function of position and time. For a two-component system, the formulation of the balance equation for the concentration of one component then requires a constitutive equation for the mass flux of that component (or, alternatively, for the migration velocity and the diffusion tensor [5]). The formulation of such an additional constitutive equation requires an understanding of the polymer dynamics and is in general a subtle problem [6]. In the presence of chemical reactions, there may even be source terms in the mass balance equations for the individual components.

The basic problem in solving a kinetic theory model is the calculation of the stresses occurring in a complex model liquid in given homogeneous flows. The calculation of tangential and normal stresses, which can all be summarized in a stress tensor $\boldsymbol{\tau}$, in homogeneous flows from molecular models is also the main subject of the second part of this book. The stress tensor or momentum flux tensor is typically obtained as an ensemble average over molecular configurations, so that one needs to understand the polymer dynamics. More precisely, one needs to find a suitable set of variables for characterizing the polymer configurations, such that a closed set of time-evolution equations can be formulated for these variables, and that the stresses can be obtained as averages of certain functions of these variables. Assume that we have found an explicit *stress calculator* for some kinetic theory model, that is an equation which makes it possible to calculate the stresses occurring in arbitrary, given, homogeneous flows as functions or functionals of the velocity gradients. What have we then learned about the behavior of polymeric liquids in general flow situations, for instance as occurring in polymer processing? The key idea is to use the relationship between velocity field and stress tensor as given by the stress calculator in the role of a constitutive equation in the *equation of motion*, which expresses the conservation of momentum,

$$\rho \left(\frac{\partial}{\partial t} + \boldsymbol{v} \cdot \boldsymbol{\nabla} \right) \boldsymbol{v} = -\boldsymbol{\nabla} p_{\mathrm{h}} - \boldsymbol{\nabla} \cdot \boldsymbol{\tau} + \rho \boldsymbol{g}, \tag{1.2}$$

where $p_{\mathrm{h}} = p_{\mathrm{h}}(\boldsymbol{r}, t)$ is the hydrostatic pressure field and $\boldsymbol{g}$ is the force per unit mass in the earth's gravitational field (as the prototype of an external force field acting on the liquid). Since (1.2) equates the substantial time derivative of the velocity times the mass density and the sum of forces acting per unit volume of the liquid it is clear that (1.2) simply expresses Newton's second law. Note that the total stress tensor has been written as the sum of an isotropic pressure term $p_h \boldsymbol{\delta}$ and a stress tensor $\boldsymbol{\tau}$ that vanishes at equilibrium.[4]

Equation (1.2), which plays an important role in the continuum mechanics description of polymeric liquids, is the generalization of the famous Navier-Stokes equation for structurally simple fluids.[5] The Navier-Stokes equation for an incompressible fluid is recovered by using the following constitutive equation for the stress tensor $\boldsymbol{\tau}$ in the general equation of motion (1.2),

$$\boldsymbol{\tau}(\boldsymbol{r},t) = -\eta \left\{ \boldsymbol{\nabla}\boldsymbol{v}(\boldsymbol{r},t) + \left[\boldsymbol{\nabla}\boldsymbol{v}(\boldsymbol{r},t)\right]^T \right\}, \tag{1.3}$$

where the parameter η is the viscosity. Incompressible liquids that can be adequately described by the simple constitutive equation (1.3) for the stress tensor in terms of the velocity field are usually referred to as *Newtonian fluids*. For Newtonian fluids, one may replace $-\boldsymbol{\nabla} \cdot \boldsymbol{\tau}$ in the equation of motion (1.2) by $\eta \boldsymbol{\nabla}^2 \boldsymbol{v}$, where $\boldsymbol{\nabla}^2$ is the Laplace operator, in order to obtain the Navier-Stokes equation in a more familiar form. For non-Newtonian fluids, such as polymer solutions or melts, the constitutive equation for the stress tensor is much more complicated (and hence more interesting). One can imagine that the stresses resulting from flow-distorted polymer molecules (i) strongly depend on the flow rate, (ii) are decisively influenced by the resulting anisotropy of the fluid, and (iii) exhibit memory effects due to the slowness of polymer relaxation. A deep understanding of the polymer dynamics is hence crucial for deriving constitutive equations, and this is why stochastic differential equations of motion are important. In continuum mechanics, the stress tensor as a functional of flow history is typically given by a nonlinear differential equation or by an integral expression. For example, for the *Maxwell fluid*, a relaxation time λ times a convected time derivative of the stress tensor is added to the left side of (1.3).

For given boundary conditions, the velocity field $\boldsymbol{v}(\boldsymbol{r},t)$ is to be calculated from the conservation laws (1.1) for mass and (1.2) for momentum supplemented by a constitutive equation such as (1.3) for the momentum flux tensor. For incompressible fluids, the above set of equations is sufficient for calculating velocity fields because the incompressibility condition derived from (1.1) can be used to eliminate the pressure from the equation of motion (1.2). For compressible fluids, one needs an additional, thermodynamic equation of state relating the density and pressure fields occurring in the conservation laws or balance equations (1.1) and (1.2). Many *examples and exercises* involving calculations of tube flows for various cross sections, flows between parallel plates, flows into dies, squeezing flows, various types of annular flows, flows around translating and rotating spheres, and flows near sharp corners for Newtonian and non-Newtonian fluids can be found in [1]. In more general cases, these partial differential equations can be solved only by numerical methods such as finite elements or finite differences. The goal of this book is not to solve the partial differential equations (1.1) and (1.2); rather we wish to obtain the functional

[4]Throughout this book, $\boldsymbol{\delta}$ denotes the unit tensor or the unit matrix.

[5]The general Navier-Stokes equation is not restricted to incompressible fluids.

relationship between the flow history and the stress tensor by solving stochastic differential equations for suitable polymer configurational variables.

Even for incompressible fluids the situation is actually more complicated than described above because the parameters occurring in the constitutive equation for the stress tensor, for instance the viscosity η in (1.3), usually depend on temperature. Since, due to viscous heating, the temperature in a flow problem is not simply a constant, one needs an additional equation for the temperature field which is obtained by combining conservation of energy, thermodynamic relationships for the internal energy, and a constitutive equation for the heat flux [1]. For example, the nonisothermal flow of a generalized Newtonian liquid through a tube with circular cross section at constant wall temperature has been discussed in Sect. 4.4 of [1]. Not only are heat effects a major problem in formulating and solving continuum mechanical equations and in interpreting experimental data [7], but also a consistent molecular approach to the energy balance equation on the kinetic theory level remains to be formulated [8]. Following a tradition in polymer kinetic theory, we here concentrate on the momentum balance and assume incompressibility and isothermal conditions. It is hoped that the power of the stochastic formulation of kinetic theory will eventually help to remove these assumptions which are very convenient but rather unrealistic in many polymer processing situations.

A serious problem in the traditional kinetic theory lies in the fact that the stress tensor is often determined only in given homogeneous flows. For an expression for the stress tensor in terms of the transposed velocity gradient tensor which characterizes a homogeneous flow, the previously mentioned term *stress calculator* has been coined in [9]. By assuming that a stress calculator may be used as a constitutive equation in the momentum balance equation of continuum mechanics, the following assumptions are made:

- The form of the expression for the stress tensor is not altered in going from homogeneous to nonhomogeneous flows.

- The stress calculator not only characterizes the stresses in given flows, but the same relationship between velocity field and stress tensor may also be used to calculate flow fields for given stresses or, more generally, to calculate stress tensors and velocity fields self-consistently.

The first assumption is plausible as long as the size of the polymers is small compared to the scale on which the velocity gradients and the composition of the fluid change noticeably. However, this important assumption should be kept in mind when one is interested in velocity profiles in narrow channels, near sharp corners, or in boundary layers. Various methods for incorporating effects due to nonhomogeneity into constitutive equations derived from molecular models have been considered in the literature [5, 6, 10].

The second of the above assumptions leads to paradoxes for models in which the interactions between the various components of a polymeric liquid, such as solvent and solute polymers, are not accounted for in a consistent manner [9, 11].

In modeling polymer solutions, one therefore needs to build in the interactions between solvent and solute components in such a manner that they are coupled to form a well-defined, coherent material, possibly of spatially varying composition.

With the above limitations and implicit assumptions in mind, from now on we assume that a stress calculator derived from a molecular model may be used as a constitutive equation in the macroscopic equation of motion (1.2). The classical goal of fluid dynamics is to go through the following sequence of steps: molecular model → constitutive equation → solution of flow problems → information on molecular configurations, where the results of each step should be compared to experiment (see p. 92 of [2] for more details). The derivation of the constitutive equation for the stress tensor in arbitrary homogeneous flows for a given kinetic theory model may be a rather formidable or even unsolvable problem. Therefore, one often evaluates the stress tensor only in special flow situations, the most prominent of which are *shear* and *extensional flows*. The material information obtained from these two types of flow discussed in the following subsections is quite different and hence very helpful in understanding more complex flow situations. In polymer kinetic theory, all relevant material properties can be obtained as suitable ensemble averages over time-dependent polymer configurations. In the second part of this book, the usefulness of stochastic simulations in generating representative ensembles of polymer configurations and thus obtaining the material properties introduced in the following is illustrated by many examples. The calculation of these material properties in given flows is very important in testing models by comparing their predictions to experimental results. Moreover, stochastic simulation techniques offer the unique possibility of avoiding the need of deriving constitutive equations and of performing the direct step molecular model → solution of flow problems, where the information on molecular configurations is automatically obtained as an intermediate result in the flow calculation; this possibility, which is a strong motivation for developing efficient simulation techniques, is briefly discussed in the final subsection of this introductory chapter (see Fig. 1.1 for the role of stochastic simulations in polymer kinetic theory and fluid dynamics).

The reader familiar with the description of various shear and extensional flows and with the material functions characterizing the stress tensor in these flow situations may skip the next two subsections. A more detailed and general discussion of shear and shear-free flows is, for example, given in Chap. 3 of [12].

1.2.1 Shear Flows

The essential feature of shear flows is that it is possible to identify a one-parameter family of material surfaces that slide relative to one another without stretching. A homogeneous simple shear flow, for which these material surfaces are planes, is given by a velocity field

$$v_1 = \dot{\gamma}\, r_2, \quad v_2 = 0, \quad v_3 = 0, \tag{1.4}$$

where the *shear rate* $\dot{\gamma}$, which is the $(1,2)$-component of the transposed velocity gradient tensor $\boldsymbol{\kappa}$ in a given Cartesian coordinate system (see Fig. 1.2), can be a function of time. Shear flows are important in many polymer processing operations, for example, in injection molding and extrusion.

For reasons of symmetry, the most general form that the stress tensor can have for the simple shear flow (1.4) is

$$\boldsymbol{\tau} = \begin{pmatrix} \tau_{11} & \tau_{12} & 0 \\ \tau_{12} & \tau_{22} & 0 \\ 0 & 0 & \tau_{33} \end{pmatrix}, \qquad (1.5)$$

where we actually display only the matrix of components characterizing the tensor $\boldsymbol{\tau}$ in the Cartesian coordinate system introduced in Fig. 1.2.[6] For incompressible fluids, one cannot distinguish between hydrostatic pressure and isotropic normal stress contributions resulting from $\boldsymbol{\tau}$ by measuring normal forces on surfaces. The occurrence of different diagonal components is a nonlinear effect arising from the anisotropic microstructure of polymeric liquids in flow.

For *steady shear flow* the shear rate is independent of time. A steady simple shear flow is easily generated between parallel plates as illustrated in Fig. 1.2, where it is assumed that the lower plate is fixed and the upper plate moves with constant velocity $\boldsymbol{V}$. For this type of flow, three independent combinations of stresses can be measured for an incompressible fluid. If it is presumed that the shear rate has been constant for such a long time that all the stresses in the fluid are time-independent, then the stress tensor can be fully characterized by three material functions depending only on the shear rate, the so-called *viscometric functions*:

$$\text{Viscosity:} \quad \eta(\dot{\gamma}) := -\tau_{12}(\dot{\gamma})/\dot{\gamma}\,, \qquad (1.6)$$
$$\text{First normal stress coefficient:} \quad \Psi_1(\dot{\gamma}) := -[\tau_{11}(\dot{\gamma}) - \tau_{22}(\dot{\gamma})]/\dot{\gamma}^2\,, \qquad (1.7)$$
$$\text{Second normal stress coefficient:} \quad \Psi_2(\dot{\gamma}) := -[\tau_{22}(\dot{\gamma}) - \tau_{33}(\dot{\gamma})]/\dot{\gamma}^2\,. \qquad (1.8)$$

An example of a time-dependent shear flow is the *start-up of steady shear flow* in which the fluid is at rest for $t < 0$ and in a state of flow described by (1.4) for $t > 0$. Then, for $t > 0$, the *stress growth coefficients* η^+, Ψ_1^+, and Ψ_2^+ depending on time and shear rate are defined by

$$\eta^+(t,\dot{\gamma}) := -\tau_{12}(t,\dot{\gamma})/\dot{\gamma}\,, \qquad (1.9)$$
$$\Psi_1^+(t,\dot{\gamma}) := -[\tau_{11}(t,\dot{\gamma}) - \tau_{22}(t,\dot{\gamma})]/\dot{\gamma}^2\,, \qquad (1.10)$$
$$\Psi_2^+(t,\dot{\gamma}) := -[\tau_{22}(t,\dot{\gamma}) - \tau_{33}(t,\dot{\gamma})]/\dot{\gamma}^2\,. \qquad (1.11)$$

The reverse experiment is *cessation of a steady shear flow* (1.4) at time $t = 0$. For $t > 0$ the *stress decay coefficients* η^-, Ψ_1^-, and Ψ_2^- are again defined by

---

[6] When a fixed coordinate system is given, we always specify a tensor by displaying the matrix of its components in that coordinate system.

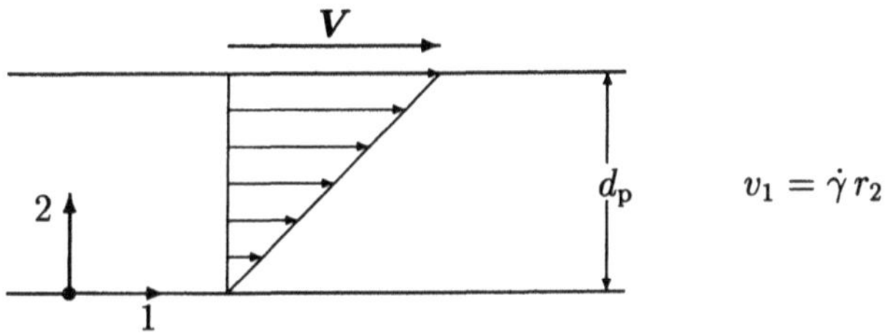

Fig. 1.2. Homogeneous simple shear flow between parallel plates with shear rate $\dot\gamma = |\boldsymbol{V}|/d_\mathrm{p}$.

$$\eta^-(t,\dot\gamma) := -\tau_{12}(t,\dot\gamma)/\dot\gamma\,, \tag{1.12}$$
$$\Psi_1^-(t,\dot\gamma) := -[\tau_{11}(t,\dot\gamma) - \tau_{22}(t,\dot\gamma)]/\dot\gamma^2\,, \tag{1.13}$$
$$\Psi_2^-(t,\dot\gamma) := -[\tau_{22}(t,\dot\gamma) - \tau_{33}(t,\dot\gamma)]/\dot\gamma^2\,. \tag{1.14}$$

Only in the limit $\dot\gamma \to 0$ are these stress decay coefficients simply related to the stress growth coefficients (1.9)–(1.11).

Another important simple shear flow with time-dependent shear rate is *oscillatory shear flow* for which the shear rate is of the form $\dot\gamma(t) = \dot\gamma_0 \cos\omega_0 t$ with fixed amplitude $\dot\gamma_0$ and frequency ω_0. In the limit of vanishing amplitude, the material behavior can be described by frequency-dependent material functions such as the *shear storage and loss moduli* (see, e.g., Appendix D of [2]). Equivalently, the material behavior in small amplitude oscillatory shear flow can be described by the *shear relaxation modulus* $G(t)$. If the shear deformations are kept small, that is, in the *linear viscoelastic regime*, the shear stress for otherwise arbitrary time-dependent shear rate can be expressed in terms of the shear relaxation modulus,

$$\tau_{12}(t) = -\int_{-\infty}^{t} G(t-t')\,\dot\gamma(t')\,\mathrm{d}t'\,. \tag{1.15}$$

For the material functions defined in (1.6), (1.9), and (1.12), this implies

$$\eta(0) = \int_0^\infty G(t')\,\mathrm{d}t'\,, \qquad \eta^+(t,0) = \eta(0) - \eta^-(t,0) = \int_0^t G(t')\,\mathrm{d}t'\,. \tag{1.16}$$

Instead of switching a constant shear rate on or off, one can also switch a constant shear stress on or off. The resulting experiments are known as *creep* and *recoil*.

For a detailed discussion of the typical, experimentally observed behavior of all the steady and time-dependent material functions mentioned above, and for a description of how they can be measured in a cone-and-plate rheometer, see §§3.3–3.4 of [1].

1.2.2 General Extensional Flows

The characteristics of shear-free or general extensional flows is that there exists a grid of orthogonally intersecting lines attached to the flowing material which always remains orthogonal in the course of the flow (see [12] for a more precise definition). Rather general shear-free flows of incompressible liquids are given by the velocity field

$$v_1 = \dot{\epsilon}\, r_1, \quad v_2 = m\, \dot{\epsilon}\, r_2, \quad v_3 = -(1+m)\, \dot{\epsilon}\, r_3, \tag{1.17}$$

where the largest principal strain rate $\dot{\epsilon}$ ($\dot{\epsilon} > 0$) and the ratio of the second largest and largest principal strain rates m ($-1/2 \leq m \leq 1$) can be functions of time. In other words, for these shear-free flows the matrix representing the transposed velocity gradient tensor $\boldsymbol{\kappa}$ in a suitable Cartesian coordinate system is diagonal. Extensional flows are found in many polymer processing operations, for example, in fiber spinning, film blowing, and sheet stretching.

For reasons of symmetry, the most general form that the stress tensor can have in that chosen coordinate system for the shear-free flow (1.17) is

$$\boldsymbol{\tau} = \begin{pmatrix} \tau_{11} & 0 & 0 \\ 0 & \tau_{22} & 0 \\ 0 & 0 & \tau_{33} \end{pmatrix} . \tag{1.18}$$

Hence, for general extensional flows of incompressible fluids, one needs only two normal stress differences to completely characterize the experimentally observable part of the stress tensor.

For *steady extensional flows*, the strain rate $\dot{\epsilon}$ and the parameter m are independent of time. The two experimentally accessible normal stress differences can be characterized by the *extensional viscosities* μ_1 and μ_2 [13]:

$$\mu_1 := -\frac{\tau_{11} - \tau_{33}}{2(2+m)\dot{\epsilon}}, \tag{1.19}$$

$$\mu_2 := -\frac{\tau_{22} - \tau_{33}}{2(1+2m)\dot{\epsilon}} . \tag{1.20}$$

If it is presumed that the strain rates have been constant for such a long time that all the stresses in the fluid are time-independent then the extensional viscosities μ_1 and μ_2 are functions of $\dot{\epsilon}$ and m only. The above definitions of the extensional viscosities have been chosen such that for a Newtonian liquid $\mu_1 = \mu_2 = \eta$.

In the same way as for simple shear flow one can investigate time-dependent *stress growth* and *stress decay* at start-up and cessation of steady shear-free flows of the form (1.17). The corresponding stress growth and stress decay coefficients are denoted by $\mu_1^{\pm}(t, \dot{\epsilon}, m)$ and $\mu_2^{\pm}(t, \dot{\epsilon}, m)$.

Several special steady or time-dependent shear-free flows are obtained for particular choices of the parameter m:

Simple (or tensile) extension:	$m = -1/2$
Equibiaxial extension:	$m = 1$
Planar extension:	$m = 0$

For equibiaxial extension, the extensional viscosities μ_1 and μ_2 are equal; for simple extension, the material function μ_2 is undefined. In both cases, the material behavior is completely characterized by the function μ_1, which is one third of the elongational or *Trouton viscosity* $\bar{\eta}$ defined in [1].

For a detailed discussion of the typical, experimentally observed behavior of the various steady and time-dependent material functions defined for general extensional flows, and for a description of experimental techniques, see [13] and §3.5 of [1]. The physical content of the various material properties in shear and extensional flows is explained in [14], where the various experimental techniques are also described in some detail.

1.2.3 The CONNFFESSIT Idea

In polymer fluid mechanics, one of the most active areas is the numerical solution of the conservation laws, along with appropriate constitutive equations. For the solution of viscoelastic fluid dynamics problems, the *finite element and finite difference methods* are of crucial importance. These numerical methods and their application to non-Newtonian flow problems have been described in detail in [15], with updates [16] and [17].

In the traditional approach, the balance equations (1.1) and (1.2) are closed by adding a differential equation or an integral expression for obtaining the stress tensor as a functional of deformation history. The numerical solution of the system of partial differential or integrodifferential equations obtained by combining conservation laws and constitutive equations is an extremely complicated mathematical problem. On top of this, there are the difficulties of deriving the closed-form constitutive equations from kinetic theory models and of formulating the proper boundary conditions which are required in this approach.

With the advent of powerful computers and of the efficient simulation techniques for the stochastic differential equations of motion obtained from kinetic theory models, which are the main subject of this book, a new approach has become possible. One can now combine simulations of a large number of coarse-grained molecules distributed over the flow region of interest with the finite element method for solving the momentum balance equation in order to solve the full flow problem. The molecules are deformed by the local flow field, and the stresses required for determining the time-evolution of that flow field are read off from the molecular configurations. No closed-form constitutive equation is required. A large number of polymer configurations evolving according to stochastic differential equations replaces the deterministic stress field as a basic dynamic variable. Like in real systems, the stresses are naturally obtained from "smart" molecules rather than from "artificial" constitutive equations. This technique, which is known as the CONNFFESSIT approach (Calculation of Non-Newtonian Flow: Finite Elements & Stochastic Simulation Techniques), was originally developed for the start-up of simple shear flow [18–20]. More recently, two-dimensional flow problems have been solved by CONNFFESSIT (axisymmetric contraction flow [21] and start-up of flow between eccentric cylinders).

Current experience shows that the CONNFFESSIT approach is numerically extremely robust. For example, the constraint that, for a Maxwell fluid, the tensor $\eta\boldsymbol{\delta} - \lambda\boldsymbol{\tau}$, which is a multiple of the so-called conformation tensor, must be positive-definite is automatically satisfied for any discretization procedure. The loss of positive-definiteness of this tensor as a consequence of inaccurate discretization procedures in the traditional approach leads to a loss of evolutionarity of the system of momentum balance and constitutive equations and hence to a breakdown of the numerical calculations.[7] The importance of this condition of positive-definiteness of the conformation tensor is discussed in Sect. 11.3.2 of [15] and in Sect. 9.8.10 of [17]. In view of the CONNFFESSIT idea, efficient simulation techniques not only provide a powerful tool for obtaining material functions from kinetic theory models in given homogeneous flows but also offer a unique possibility for solving flow problems (see Fig. 1.1 for the role of stochastic simulations in polymer kinetic theory and fluid dynamics).

Kinetic theory models for which only approximate or no constitutive equations are known can be treated rigorously by means of the CONNFFESSIT approach. This approach bridges the gap between the molecular models of polymer kinetic theory and the important problem of viscoelastic flow calculation. One can easily switch models by exchanging simulation subroutines, and the maximum possible information about the molecular stretch and orientation affecting the mechanical properties of finished plastic products is automatically available. These promising possibilities are here sketched as a motivation for learning more about stochastic simulations in polymer dynamics and about the required background from the theory of stochastic processes and stochastic differential equations. At least for the author, these exciting perspectives have been the motivation for summarizing and further developing state-of-the-art stochastic simulation techniques in this book.

References

1. Bird RB, Armstrong RC, Hassager O (1987) *Dynamics of Polymeric Liquids, Vol 1, Fluid Mechanics*, 2nd Edn. Wiley-Interscience, New York Chichester Brisbane Toronto Singapore
2. Bird RB, Curtiss CF, Armstrong RC, Hassager O (1987) *Dynamics of Polymeric Liquids, Vol 2, Kinetic Theory*, 2nd Edn. Wiley-Interscience, New York Chichester Brisbane Toronto Singapore
3. Bird RB, Wiest JM (1995) Annu. Rev. Fluid Mech. 27: 169
4. Bird RB, Öttinger HC (1992) Annu. Rev. Phys. Chem. 43: 371

[7] This loss of evolutionarity is caused by an instability of the Hadamard type in which short-wave disturbances sharply increase in amplitude.

5. Öttinger HC (1992) Rheol. Acta 31: 14
6. Beris AN, Mavrantzas VG (1994) J. Rheol. 38: 1235
7. Schneider J (1988) in: Giesekus H, Hibberd MF (eds) *Progress and Trends in Rheology II*: Proceedings of the Second Conference of European Rheologists. Steinkopff, Darmstadt, p 157 (Supplement to Rheologica Acta)
8. Curtiss CF, Bird RB (1996) to appear in Advances in Polymer Science
9. Lodge AS (1988) J. Rheol. 32: 93
10. Bhave AV, Armstrong RC, Brown RA (1991) J. Chem. Phys. 95: 2988
11. Öttinger HC, Rabin Y (1989) J. Rheol. 33: 725
12. Lodge AS (1974) *Body Tensor Fields in Continuum Mechanics*. Academic, New York San Francisco London
13. Demarmels A, Meissner J (1986) Coll. & Polym. Sci. 264: 829
14. Barnes HA, Hutton JF, Walters K (1989) *An Introduction to Rheology*. Elsevier, Amsterdam Oxford New York Tokyo (Rheology Series, Vol 3)
15. Crochet MJ, Davies AR, Walters K (1984) *Numerical Simulation of Non-Newtonian Flow*. Elsevier, Amsterdam Oxford New York Tokyo (Rheology Series, Vol 1)
16. Crochet MJ (1989) Rubber Chem. Technol. 62: 426
17. Keunings R (1989) in: Tucker CL (ed) *Fundamentals of Computer Modeling for Polymer Processing*. Hanser, Munich Vienna New York, p 403 (Computer Aided Engineering for Polymer Processing, Vol 2)
18. Öttinger HC, Laso M (1992) in: Moldenaers P, Keunings R (eds) *Theoretical and Applied Rheology, Vol 1*: Proceedings of the XIth International Congress on Rheology. Elsevier, Amsterdam London New York Tokyo, p 286
19. Laso M, Öttinger HC (1993) J. Non-Newtonian Fluid Mech. 47: 1
20. Öttinger HC, Laso M (1994) in: López de Haro M, Varea C (eds) *Lectures on Thermodynamics and Statistical Mechanics*: Proceedings of the XXII Winter Meeting on Statistical Physics. World Scientific, Singapore New Jersey London Hong Kong, p 139
21. Feigl K, Laso M, Öttinger HC (1995) Macromolecules 28: 3261

Part I

Stochastic Processes

One might ask why it is necessary to read more than a hundred pages on mathematical stochastics if one is interested in gaining a fundamental understanding of the flow behavior of polymeric liquids? As pointed out in the preface, the basic equations describing the polymer dynamics in the coarse-grained models of kinetic theory are stochastic in character. In order to recognize and understand the stochastic nature and content of these kinetic theory equations it is crucial first to develop the underlying theory of stochastics. To reassure the hesitant reader it should be pointed out that only those concepts of stochastics are developed here that are really useful in understanding polymeric fluids. A considerable amount of space is devoted to elucidating and illustrating these concepts.

For the mathematician it may be interesting and surprising to see how many advanced concepts and results of mathematical stochastics are actually useful, if not even indispensable, in the kinetic theory of polymeric fluids. The most fundamental concept in describing dynamical systems subject to random noise is that of a *stochastic process*. On the way to introducing stochastic processes, the notions of events, probabilities, and random variables need to be introduced first. In understanding and formulating the dynamics of a system which is modeled in terms of a stochastic process, stochastic differential equations play a crucial role. Therefore, a broad background from the theory of stochastic processes must be provided. Since these mathematical foundations are of a very general and generic nature, the *stepping-stone to stochastics* offered in the first part of this book should be helpful in many other branches of science and engineering as well.

In applied textbooks, random variables and stochastic processes are defined through their probability distributions. In traditional polymer kinetic theory, for example, diffusion equations which describe the time evolution of probability densities in polymer configuration space are ubiquitous. Working on the level of distributions is clearly unsatisfactory or even insufficient if one wishes to perform computer simulations: one creates trajectories, that is realizations of the random objects themselves rather than their distributions. It is hence mandatory to have a clear definition of "trajectories," not only to dispel all doubts about the fundamental soundness of the theory of stochastic differential equations and of stochastic simulation techniques, but also as a starting point for developing more efficient simulation algorithms and for gaining a feeling for the effects of approximations or changes in some model parameter on the level of trajectories. For example, the effect of an approximation on the time-evolution of a system may be much more easily grasped by looking at some typical trajectories rather than at the time evolution of an average. The precise understanding of the concept of trajectories may also be important for introducing constraints, implementing stochastic time-step control or, quite generally, for exploiting advanced mathematical results in solving applied problems. In view of the highly developed state of mathematical stochastics, bridging the gap between this branch of mathematics and the natural and applied sciences is very important.

2. Basic Concepts from Stochastics

Stochastics is concerned with the mathematical analysis of intuitive notions like "chance," "randomness," "fluctuations," or "noise." In order to develop a *sound mathematical theory of stochastics* it is necessary to give rigorous definitions for the following basic concepts, which have an intuitive meaning based on everyday experience, especially from games of chance: event, probability, random variable, expectation, stochastic independence, conditioning, convergence, and stochastic process. The formal definitions of these objects in this chapter are made clear by many examples. The examples and exercises of this chapter play a most important role in this book because they not only illustrate the meaning of abstract concepts, but also contain important information required in the subsequent chapters.

There are many recommendable basic textbooks on mathematical stochastics; one of the oldest and most popular, especially among natural scientists and engineers, is the comprehensive two-volume book by W. Feller [1]. The explanation of the basic concepts given below follows a textbook by M. Loève [2]. Rigorous proofs of all the theorems and results cited in this chapter can, for example, be found in a very systematic and careful introduction to stochastics by H. Bauer [3]. A clear and concise introduction to the basic concepts is given in a book by L. Arnold [4].

2.1 Events and Probabilities

As a sound foundation for the entire structure of probability theory we first need to develop axiomatic definitions for two elementary notions, namely, event and probability. This can be done by analyzing the implications of the intuitive perception of these notions. The rigorous formulation of mathematical stochastics based on the axiomatization of events and probabilities was developed by the versatile Russian scientist Andrei Nikolaevich Kolmogorov (1903–1987) in the early 1930s. The amazingly modern approach of his pioneering textbook [5], which was first published in 1933 (in German), underlines Kolmogorov's immense influence on the formulation and development of modern stochastics.

2.1.1 Events and σ-Algebras

In order to develop a rigorous, axiomatic definition of the concept of *event* we first analyze the intuitive meaning of this notion. Since the notion of event is related to the occurrence or non-occurrence of certain phenomena or outcomes in an experiment one intuitively expects that various ways of comparing and combining events must be possible. If A and B are events, it should certainly be possible to give well-defined meanings to the following intuitive notions:

"A implies B"	$A \subset B$
"A and B are equivalent"	$A = B$
"A and B"	$A \cap B$
"A or B"	$A \cup B$
"not A"	A^C

For example, "A implies B" means that whenever the event A occurs, then the event B occurs necessarily. "A and B are equivalent" if and only if "A implies B" and "B implies A." "A and (or) B" is an event which occurs if and only if both the event A and the event B occur (at least one of the events A or B occurs). Finally, the event "not A" occurs if and only if A does not occur. The formal notation used for these intuitively explained relations and operations for events is taken from set theory; as a matter of fact, Kolmogorov's axiomatic definition of events (see Definition 2.1 below) relies on the framework of set theory.

Once the above relations and operations are defined, special events which always or never occur can easily be introduced, and further relationships between events can be defined:

"sure event Ω"	$\Omega = A \cup A^C$
"impossible event $\emptyset$"	$\emptyset = A \cap A^C$
"A and B are incompatible"	$A \cap B = \emptyset$

With the interpretation of the symbols $\subset$, $=$, $\cap$, $\cup$, and ^{C} in terms of occurrence and non-occurrence of certain phenomena in mind, one expects the following relationships to be true:

$$\begin{aligned} A \cap B &= B \cap A, & A \cup B &= B \cup A\,; \\ (A \cap B) \cap C &= A \cap (B \cap C), & (A \cup B) \cup C &= A \cup (B \cup C)\,; \\ A \cap (B \cup C) &= (A \cap B) \cup (A \cap C), & A \cup (B \cap C) &= (A \cup B) \cap (A \cup C)\,; \\ (A \cap B)^C &= A^C \cup B^C, & (A \cup B)^C &= A^C \cap B^C\,; \end{aligned}$$
$$(A^C)^C = A\,.$$

In other words, the operations $\cap$ and $\cup$ are commutative, associative, and distributive, and the operation $\cap$ can be defined in terms of the operations ^{C} and $\cup$ (or, alternatively, $\cup$ can be defined in terms of ^{C} and $\cap$). As an example, the distributive law $A \cup (B \cap C) = (A \cup B) \cap (A \cup C)$ is illustrated in Fig. 2.1.

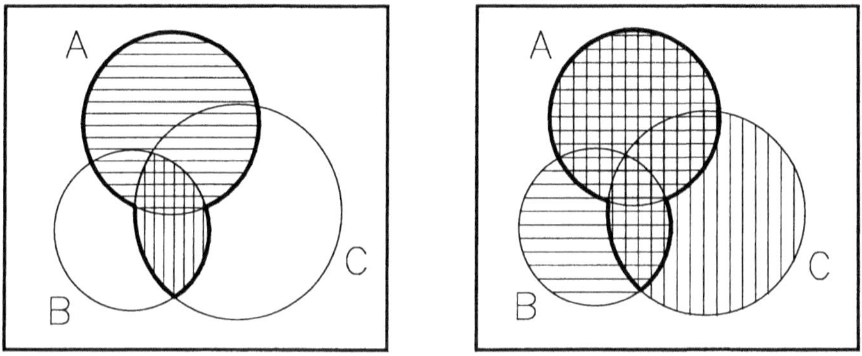

Fig. 2.1. Illustration of the distributive law $A \cup (B \cap C) = (A \cup B) \cap (A \cup C)$

In the left diagram, the occurrence of the events A and $B \cap C$ is indicated by horizontal and vertical hatching, respectively; the event $A \cup (B \cap C)$ occurs for the outcomes in the region bounded by the thick curve. In the right diagram, the events $A \cup B$ and $A \cup C$ are indicated by horizontal and vertical hatching, respectively; the event $(A \cup B) \cap (A \cup C)$ occurs again for the outcomes in the region bounded by the thick curve. A similar illustration of the other rules for combining events is left as an exercise to the reader.

In summary, by working out some implications of the intuitive notion of event we discovered the rules for operations on sets. This suggests interpreting *events as subsets* of a given set Ω which contains all possible outcomes of an experiment and thus obviously depends on the situation to be described. For example, in modeling a game of dice we can choose

$$\Omega = \{\text{⚀}, \text{⚁}, \text{⚂}, \text{⚃}, \text{⚄}, \text{⚅}\}$$

or, more conveniently, $\Omega = \{1, 2, 3, 4, 5, 6\}$ as the set of all possible outcomes (or sure event).[1] In all the examples occurring in the second part of this book, Ω is some set of real numbers ($\Omega \subset \mathbb{R}$) or of arrays of real numbers characterizing polymer configurations ($\Omega \subset \mathbb{R}^d$). For flexible dumbbell models of polymers, for example, $\Omega = \mathbb{R}^3$ represents all possible internal configurations. In terms of sets, $A \subset B$ means that A is a *subset* of B, $A \cap B$ is the *intersection* and $A \cup B$ the *union* of two sets A and B, A^C is the set *complementary* to the set A in Ω, and $\emptyset$ is the *empty set*.

For a given set Ω of all possible outcomes of an experiment, one could admit all subsets of Ω as possible events. In many applications, however, this set of

[1]Following standard notation, we use braces whenever we specify a set by listing its elements.

events would be unnecessarily large and much effort could be saved by introducing some smaller set of events containing only the subsets "of interest." In the case of infinite sets Ω, the set of all subsets of Ω may furthermore be an extremely large set (of higher order infinity) and mathematical difficulties can arise later when introducing probabilities. For example, the continuous probability measures on $\mathbb{R}$ introduced in the next subsection cannot be defined if all subsets of $\mathbb{R}$ are admitted as events.

Hence, instead of using the set of all subsets of Ω, denoted as $\mathcal{P}(\Omega)$, one often uses a smaller set $\mathcal{A} \subset \mathcal{P}(\Omega)$ as the set of possible events in a given situation (all the subsets of Ω contained in the given set $\mathcal{A}$ are then referred to as "events"). One would certainly always want the "sure event Ω" to be in the set of events, $\mathcal{A}$. Furthermore, if A and B are events in $\mathcal{A}$, then "not A" and "A or B" should also be events in $\mathcal{A}$. These considerations lead to the following axiomatic definition of σ-algebras $\mathcal{A}$ as possible sets of events.

Definition 2.1 *A set $\mathcal{A}$ of subsets of Ω, $\mathcal{A} \subset \mathcal{P}(\Omega)$, is a* σ-algebra *(in Ω) if and only if $\mathcal{A}$ satisfies the following axioms:*

$$\Omega \in \mathcal{A}, \tag{2.1}$$

$$\text{for every } A \in \mathcal{A}: \quad A^C \in \mathcal{A}, \tag{2.2}$$

$$\text{for every sequence } (A_j)_{j \in \mathbb{N}} \text{ in } \mathcal{A}: \quad \bigcup_{j=1}^{\infty} A_j \in \mathcal{A}. \tag{2.3}$$

The sets contained in a σ-algebra are referred to as events. *A set Ω in which a σ-algebra $\mathcal{A}$ is selected is called a* measurable space $(\Omega, \mathcal{A})$.

It should be pointed out for clarity that $\mathcal{A}$ may have uncountably many elements and that the sets A_j in axiom (2.3) are not necessarily distinct. In some cases, in particular in formal developments and as long as no probability measure is introduced on a measurable space, we also refer to the elements of σ-algebras as *measurable sets*.

At first glance, the admission of infinite sequences in the third of the above axioms for σ-algebras may be somewhat surprising. From the above explanation, one would expect the union of two and hence of any finite number of sets in $\mathcal{A}$ to be contained in $\mathcal{A}$, too (in which case $\mathcal{A}$ would be referred to as an algebra rather than a σ-algebra). In the derivation of many profound results of probability theory it is, however, crucial to admit countably infinite sequences of subsets in the axiom (2.3). It is, for example, easy to imagine why consideration of infinite sequences of events is important in proving limit theorems or in handling stochastic processes.

The axioms (2.1)–(2.3) have several immediate consequences. According to (2.1), a σ-algebra is a non-empty set. The axioms (2.1) and (2.2) imply that the impossible event, $\emptyset = \Omega^C$, must be contained in every σ-algebra; the "Ω" in axiom (2.1) may hence equivalently be replaced by "$\emptyset$." The identity

$$\bigcap_{j=1}^{\infty} A_j = \Big(\bigcup_{j=1}^{\infty} A_j^C\Big)^C$$

in combination with the axioms (2.2) and (2.3) guarantees that the intersection of a sequence of sets $(A_j)_{j\in\mathbb{N}}$ in $\mathcal{A}$ is also contained in $\mathcal{A}$; the union in axiom (2.3) may hence equivalently be replaced by the intersection.

Exercise 2.2 Show that the union and intersection of any finite number of sets in $\mathcal{A}$ is also an element of $\mathcal{A}$.

In an arbitrary set Ω, the set $\mathcal{A} = \mathcal{P}(\Omega)$ satisfies the axioms (2.1)–(2.3), and $\mathcal{P}(\Omega)$ is hence a σ-algebra. In general, however, $\mathcal{P}(\Omega)$ "resolves too-detailed information" which is often not actually required in modeling a certain experiment or certain phenomena. The following example should help to explain such vague statements and to illustrate the general idea.

Example 2.3 Events in a Game of Dice
When tossing a die once, the natural set of all possible outcomes is $\Omega = \{1, 2, 3, 4, 5, 6\}$. We consider two examples of σ-algebras in Ω in order to illustrate this concept:

1. In general, when tossing a die, one would like each set $A_j = \{j\}$, $j = 1, \ldots 6$ containing only a single element to be a possible event ("the event of throwing j"). Since any non-empty set $A \subset \Omega$ can be expressed as $A = \bigcup_{j\in A} A_j$, the only σ-algebra containing all sets A_j is $\mathcal{A}_1 = \mathcal{P}(\Omega)$ (remember that in every set Ω, the set $\mathcal{P}(\Omega)$ is a σ-algebra).

2. In some games of dice, one is interested only in throwing a "6." The smallest σ-algebra containing $\{6\}$ as an event is $\mathcal{A}_2 = \{\Omega, \emptyset, \{6\}, \{1, 2, 3, 4, 5\}\}$. The reader should verify that $\mathcal{A}_2$ indeed satisfies the axioms (2.1)–(2.3). When the σ-algebra $\mathcal{A}_2$ is used, the sets A_j for $j \neq 6$ are not events! When interested only in throwing a "6," one does not need to "resolve any further details" of the event $\{1, 2, 3, 4, 5\}$.

For a finite or countable set Ω, one in general chooses $\mathcal{P}(\Omega)$ as the most natural σ-algebra. Thus, the natural σ-algebra for a game of dice is $\mathcal{A}_1$. The σ-algebra $\mathcal{A}_2$, on the other hand, seems to be rather artificial. We leave it as a problem to the reader to figure out situations in which any normal person would choose $\mathcal{A}_2$ as the most natural σ-algebra and where $\mathcal{A}_1$ would be a rather eccentric choice. (Hint: Imagine a game of Russian roulette with a six-shot revolver; then, the issue of interest can be transformed from "6 or not 6" into "to be or not to be.") □

Exercise 2.4 What is Ω in a game of roulette? Think of different gambling strategies and how they are reflected in the choice of an appropriate σ-algebra. List some possible σ-algebras in Ω.

In order to specify a σ-algebra it is very useful to introduce the notion of a generating system for a σ-algebra. In the example of a game of dice, the σ-algebra $\mathcal{A}_1$ was obtained by the requirement that all one-element sets A_j should be events, and $\mathcal{A}_2$ was the minimum choice for having $\{6\}$ as an event; respective generating systems are $\mathcal{E}_1 = \{A_j \mid j = 1, \dots 6\}$ and $\mathcal{E}_2 = \big\{\{6\}\big\}$.[2] The general concept of a generating system is introduced in the following definition.

Definition 2.5 *Let $\mathcal{E} \subset \mathcal{P}(\Omega)$ be a non-empty set of subsets of Ω. The smallest σ-algebra containing $\mathcal{E}$ is referred to as the* σ-algebra generated by *$\mathcal{E}$ and is denoted by $\mathcal{A}(\mathcal{E})$. The set $\mathcal{E}$ is a* generating system *for the σ-algebra $\mathcal{A}$ if and only if $\mathcal{A}(\mathcal{E}) = \mathcal{A}$.*

The existence of a smallest σ-algebra containing $\mathcal{E}$ is not obvious. Notice, however, that the intersection of an arbitrary number of σ-algebras is again a σ-algebra. Therefore, the smallest σ-algebra containing $\mathcal{E}$ can be obtained as the intersection of all σ-algebras containing $\mathcal{E}$.

Exercise 2.6 Give convenient generating systems for the σ-algebras which, in Exercise 2.4, you found appropriate for different gambling strategies in a game of roulette.

For the σ-algebra $\mathcal{A}_2$ of Example 2.3, we can now write $\mathcal{A}_2 = \mathcal{A}(\mathcal{E}_2) = \mathcal{A}\left(\big\{\{6\}\big\}\right)$. A most important generating system and the σ-algebra generated by that system are discussed in the following example. The σ-algebra defined in that example is implicitly assumed whenever stochastics in $\mathbb{R}^d$ is used in the applied literature without going into mathematical details.

Example 2.7 Borel σ-Algebra

For $\Omega = \mathbb{R}$, one would like all open intervals to constitute possible events. We hence consider the σ-algebra $\mathcal{B} := \mathcal{A}(\mathcal{E}_1)$ generated by the system $\mathcal{E}_1 = \{A \subset \mathbb{R} \,|A =]a, b[\text{ with } a, b \in \mathbb{R},\ a \le b\}$. The σ-algebra $\mathcal{B}$ generated by the open intervals $]a, b[$ is known as the *Borel σ-algebra*.

Since any one-element subset $\{a\} \subset \mathbb{R}$ can be written in the form

$$\{a\} = \bigcap_{j=1}^{\infty} \left]a - \frac{1}{j}, a + \frac{1}{j}\right[, \tag{2.4}$$

one concludes that $\{a\} \in \mathcal{B}$ for every set containing only a single element $a \in \mathbb{R}$. For that reason, every interval of the form $[a, b]$, $[a, b[$, or $]a, b]$ is contained in $\mathcal{B}$, too, and it does not matter which type of interval is used in the generating system $\mathcal{E}_1$ (notice that the analog of (2.4) holds for each of the above types of intervals). An even more convenient generating system consists of the intervals $]-\infty, a]$ for $a \in \mathbb{R}$.

For the system $\mathcal{E}_2 = \{\, A \subset \mathbb{R} \mid A \text{ is an open set } \}$, which is obviously larger than $\mathcal{E}_1$, one also obtains $\mathcal{B} = \mathcal{A}(\mathcal{E}_2)$. This is a consequence of the fact that every open set in $\mathbb{R}$ can be written as the union of a countable system of open

[2] Rather than listing all the elements of a set we often characterize its elements by a property that is specified after a vertical bar; the vertical bar may be read as "for" or "for which."

intervals and must hence be contained in $\mathcal{A}(\mathcal{E}_1)$. Since the Borel σ-algebra can alternatively be defined as the σ-algebra generated by the system of open (or closed, or compact) sets this concept can immediately be generalized to any topological space Ω. The Borel σ-algebra is the most natural σ-algebra for a topological space Ω. □

Example 2.8 Product σ-Algebra

For later reference we here develop the concept of *product σ-algebras*. On a first reading, this example may look rather formal, but the ideas should become clear when they are used later in this chapter (in particular, in connection with the concept of stochastic independence).

Since $\mathbb{R}^d$ is a topological space, there is a Borel σ-algebra $\mathcal{B}^d$ defined on $\Omega = \mathbb{R}^d$. This σ-algebra is generated by open "d-dimensional intervals" of the form $]a_1, b_1[\times]a_2, b_2[\times \ldots \times]a_d, b_d[$. This is an example of the more general construction of the product σ-algebra $\mathcal{A}_1 \otimes \mathcal{A}_2 \otimes \ldots \otimes \mathcal{A}_d$ of d given σ-algebras on the product set $\Omega_1 \times \Omega_2 \times \ldots \times \Omega_d$ when d measurable spaces $(\Omega_j, \mathcal{A}_j)$ are given. If, for each $j = 1, \ldots d$, $\mathcal{E}_j$ is a generating system of $\mathcal{A}_j$ containing Ω_j (or, more generally, containing an increasing sequence $A_{j1} \subset A_{j2} \subset A_{j3} \subset \ldots$ with $A_{j1} \cup A_{j2} \cup A_{j3} \cup \ldots = \Omega_j$), then $\mathcal{A}_1 \otimes \mathcal{A}_2 \otimes \ldots \otimes \mathcal{A}_d$ is the σ-algebra generated by the set of all products $A_1 \times A_2 \times \ldots \times A_d$ with $A_j \in \mathcal{E}_j$ (this construction is independent of the choice of the generating systems; in particular, one may choose $\mathcal{E}_j = \mathcal{A}_j$).

The σ-algebra $\mathcal{A}_1 \otimes \mathcal{A}_2 \otimes \ldots \otimes \mathcal{A}_d$ can also be generated by all products of the form $A_1 \times A_2 \times \ldots \times A_d$ where at most one $A_j \in \mathcal{A}_j$ is not equal to Ω_j; the general set $A_1 \times A_2 \times \ldots \times A_d$ can be obtained as the intersection of d sets with only one factor not being the full space. This construction motivates the following generalization: for an infinite product of measurable spaces, we define the product σ-algebra as the σ-algebra generated by all infinite products of sets in which at most one factor is not the full space. The same σ-algebra is obtained when a finite or countable number of factors are allowed to be arbitrary events from the corresponding σ-algebras. □

The Borel σ-algebra $\mathcal{B}^d$ is of great importance in the subsequent chapters because in all kinetic theory models of polymer solutions and melts the polymer configurations are represented as column vectors in some space $\mathbb{R}^d$. In the remainder of this book, it is understood that $\Omega = \mathbb{R}$ or $\mathbb{R}^d$ is always considered together with the corresponding natural σ-algebra, $\mathcal{B}$ or $\mathcal{B}^d$. When Ω is a subset of $\mathbb{R}^d$, we use $\mathcal{A} := \{B \cap \Omega \mid B \in \mathcal{B}^d\}$ as the natural σ-algebra in Ω. Note that this construction of a σ-algebra $\mathcal{A}$ for a subset $\Omega \subset \Omega'$ works for every measurable space $(\Omega', \mathcal{A}')$, even if $\Omega \notin \mathcal{A}'$.

Exercise 2.9 Have you ever encountered a subset of $\mathbb{R}$ not contained in $\mathcal{B}$?

2.1.2 Probability Axioms

The intuitive meaning of the notion of *probability* is associated with the relative frequency of the occurrence of certain events in repeated trials. Suppose that all possible outcomes for a given experiment are summarized in a set Ω, and that the same experiment can be performed n times where the outcome of each of these n trials is not affected by the outcomes of all the other $n-1$ trials ("completely independent, identical trials"). The possibility of repeated trials is a fundamental assumption in science and in games of chance. One assumes that every experiment can be performed again and again, the knowledge of past and present outcomes having no influence on future ones.

Let the number of occurrences of the events A and B in n repeated trials be n_A and n_B, respectively. Then, $f(A) = n_A/n$ is the *relative frequency* of the event A. For sufficiently large numbers of trials n, one expects the relative frequency $f(A) \in [0,1]$ to approach a well-defined number, which is called the probability $P(A)$ of the event A in the given experiment. If the events A and B are incompatible, $A \cap B = \emptyset$, one has $f(A \cup B) = (n_A + n_B)/n = f(A) + f(B)$. Furthermore, the relative frequency of the set of all possible outcomes or sure event must be unity. We are hence led to the following axiomatic definition of probabilities.

Definition 2.10 *Let $(\Omega, \mathcal{A})$ be a measurable space. A function $P : \mathcal{A} \to \mathbb{R}$ is a* probability measure *if, and only if, P satisfies the following axioms:*[3]

$$\text{for every } A \in \mathcal{A}: \quad P(A) \geq 0\,, \tag{2.5}$$

$$P(\Omega) = 1\,, \tag{2.6}$$

$$\begin{array}{r} \text{for every sequence of non-} \\ \text{overlapping sets } (A_j)_{j\in\mathbb{N}} \text{ in } \mathcal{A} \\ (\text{i.e., } A_j \cap A_k = \emptyset \text{ for all } j \neq k): \end{array} \quad P\Big(\bigcup_{j=1}^{\infty} A_j\Big) = \sum_{j=1}^{\infty} P(A_j)\,. \tag{2.7}$$

The triple $(\Omega, \mathcal{A}, P)$ is referred to as a probability space.

The axioms (2.5)–(2.7) have several immediate consequences: (i) $P(\emptyset) = 0$; a set A with $P(A) = 0$ is referred to as a *null set*, and the impossible event $\emptyset$ is thus an example of a null set. (ii) The additivity property (2.7) holds also for any finite number of events A_j. (iii) $P(A^C) = 1 - P(A)$. (iv) $P(A) \leq P(B)$ for $A \subset B$.

Exercise 2.11 Prove the above properties (i)–(iv).

Example 2.12 Psychology of Throwing Dice
In order to model a game of dice, we use $\Omega = \{1,2,3,4,5,6\}$ and $\mathcal{A} = \mathcal{A}_2 =$

[3] In our notation for functions, P is the name of the function, $\mathcal{A}$ is its domain, and $\mathbb{R}$ is its range. In addition, the concrete form of a function can be specified by means of the symbol "$\mapsto$"; for example, we write $g : \mathbb{R} \to [-1,1]$, $x \mapsto \sin x$ for the sine function.

$\mathcal{A}\left(\{\{6\}\}\right)$ as introduced in Example 2.3. One can check that the function $P : \mathcal{A}_2 \to \mathbb{R}$ defined by $P(\Omega) = 1$, $P(\emptyset) = 0$, $P(\{6\}) = \frac{1}{7}$, and $P(\{1,2,3,4,5\}) = \frac{6}{7}$ satisfies the axioms (2.5)–(2.7) and hence constitutes a probability measure on $(\Omega, \mathcal{A}_2)$.

The definition $P(\{6\}) = \frac{1}{7}$, which unambiguously characterizes the probability measure P (cf. a general theorem cited in Example 2.16), illustrates that the axioms (2.5)–(2.7) specify only the general, mathematical properties of probability measures, whereas the determination of a reasonable measure for a concrete problem requires intuition or plausible assumptions. The unexpected value $\frac{1}{7}$, rather than the usual $\frac{1}{6}$, is based on bad experience; it seems to be "observed" in situations where one is impatiently waiting for a "6." For the always lucky opponents in the same situation, the $\frac{1}{7}$ should, again according to experience, be replaced by $\frac{1}{5}$. □

Example 2.13 Probability Measures in Discrete Spaces
Let $\Omega = \{\omega_1, \omega_2, \ldots\}$ be a countable set; we assume here that Ω is infinite, although exactly the same procedure can be used for finite Ω. Given a sequence of nonnegative numbers $(p_j)_{j\in\mathbb{N}}$ with $\sum_{j=1}^{\infty} p_j = 1$, a probability measure can be defined for an arbitrary σ-algebra $\mathcal{A}$ in Ω as follows:

$$P(A) := \sum_{j\in I(A)} p_j \quad \text{for every } A \in \mathcal{A}\,, \tag{2.8}$$

where $I(A) := \{j \mid \omega_j \in A\}$. For the natural σ-algebra in a discrete space Ω, $\mathcal{A} = \mathcal{P}(\Omega)$, this probability measure is completely characterized by fixing the values

$$P\left(\{\omega_j\}\right) = p_j\,. \tag{2.9}$$

In general, it is not clear that $\{\omega_j\} \in \mathcal{A}$. □

Example 2.14 Poissonian Probability Measure
For $\Omega = \{0, 1, 2, \ldots\}$ and $\mathcal{A} = \mathcal{P}(\Omega)$, the Poissonian probability measure (or Poissonian distribution) is defined by

$$P(\{j\}) := p_j = \mathrm{e}^{-\alpha}\,\frac{\alpha^j}{j!}\,,$$

where $\alpha \in \mathbb{R}$ is assumed to be positive. The Poissonian distribution plays a central role in the study of random variables which take on nonnegative integer values. □

Example 2.15 Measures Concentrated at a Single Point
Given a measurable space $(\Omega, \mathcal{A})$ and $\omega \in \Omega$, the function

$$P_\omega : \mathcal{A} \to \mathbb{R}, \quad A \mapsto P_\omega(A) = \begin{cases} 1 & \text{if } \omega \in A \\ 0 & \text{if } \omega \notin A \end{cases}$$

constitutes a probability measure. The entire probability is located at the single point ω; the deterministic outcome of an experiment described by P_ω is ω. □

Example 2.16 Continuous Probability Measures
Let $(\Omega, \mathcal{A}) = (\mathbb{R}, \mathcal{B})$, $p : \mathbb{R} \to \mathbb{R}$, $x \mapsto p(x)$ be an integrable (for example, a piecewise continuous) function with $p(x) \geq 0$ and $\int p(x)\, dx = 1$ (when no limits of integration are specified, it is always assumed that the integration extends over the full domain of the integrand). One can show that there exists a unique probability measure P on $(\mathbb{R}, \mathcal{B})$ with

$$\boxed{P(]a, b[) := \int_a^b p(x)\, dx \quad \text{for} \quad a \leq b.} \tag{2.10}$$

Then, P is referred to as a *continuous probability measure* with *probability density* p. The uniqueness of P, which is so far defined only on the elements of the generating system $\mathcal{E}_1$ of Example 2.7, is a consequence of the following theorem: If $\mathcal{E} \subset \mathcal{P}(\Omega)$ is a $\cap$-stable generating system of $\mathcal{A}$ (i.e., $A, B \in \mathcal{E}$ implies $A \cap B \in \mathcal{E}$), then every probability measure on $(\Omega, \mathcal{A})$ is determined by its values on $\mathcal{E}$. For proving the existence of the measure P, it is crucial that the additivity of integrals guarantees the validity of the condition (2.7). The validity of (2.5) and (2.6) trivially follows from the assumed properties of p. In many applied textbooks using stochastics, the concept of probability density defined through (2.10) is introduced more vaguely by saying something like "$p(x)\, dx$ is the probability of finding a value of x in an interval of length dx around x."

A convenient means for roughly characterizing a continuous probability measure with density p is the *mean* α and the *width* σ defined as follows,

$$\alpha := \int x\, p(x)\, dx\,, \tag{2.11}$$

$$\sigma^2 := \int (x-\alpha)^2 p(x)\, dx = \int x^2 p(x)\, dx - \alpha^2\,, \tag{2.12}$$

provided that the corresponding integrals exist. The full information about the probability density p is contained in the Fourier transform

$$\tilde{p}(q) := \int e^{iqx}\, p(x)\, dx\,. \tag{2.13}$$

The Fourier transform of a probability density is often referred to as the *characteristic function* of the corresponding probability measure. The mean and width can be obtained from the first two derivatives of $\tilde{p}(q)$ with respect to q, evaluated at $q = 0$ (a more general result is given in Theorem 2.45 below). In the subsequent sections, characteristic functions will turn out to be a very powerful tool for handling Gaussian probability measures, marginal distributions, sums of independent random variables, and the convergence of random variables.

The probability measure P_ω on $(\mathbb{R}, \mathcal{B})$ concentrated at a single point $\omega \in \mathbb{R}$ as defined in Example 2.15 is sometimes treated as if it were a continuous probability measure with density $p_\omega(x) = \delta(x - \omega)$, where $\delta(x)$ is the Dirac δ-function (which is a generalized function).

The generalization of the above construction of continuous probability measures to the measurable space $(\mathbb{R}^d, \mathcal{B}^d)$ is straightforward. In that case, the variables x and q are replaced by d-dimensional column vectors, and all integrations must be performed in d dimensions. Also the mean α must be replaced with a column vector, and σ^2 must be generalized to a $d \times d$-matrix. For example, the characteristic function of a d-dimensional probability measure with density $p(\boldsymbol{x})$ is

$$\tilde{p}(\boldsymbol{q}) := \int e^{i\boldsymbol{q}\cdot\boldsymbol{x}}\, p(\boldsymbol{x})\, \mathrm{d}^d x\,, \tag{2.14}$$

where $\int \mathrm{d}^d x := \int \mathrm{d}x_1 \int \mathrm{d}x_2 \ldots \int \mathrm{d}x_d$, and $\boldsymbol{q}\cdot\boldsymbol{x}$ is the standard scalar product (or the contraction) of the column vectors $\boldsymbol{q}, \boldsymbol{x} \in \mathbb{R}^d$. □

Exercise 2.17 Calculate $\boldsymbol{\alpha}$, $\boldsymbol{\sigma}^2$ and $\tilde{p}(\boldsymbol{q})$ for the probability measure with density $p(x_1, x_2) = 1$ for $(x_1, x_2) \in [0,1] \times [0,1]$ and $p(x_1, x_2) = 0$ otherwise.

In many applied textbooks, the entire theory of stochastics is built on the discrete and continuous probability measures introduced separately in Examples 2.13 and 2.16 without specifying any σ-algebras. In a more general and rigorous approach, the concept of probability space $(\Omega, \mathcal{A}, P)$ is the common basis of stochastics.

The interpretation of probability in terms of relative frequency given in the beginning of this subsection immediately suggests definitions of conditional probabilities and stochastic independence of events for a given probability measure P on a measurable space $(\Omega, \mathcal{A})$. When A and B are arbitrary events $(A, B \in \mathcal{A})$, the relative frequency of A in the n_B trials in which B occurs is n_{AB}/n_B, where n_{AB} is the number of occurrences of the combined event "A and B." Since $n_{AB}/n_B = f(A \cap B)/f(B)$, the *conditional probability* of A under the condition that the event B with $P(B) > 0$ occurs is naturally defined as

$$P(A|B) := \frac{P(A \cap B)}{P(B)}\,. \tag{2.15}$$

Exercise 2.18 Show that the function $A \mapsto P(A|B)$ is a probability measure.

For stochastically independent events A and B, the probability of A should be independent of whether or not B occurs, $P(A|B) = P(A)$. One can hence define the stochastic *independence* of two arbitrary events A and B for a given probability measure P by the following, symmetric condition,

$$P(A \cap B) = P(A)P(B)\,.$$

Independence is intimately related to the factorization of the probabilities of combined events. More generally, we adopt the following definitions in which $\mathbb{T}$ represents an arbitrary set of labels and the underlying probability space $(\Omega, \mathcal{A}, P)$ is given.

Definition 2.19 *The events $(A_t)_{t\in\mathbb{T}}$ are said to be* independent *if and only if for every finite subset $\{t_1, t_2, \ldots t_n\} \subset \mathbb{T}$*

$$P(A_{t_1} \cap A_{t_2} \cap \ldots \cap A_{t_n}) = P(A_{t_1})\, P(A_{t_2}) \ldots P(A_{t_n})\,. \tag{2.16}$$

From this definition it is obvious that the concept of independence crucially depends on the underlying probability measure. This concept is of great importance in many practical applications because the independence of events can often be justified by plausible arguments, and because it has far-reaching consequences. Particularly important results are the strong law of large numbers, which is the basis of many stochastic simulation techniques, and the central limit theorem, which explains the importance of Gaussian probability measures.

Exercise 2.20 Does the property $P(A_1 \cap A_2 \cap A_3) = P(A_1)\,P(A_2)\,P(A_3)$ imply the independence of the events A_1, A_2, A_3?

Definition 2.21 *The sets of events* $(\mathcal{E}_t)_{t\in\mathbb{T}}$, *with* $\mathcal{E}_t \subset \mathcal{A}$, *are said to be* independent *if and only if for every finite subset* $\{t_1, t_2, \ldots t_n\} \subset \mathbb{T}$ *and for every choice of events* $A_{t_j} \in \mathcal{E}_{t_j}$, *for* $j = 1, \ldots n$, *the property (2.16) holds.*

2.1.3 Gaussian Probability Measures

In Example 2.12 we have seen that it is very important to find the probability measure which adequately represents a given problem. A measure which naturally occurs in very many practical situations is the Gaussian probability measure (often referred to as Gaussian, or normal, distribution). The great importance of the Gaussian probability measure arises from the central limit theorem which, roughly speaking, asserts that sums of large numbers of independent outcomes of an experiment under very general conditions tend to be distributed according to a Gaussian probability measure (a more rigorous formulation of an important special case of the central limit theorem is given in Example 2.71 below). Gaussian probability measures appear in many places throughout this book, for example, in characterizing Brownian forces, in the Rouse and Zimm models, in perturbation theory, and in various approximation schemes.

Gaussian distributions are continuous probability measures. The simplest one-dimensional Gaussian probability density (see Fig. 2.2) is,

$$p_{0\delta}(x) = \frac{1}{\sqrt{2\pi}} \exp\left\{-\frac{1}{2}x^2\right\}.$$

By multiplying d one-dimensional Gaussian probability densities one obtains the simplest Gaussian probability density in d-dimensional space:

$$p_{0\delta}(\boldsymbol{x}) = \frac{1}{\sqrt{(2\pi)^d}} \exp\left\{-\frac{1}{2}\sum_{j=1}^{d} x_j^2\right\} = \frac{1}{\sqrt{(2\pi)^d}} \exp\left\{-\frac{1}{2}\boldsymbol{x}^2\right\}. \tag{2.17}$$

Exercise 2.22 Check the correct normalization of the positive functions $p_{0\delta}(\boldsymbol{x})$ for arbitrary d. (Hint: For $d = 2$, this can be done by means of a transformation to polar coordinates.)

Fig. 2.2. The German mathematician, astronomer and physicist Carl Friedrich Gauß [or, Gauss] (1777–1855) together with a plot of a one-dimensional Gaussian probability density (and historical buildings in Göttingen, where he studied and lived during almost his entire career).

According to Example 2.16, the positive function $p_{0\delta}(\boldsymbol{x})$ defines a Gaussian probability measure on $(\mathbb{R}^d, \mathcal{B}^d)$. The subscripts "$0\delta$" on the density p indicate that the following expressions for the d-dimensional generalizations of the mean (2.11) and the square of the width (2.12) hold:

$$\int x_j \, p_{0\delta}(\boldsymbol{x}) \, \mathrm{d}^d x = 0 , \tag{2.18}$$

$$\int x_j x_k \, p_{0\delta}(\boldsymbol{x}) \, \mathrm{d}^d x = \delta_{jk} , \tag{2.19}$$

where δ_{jk} is Kronecker's δ-symbol (the quantities δ_{jk} may be considered as the elements of the unit matrix $\boldsymbol{\delta}$, i.e., $\delta_{jk} = 1$ for $j = k$, $\delta_{jk} = 0$ for $j \neq k$).

Exercise 2.23 The fact that the d-dimensional probability density $p_{0\delta}(\boldsymbol{x})$ is obtained by multiplying d one-dimensional densities implies that the d components of $\boldsymbol{x}$ are distributed independently. In order to make this statement more precise we introduce the d σ-algebras

$$\mathcal{A}_j = \{B_1 \times B_2 \times \ldots \times B_d \mid B_j \in \mathcal{B}, \, B_k = \mathbb{R} \text{ for } k \neq j\}, \quad j = 1, \ldots d,$$

containing those events for which only the jth component of a possible outcome is "of interest." Check that the σ-algebras $\mathcal{A}_1, \ldots \mathcal{A}_d$ are independent according to Definition 2.21 when $(\mathbb{R}^d, \mathcal{B}^d)$ is equipped with the Gaussian probability measure given by the probability density (2.17).

In general, starting from a given probability density $p(\boldsymbol{x})$ one can construct another probability density by means of a smooth one-to-one transformation from $\boldsymbol{x}$ to another variable $\boldsymbol{y}$. Namely, after taking the determinant of the Jacobian of the transformation from $\boldsymbol{x}$ to $\boldsymbol{y}$ into account, the transformation yields another nonnegative, normalized function (that is, another probability density). The most general Gaussian density, which is proportional to the exponential of a quadratic form, can be constructed from (2.17) by means of the affine linear transformation

$$y_j = \alpha_j + \sum_{n=1}^{d} \sigma_{jn} x_n, \qquad x_j = \sum_{n=1}^{d} \sigma^{-1}_{jn} (y_n - \alpha_n) , \tag{2.20}$$

where $\boldsymbol{\sigma}^{-1}$ is the inverse of a regular $d \times d$-matrix $\boldsymbol{\sigma}$, and $\boldsymbol{\alpha} \in \mathbb{R}^d$ is an arbitrary column vector. With this transformation, one obtains

$$p_{\alpha\Theta}(\boldsymbol{y}) = \frac{1}{\sqrt{(2\pi)^d \det(\boldsymbol{\Theta})}} \exp\left\{ -\frac{1}{2} \sum_{l,n=1}^{d} (y_l - \alpha_l) \, \Theta^{-1}_{ln} \, (y_n - \alpha_n) \right\} ,$$

or

$$\boxed{p_{\alpha\Theta}(\boldsymbol{y}) = \frac{1}{\sqrt{(2\pi)^d \det(\boldsymbol{\Theta})}} \exp\left\{ -\frac{1}{2} (\boldsymbol{y} - \boldsymbol{\alpha}) \cdot \boldsymbol{\Theta}^{-1} \cdot (\boldsymbol{y} - \boldsymbol{\alpha}) \right\} ,} \tag{2.21}$$

where

$$\Theta_{jk} = \sum_{n=1}^{d} \sigma_{jn}\sigma_{kn}\,, \tag{2.22}$$

and the correct normalization can easily be checked by applying the transformation formula for integrals. The invariance of the class of Gaussian probability measures under linear transformations is a very important feature of these probability measures. The most general d-dimensional Gaussian probability density is characterized by a column vector $\boldsymbol{\alpha} \in \mathbb{R}^d$ and a $d \times d$-matrix $\boldsymbol{\Theta}$ which, as a consequence of (2.22) and of the regularity of $\boldsymbol{\sigma}$, must be symmetric and positive-definite (that is, $\boldsymbol{x} \cdot \boldsymbol{\Theta} \cdot \boldsymbol{x} > 0$ for all $\boldsymbol{x} \in \mathbb{R}^d$, $\boldsymbol{x} \neq \boldsymbol{0}$). By means of the transformation formula for integrals, one obtains from (2.18) and (2.19)

$$\int y_j \, p_{\alpha\Theta}(\boldsymbol{y}) \, \mathrm{d}^d y = \alpha_j\,, \tag{2.23}$$

$$\int (y_j - \alpha_j)(y_k - \alpha_k) \, p_{\alpha\Theta}(\boldsymbol{y}) \, \mathrm{d}^d y = \Theta_{jk}\,, \tag{2.24}$$

and, by performing the linear transformation (2.20), one can also evaluate the characteristic function (2.14) of the Gaussian probability measure given by $p_{\alpha\Theta}$,

$$\begin{aligned} \tilde{p}_{\alpha\Theta}(\boldsymbol{q}) &= \frac{1}{\sqrt{(2\pi)^d}} \int \exp\left\{ i\boldsymbol{q} \cdot (\boldsymbol{\alpha} + \boldsymbol{\sigma} \cdot \boldsymbol{x}) - \frac{1}{2}\boldsymbol{x}^2 \right\} \mathrm{d}^d x \\ &= \frac{1}{\sqrt{(2\pi)^d}} \mathrm{e}^{i\boldsymbol{q} \cdot \boldsymbol{\alpha}} \int \exp\left\{ -\frac{1}{2}\left(\boldsymbol{x} - i\boldsymbol{\sigma}^T \cdot \boldsymbol{q}\right)^2 - \frac{1}{2}\boldsymbol{q} \cdot \boldsymbol{\sigma} \cdot \boldsymbol{\sigma}^T \cdot \boldsymbol{q} \right\} \mathrm{d}^d x\,. \end{aligned}$$

Since the integral of the squared term in the exponential yields $\sqrt{(2\pi)^d}$ and thus cancels the normalization factor, we finally obtain:

$$\boxed{\tilde{p}_{\alpha\Theta}(\boldsymbol{q}) = \exp\left\{ i\boldsymbol{q} \cdot \boldsymbol{\alpha} - \frac{1}{2}\boldsymbol{q} \cdot \boldsymbol{\Theta} \cdot \boldsymbol{q} \right\}.} \tag{2.25}$$

The characteristic function of a Gaussian probability measure is also an exponential of a quadratic form with $\boldsymbol{\Theta}$ replaced by $\boldsymbol{\Theta}^{-1}$.

The most important conclusion of this subsection on Gaussian probability measures is that, in order to specify such a probability measure in d dimensions, it suffices to give a column vector $\boldsymbol{\alpha}$ and a positive-definite symmetric matrix $\boldsymbol{\Theta}$. For specified $\boldsymbol{\alpha}$ and $\boldsymbol{\Theta}$, the corresponding Gaussian probability density is given by the exponential of a quadratic form shown in (2.21). In an even simpler manner (without inverting $\boldsymbol{\Theta}$ and without calculating the determinant of $\boldsymbol{\Theta}$ occurring in the normalization factor of the probability density), one can write down the characteristic function (2.25) of the Gaussian probability measure characterized by $\boldsymbol{\alpha}$ and $\boldsymbol{\Theta}$. According to (2.23) and (2.24), the quantities $\boldsymbol{\alpha}$ and $\boldsymbol{\Theta}$ are the d-dimensional generalizations of the mean and the square of the width, written here only for Gaussian probability measures. For a one-dimensional Gaussian probability measure with parameters α and Θ, the width

$\sigma = \sqrt{\Theta}$ is often referred to as the standard deviation, and the probability of the interval $[\alpha - \sigma, \alpha + \sigma]$ is approximately equal to 0.6827.

Exercise 2.24 Verify the correct normalization of $p_{\alpha\Theta}$, and prove (2.23) and (2.24).

Exercise 2.25 Assume that intelligence, as described by the intelligence quotient (IQ), is distributed according to a Gaussian probability measure with mean $\alpha = 100$ and standard deviation $\sigma = 16$. How many people in the world do you expect to have a negative IQ? What is the percentage of geniuses (IQ$\geq$ 140) amongst us?

2.2 Random Variables

After introducing the fundamental notions of event and probability in an axiomatic manner we can now define more advanced concepts. One of the most widely used concepts in mathematical stochastics is the notion of a random variable. We first present the general definition of random variables and characterize their distributions. Subsequently, we describe and investigate possible relations between different random variables which will be important in the theory of stochastic dynamical systems, or stochastic processes, in which one has to deal with a family of random variables labeled by a time parameter.

2.2.1 Definitions and Examples

The concept of a random variable is closely related to the consequences of the random occurrence of certain events or outcomes in an experiment. For example, the money a gambler wins or loses in a game of dice is a "random variable" because the amount of money won or lost takes values depending on the possible random outcomes $\omega \in \{1, \ldots, 6\}$. For a real-valued random variable X, in general, the real number X is a dependent variable which varies with $\omega \in \Omega$ where $(\Omega, \mathcal{A}, P)$ is the underlying probability space used to model randomness. In other words, a real-valued random variable is a function, $X : \Omega \to \mathbb{R}$. For many purposes it is sufficient to know the probability measure P^X describing with which probability the random variable X takes on certain values in its range (P^X is usually called the distribution of the random variable X). However, many properties of probability measures can be formulated more conveniently in terms of random variables, in particular, properties related to the transformation of probability measures or to conditional probabilities. Furthermore, the concept of a random variable is very useful because (pseudo-)random variables play an important role in computer simulations.

In order to make sure that the distribution P^X of a random variable X can be defined we first introduce the concept of measurable functions.

Definition 2.26 *Let* $(\Omega, \mathcal{A})$ *and* $(\Omega', \mathcal{A}')$ *be measurable spaces. A function* $X : \Omega \to \Omega', \ \omega \mapsto X(\omega)$ *is a* measurable function *if and only if*

$$X^{-1}(A') := \{\omega \in \Omega \mid X(\omega) \in A'\} \in \mathcal{A} \quad \textit{for all } A' \in \mathcal{A}'. \tag{2.26}$$

In the following, the notation $X : (\Omega, \mathcal{A}) \to (\Omega', \mathcal{A}')$ *implies that* $X : \Omega \to \Omega'$ *is a measurable function from* $(\Omega, \mathcal{A})$ *to* $(\Omega', \mathcal{A}')$.

For measurable functions, the inverse images of measurable sets are again measurable sets. This property is very important for the following reason. For a measurable function X on a probability space $(\Omega, \mathcal{A}, P)$ it is possible to determine the probability that X takes on a value from $A' \in \mathcal{A}'$; this is so because $X^{-1}(A') \in \mathcal{A}$ is a measurable set and hence the probability $P(X^{-1}(A'))$ is defined.

Definition 2.27 *A measurable function* $X : (\Omega, \mathcal{A}) \to (\Omega', \mathcal{A}')$ *on a probability space* $(\Omega, \mathcal{A}, P)$ *is called a* random variable. *The probability measure on* $(\Omega', \mathcal{A}')$ *defined by*

$$P^X(A') := P(X^{-1}(A')) \quad \textit{for all } A' \in \mathcal{A}' \tag{2.27}$$

is referred to as the distribution of the random variable X.

We occasionally refer to measurable functions as random variables even if no probability measure P is explicitly specified. For real-valued random variables, the image space $\mathbb{R}$ is always assumed to be equipped with the Borel σ-algebra $\mathcal{B}$. If the probability measure P^X on $(\Omega', \mathcal{A}') = (\mathbb{R}, \mathcal{B})$ has a probability density $p^X(x)$ one can heuristically describe a real-valued random variable by saying that "the probability of finding a value of the random variable in an interval of length $\mathrm{d}x$ around x is $p^X(x)\,\mathrm{d}x$," a statement that is often found when random variables are introduced in the applied literature. This seemingly simple introduction of random variables through their distributions gets quite involved when families or transformations of random variables need to be considered. Some aspects of random variables, like coarse-grained random variables (conditional expectations), are almost impossible to discuss without carefully introducing random variables as measurable functions. Also, for understanding the various concepts of convergence for random variables it is insufficient to introduce random variables only through their distributions.

It is often convenient and illuminating to use the following suggestive notation for characterizing distributions of random variables,

$$P(X \in A') := P^X(A'). \tag{2.28}$$

The general idea behind this notation becomes more obvious when we use (2.26) and (2.27) for the definition of distributions in order to obtain the identity $P(X \in A') = P(\{\omega \mid X(\omega) \in A'\})$.

Example 2.28 Indicator Functions

Given a measurable space $(\Omega, \mathcal{A})$, the indicator function of an event $A \in \mathcal{A}$ defined as

$$\chi_A : \Omega \to \mathbb{R}, \quad \omega \mapsto \begin{cases} 1 & \text{if } \omega \in A \\ 0 & \text{if } \omega \notin A \end{cases}$$

is a real-valued measurable function on $(\Omega, \mathcal{A})$. For every $B \in \mathcal{B}$, the set $\chi_A^{-1}(B)$ is one of the four measurable sets A, A^C, Ω, $\emptyset$ depending on whether or not 1, or 0, or both, or neither are contained in B. □

Example 2.29 Emotions in a Game of Dice
Once more we consider a game of dice with $\Omega = \{1, 2, 3, 4, 5, 6\}$ (cf. Example 2.3). For $\mathcal{A}_1 = \mathcal{P}(\Omega)$, every function on Ω taking on values in a measurable space is a random variable (this is of course true for any Ω when it is equipped with the σ-algebra $\mathcal{A} = \mathcal{P}(\Omega)$).

For $\Omega' = \{$"wow!", "damn!"$\}$ and $\mathcal{A}' = \mathcal{P}(\Omega')$, the function

$$X : \Omega \to \Omega', \quad \omega \mapsto \begin{cases} \text{"wow!"} & \text{if } \omega = 6 \\ \text{"damn!"} & \text{if } \omega \neq 6 \end{cases}$$

is a random variable on the measurable space $(\Omega, \mathcal{A}_2)$ with values in $(\Omega', \mathcal{A}')$, where $\mathcal{A}_2 = \mathcal{A}\left(\left\{\{6\}\right\}\right)$. Notice that X is a random variable in a mathematically rigorous sense. In general, a function on Ω that is not constant on $\{1, 2, 3, 4, 5\}$ cannot be a random variable with respect to $\mathcal{A}_2$ because "it resolves information not contained in the σ-algebra $\mathcal{A}_2$." □

Exercise 2.30 For the game of roulette considered in Exercise 2.4, give an example of a rigorous and complete definition of a random variable taking on values in the measurable space of emotions $(\Omega', \mathcal{A}')$ of Example 2.29.

It follows from the properties of functions that for every function from a set Ω to a measurable space $(\Omega', \mathcal{A}')$ the set $\mathcal{A}^X := \{X^{-1}(A') \mid A' \in \mathcal{A}'\}$ of subsets of Ω is a σ-algebra. This σ-algebra is referred to as the *σ-algebra induced by X*; the function X is measurable if $\mathcal{A}^X \subset \mathcal{A}$. Notice that in order to guarantee that a function is measurable it is sufficient if the property (2.26) is restricted to all A' from a generating system of $\mathcal{A}'$. The σ-algebra $\mathcal{A}^X$ represents "the information resolved by X." If the function X does not resolve any more information than another measurable function Y that is defined on the same measurable space, that is $\mathcal{A}^X \subset \mathcal{A}^Y$, we would intuitively expect that X can be expressed as a function of Y. This idea is important for the construction of conditional probabilities and conditional expectations. Furthermore, the fact that intuitive ideas such as "a random variable X depends only on (or is a function of) another random variable Y" can so conveniently and elegantly be expressed in terms of the induced σ-algebras is very useful for an intuitive understanding of formal assumptions in the theory of stochastic integrals and stochastic differential equations. A more precise statement is made in the following theorem.

Theorem 2.31 *Let $X : (\Omega, \mathcal{A}) \to (\Omega', \mathcal{A}')$ and $Y : (\Omega, \mathcal{A}) \to (\Omega'', \mathcal{A}'')$ be measurable functions on the same measurable space, where $(\Omega'', \mathcal{A}'')$ is chosen* *such that $Y(\Omega) = \Omega''$. If all one-element subsets of Ω' are measurable and if $\mathcal{A}^X \subset \mathcal{A}^Y$, then there exists a measurable function $g : (\Omega'', \mathcal{A}'') \to (\Omega', \mathcal{A}')$ such that X is the composite of g and Y, that is $X = g(Y)$.*

The assumption $\mathcal{A}^X \subset \mathcal{A}^Y$ implies that for any $A' \in \mathcal{A}'$ there exists $A'' \in \mathcal{A}''$ such that $X^{-1}(A') = Y^{-1}(A'')$. The proof of this theorem is based on the fact that for each $\omega \in \Omega'$ one can select a set $A_\omega \in \mathcal{A}''$ such that $X^{-1}(\{\omega\}) = Y^{-1}(A_\omega)$ (here the assumption $\{\omega\} \in \mathcal{A}'$ is crucial for guaranteeing that $A_\omega \in \mathcal{A}''$). Then, the sets A_ω form a partition of Ω'' consisting of mutually disjoint sets, and the function g is constructed such that it takes the value ω on A_ω. The requirement that all one-element subsets of Ω' be measurable is automatically satisfied when we consider $\Omega' \subset \mathbb{R}^d$ together with the corresponding Borel σ-algebra (cf. Example 2.7).

Exercise 2.32 Consider the measurable space $(\Omega, \mathcal{A}_1)$ and the random variable X of Example 2.29, where we here regard X as a random variable on $(\Omega, \mathcal{A}_1)$. Another random variable $Y : (\Omega, \mathcal{A}_1) \to (\{0,1,2\}, \mathcal{P}(\{0,1,2\}))$ is defined by $Y(1) = Y(2) = Y(3) = Y(4) = 0$, $Y(5) = 1$, $Y(6) = 2$. Determine the induced σ-algebras $\mathcal{A}^X$ and $\mathcal{A}^Y$, and verify the relation $\mathcal{A}^X \subset \mathcal{A}^Y$. Construct the measurable function g such that $X = g(Y)$ (the existence of g is guaranteed by Theorem 2.31).

Example 2.33 Continuous Functions
Let Ω and Ω' be topological spaces with the corresponding Borel σ-algebras $\mathcal{B}$ and $\mathcal{B}'$ (cf. Example 2.7). Any continuous function from Ω to Ω' is a measurable function. This is an immediate consequence of the close, formal similarity between the definitions of continuous and measurable functions. The inverse images of the open subsets of Ω' are open and hence measurable subsets of Ω; since the open subsets of Ω' form a generating system of $\mathcal{B}'$, this proves the measurability of continuous functions. □

The above-mentioned analogy between continuous and measurable functions suggests that sums, differences, scalar multiples, products and, if well-defined, quotients of real-valued random variables are also random variables. For example, the sum of two random variables X and Y can be shown to be a random variable by means of the identity

$$(X+Y)^{-1}(]-\infty, a[) = \bigcup_{b \text{ rational}} X^{-1}(]-\infty, b[) \cap Y^{-1}(]-\infty, a-b[)\,.$$

Furthermore, it is clear from Example 2.8 that a function with range $(\Omega', \mathcal{A}') = (\mathbb{R}^d, \mathcal{B}^d)$ is measurable if its d component functions are measurable, real-valued functions. A vector-valued function is a random variable if all its component functions are real-valued random variables. In short, we are free to do all kinds of elementary calculations with real- and vector-valued random variables—the result will always be another random variable.

Example 2.34 Random Variables with Given Distributions
For a given probability measure P' on $(\Omega', \mathcal{A}')$ there always exists a random variable X for which $P^X = P'$. In order to see this, one can choose $(\Omega, \mathcal{A}, P) = (\Omega', \mathcal{A}', P')$ and $X: \Omega \to \Omega'$, $\omega \mapsto \omega$. This trivial construction justifies later formulations like "let X be a random variable with distribution P'." For a given distribution, the random variable X is of course not unique. Since, according to this construction, any probability measure can be considered as the distribution of a suitable random variable the term probability distribution is often used instead of probability measure (for example, "Gaussian distribution" sounds much more familiar and less awkward than "Gaussian probability measure"). □

Example 2.35 Composed Functions
Let $X: (\Omega, \mathcal{A}) \to (\Omega', \mathcal{A}')$ and $X': (\Omega', \mathcal{A}') \to (\Omega'', \mathcal{A}'')$ be measurable functions. Then the *composed function* or *composite* $X'(X): \Omega \to \Omega''$, $\omega \mapsto X'(X(\omega))$ is a measurable function from $(\Omega, \mathcal{A})$ to $(\Omega'', \mathcal{A}'')$.

For a random variable $\boldsymbol{X}: (\Omega, \mathcal{A}) \to (\mathbb{R}^d, \mathcal{B}^d)$ with a continuous distribution characterized by the density p^X and a differentiable, one-to-one mapping, $\boldsymbol{g}: \mathbb{R}^d \to \mathbb{R}^d$ with continuous derivatives, the function $\boldsymbol{Y} = \boldsymbol{g}(\boldsymbol{X})$ is a random variable with continuous distribution. The probability that $\boldsymbol{Y}$ will assume values in a neighborhood of $\boldsymbol{y} \in \mathbb{R}^d$ is equal to the probability that $\boldsymbol{X}$ will assume values in the inverse image of this neighborhood under the function $\boldsymbol{g}$ (this inverse image is a neighborhood of $\boldsymbol{x} = \boldsymbol{g}^{-1}(\boldsymbol{y})$). In order to compare the corresponding probability densities one has to account for the difference in the size of the volume elements $\mathrm{d}^d x$ and $\mathrm{d}^d y$, which is given by the absolute value of the determinant of the Jacobian of the transformation from $\boldsymbol{y}$ to $\boldsymbol{x}$,

$$p^Y(\boldsymbol{y}) = p^X(\boldsymbol{g}^{-1}(\boldsymbol{y})) \left| \det \frac{\partial \boldsymbol{g}^{-1}(\boldsymbol{y})}{\partial \boldsymbol{y}} \right| . \tag{2.29}$$

The construction of the general Gaussian probability density $p_{\alpha\Theta}$ from the special case $p_{0\delta}$ described in Sect. 2.1.3 may be regarded as an example of the application of (2.29), with the linear transformation $\boldsymbol{g}$ defined in (2.20).

We next discuss an important, explicit example for $d = 2$. We consider $\Omega' =]0,1[\times]0,1[$ together with the natural σ-algebra $\mathcal{A}' = \{B \cap \Omega' \mid B \in \mathcal{B}^2\}$ (see the remarks after Example 2.8). The probability density for the distribution of $\boldsymbol{X}$ is assumed to be $p^X(\boldsymbol{x}) = 1$, that is uniform, and we consider the mapping

$$\begin{aligned} y_1 = g_1(x_1, x_2) &= \sqrt{-2 \ln x_1}\, \sin(2\pi x_2)\,, \\ y_2 = g_2(x_1, x_2) &= \sqrt{-2 \ln x_1}\, \cos(2\pi x_2)\,, \end{aligned}$$

which implies $-(y_1^2 + y_2^2)/2 = \ln x_1$. For $\boldsymbol{g}$ to be a one-to-one mapping we take $\Omega'' = \mathbb{R}^2 - \{(x_1, 0) \mid x_1 \in [0, \infty[\}$ together with the natural σ-algebra $\mathcal{A}'' = \{B \cap \Omega'' \mid B \in \mathcal{B}^2\}$. As an auxiliary quantity we calculate

$$\det \frac{\partial \boldsymbol{g}(\boldsymbol{x})}{\partial \boldsymbol{x}} = \frac{2\pi}{x_1} .$$

From the general transformation rule for probability densities (2.29) one thus obtains for $\boldsymbol{y} \in \Omega''$

$$p^{Y}(\boldsymbol{y}) = \left|\det \frac{\partial \boldsymbol{g}^{-1}(\boldsymbol{y})}{\partial \boldsymbol{y}}\right| = \frac{x_1(\boldsymbol{y})}{2\pi} = \frac{1}{2\pi}\exp\left\{-\frac{1}{2}\left(y_1^2 + y_2^2\right)\right\}.$$

In other words, the distribution of $\boldsymbol{Y} = \boldsymbol{g}(\boldsymbol{X})$ is Gaussian. This example and the following exercise will later be of great importance for carrying out computer simulations in which one needs to construct random numbers with a Gaussian distribution from random numbers with a uniform distribution. □

Exercise 2.36 Let $\boldsymbol{X}$ be a random variable with uniform distribution in the unit circle $\Omega' = \{(x_1, x_2) \mid 0 < x_1^2 + x_2^2 < 1\}$. For

$$y_j = g_j(x_1, x_2) = \sqrt{\frac{-2\ln(x_1^2 + x_2^2)}{x_1^2 + x_2^2}}\, x_j \qquad \text{for } j = 1, 2,$$

determine the distribution of $\boldsymbol{Y} = \boldsymbol{g}(\boldsymbol{X})$.

Exercise 2.37 Let $\boldsymbol{X}$ be a random variable with uniform distribution in the semicircle $\Omega' = \{(x_1, x_2) \mid 0 < x_1,\ x_1^2 + x_2^2 < 1\}$. For

$$y_1 = g_1(x_1, x_2) = 1 - 2(x_1^2 + x_2^2), \qquad y_2 = g_2(x_1, x_2) = \arctan\frac{x_2}{x_1},$$

determine the distribution of $\boldsymbol{Y} = \boldsymbol{g}(\boldsymbol{X})$.

Example 2.38 More About Composed Functions

In the preceding example, the assumption that $\boldsymbol{g}$ is a one-to-one mapping is not essential for guaranteeing that $\boldsymbol{g}(\boldsymbol{X})$ is a random variable but, in general, it simplifies the calculation of the distribution of $\boldsymbol{g}(\boldsymbol{X})$ considerably. We next consider the complications arising when the function $\boldsymbol{g}$ is not a one-to-one mapping.

Assume that X is a real-valued random variable taking on values in $[0, 1]$ with a uniform distribution, that is, $p^X(x) = 1$. For arbitrary constants c_1, c_2, the function $g(x) = \left(x - \frac{1}{2}\right)\left(c_1\left|x - \frac{1}{2}\right| + c_2\right)$ has a continuous derivative. For suitably chosen c_1, c_2, the function g is non-monotonic; we here choose $c_1 = 10.7$ and $c_2 = -0.7$ (see Fig. 2.3).

The range of the composed random variable $Y = g(X)$ is $[-b, b]$, where $b = \frac{1}{4}c_1 + \frac{1}{2}c_2 = 2.325$. For $|y| > a$ with $a = c_2^2/(4c_1) \approx 0.01$, there is a unique inverse image $x = g^{-1}(y)$, and the probability density can be calculated as in the preceding example,

$$p^Y(y) = \left|\frac{\mathrm{d}g^{-1}(y)}{\mathrm{d}y}\right| = \frac{1}{\sqrt{c_2^2 + 4c_1|y|}} \qquad \text{for } \ a < |y| \le b\,.$$

For $|y| < a$, each y has three inverse images, each of which contributes to the probability density $p^Y(y)$. After evaluating $\mathrm{d}y/\mathrm{d}x$ at each of these inverse images of a given y we obtain

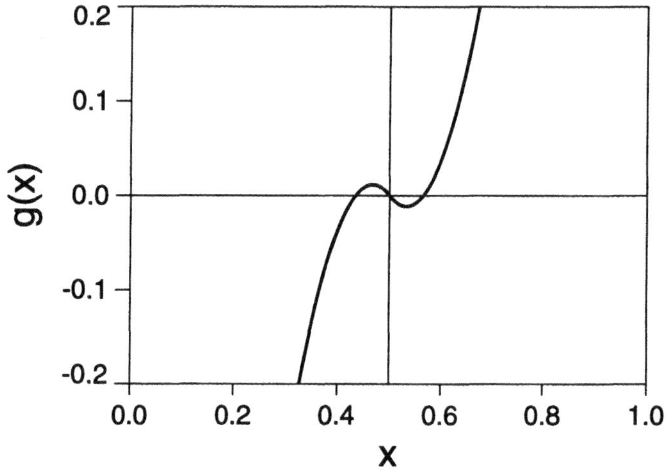

Fig. 2.3. The non-monotonic function $g(x)$ used in Example 2.38 to transform a random variable with uniform distribution.

$$p^Y(y) = \frac{1}{\sqrt{c_2^2 + 4c_1|y|}} + \frac{2}{\sqrt{c_2^2 - 4c_1|y|}} \qquad \text{for} \quad |y| < a\,. \tag{2.30}$$

The probability density $p^Y(y)$ in the neighborhood of the origin is shown in Fig. 2.4; the density jumps to zero at $y = \pm b$. Notice that the second term in (2.30) has an integrable square-root singularity for $|y| \to a$. This is the generic situation for a function g possessing maxima and minima of the quadratic type. The contribution of the singular part of the probability density to the total probability is $2\,|c_2|/c_1 \approx 0.13$, which is simply the distance between the smallest and largest zeros of the function g. □

2.2.2 Expectations and Moments

In this section, we consider random variables taking on values only in $\mathbb{R}$ or $\mathbb{R}^d$. The coarsest probabilistic information concerning a real-valued random variable X is its average over the entire probability space, usually referred to as the *expectation* $\langle X \rangle$. The evaluation of expectations is the primary goal in many applications; for example, in the kinetic theory of polymeric fluids, stresses are obtained as expectations of polymer configurations. We introduce the expected or mean value of a real-valued random variable on a given probability space $(\Omega, \mathcal{A}, P)$ in three steps.

Step 1: We first consider *simple* random variables which, by definition, are linear combinations of indicator functions,

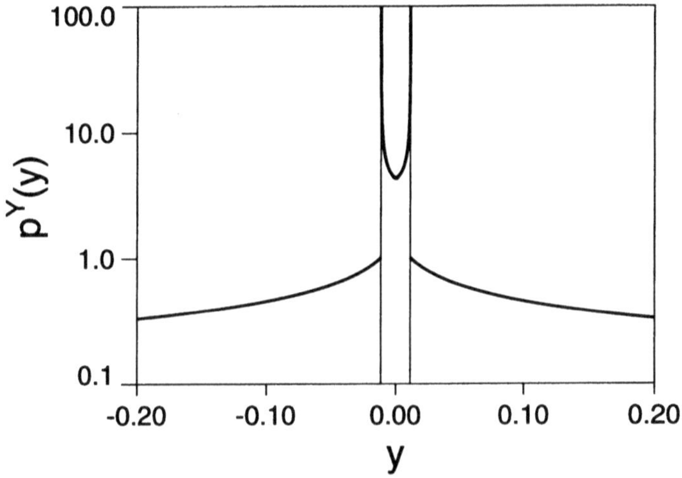

Fig. 2.4. The singular probability density obtained via a non-monotonic transformation from a uniform distribution; see Example 2.38 for details.

$$X = \sum_{j=1}^{n} c_j \, \chi_{A_j} \, , \tag{2.31}$$

with $c_j \in \mathbb{R}$ and $A_j \in \mathcal{A}$. The expectation of the simple random variable (2.31) is defined by

$$\langle X \rangle := \sum_{j=1}^{n} c_j \, P(A_j) \, . \tag{2.32}$$

This definition is independent of the particular representation of a given simple random variable in the form (2.31).

Example 2.39 Expectations in a Game of Dice
Consider the probability space of Example 2.12, that is $\Omega = \{1, 2, 3, 4, 5, 6\}$, $\mathcal{A} = \{\Omega, \emptyset, \{6\}, \{6\}^C\}$, and the probability measure characterized by $P(\{6\}) = \frac{1}{7}$. The real-valued random variable

$$\omega \mapsto X(\omega) = \begin{cases} 5 & \text{if } \omega = 6 \\ -1 & \text{if } \omega \neq 6 \end{cases}$$

which is a slightly more serious version of the function considered in Example 2.29, is a simple random variable. For each of the following three different representations of X in the form (2.31),

$$X = -\chi_\Omega + 6\,\chi_{\{6\}} = 5\,\chi_{\{6\}} - \chi_{\{6\}^C} = 5\,\chi_\Omega - 6\,\chi_{\{6\}^C} \, ,$$

the definition (2.32) gives $\langle X \rangle = -\frac{1}{7}$. The first of these representations of X can be considered as a description of the rules for a particular game of dice: after paying one dollar to take part in the game, a gambler gets six dollars back when the event $\{6\}$ occurs. The second representation shows more clearly that in the event $\{6\}$ one has a net win of five dollars and otherwise one dollar is lost. The third representation, corresponding to an optimist's point of view, counts on a net win of five dollars; however, when the event $\{6\}$ does not occur, the net win is actually six dollars less (and even the biggest optimist then has to admit a loss of one dollar). □

Step 2: For a nonnegative random variable X we define

$$\langle X \rangle := \sup_{\substack{Y \leq X \\ Y \text{ simple}}} \langle Y \rangle \,, \tag{2.33}$$

provided that the supremum is finite; otherwise, we say that the expectation of X does not exist. The definition (2.33) corresponds to an approximation of a nonnegative random variable by simple random variables from below. The expectation can also be obtained by considering a sequence of simple functions $(Y_n)_{n \in \mathbb{N}}$ with $Y_n \leq Y_{n+1}$ for all $n \in \mathbb{N}$ and $\sup_{n \in \mathbb{N}} Y_n = Y$. When applied to nonnegative simple functions, the definition (2.33) is consistent with (2.32).

Step 3: For an arbitrary real-valued random variable X we introduce the measurable set $A = X^{-1}([0, \infty[)$ and define

$$\langle X \rangle := \langle X \, \chi_A \rangle - \langle -X \, \chi_{A^c} \rangle \,, \tag{2.34}$$

where the indicator functions make sure that the two expectations on the right side can be evaluated according to Step 2. For nonnegative functions, one has $A = \Omega$, and only the first term contributes. The expectation of a random variable X exists if both expectations on the right side of (2.34) are finite. When applied to simple functions, the definition (2.34) is consistent with (2.32).

Example 2.40 Expectations for Discrete Probability Measures
Consider the discrete probability space $(\Omega, \mathcal{A}, P)$ discussed in Example 2.13 with $\Omega = \{\omega_1, \omega_2, \ldots\}$, with $\mathcal{A} = \mathcal{P}(\Omega)$, and with $P(\{\omega_j\}) = p_j$. Let X be a real-valued random variable defined on $(\Omega, \mathcal{A}, P)$ such that $\langle X \rangle$ exists. In this situation, one has

$$\langle X \rangle = \sum_{j=1}^{\infty} X(\omega_j) \, p_j \,. \tag{2.35}$$

If the number of values taken by X is finite, X is a linear combination of indicator functions, and (2.35) follows immediately from the definitions (2.8) and (2.32). If X takes on infinitely many different values, Steps 2 and 3 of the above definition of expectations have to be followed in order to arrive at (2.35).
□

Example 2.41 Expectations for Continuous Probability Measures
Consider the continuous probability space $(\Omega, \mathcal{A}, P)$ with $\Omega = \mathbb{R}$, with $\mathcal{A} = \mathcal{B}$,

and with P given by a probability density $p(x)$ as discussed in Example 2.16. Let $g : (\mathbb{R}, \mathcal{B}) \to (\mathbb{R}, \mathcal{B})$ be a measurable function such that the product $g(x)p(x)$ is integrable (for example, piecewise continuous).[4] In this situation, one finds

$$\langle g \rangle = \int g(x)\, p(x)\, \mathrm{d}x \,. \tag{2.36}$$

For a simple function for which the sets A_j are open intervals, this result follows immediately from the definitions (2.10) and (2.32). In following Step 2 of the above procedure for defining expectations, one needs the theorem of monotone convergence for the integrals on the right side of (2.36) when approximating g by increasing simple functions (see, e.g., Theorem 2.3.4 of [3]). Step 3 of the above construction of expectations is the analog of a similar rule for decomposing the integral of an integrable function into the integrals of the positive and negative parts of that function. □

Example 2.41 suggests that evaluating the expectation of a random variable is very closely related to integration. For that reason, the notation

$$\int X \,\mathrm{d}P := \langle X \rangle \tag{2.37}$$

is often used in the mathematical literature. We use this notation only when it is important to point out which probability measure P is to be used in evaluating an expectation.

Exercise 2.42 Consider the probability space given by $\Omega = [0, 1]$, the Borel σ-algebra on $[0, 1]$, and the continuous probability measure with density $p(x) = 1$. Two random variables are introduced by $g(x) = x$ and $\tilde{g}(x) = x^2$. Evaluate $\langle g \rangle$ and $\langle \tilde{g} \rangle$ both by means of Example 2.41 and according to Steps 1 and 2 of the definition.

Example 2.43 Expectations of Composed Functions
Let $X : (\Omega, \mathcal{A}) \to (\Omega', \mathcal{A}')$ and $X' : (\Omega', \mathcal{A}') \to (\mathbb{R}, \mathcal{B})$ be measurable functions on $(\Omega, \mathcal{A}, P)$ and $(\Omega', \mathcal{A}', P^X)$, respectively. One then finds for the expectation of the composite $X'(X)$,

$$\int X'(X) \,\mathrm{d}P = \int X' \,\mathrm{d}P^X \,, \tag{2.38}$$

where the expectation on the right side is evaluated with P^X on $(\Omega', \mathcal{A}')$ rather than on the standard probability space $(\Omega, \mathcal{A}, P)$. Equation (2.38) can be proved by following all the steps of the definition of expectations and by exploiting the definition of the distribution P^X of X (if X' is a simple random variable, then $X'(X)$ is also a simple random variable). □

[4] It is assumed that the reader has an a priori understanding of integrals. A very general theory of integration can be developed by following the three steps of the definition of expectations presented in this section for more general measures (when no normalization condition is required).

Exercise 2.44 Show that (2.38) contains an abstract formulation of the transformation formula for integrals. (Hint: The familiar form of the transformation formula can be recovered by choosing $\Omega = \Omega' = \mathbb{R}$ and assuming that P and P^X are continuous probability measures.)

In summary, we now have the simple formulas (2.35) and (2.36) for evaluating the expectations of random variables on probability spaces with discrete and continuous probability measures as sums or integrals, respectively, and we have seen how to define expectations for more general probability measures. As for sums and integrals, one can in general derive two basic properties of expectations,

$$\begin{aligned} \langle c_1 X + c_2 Y \rangle &= c_1 \langle X \rangle + c_2 \langle Y \rangle \,, & (2.39) \\ X \leq Y &\Rightarrow \langle X \rangle \leq \langle Y \rangle \,, & (2.40) \end{aligned}$$

where X and Y are random variables with finite expectations and c_1, c_2 are arbitrary real constants. The validity of these properties which one expects for any type of integral is a further justification of the notation (2.37). By choosing X' in Example 2.43 as the identity, one can conclude that the expectation of a random variable depends only on the distribution P^X of X.

After having the general definition of expectations available, we can now generalize some important concepts previously introduced in Example 2.16 only for continuous probability densities. In order to establish the connection, we discuss certain special cases of expectations. If X is a real-valued random variable, we consider the expectations of the random variables e^{iqX} for arbitrary $q \in \mathbb{R}$ and of X^n for arbitrary $n \in \mathbb{N}$. If X has a continuous distribution, we obtain from Examples 2.41 and 2.43

$$\left\langle \mathrm{e}^{iqX} \right\rangle = \int \mathrm{e}^{iqx}\, p^X(x)\, \mathrm{d}x = \tilde{p}^X(q) \,, \tag{2.41}$$

which by (2.13) is the characteristic function of the probability distribution of X or, simply, the *characteristic function* of X. For the expectation of X^n, that is the nth *moment* of the random variable X, one similarly has

$$\langle X^n \rangle = \int x^n\, p^X(x)\, \mathrm{d}x \,. \tag{2.42}$$

The first moment of the random variable X is the mean of its continuous probability distribution, which was introduced in Example 2.16. Furthermore, the square of the width of the distribution defined in Example 2.16 is equal to the *variance* of the random variable X,

$$\mathrm{Var}(X) := \left\langle (X - \langle X \rangle)^2 \right\rangle = \langle X^2 \rangle - \langle X \rangle^2 \,. \tag{2.43}$$

Since the expectations in (2.41) and (2.42) can also be defined for random variables X for which the probability distributions are not continuous, the concepts of characteristic functions and moments of random variables are very general (assuming that the moments exist). All these quantities depend only on the distribution of the random variable X. As a matter of fact, the characteristic function $\tilde{p}^X$ uniquely characterizes the distribution of the random variable X. The nth moment of a random variable can be obtained from the nth derivative of its characteristic function with respect to q after setting $q = 0$. A more precise statement is given in the following theorem (see Sect. 8.3 of [3]).

Theorem 2.45 *Let X be a random variable with finite moment $\langle X^n \rangle$ for some $n \in \mathbb{N}$. Then, for each $k = 1, \ldots, n$, the moment $\langle X^k \rangle$ is finite, the characteristic function $\tilde{p}^X$ has bounded and continuous derivatives of orders $1, \ldots, n$, and one has the following expansion for small q,*

$$\tilde{p}^X(q) = \sum_{k=0}^{n} \frac{(iq)^k}{k!} \left\langle X^k \right\rangle + o(q^n) . \tag{2.44}$$

The first statement is a consequence of the inequality

$$\left\langle |X|^k \right\rangle = \left\langle |X|^k \chi_{\{\omega||X(\omega)|\leq 1\}} \right\rangle + \left\langle |X|^k \chi_{\{\omega||X(\omega)|>1\}} \right\rangle \leq 1 + \langle |X|^n \rangle .$$

The rest follows by induction, where the theorem of dominated convergence needs to be used (see, e.g., Theorem 2.7.4 of [3]), and by Taylor expansion.

The generalization of all the concepts developed in this section to the case of vector-valued random variables is obvious. Expectations can be defined componentwise, and in the definition of characteristic functions the argument q has to be replaced with a column vector of the same length as the random variable X. The conclusions of Theorem 2.45 can be generalized when for each component the nth moment is finite, where (2.44) is replaced with the multivariate Taylor expansion of the characteristic function. Instead of the variance, one introduces the *covariance matrix* with elements

$$\left\langle (X_j - \langle X_j \rangle)(X_k - \langle X_k \rangle) \right\rangle = \langle X_j X_k \rangle - \langle X_j \rangle \langle X_k \rangle . \tag{2.45}$$

The diagonal components of the covariance matrix are the variances for each component. An off-diagonal component $\mathrm{Cov}(X_j, X_k) := \langle X_j X_k \rangle - \langle X_j \rangle \langle X_k \rangle$ is usually referred to as the *covariance* of X_j and X_k.

2.2.3 Joint Distributions and Independence

After defining random variables and characterizing their distributions, we next describe and discuss relationships between different random variables. Let $X : (\Omega, \mathcal{A}) \to (\Omega', \mathcal{A}')$ and $Y : (\Omega, \mathcal{A}) \to (\Omega'', \mathcal{A}'')$ be two random variables defined on the same probability space $(\Omega, \mathcal{A}, P)$. The *joint distribution* of X and Y is defined as the distribution of the product mapping which maps each ω into the pair of values $(X(\omega), Y(\omega))$, more precisely $(X, Y) : (\Omega, \mathcal{A}) \to (\Omega' \times \Omega'', \mathcal{A}' \otimes \mathcal{A}'')$, $\omega \mapsto (X(\omega), Y(\omega))$ (cf. the definition of product spaces and product σ-algebras in Example 2.8). Even though the joint probability distribution simply reduces to the distribution of a more complex random variable, this additional concept is often useful when considering several different random variables (the generalization of the concept of joint distributions from two to any finite number of random variables is obvious). In particular, joint distributions play a key role in constructing stochastic processes. This concept is even more useful if one extends the notation (2.28),

$$\begin{aligned} P(X \in A', Y \in A'') &:= P^{(X,Y)}(A' \times A'') \\ &= P(\{\omega \mid X(\omega) \in A', Y(\omega) \in A''\}) \qquad (2.46) \\ &= P\left(X^{-1}(A') \cap Y^{-1}(A'')\right), \end{aligned}$$

for all $A' \in \mathcal{A}'$ and all $A'' \in \mathcal{A}''$. The first equality in (2.46) is a straightforward generalization of the notation (2.28), the second equality follows by applying the definition of distributions to product mappings, and the third equality is obtained by representing the set in braces in an equivalent manner. If the joint distribution of a set of random variables is considered together with the joint distribution of a subset of these variables, the latter distribution is often referred to as a *marginal distribution*. For example, the distribution of X is a marginal distribution of the joint distribution of X and Y considered in (2.46).

According to the general definition of joint distributions, the probability distribution of a vector-valued random variable constitutes the joint distribution of all the component functions.

Exercise 2.46 Determine the joint distribution of the random variables X and Y of Exercise 2.32 (a probability measure on the underlying measurable space is given by $p(\{j\}) = 1/6$ for $j \in \Omega$).

In general, joint distributions characterize the relationship between random variables which are defined on the same probability space. The simplest possible relationship between random variables occurs for independent random variables. In order to express this relationship in a rigorous manner we first need to generalize the concept of independence, previously introduced for events (Definition 2.19) and for sets of events (Definition 2.21), to random variables. Let $\mathbb{T}$ represent an arbitrary set of labels and let a probability space $(\Omega, \mathcal{A}, P)$ be given.

Definition 2.47 *The random variables $(X_t)_{t\in\mathbb{T}}$ defined on $(\Omega, \mathcal{A}, P)$ are* independent *if and only if the induced σ-algebras $(\mathcal{A}^{X_t})_{t\in\mathbb{T}}$ are independent sets of events.*

Notice that the independence of random variables is not a matter of which values those random variables take but rather is determined by the induced σ-algebras ("the resolved information"). The independence of random variables can be recognized from the properties of the joint distribution.

Theorem 2.48 *The random variables* $X: (\Omega, \mathcal{A}) \to (\Omega', \mathcal{A}')$ *and* $Y: (\Omega, \mathcal{A}) \to (\Omega'', \mathcal{A}'')$ *on* $(\Omega, \mathcal{A}, P)$ *are independent if and only if for all* $A' \in \mathcal{A}'$ *and* $A'' \in \mathcal{A}''$ *the joint distribution has the property*

$$P(X \in A', Y \in A'') = P(X \in A')\, P(Y \in A'') . \tag{2.47}$$

This theorem follows immediately from the definition of the stochastic independence of X and Y, which is equivalent to the independence of all pairs of sets of the form $X^{-1}(A')$ and $Y^{-1}(A'')$. For any finite number of random variables, independence is equivalent to the complete factorization (one factor for each random variable) of the joint distribution. In the spirit of Definitions 2.19 and 2.21, one can generalize Theorem 2.48 even to an infinite number of random variables (in that case, all finite-dimensional marginal distributions need to be of the product form).

In the special case of random variables assuming values in $\mathbb{R}$ or $\mathbb{R}^d$, independence allows us to simplify expectations.

Theorem 2.49 *Let* X *and* Y *be independent random variables, and* $\boldsymbol{g}_X$ *and* $\boldsymbol{g}_Y$ *be real-valued or vector-valued measurable functions on the respective image spaces. Then, the real-valued or vector-valued random variables* $\boldsymbol{g}_X(X)$ *and* $\boldsymbol{g}_Y(Y)$ *are also independent and, if all involved expectations exist, one has*

$$\langle \boldsymbol{g}_X(X)\, \boldsymbol{g}_Y(Y) \rangle = \langle \boldsymbol{g}_X(X) \rangle \, \langle \boldsymbol{g}_Y(Y) \rangle . \tag{2.48}$$

Note that Definition 2.47 implies that with X and Y the random variables $\boldsymbol{g}_X(X)$ and $\boldsymbol{g}_Y(Y)$ are also independent. After applying the transformation formula (2.38), the factorization of expectations is an immediate consequence of the product form (2.47) of the joint distribution. Only if (2.48) holds for a sufficiently large class of functions $\boldsymbol{g}_X$, $\boldsymbol{g}_Y$, can the independence of X and Y be inferred from the factorization of expectations. The factorization of expectations is an extremely useful property when dealing with independent random variables; this property is so intuitive that we will often use it in the following without explicitly referring to Theorem 2.49.

For a random variable $\boldsymbol{X}$ taking on values in $\mathbb{R}^d$, (2.48) implies that the covariance matrix defined in (2.45) is diagonal whenever the components of $\boldsymbol{X}$ are independent real-valued random variables. If X and Y are independent real-valued random variables, then (2.48) implies that $\mathrm{Cov}(X,Y) = 0$. Random variables with vanishing covariance are sometimes referred to as *uncorrelated*. While independent random variables are uncorrelated, vanishing covariance does not generally imply independence, as the following example shows.

Example 2.50 Dependent Random Variables with Zero Covariance
Consider $\Omega = \{-1, 0, 1\}$ with $\mathcal{A} = \mathcal{P}(\Omega)$ and the discrete probability measure defined by $P(\{-1\}) = P(\{0\}) = P(\{1\}) = \frac{1}{3}$. Let X and Y be real-valued

random variables defined via $X(-1) = -1$, $X(0) = 0$, $X(1) = 1$ and $Y(-1) = 1$, $Y(0) = -2$, $Y(1) = 1$. On the one hand, by applying Example 2.40, we obtain $\langle X \rangle = \langle Y \rangle = \langle XY \rangle = 0$ so that $\mathrm{Cov}(X, Y) = 0$. On the other hand, Y can be regarded as a function of X because X is the identity mapping and, in this situation, X and Y can be independent only if Y is constant (except on a null set). Indeed, for $A' = A'' = \{1\}$ the criterion (2.47) for independence is violated because the left-hand side is $P(\{1\}) = \frac{1}{3}$, the right-hand side is $P(\{1\})P(\{-1,1\}) = \frac{2}{9}$. □

A further important factorization property for independent random variables can be obtained if the joint distribution and hence also the marginal distributions of the vector-valued random variables $\boldsymbol{X}$ and $\boldsymbol{Y}$ are continuous probability measures. Then, (2.47) implies that the joint probability density is of the product form, and vice versa. All these factorization properties for independent random variables can immediately be generalized from two to any finite number of random variables.

2.2.4 Conditional Expectations and Probabilities

In the preceding subsection, we discussed joint probability distributions as a concept for characterizing the relationship between random variables, and we established their form for the particularly simple case of independent random variables. In this subsection, we introduce another concept for characterizing the relationship between real-valued or vector-valued random variables, namely conditional expectations. The concept of conditional expectations is crucial in understanding important classes of stochastic processes, such as Markov processes and martingales. For a real-valued random variable X, a conditional expectation of X is another real-valued random variable which, loosely speaking, is a coarse-grained version of X. Such a conditional expectation of X resolves less information than X itself, for example, only the information resolved by some other random variable.

In order to clarify these rather vague statements we first discuss an intuitive, but at the same time rigorous and typical, example before giving the general definition of conditional expectations. We consider $\Omega = [0, 6[$ together with the Borel σ-algebra $\mathcal{B}$ and the continuous probability measure with density $p(\omega) = \frac{1}{6}$ for $\omega \in \Omega$. As a random variable X we consider the third-order polynomial

$$X\colon [0,6[\to \mathbb{R}, \quad \omega \mapsto X(\omega) := \omega^3 - 10\omega^2 + 27\omega - 10\,,$$

with zeros at 5, $(5 \pm \sqrt{17})/2$, a maximum at $(10 - \sqrt{19})/3$, and a minimum at $(10 + \sqrt{19})/3$.

Figure 2.5 shows three coarse-grained versions of this random variable X. The coarsest version, represented by three dashed horizontal lines of length two, resolves only information contained in the σ-algebra $\mathcal{A}_3$ generated by $\{[0,2[, [2,4[, [4,6[\}$. This coarse-grained random variable is a conditional expectation of X, denoted by $E(X|\mathcal{A}_3)$. On the coarse events contained in $\mathcal{A}_3$,

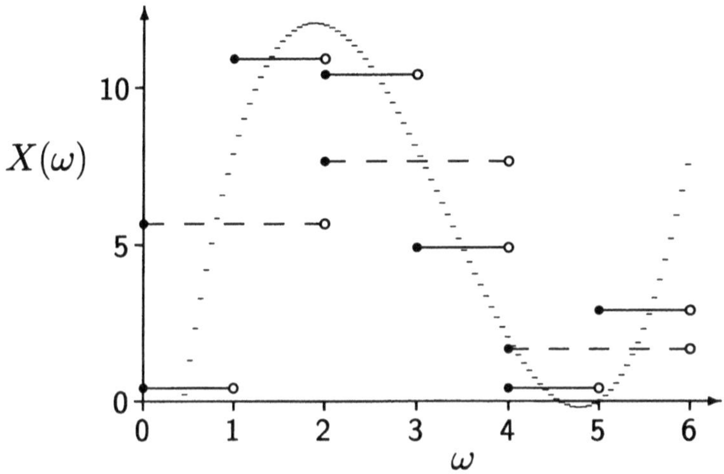

Fig. 2.5. Various coarse-grained versions or conditional expectations of a continuous random variable (see text for details). A *filled (open) circle* at the end of a horizontal line indicates that the function at the corresponding ω assumes (does not assume) the value given by the horizontal line.

$E(X|\mathcal{A}_3)$ and X have the same averages (this is the precise sense in which we speak of coarse graining).

A less coarse-grained version of X is given by the random variable $E(X|\mathcal{A}_6)$ where $\mathcal{A}_6$ is the σ-algebra generated by the six intervals $[j-1, j[$, $j = 1, \ldots 6$ (see the curve consisting of six steps of length one in Fig. 2.5). We can regard $E(X|\mathcal{A}_3)$ as a coarse-grained version of X or of $E(X|\mathcal{A}_6)$. If one resolves even more information by considering $E(X|\mathcal{A}_{100})$, where $\mathcal{A}_{100}$ is the σ-algebra obtained by partitioning Ω into 100 intervals of equal length (short dashes in Fig. 2.5), one can almost guess the third-order-polynomial character of the original random variable X. The example illustrated in Fig. 2.5 shows all the essential features of conditional expectations. We are now ready for a formal introduction of this concept.

Definition 2.51 *Let X be a real-valued random variable on $(\Omega, \mathcal{A}, P)$ for which $\langle X \rangle$ exists, and let $\mathcal{A}' \subset \mathcal{A}$ be a sub-σ-algebra of $\mathcal{A}$. A random variable $Y : (\Omega, \mathcal{A}') \to (\mathbb{R}, \mathcal{B})$ is a* conditional expectation *of X given $\mathcal{A}'$ if and only if*

$$\langle X \chi_{A'} \rangle = \langle Y \chi_{A'} \rangle \qquad \textit{for all } A' \in \mathcal{A}' . \tag{2.49}$$

We then write $E(X|\mathcal{A}') := Y$.

In words, conditional expectations are random variables defined on coarser σ-algebras that produce the same averages on the events contained in the coarse σ-algebras. Even though obvious in our example, in general, the proof of the existence of conditional expectations is quite involved (see, for example, Theorem 10.1.1 of [3]). One can furthermore prove the uniqueness of conditional expectations in the following sense: if two random variables Y and Y' have the properties listed in Definition 2.51, then $\{\omega \mid Y(\omega) \neq Y'(\omega)\}$ is a null set. In the usual language of probability theory, Y is *almost surely* equal to Y', or a conditional expectation is almost surely unique. In theorems and definitions we always state explicitly when the qualifier "almost surely" must be applied; in the text it is understood that statements like "two random variables are equal," "a random variable is unique," or "all values of a random variable have a certain property" in general only hold outside a null set.[5]

Exercise 2.52 Let $\boldsymbol{X}'$ be a random variable with uniform distribution in the unit circle $\{(x_1, x_2) \mid x_1^2 + x_2^2 < 1\}$. If we wish to make use of the results of Exercise 2.36, for example in designing a Gaussian random number generator (cf. Exercise 4.9), we can define a modified random variable $\boldsymbol{X}$ which does not take on the value $(0,0)$,

$$\boldsymbol{X}(\omega) := \begin{cases} (0, 1/2) & \text{if } \boldsymbol{X}'(\omega) = (0,0) \\ \boldsymbol{X}'(\omega) & \text{otherwise} \end{cases} .$$

Show that $\boldsymbol{X}$ is almost surely equal to $\boldsymbol{X}'$.

For understanding the idea behind conditional expectations it is crucial to have the concept of σ-algebra in order to characterize the "information content" of a random variable and to compare various coarse-grained random variables. Since, in many popular textbooks applying stochastics to the natural sciences or engineering, random variables are introduced in a very naive way through their distributions without referring to σ-algebras, the role of conditional expectations in applied stochastics is often not well understood. This fact causes severe limitations because conditional expectations are an important concept in understanding the general theory of stochastic calculus. From our intuitive interpretation of conditional expectations it is clear that these coarse-grained random variables, for example, play an important role as optimal estimates in filtering problems when only incomplete information is available (see Chap. 12 of [4]).

Exercise 2.53 Consider the random variable Y of Exercise 2.32, which may be regarded as a random variable with values in $(\mathbb{R}, \mathcal{B})$. Evaluate the conditional expectation $E(Y|\mathcal{A}(\{\{6\}\}))$. Did you have to make assumptions about the underlying probability measure?

Several important properties of conditional expectations follow immediately from the Definition 2.51. By applying (2.49) to $A' = \Omega$, we obtain

[5]Knowing that conditional expectations are almost surely unique is good enough for all practical applications, because exceptional cases occurring with zero probability are of no concern. We always have the freedom to change random variables on an event with zero probability without changing the situation or "physics" modeled by the random variable and, whenever it is convenient, we make use of this freedom.

$$\langle X \rangle = \langle E(X|\mathcal{A}') \rangle \,. \tag{2.50}$$

The corresponding properties of expectations, (2.39) and (2.40), lead to the following equations for conditional expectations,

$$E(c_1 X + c_2 Y|\mathcal{A}') \;=\; c_1\, E(X|\mathcal{A}') + c_2\, E(Y|\mathcal{A}')\,, \tag{2.51}$$

$$X \leq Y \;\Rightarrow\; E(X|\mathcal{A}') \leq E(Y|\mathcal{A}')\,, \tag{2.52}$$

which hold pointwise outside a null set (since conditional expectations can be arbitrarily modified on null sets).

If a random variable X is measurable with respect to the coarser σ-algebra $\mathcal{A}'$, one immediately verifies that $E(X|\mathcal{A}') = X$ (quite generally, $\mathcal{A}'$-measurable factors in conditional expectations with respect to $\mathcal{A}'$ can be pulled out of conditional expectations). In particular, if $\mathcal{A}' = \mathcal{A}^Y$, that is if X resolves no more information than another random variable Y, then conditioning has no effect, $E(X|\mathcal{A}^Y) = X$. If, on the other hand, two random variables X and Y are independent one finds an almost surely constant conditional expectation,

$$E(X|\mathcal{A}^Y) = \langle X \rangle \,. \tag{2.53}$$

In order to prove this equality, by definition, one has to show that $\langle X\, \chi_A \rangle = \langle \langle X \rangle\, \chi_A \rangle$ for all $A \in \mathcal{A}^Y$, which follows from (2.48). Loosely speaking, the result (2.53) can be stated as follows: in the case of independence, the conditioning σ-algebra in a conditional expectation can be dropped, so that the conditional expectation is simply given by the expectation.

Exercise 2.54 Show that if the coarse-graining procedure is performed in several steps then the final result does not depend on intermediate levels of coarse graining but only on the coarsest σ-algebra used for conditioning; that is, for $\mathcal{A}'' \subset \mathcal{A}' \subset \mathcal{A}$,

$$E\Big(E(X|\mathcal{A}')\Big|\mathcal{A}''\Big) = E(X|\mathcal{A}'')\,.$$

Example 2.55 From Conditional Expectations to Probabilities
For a given $(\Omega, \mathcal{A}, P)$, let $A, B \in \mathcal{A}$ with $0 < P(B) < 1$, and let $\mathcal{A}' = \{\emptyset, B, B^C, \Omega\}$. One then has

$$E(\chi_A|\mathcal{A}')(\omega) = \begin{cases} P(A|B) & \text{if } \omega \in B \\ P(A|B^C) & \text{if } \omega \in B^C \,. \end{cases}$$

This relationship, which can be proved by verifying the defining equation (2.49) for all four $A' \in \mathcal{A}'$, establishes a relationship between the conditional probabilities defined in (2.15) and the conditional expectations introduced in this subsection. By averaging we obtain the formula

$$P(A) = \langle \chi_A \rangle = P(A|B)\, P(B) + P(A|B^C)\, P(B^C)\,,$$

which is sometimes referred to as the *total probability rule.* □

Example 2.55 can be used as a guideline for constructing the conditional probability that a random variable X takes on a value in A given that the random variable Y takes on the value y, that is $P(X \in A|Y = y)$. The basic idea is to obtain conditional probabilities as the values of a conditional expectation on the conditioning event. In order to define this quantity, which is a very useful tool for describing the relationship between two random variables X and Y, we consider $E\left(\chi_{\{\omega|X(\omega)\in A\}}|\mathcal{A}^Y\right)$, which is a measurable function on $(\Omega, \mathcal{A}^Y)$. We next need to find the value of this function when Y takes on the value y. According to Theorem 2.31, this function can be decomposed in the following way,

$$E\left(\chi_{\{\omega|X(\omega)\in A\}}|\mathcal{A}^Y\right) = g(Y) \qquad \text{for some function } g\,,$$

and we then define the *conditional probability*

$$P(X \in A|Y = y) := g(y)\,. \tag{2.54}$$

For given A, this function is unique except on a set of y-values with zero measure P^Y. This completes our general definition of conditional probabilities which can be applied to random variables X and Y with arbitrary discrete, continuous, and mixed distributions. The definition has been motivated by Example 2.55 for the discrete case, and in Example 2.56 we verify that our abstraction is reasonable also for random variables with continuous distributions. Before doing so, we give an integral equation characterizing conditional probabilities in a more intuitive manner, and we show how conditional probabilities can be used to construct conditional expectations by straightforward integration.

In order to show that some given function $\tilde{g}(y)$ is equal to $P(X \in A|Y = y)$, according to the definition of conditional expectations (2.49) and the transformation formula (2.38), one needs to show that the following equation holds,

$$P(X \in A, Y \in B) = \int_B \tilde{g}\,dP^Y := \int \tilde{g}\,\chi_B\,dP^Y \qquad \text{for all } B \in \mathcal{B}\,. \tag{2.55}$$

Conditional probabilities play an important role in the theory of general Markov processes. They are particularly useful in the construction of variance reduced numerical integration schemes for stochastic differential equations. Furthermore, the conditional probability $P(X \in A|Y = y)$ is a very useful quantity because it has an intuitive interpretation and, at the same time, it can be used as a basis for constructing various conditional expectations, which is the other fundamental concept developed in this subsection. In fact, knowing the conditional probabilities $P(X \in A|Y = y)$ for all $A \in \mathcal{B}$, we can evaluate all conditional expectations of the form $E(g(X)|\mathcal{A}^Y)$ by straightforward integration. In order to understand this procedure it is important to realize that the collection of conditional probabilities $P(X \in A|Y = y)$ for all different $A \in \mathcal{B}$ can be chosen such that for each $y \in \mathbb{R}$ the function $P(X \in \cdot|Y = y)\colon \mathcal{B} \to \mathbb{R}$, $A \mapsto P(X \in A|Y = y)$ is a probability measure (see Theorem 10.3.5 of [3]; here the dot ("$\cdot$") indicates the point at which events are to be inserted as arguments of the function). For each $y \in \mathbb{R}$, we can then define the expectation,

$$E(g(X)|Y=y) := \int g \, \mathrm{d}P(X \in \cdot|Y=y)\,, \tag{2.56}$$

with respect to the measure $P(X \in \cdot|Y=y)$. Then, the conditional expectation $E(g(X)|\mathcal{A}^Y)$ is obtained as the composite of the functions $E(g(X)|Y=y)$ and Y. For simple functions g, this statement follows immediately from the definition of the integral on the right side of (2.56) according to Step 1 in Sect. 2.2.2 and from the definition (2.54) of conditional probabilities. The proof of the general result can then be based on the theorem of monotone convergence (see, e.g., Theorem 2.3.4 and p. 314 of [3]).

If the distribution of Y is a continuous probability measure with density $p^Y(y)$ then, by taking the expectation of the expression for $E(g(X)|\mathcal{A}^Y)$ constructed from (2.56), we obtain

$$\boxed{\langle g(X)\rangle = \int E(g(X)|Y=y)\, p^Y(y)\, \mathrm{d}y\,.} \tag{2.57}$$

This formula is a helpful tool for evaluating expectations by "distinguishing cases." For example, this formula can be applied to reformulate the stress tensor in reptation models (see Sect. 6.1.2). Equation (2.57) may be regarded as a generalization of the total probability rule of Example 2.55.

Example 2.56 Conditional Probability Densities

Since the above definition of conditional probabilities is rather formal, we want to show that this definition indeed captures the intuitive meaning of conditional probability by considering the illustrative case of random variables with continuous distributions. This example also provides a valuable background for the theory of Markov processes developed in the section on stochastic processes.

If the joint distribution of X and Y is a continuous probability measure with density $p^{(X,Y)}(x,y)$, and $p^Y(y) = \int p^{(X,Y)}(x,y)\mathrm{d}x$ is the marginal probability density for the distribution of Y, then $P(X \in \cdot|Y=y)$ is also a continuous probability measure, and one has

$$P(X \in A|Y=y) = \int_A p(X=x|Y=y)\, \mathrm{d}x\,, \tag{2.58}$$

with

$$p(X=x|Y=y) = \begin{cases} \dfrac{p^{(X,Y)}(x,y)}{p^Y(y)} & \text{if } p^Y(y) \neq 0 \\ 0 & \text{otherwise}\,. \end{cases} \tag{2.59}$$

This explicit representation can be proved by verifying the condition (2.55). The result (2.59) for the conditional probability density is very similar in form to the definition of the conditional probability in (2.15). If one views a plot of the joint probability density $p^{(X,Y)}(x,y)$ as a function of the two variables x and y as a "landscape," the conditional probability density $p(X=x|Y=y)$ can be interpreted as the profile of the landscape for a cut along the fixed "latitude" y, which is then properly normalized (see Fig. 2.6). For the conditional

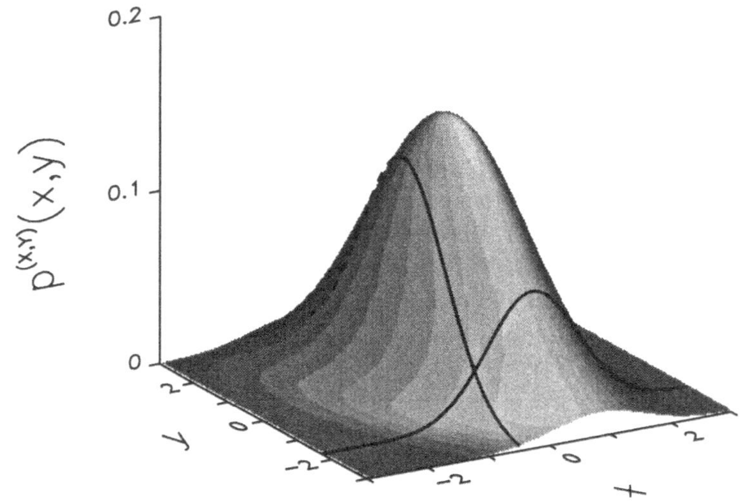

Fig. 2.6. Conditional probability densities for a pair of dependent Gaussian random variables. After suitable normalization, the *thick curves* represent $p(X = x|Y = -1.7)$ and $p(Y = y|X = -0.5)$.

probabilities defined in (2.58) and (2.59) it is obvious that, for each $y \in \mathbb{R}$, the quantity $P(X \in A|Y = y)$ as a function of A is a continuous probability measure. Explicit examples of this procedure are considered in the next subsection on Gaussian random variables.

In the situation of this example, the formula (2.57) takes the form

$$\langle g(X) \rangle = \int\int g(x)\, p(X = x|Y = y)\, p^Y(y)\, \mathrm{d}y\, \mathrm{d}x\,, \tag{2.60}$$

which is obvious from (2.59). □

Conditional expectations for vector-valued random variables can be defined componentwise. The construction of the conditional probability (2.54) need not be modified for vector-valued random variables; if $\boldsymbol{Y}$ is a random variable taking values in $\mathbb{R}^d$, we occasionally use the notation $P(X \in A|Y_1 = y_1, \ldots, Y_d = y_d) := P(X \in A|\boldsymbol{Y} = \boldsymbol{y})$. In introducing conditional probability densities, multi-dimensional integrals are to be used in (2.58).

2.2.5 Gaussian Random Variables

We here extend the discussion of Gaussian probability measures started in Sect. 2.1.3. The results compiled in this section allow a simple handling of Gaussian probability measures in most situations occurring in applications. Gaussian probability measures can, according to Example 2.34, be regarded as distributions of suitable vector-valued random variables. We refer to such random variables as *Gaussian random variables*, and we no longer distinguish between Gaussian probability measures and Gaussian distributions. We speak of a set of Gaussian random variables if the joint distribution of all these random variables is Gaussian.

Consider a vector-valued Gaussian random variable $\boldsymbol{X}$ whose distribution has a density $p_{\alpha\Theta}$ of the form (2.21). Equations (2.23) and (2.24) in combination with Example 2.41 imply that the first moment of $\boldsymbol{X}$ is given by $\langle X_j \rangle = \alpha_j$, and that the components of the covariance matrix are given by $\langle X_j X_k \rangle - \langle X_j \rangle \langle X_k \rangle = \Theta_{jk}$. Therefore, the first and second moments of a Gaussian random variable determine all the parameters in the probability density; consequently, all higher moments can be expressed in terms of the first and second moments. This observation is often used to truncate hierarchies of equations for increasingly higher moments by assuming Gaussian distributions. Since the joint distribution of a family of Gaussian random variables is also characterized by its first and second moments, such a joint distribution is fully determined by at most pairwise correlations between the random variables.

Among applied scientists, the set of explicit rules for decomposing higher moments of Gaussian distributions in terms of first and second moments is known as "Wick's theorem." The derivation of such decomposition rules can be based on the identity

$$\langle \boldsymbol{X} g(\boldsymbol{X}) \rangle = \langle \boldsymbol{X} \rangle \langle g(\boldsymbol{X}) \rangle + \Big(\langle \boldsymbol{X}\boldsymbol{X} \rangle - \langle \boldsymbol{X} \rangle \langle \boldsymbol{X} \rangle \Big) \cdot \left\langle \frac{\partial}{\partial \boldsymbol{x}} g(\boldsymbol{X}) \right\rangle , \tag{2.61}$$

which can be verified by expressing all expectations as integrals, inserting the explicit expression for the Gaussian probability density, and integrating by parts. Of course, the otherwise arbitrary differentiable function g must be such that the expectations in (2.61) exist and that the boundary terms at infinity vanish in the integration by parts. Equation (2.61) is particularly simple when $\langle \boldsymbol{X} \rangle = \boldsymbol{0}$. For example, if we apply it in that case to the third-order product $g(\boldsymbol{x}) = x_{j_1} x_{j_2} x_{j_3}$, we obtain from the j_4-component of (2.61)

$$\begin{aligned} \langle X_{j_1} X_{j_2} X_{j_3} X_{j_4} \rangle &= \langle X_{j_1} X_{j_2} \rangle \langle X_{j_3} X_{j_4} \rangle \\ &+ \langle X_{j_1} X_{j_3} \rangle \langle X_{j_2} X_{j_4} \rangle + \langle X_{j_1} X_{j_4} \rangle \langle X_{j_2} X_{j_3} \rangle . \end{aligned} \tag{2.62}$$

If the components of a general vector-valued random variable $\boldsymbol{X}$ are independent, the covariance matrix is diagonal. For Gaussian random variables, the reverse of this statement is also true. This is because a diagonal $\boldsymbol{\Theta}$ is equivalent to a probability density $p_{\alpha\Theta}$ of the product form. In particular, for the special case $p_{\alpha\Theta} = p_{0\delta}$, the d components of the random variable are independent real-valued Gaussian random variables with vanishing first moments and variances of unity. It is a remarkable property of vector-valued Gaussian random variables that a diagonal covariance matrix (representing uncorrelated components) is equivalent to the independence of the components. This is not true for general random variables, as we learned from Example 2.50.

Example 2.35 implies that any regular linear transformation of a given Gaussian random variable $\boldsymbol{X}$ leads to another Gaussian random variable $\boldsymbol{Y}$. If $\boldsymbol{Y} = \mathrm{M} \cdot \boldsymbol{X} + \boldsymbol{c}$ where M is a regular matrix and $\boldsymbol{c}$ is an arbitrary column vector, we can obtain the expectation and covariance of the Gaussian random variable $\boldsymbol{Y}$ from the corresponding quantities $\boldsymbol{\alpha}$ and $\boldsymbol{\Theta}$ for the random variable $\boldsymbol{X}$ according to

$$\langle \boldsymbol{Y} \rangle = \mathrm{M} \cdot \boldsymbol{\alpha} + \boldsymbol{c}, \qquad \langle \boldsymbol{Y}\boldsymbol{Y} \rangle - \langle \boldsymbol{Y} \rangle \langle \boldsymbol{Y} \rangle = \mathrm{M} \cdot \boldsymbol{\Theta} \cdot \mathrm{M}^T . \tag{2.63}$$

This result can be obtained either from the transformation of probability densities discussed in Example 2.35 or from the linearity (2.39) of expectations.

Exercise 2.57 Let a Gaussian random variable $\boldsymbol{X}$ with covariance matrix $\boldsymbol{\Theta}$ be given. Establish a linear transformation between $\boldsymbol{X}$ and another Gaussian random variable $\boldsymbol{Y}$ with independent components by diagonalizing the symmetric matrix $\boldsymbol{\Theta}$. (Remark: This construction or similar ideas can be exploited in simulations when vector-valued Gaussian random variables $\boldsymbol{X}$ with given covariance matrices $\boldsymbol{\Theta}$ are constructed by means of linear transformations from independent real-valued Gaussian random variables (the components of $\boldsymbol{Y}$) which are much easier to simulate.)

We next construct the marginal distributions of Gaussian random variables. If we select $n \leq d$ of the components of a vector-valued random variable $\boldsymbol{X}$, say $X_{j_1}, \ldots, X_{j_n}$ with $1 \leq j_1 < \ldots < j_n \leq d$, and we write $\boldsymbol{X}' = (X_{j_1}, \ldots, X_{j_n})^T$, the characteristic functions of $\boldsymbol{X}$ and $\boldsymbol{X}'$ are closely related. Namely, for arbitrary $\boldsymbol{q}' \in \mathbb{R}^n$ we choose $\boldsymbol{q} \in \mathbb{R}^d$ such that $q_{j_1} = q'_1, \ldots, q_{j_n} = q'_n$, and the remaining components of $\boldsymbol{q}$ are zero; we then obtain

$$\tilde{p}^{X'}(\boldsymbol{q}') = \left\langle \mathrm{e}^{i\boldsymbol{q}' \cdot \boldsymbol{X}'} \right\rangle = \left\langle \mathrm{e}^{i\boldsymbol{q} \cdot \boldsymbol{X}} \right\rangle = \tilde{p}^{X}(\boldsymbol{q}) .$$

In general, the characteristic function of a marginal distribution is obtained by keeping only those components of $\boldsymbol{q}$ corresponding to the components of $\boldsymbol{X}$ that are considered in the marginal distribution while setting the remaining components of $\boldsymbol{q}$ equal to zero. From the explicit form (2.25) of the characteristic function for a Gaussian random variable we find that the distribution of $\boldsymbol{X}'$ is also Gaussian. The vanishing components of $\boldsymbol{q}$ suppress the corresponding contributions in the quadratic exponential of (2.25), and the remaining contributions for the marginal distribution are of course consistent with $\langle X'_k \rangle = \langle X_{j_k} \rangle$ and $\mathrm{Cov}(X'_k, X'_l) = \mathrm{Cov}(X_{j_k}, X_{j_l})$. In summary: the marginal distributions of Gaussian distributions are also Gaussian; they are obtained by deleting the appropriate entries from the first-moment vector and the appropriate rows and columns from the covariance matrix.

So far, we have assumed that the covariance matrices of Gaussian distributions are positive-definite. It is often convenient to relax that condition to positive-semidefinite (that is, $\boldsymbol{x} \cdot \boldsymbol{\Theta} \cdot \boldsymbol{x} \geq 0$ for all $\boldsymbol{x} \in \mathbb{R}^d$; in other words, the eigenvalues of the symmetric matrix $\boldsymbol{\Theta}$ are assumed to be only nonnegative instead of strictly positive). For example, one can then use (2.63) for singular matrices $\mathbf{M}$ also. While a covariance matrix $\boldsymbol{\Theta}$ which is positive-semidefinite but not positive-definite does not cause any problems in the characteristic function (2.25), the resulting singular nature of $\boldsymbol{\Theta}$ makes the expression (2.21) for the probability density meaningless. If we assume that the covariance matrix $\boldsymbol{\Theta}$ is diagonal (without loss of generality, because a diagonal $\boldsymbol{\Theta}$ can be achieved by a non-singular linear transformation), $\boldsymbol{\Theta}$ has at least one vanishing diagonal element. Those factors of the fully factorized distribution function which correspond to zero eigenvalues of $\boldsymbol{\Theta}$ describe one-dimensional distributions of zero width and may be thought of as δ-functions. These δ-function factors correspond to distributions concentrated at single points or to deterministic outcomes for the respective components. The validity of this δ-function interpretation is best seen from the corresponding factors in the characteristic functions. The considerations of this paragraph show that if the covariance matrix has zero eigenvalues then the corresponding Gaussian distribution is concentrated in a lower-dimensional subspace.

After admitting covariance matrices with zero eigenvalues and discussing marginal distributions it is clear that transformations of the form $\boldsymbol{Y} = \mathbf{M} \cdot \boldsymbol{X} + \boldsymbol{c}$ lead from one Gaussian random variable to another one, even if the number of components of $\boldsymbol{Y}$ is less than or greater than the number of components of $\boldsymbol{X}$. Therefore, the class of Gaussian random variables is invariant under general affine-linear transformations. In particular, any linear combination of Gaussian random variables is Gaussian.

For a pair of arbitrary vector-valued Gaussian random variables $\boldsymbol{X}$ and $\boldsymbol{Y}$, the conditional probability density $p(\boldsymbol{X} = \boldsymbol{x} | \boldsymbol{Y} = \boldsymbol{y})$ is Gaussian. This follows from (2.59) and the fact that the ratio of two exponentials of quadratic forms is again an exponential of a quadratic form. More intuitively, we can say that any cut through a Gaussian density along a hyperplane produces another Gaussian distribution. In the following, we consider an explicit example.

Example 2.58 Reconstruction of Intermediate Positions

Let X_1 and X_2 be independent Gaussian random variables with vanishing first moment and variance σ^2. If we want to calculate the conditional probability density $p(X_1 = x | X_1 + X_2 = y)$, we first need to construct the density of the joint distribution of $X = X_1$ and $Y = X_1 + X_2$. From the above considerations we know that this joint distribution of X and Y is also Gaussian, and we find for the first and second moments $\langle X \rangle = \langle Y \rangle = 0$, $\langle X^2 \rangle = \langle XY \rangle = \sigma^2$, and $\langle Y^2 \rangle = 2\sigma^2$ where $\langle X_1 X_2 \rangle = 0$ has been used (which is a consequence of the independence of X_1 and X_2). By inserting the inverted 2×2 covariance matrix into (2.21) we thus obtain

$$p^{(X,Y)}(x,y) = \frac{1}{2\pi\sigma^2} \exp\left\{-\frac{2x^2 - 2xy + y^2}{2\sigma^2}\right\}.$$

The marginal probability density for the one-dimensional Gaussian variable Y with $\langle Y \rangle = 0$ and $\langle Y^2 \rangle = 2\sigma^2$ is

$$p^Y(y) = \frac{1}{2\sqrt{\pi\sigma^2}} \exp\left\{-\frac{y^2}{4\sigma^2}\right\}.$$

After inserting these probability densities into (2.59), we obtain the final result

$$p(X_1 = x | X_1 + X_2 = y) = \frac{1}{\sqrt{\pi\sigma^2}} \exp\left\{-\frac{[x-(y/2)]^2}{\sigma^2}\right\}. \tag{2.64}$$

This is a Gaussian probability density with mean $\frac{1}{2}y$ and squared width $\frac{1}{2}\sigma^2$.

The result (2.64) may be interpreted as follows. If X_1 and X_2 are regarded as independent successive Gaussian steps, then the conditioning event is that the outcome of the two steps when starting at 0 is $X_1 + X_2 = y$. Under this condition, one can ask where the first step ended. The answer given in (2.64) is that the average position after the first step is the midpoint between 0 and y, that the fluctuations around this position are Gaussian, and that the variance of the intermediate position is $\frac{1}{2}\sigma^2$, which is only $\frac{1}{2}$ of the variance of X_1. □

Exercise 2.59 Calculate the conditional expectations $E(X|\mathcal{A}^Y)$ and $E(Y|\mathcal{A}^X)$ for the random variables X and Y of Example 2.58.

Exercise 2.60 Let $(\boldsymbol{X}, \boldsymbol{Y})$ be a vector-valued Gaussian random variable, and introduce the matrices $\boldsymbol{\Theta} := \langle \boldsymbol{YY} \rangle - \langle \boldsymbol{Y} \rangle \langle \boldsymbol{Y} \rangle$ and $\boldsymbol{\Theta}' := \langle \boldsymbol{XY} \rangle - \langle \boldsymbol{X} \rangle \langle \boldsymbol{Y} \rangle$. Derive the following identities,

$$\begin{aligned} E(\boldsymbol{X}|\mathcal{A}^Y) &= \langle \boldsymbol{X} \rangle + \boldsymbol{\Theta}' \cdot \boldsymbol{\Theta}^{-1} \cdot (\boldsymbol{Y} - \langle \boldsymbol{Y} \rangle), \\ E(\boldsymbol{XX}|\mathcal{A}^Y) - E(\boldsymbol{X}|\mathcal{A}^Y)E(\boldsymbol{X}|\mathcal{A}^Y) &= \langle \boldsymbol{XX} \rangle - \langle \boldsymbol{X} \rangle \langle \boldsymbol{X} \rangle - \boldsymbol{\Theta}' \cdot \boldsymbol{\Theta}^{-1} \cdot \boldsymbol{\Theta}'^T. \end{aligned}$$

Verify the results of Exercise 2.59.

2.2.6 Convergence of Random Variables

The convergence of sequences of random variables plays an important role in stochastics. On the one hand, many concepts of stochastics are first introduced for certain simple functions and then generalized by using sequences of such simple functions converging to more general functions (this is also a typical procedure for proving theorems). On the other hand, limit theorems for sums of independent random variables are among the most important results of stochastics. For the purpose of this book, a careful discussion of the convergence of random variables is important for understanding the behavior of the solutions of time-discretized stochastic differential equations in the limit of vanishing time-step width. On another level, the strong law of large numbers provides the justification for estimating expectations by ensemble averages.

Consider a sequence $(X_n)_{n\in\mathbb{N}}$ of real-valued random variables. Since the X_n are real-valued functions, the usual concept of pointwise convergence can be applied. As previously pointed out, for the purposes of stochastics it is sufficient if functions converge almost surely. In the following definition, we consider two further important concepts of convergence which are related to the convergence of moments and to the probability of having small deviations.

Definition 2.61 *Suppose that the sequence $(X_n)_{n\in\mathbb{N}}$ of real-valued random variables and the real-valued random variable X are defined on the same probability space $(\Omega, \mathcal{A}, P)$.*
The random variable X is the almost sure limit *of the sequence $(X_n)_{n\in\mathbb{N}}$, or $X = \text{as-}\lim_{n\to\infty} X_n$, if and only if*

$$P(\{\omega \mid |X_n(\omega) - X(\omega)| \to 0 \text{ as } n \to \infty\}) = 1. \tag{2.65}$$

The random variable X is the mean-square limit *of the sequence $(X_n)_{n\in\mathbb{N}}$, or $X = \text{ms-}\lim_{n\to\infty} X_n$, if and only if*

$$\left\langle |X_n - X|^2 \right\rangle \to 0 \text{ as } n \to \infty. \tag{2.66}$$

The random variable X is the stochastic limit *of the sequence $(X_n)_{n\in\mathbb{N}}$, or $X = \text{st-}\lim_{n\to\infty} X_n$, if and only if for all $\epsilon > 0$*

$$P(\{\omega \mid |X_n(\omega) - X(\omega)| \geq \epsilon\}) \to 0 \text{ as } n \to \infty. \tag{2.67}$$

For all these modes of convergence, the limit X is almost surely unique. The ms-convergence plays an important role in defining stochastic integrals and in solving stochastic differential equations by discretization. Of course, it is important to establish relationships between as-, ms-, and st-convergence. The basic results are summarized in the following two theorems.

Theorem 2.62 *Suppose that the sequence $(X_n)_{n\in\mathbb{N}}$ of real-valued random variables and the real-valued random variable X are defined on the same probability space $(\Omega, \mathcal{A}, P)$. Then:*

$$X = \text{as-}\lim_{n\to\infty} X_n \implies X = \text{st-}\lim_{n\to\infty} X_n\,,$$
$$X = \text{ms-}\lim_{n\to\infty} X_n \implies X = \text{st-}\lim_{n\to\infty} X_n\,.$$

This means that stochastic convergence is the weakest of the three modes of convergence introduced in Definition 2.61. The first statement of Theorem 2.62 is trivial (assume that the statement is not true to see this); the second implication follows from the following chain of inequalities, which are based on the monotony of expectations (2.40),

$$\begin{aligned}\langle |X_n - X|^2 \rangle &\geq \langle |X_n - X|^2 \chi_{\{\omega | |X_n(\omega) - X(\omega)| \geq \epsilon\}} \rangle \\ &\geq \epsilon^2 P(\{\omega \mid |X_n(\omega) - X(\omega)| \geq \epsilon\}) .\end{aligned}$$

In the special situation of so-called dominated convergence, the following theorem states that as-convergence is the strongest mode of convergence, and that the other two modes are equivalent.

Theorem 2.63 *Suppose that the sequence $(X_n)_{n \in \mathbb{N}}$ of real-valued random variables and the real-valued random variable X are defined on the same probability space $(\Omega, \mathcal{A}, P)$. If there exists a random variable Y on $(\Omega, \mathcal{A}, P)$ with finite expectation such that $X_n^2 \leq Y$ for all $n \in \mathbb{N}$, then:*

$$\begin{aligned} X = \text{as-}\lim_{n\to\infty} X_n &\implies X = \text{ms-}\lim_{n\to\infty} X_n \text{ and } X = \text{st-}\lim_{n\to\infty} X_n , \\ X = \text{ms-}\lim_{n\to\infty} X_n &\iff X = \text{st-}\lim_{n\to\infty} X_n . \end{aligned}$$

Only the reverse direction of the second assertion remains to be shown (the rest follows from Theorem 2.62). A full proof is not particularly illuminating; it requires lengthy discussions of inequalities (see Theorem 2.12.4 of [3]).

Exercise 2.64 Consider $\Omega = [0,1[$ together with the Borel σ-algebra on $[0,1[$ and the continuous probability measure with density $p(x) = 1$. Let $X_n := \sqrt{n}\,\chi_{]0,1/n[}$ for $n = 1, 2, \ldots$ be a sequence of random variables. For which $\$\$ \in \{\text{as,ms,st}\}$ is $\$\$\text{-}\lim_{n\to\infty} X_n = 0$? Can one make use of Theorems 2.62 and 2.63?

Exercise 2.65 Consider $\Omega = [0,1[$ together with the Borel σ-algebra on $[0,1[$ and the continuous probability measure with density $p(x) = 1$. Decompose $n = 1, 2, \ldots$ in the form $n = 2^j + k$, $0 \leq k < 2^j$, with nonnegative integers j and k, and let $X_n := \chi_{[k\,2^{-j},(k+1)2^{-j}[}$. Answer the questions of Exercise 2.64.

Definition 2.61 allows us to check whether a sequence of random variables converges in one or more of the given senses to a given random variable X. In many applications, only a sequence of random variables is given, and it is important to find out whether or not this sequence converges in some sense to some (unknown) limiting random variable. In that situation, the analog of the Cauchy criterion for sequences of real numbers allows us to come to a decision about the convergence of a sequence of random variables for each of the three modes of convergence introduced in Definition 2.61 (see Theorem 2.7.5 and Remark (6) on p. 100 of [3]).

Theorem 2.66 (Cauchy Criterion). *Suppose that all the real-valued random variables $(X_n)_{n \in \mathbb{N}}$ are defined on the same probability space $(\Omega, \mathcal{A}, P)$. Then, the condition*

$$\$\$\text{-}\lim_{n,l\to\infty} |X_n - X_l| = 0 ,$$

is equivalent to the existence of a random variable X defined on $(\Omega, \mathcal{A}, P)$ such that

$$X = \$\$\text{-}\lim_{n\to\infty} X_n\,,$$

where \$\$ represents any of the three modes of convergence in Definition 2.61, that is \$\$ $\in$ {as,ms,st}.

All the definitions and theorems of this subsection can immediately be generalized to the case of vector-valued random variables by interpreting all absolute values as lengths of column vectors.

Example 2.67 Strong Law of Large Numbers
Let $(X_n)_{n\in\mathbb{N}}$ be a sequence of independent, identically distributed, real-valued random variables on the same probability space $(\Omega, \mathcal{A}, P)$.[6] If the common expectation $\langle X_n \rangle = \alpha$ is finite, one has

$$\text{as-}\lim_{n\to\infty} \frac{1}{n}\sum_{j=1}^{n} X_j = \alpha\,. \tag{2.68}$$

Note that the as-lim in (2.68) is the strongest mode of convergence, and this fact explains the name "strong law of large numbers." This important result states that we can approximate the common expectation α to arbitrary accuracy by averaging $X_1, \ldots, X_n$ for sufficiently large n (except on a null set). The limit theorem (2.68) is the mathematical basis and justification for estimating probabilities from relative frequencies. The proof of this deep theorem requires several pages of careful estimates (see Sect. 6.4 of [3]). □

Exercise 2.68 Let $(X_n)_{n\in\mathbb{N}}$ be a sequence of independent, identically distributed, real-valued random variables on the same probability space $(\Omega, \mathcal{A}, P)$ with finite mean α and variance Θ. Show that

$$\text{as-}\lim_{n\to\infty}\left[\frac{1}{n-1}\sum_{j=1}^{n} X_j^2 - \frac{1}{n(n-1)}\left(\sum_{j=1}^{n} X_j\right)^2\right] = \Theta\,.$$

Show that, for any $n \geq 2$, the random variable in square brackets is nonnegative and that its expectation is equal to Θ. This is the reason why the nth term is frequently used in estimating the variance Θ on the basis of n independent realizations of a random variable.

All the modes of convergence introduced in Definition 2.61 depend on the values of the functions X_n. It is often convenient to have a further convergence mode which involves only the distributions of the random variables X_n. Such a mode of convergence offers the possibility of relating random variables defined on

[6] The existence of sequences of independent random variables with given distributions seems to be intuitively clear, however, a formal proof of existence is actually quite involved (see Sect. 5.4 of [3]).

different probability spaces (even though it is known to be difficult to compare apples and oranges, it might be useful to compare their sizes and shapes). For us, the most important aspect of the convergence of distributions is the convergence of certain expectations. Such a situation occurs, for example, when analyzing the convergence of a series of results from computer simulations with different parameter settings. The above remarks motivate the following definition:

Definition 2.69 *Suppose that the sequence $(X_n)_{n\in\mathbb{N}}$ of real-valued random variables and the real-valued random variable X are given (all these random variables may be defined on different probability spaces).*
The random variable X is a limit in distribution *of the sequence $(X_n)_{n\in\mathbb{N}}$, or $X = \text{db-}\lim\limits_{n\to\infty} X_n$, if and only if*

$$\lim_{n\to\infty} \langle g(X_n)\rangle = \langle g(X)\rangle \tag{2.69}$$

for all bounded continuous functions $g\colon \mathbb{R} \to \mathbb{R}$.

Of course, the limit X is not unique; only the distribution of X is uniquely determined by the converging sequence $(X_n)_{n\in\mathbb{N}}$. The convergence in distribution is weaker than all three modes of convergence introduced in Definition 2.61, that is, all those modes of convergence (as, ms, st) imply the convergence in distribution. The convergence in distribution does not automatically imply the convergence of moments because the corresponding functions are not bounded. Under rather general conditions (uniform integrability), however, the convergence of moments can be established (see Sect. 11.4 of [2]). A very important and convenient criterion for verifying the convergence in distribution based on a smaller but still sufficiently rich class of functions in Definition 2.69 is given by the following theorem.

Theorem 2.70 *Suppose that the sequence $(X_n)_{n\in\mathbb{N}}$ of real-valued random variables and the real-valued random variable X are given. Then, $X = \text{db-}\lim\limits_{n\to\infty} X_n$ if and only if the characteristic functions of X_n converge pointwise to the characteristic function of X, that is,*

$$\lim_{n\to\infty} \tilde{p}^{X_n}(q) = \tilde{p}^{X}(q) \qquad \textit{for all } q \in \mathbb{R}\,. \tag{2.70}$$

Example 2.71 Central Limit Theorem
Let $(X_n)_{n\in\mathbb{N}}$ be a sequence of independent, identically distributed, real-valued random variables on the same probability space $(\Omega, \mathcal{A}, P)$. If the expectations $\langle X_n\rangle$ vanish (this assumption can be made without loss of generality since one can otherwise subtract the expectation from each X_n), and if the variances $\langle X_n^2\rangle = \Theta$ are finite, (2.44) yields the following Taylor expansion (in q) for the identical characteristic functions,

$$\tilde{p}^{X_n}(q) = 1 - \frac{1}{2}\,\Theta q^2 + o(q^2)\,.$$

We next consider the variances of the random variables

$$Y_n = \frac{1}{\sqrt{n}} \sum_{j=1}^{n} X_j \, , \tag{2.71}$$

which are

$$\left\langle Y_n^2 \right\rangle = \frac{1}{n} \sum_{j,k=1}^{n} \langle X_j X_k \rangle = \Theta \, ,$$

where the fact that the covariances of independent random variables vanish has been used. The motivation for dividing the sum in (2.71) by $\sqrt{n}$ is that this normalization leads to a variance $\langle Y_n^2 \rangle$ which is independent of n. Using the definition of characteristic functions as well as the factorization (2.48) of expectations for products of independent random variables, we obtain the characteristic function of Y_n,

$$\begin{aligned} \tilde{p}^{Y_n}(q) &= \left\langle \exp\left\{ iq \frac{1}{\sqrt{n}} \sum_{j=1}^{n} X_j \right\} \right\rangle \\ &= \prod_{j=1}^{n} \left\langle \exp\left\{ i \frac{q}{\sqrt{n}} X_j \right\} \right\rangle = \prod_{j=1}^{n} \tilde{p}^{X_j}\left(\frac{q}{\sqrt{n}}\right) \\ &= \left[1 - \frac{1}{2} \Theta q^2 \frac{1}{n} + o\left(\frac{1}{n}\right) \right]^n \underset{n\to\infty}{\longrightarrow} \mathrm{e}^{-\Theta q^2/2} = \tilde{p}_{0\Theta}(q) \, . \end{aligned}$$

In conclusion, the normalized sums of independent random variables (2.71) converge in distribution to a Gaussian random variable with probability density $p_{0\Theta}$. The result just proved in three lines is an important special case of the central limit theorem which establishes the db-convergence of sums of independent random variables to Gaussian random variables under much more general conditions (see Sect. 9.2 of [3]). In particular, a closer examination of the above proof shows that the assumption of identically distributed random variables is not crucial. The central limit theorem explains the immense importance of Gaussian random variables as it states that the accumulated effects of many independent random outcomes with finite variance always tend to become Gaussian. The theorem also offers a possibility for generating approximate Gaussian random variables by adding independent random variables, say with a uniform distribution. □

2.3 Basic Theory of Stochastic Processes

Random variables describe the consequences of the random occurrence of certain outcomes in an experiment. For example, the configuration of a polymer molecule as a result of complex interactions with its surroundings can be regarded as a random variable. Of course, such a polymer configuration changes in time, and it is hence natural to consider time-dependent random variables. The discussion of time-dependent random variables, or stochastic processes, completes this chapter on the basic concepts of stochastics. We first present some general definitions and theorems, and we then discuss three important classes of stochastic processes (Gaussian processes, Markov processes, martingales) together with the respective tools for handling them. A comprehensive discussion of the general theory of stochastic processes and of the particular properties of the most important classes of stochastic processes can be found in the pioneering book by J. L. Doob [6].

2.3.1 Definitions and Distributions

Throughout this section on stochastic processes we consider a range of time $\mathbb{T}$ which is either the closed interval $[0, t_{\max}]$ or the nonnegative axis $[0, \infty[$.

Definition 2.72 *A family of random variables* $\boldsymbol{X} = (\boldsymbol{X}_t)_{t \in \mathbb{T}}$ *from a probability space* $(\Omega, \mathcal{A}, P)$ *into* $(\mathbb{R}^d, \mathcal{B}^d)$ *is called a vector-valued* stochastic process. *For* $d = 1$, $X = (X_t)_{t \in \mathbb{T}}$ *is referred to as a real-valued stochastic process. The joint distributions of* $\boldsymbol{X}_{t_1}, \boldsymbol{X}_{t_2}, \ldots, \boldsymbol{X}_{t_n}$ *for all* $n \in \mathbb{N}$ *and* $t_1, t_2, \ldots, t_n \in \mathbb{T}$ *are referred to as the* finite-dimensional marginal distributions *of the stochastic process* $\boldsymbol{X}$.

For simplicity of notation only, we assume for most of the following considerations that X is a real-valued stochastic process. A stochastic process is a function of two variables, $X_t(\omega)$. The usual notation suppresses the probability space variable ω. When the process X carries a subscript, for example when X is a component of a vector-valued process, for clarity, we include the time-argument in the list of variables in parentheses and write $X_j(t, \omega)$ or $X_j(t)$ (for example, the latter notation is used throughout Chapter 4 in the discussion of bead-spring chain models of dilute solutions because there the bead position vectors and connector vectors need to be labeled). For each fixed $t \in \mathbb{T}$, the function $\omega \mapsto X_t(\omega)$ is a real-valued random variable defined on $(\Omega, \mathcal{A}, P)$. For each fixed $\omega \in \Omega$, the function $t \mapsto X_t(\omega)$ is a real function defined on $\mathbb{T}$. Such a time-dependent function for given ω is called a *trajectory* (or *sample path*) of the process X (see Fig. 2.7).

A stochastic process can be thought of as a random variable for which the range is the set of real functions on $\mathbb{T}$; the result of an outcome $\omega \in \Omega$ is the function $t \mapsto X_t(\omega)$. In order to justify this picture of stochastic processes as trajectory-valued random variables, one must equip the space of functions with a σ-algebra and to check whether the inverse image of that σ-algebra under X is contained in $\mathcal{A}$. Since the set of real functions on $\mathbb{T}$ can be regarded as an uncountable product of copies of $\mathbb{R}$, one factor for each $t \in \mathbb{T}$, it is natural to consider that uncountable product space together with the product σ-algebra introduced in Example 2.8. We denote the resulting measurable space of real functions by $(\mathbb{R}^{\mathbb{T}}, \mathcal{B}^{\mathbb{T}})$. Indeed, for that product σ-algebra, the fact that each X_t is a real-valued random variable immediately implies that the stochastic process X is a trajectory-valued random variable, that is, a measurable function X :

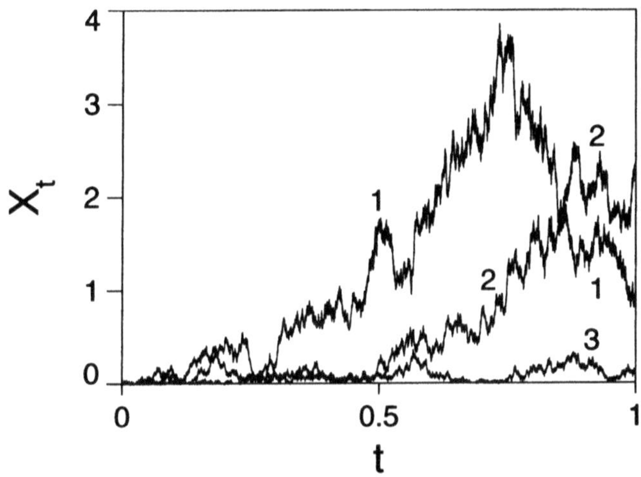

Fig. 2.7. Three trajectories $X_t(\omega_j)$, $j = 1, 2, 3$ of a real-valued stochastic process (the trajectories are labeled by j).

$(\Omega, \mathcal{A}) \to (\mathbb{R}^{\mathbb{T}}, \mathcal{B}^{\mathbb{T}})$. Therefore, there exists a distribution P^X of X on $(\mathbb{R}^{\mathbb{T}}, \mathcal{B}^{\mathbb{T}})$. According to the theorem cited in Example 2.16, the distribution of a stochastic process is uniquely characterized by its values on a $\cap$-stable generating system and, therefore, by the values of the finite-dimensional marginal distributions on product sets,

$$\overline{P}_{t_1 \ldots t_n}(B_1 \times \ldots \times B_n) := P(X_{t_1} \in B_1, \ldots, X_{t_n} \in B_n), \qquad (2.72)$$

for all $n \in \mathbb{N}$, for all $B_1, \ldots, B_n \in \mathcal{B}$ and for all $t_1, \ldots, t_n \in \mathbb{T}$ (we assume $t_1 < t_2 < \ldots < t_n$ in order to avoid redundant information). Knowing that the family of measures $\overline{P}_{t_1 \ldots t_n}$ uniquely characterizes the distribution of a given stochastic process we can next ask, for a given family of measures $\overline{P}_{t_1 \ldots t_n}$, under what conditions there exists a stochastic process with the corresponding finite-dimensional marginal distributions. There is a compatibility condition,

$$\begin{aligned} \overline{P}_{t_1 \ldots t_{k-1} t_{k+1} \ldots t_n}&(B_1 \times \ldots \times B_{k-1} \times B_{k+1} \times \ldots \times B_n) = \\ &\overline{P}_{t_1 \ldots t_n}(B_1 \times \ldots \times B_{k-1} \times \mathbb{R} \times B_{k+1} \times \ldots \times B_n), \qquad (2.73) \end{aligned}$$

for all $k \in \{1, \ldots, n\}$, which is obvious from (2.72) and the fact that X_{t_k} must assume some value in $\mathbb{R}$, that is $X_{t_k}^{-1}(\mathbb{R}) = \Omega$. The following theorem states that this compatibility condition is sufficient for the existence of a process with marginal distributions $\overline{P}_{t_1 \ldots t_n}$.

Theorem 2.73 (Kolmogorov Extension Theorem). *Suppose $\overline{P}_{t_1 \ldots t_n}$ is a family of measures on $(\mathbb{R}^n, \mathcal{B}^n)$, one for each $n \in \mathbb{N}$ and $t_1, \ldots, t_n \in \mathbb{T}$*

with $t_1 < t_2 < \ldots < t_n$. If the members of this family satisfy the compatibility condition (2.73), then there exists a stochastic process X whose family of finite-dimensional marginal distributions coincides with the given family of measures, as expressed in (2.72).

The proof of this deep theorem is highly nontrivial (see Sect. 12.1 of [3]). Whenever we are interested only in the distribution of a stochastic process rather than in its trajectories, the Kolmogorov extension theorem simplifies the situation greatly because we have to specify only a family of measures $\overline{P}_{t_1 \ldots t_n}$ on finite-dimensional product spaces and to verify the compatibility condition. The constructions of most stochastic processes, particularly in the applied literature, are based on the Kolmogorov extension theorem.

Exercise 2.74 Let the probability densities $p_t(x)$ for $t \in \mathbb{T}$ be given. Consider the family of continuous measures with probability densities $\overline{p}_{t_1 \ldots t_n}(x_1, \ldots, x_n) = \prod_{j=1}^{n} p_{t_j}(x_j)$. Show that the compatibility condition (2.73) is satisfied. Describe the corresponding stochastic process whose existence is guaranteed by the Kolmogorov extension theorem. (Remark: This exercise shows how the Kolmogorov extension theorem may be used to establish the existence of sequences of independent, identically distributed, real-valued random variables assumed in Example 2.67.)

Exercise 2.75 Let X_0 be a real-valued random variable whose distribution is the measure P_0 concentrated at 0, and define $X_t = X_0 + ct$ where $c \in \mathbb{R}$ is a constant. What kind of dynamical system is described by this stochastic process? Determine the finite-dimensional marginal distributions.

We say that a process is *stationary* if its distribution is invariant under time displacements, that is, if $\overline{P}_{t_1+t \ldots t_n+t} = \overline{P}_{t_1 \ldots t_n}$ for all the finite-dimensional marginal distributions (2.72).

In many situations, for example in constructing solutions of stochastic differential equations, one is interested not only in the distributions of processes but also in their trajectories. Since we can regard stochastic processes as trajectory-valued random variables, as for all random variables, it does not matter if we change the trajectories on a null set. We hence call two processes *indistinguishable* if, with the possible exception of a null set A, all their trajectories are equal. For indistinguishable processes X and Y and any given $t \in \mathbb{T}$, the two random variables X_t and Y_t are equal (except on the null set A). In general, if for each t the random variables X_t and Y_t are equal (as always, that means that $X_t(\omega) = Y_t(\omega)$ for $\omega \in N_t^C$ where N_t is now a null set that may depend on t), we call the processes X and Y *equivalent*. In this case, each process is referred to as a *version* of the other. For equivalent processes X and Y, the finite collections of random variables $(X_t)_{t\in\{t_1,\ldots,t_n\}}$ and $(Y_t)_{t\in\{t_1,\ldots,t_n\}}$ are equal except on the null set $A = N_{t_1} \cup \ldots \cup N_{t_n}$. Then, all finite-dimensional marginal distributions and consequently also the distributions of the processes X and Y coincide.

In summary, indistinguishable processes are equivalent, and equivalent processes have the same distribution. Indistinguishability is a much stronger requirement than equivalence because, for equivalent processes X and Y, the union of the uncountable number of sets N_t for which the trajectories differ at time t is generally not a null set (or, if not measurable, it is not contained in a null set). If the processes have continuous (or only right continuous) paths, equivalence implies indistinguishability, and the two concepts thus coincide.

In modeling physical situations by stochastic processes it is often crucial that the trajectories be continuous. There is an important criterion for processes with continuous trajectories (see Theorem 12.2.7 of [3]).

Theorem 2.76 (Kolmogorov-Prokhorov Theorem). *Suppose that $X = (X_t)_{t \in \mathbb{T}}$ is a stochastic process on $(\Omega, \mathcal{A}, P)$. If there are positive constants a, b, c such that for all $t, t' \in \mathbb{T}$*

$$\langle |X_t - X_{t'}|^a \rangle \leq c\,|t - t'|^{1+b},$$

then there exists a version of X which has continuous trajectories.

One might be tempted to believe that for a process X satisfying the assumptions of Theorem 2.76 the probability of the set of all continuous functions, $C^{\mathbb{T}}$, is equal to unity, that is, $P^X(C^{\mathbb{T}}) = 1$. However, the set $C^{\mathbb{T}}$ is not contained in $\mathcal{B}^{\mathbb{T}}$. The next idea would be to look for a measurable subset $A \subset C^{\mathbb{T}}$ with $P^X(A) = 1$. In general terms, $C^{\mathbb{T}}$ would then be contained in the completion of the σ-algebra $\mathcal{B}^{\mathbb{T}}$ for the given distribution P^X.[7] It can be shown that the only subset of $C^{\mathbb{T}}$ contained in $\mathcal{B}^{\mathbb{T}}$ is the empty set, so that a measurable subset $A \subset C^{\mathbb{T}}$ with $P^X(A) = 1$ cannot exist. In the situation of the Theorem 2.76, we conclude that $C^{\mathbb{T}}$ is not even measurable when considering the completion of $(\mathbb{R}^{\mathbb{T}}, \mathcal{B}^{\mathbb{T}}, P^X)$. However, Theorem 2.76 guarantees that there is a version of X such that $X^{-1}(C^{\mathbb{T}}) = \Omega$, that is $P(X^{-1}(C^{\mathbb{T}})) = 1$.[8]

All the definitions and theorems of this subsection are readily generalized from real-valued to vector-valued stochastic processes.

2.3.2 Gaussian Processes

Like Gaussian probability measures and Gaussian random variables, Gaussian stochastic processes also deserve special attention because they occur naturally in many applications. Gaussian processes also serve as illustrative examples of the general concepts and the application of the theorems discussed in the preceding subsection. This class of processes can be characterized and handled in a particularly simple manner.

---

[7] The *completion* of a probability space $(\Omega, \mathcal{A}, P)$ is another probability space $(\Omega, \mathcal{A}^*, P^*)$ for which the σ-algebra $\mathcal{A}^*$ contains all sets of the form $A \cup B$ with $A \in \mathcal{A}$, $B \subset C \in \mathcal{A}$, $P(C) = 0$, and the measure P^* is obtained by setting $P^*(A \cup B) = P(A)$. Such a completion is a *complete* probability space, that is, by definition, every subset of a null set is measurable.

[8] The discussion in this paragraph is similar to Sect. 1.2 of [7].

Definition 2.77 *A real-valued stochastic process $X = (X_t)_{t\in\mathbb{T}}$ on $(\Omega, \mathcal{A}, P)$ is called a* Gaussian process *if and only if all finite-dimensional marginal distributions of X are Gaussian.*

It was pointed out in Sect. 2.2.5 that all finite-dimensional Gaussian distributions are characterized by their first and second moments. Therefore, the functions

$$\alpha_t = \langle X_t \rangle , \tag{2.74}$$

$$\Theta_{tt'} = \langle X_t X_{t'} \rangle - \langle X_t \rangle \langle X_{t'} \rangle , \tag{2.75}$$

of one or two variables $t, t' \in \mathbb{T}$, respectively, fully determine the distribution of a given Gaussian process X. Since a Gaussian process is characterized by the functions (2.74) and (2.75) we may ask the following question: what conditions must the given functions $\alpha : \mathbb{T} \to \mathbb{R}$, $t \mapsto \alpha_t$ and $\Theta : \mathbb{T} \times \mathbb{T} \to \mathbb{R}$, $(t, t') \mapsto \Theta_{tt'}$ satisfy in order to guarantee the existence of a Gaussian process X for which (2.74) and (2.75) hold? The function α_t is arbitrary since one can easily adjust it by adding a suitable deterministic function to a given Gaussian process without affecting (2.75). Concerning the function $\Theta_{tt'}$, we have to make sure that the covariance matrices of all finite-dimensional marginal distributions are symmetric and positive-semidefinite, that is

$$\Theta_{tt'} = \Theta_{t't} \quad \text{for all } t, t' \in \mathbb{T}, \tag{2.76}$$

and

$$\sum_{j,k=1}^{n} \Theta_{t_j t_k} x_j x_k \geq 0 \quad \text{for all } n \in \mathbb{N};\ t_1, \ldots, t_n \in \mathbb{T};\ x_1, \ldots, x_n \in \mathbb{R}. \tag{2.77}$$

The conditions (2.76) and (2.77) allow us to define a family of Gaussian measures $\overline{P}_{t_1 \ldots t_n}$. The validity of the compatibility condition (2.73) follows from the discussion of marginal distributions in Sect. 2.2.5. By applying the Kolmogorov extension theorem we have thus proved the following result.

Theorem 2.78 *Suppose that the functions $\alpha : \mathbb{T} \to \mathbb{R}$, $t \mapsto \alpha_t$ and $\Theta : \mathbb{T} \times \mathbb{T} \to \mathbb{R}$, $(t, t') \mapsto \Theta_{tt'}$ are given. If $\Theta_{tt'}$ satisfies the conditions (2.76) and (2.77), then there exists a Gaussian process X for which (2.74) and (2.75) hold.*

This theorem is very important because it allows us to construct Gaussian processes in a very simple manner. As an example we discuss a Gaussian process that was suggested as a model of "Brownian motion" (a detailed discussion of that phenomenon is postponed to Sect. 3.1, where the dynamics of a Brownian particle is described by a stochastic differential equation). This process is named after the US-American mathematician Norbert Wiener (1894–1964).

Example 2.79 Wiener Process
In order to construct a particular Gaussian process that is known as the Wiener

process (see, e.g., §41 of [2], Sect. 3.1 of [4] or Chap. I of [8]), we specify the following functions for $\mathbb{T} = [0, \infty[$,

$$\alpha_t = 0\,, \quad \Theta_{tt'} = \min(t, t') \quad \text{for all } t, t' \in \mathbb{T}\,. \tag{2.78}$$

In order to verify the criterion (2.77) for this symmetric function $\Theta_{tt'}$, we write

$$\begin{aligned}\sum_{j,k=1}^{n} \Theta_{t_j t_k}\, x_j x_k &= \sum_{j,k=1}^{n} x_j x_k \int \chi_{[0,t_j]}(t)\chi_{[0,t_k]}(t)\, dt \\ &= \int \left(\sum_{j=1}^{n} x_j \chi_{[0,t_j]}(t) \right)^2 dt \geq 0\,.\end{aligned}$$

Theorem 2.78 hence guarantees the existence of a Gaussian process $W = (W_t)_{t \in \mathbb{T}}$, the Wiener process, with

$$\langle W_t \rangle = 0\,, \qquad \langle W_t W_{t'} \rangle = \min(t, t')\,. \tag{2.79}$$

Since the moment $\langle W_0^2 \rangle$ vanishes, W_0 is almost surely equal to zero, and we always assume that W_0 vanishes identically. The increments $W_t - W_{t'}$ of the Wiener process are Gaussian random variables for which we obtain from (2.79)

$$\langle W_t - W_{t'} \rangle = 0\,, \qquad \left\langle (W_t - W_{t'})^2 \right\rangle = |t - t'|\,. \tag{2.80}$$

The fact that the mean-square increment increases linearly in time suggests that the Wiener process is closely related to diffusive motion. Three trajectories of the process $X_t = W_t^2$ are shown in Fig. 2.7 in which the idea of stochastic processes as trajectory-valued random variables is illustrated. If we consider $t_1 \leq t_2 \leq t_3 \leq t_4$, application of (2.79) gives

$$\langle (W_{t_4} - W_{t_3})(W_{t_2} - W_{t_1}) \rangle = t_2 - t_1 - t_2 + t_1 = 0\,,$$

which shows that the increments of the Wiener process for disjoint time intervals are uncorrelated and, since they are Gaussian random variables, the increments are thus independent. The Wiener process is the prototype of a stochastic process with independent increments. Under quite general assumptions (finite $\langle X_t^2 \rangle$ for all $t \in \mathbb{T}$, continuous trajectories, Gaussian initial condition X_0), stochastic processes with independent increments are Gaussian. The Gaussian nature of processes with independent increments is closely related to the central limit theorem, and it explains why Gaussian processes are particularly important.

After evaluating the fourth moments of the increments by means of (2.62),

$$\left\langle (W_t - W_{t'})^4 \right\rangle = 3 \left\langle (W_t - W_{t'})^2 \right\rangle^2 = 3|t - t'|^2\,,$$

we can apply the Kolmogorov-Prokhorov Theorem 2.76 in order to prove that the Wiener process has a version with continuous trajectories. Henceforth, we consider only continuous versions of the Wiener process. The second equation in (2.80) shows that typical increments of the Wiener process in a time interval of size Δt are of the order of $\sqrt{\Delta t}$ and hence suggests that the trajectories of the Wiener process are not differentiable. Indeed, it can rigorously be shown that the trajectories of the Wiener process are nowhere differentiable (with the possible exception of a null set). Therefore, the trajectories of the Wiener process provide plenty of examples for functions which are continuous but nowhere differentiable. The wild fluctuations of the Wiener process of order $\sqrt{\Delta t}$ within a short time interval Δt also suggest that the length of a trajectory over any finite time interval must be infinite; more rigorously, the trajectories of the Wiener process are almost surely of unbounded variation. These properties of the Wiener process lead to the peculiarities of stochastic calculus, in which the Wiener process plays a prominent role. □

In view of the fundamental importance of the Wiener process in introducing noise, formulating stochastic differential equations and developing simulation algorithms we supplement the rigorous definition of the preceding example by some heuristic remarks.

In defining stochastic integrals and in developing an Euler integration scheme for stochastic differential equations one introduces partitions $0 = t_0 < t_1 \cdots < t_{n-1} < t_n$ of the time range of interest. Then, one needs only the n random variables $W_{t_j} - W_{t_{j-1}}$, for $j = 1, 2, \ldots, n$, where $W_0 = 0$, in order to evaluate stochastic integrals or solutions of stochastic difference equations. According to Example 2.79, these increments are n independent Gaussian random variables with variances $t_j - t_{j-1}$. The understanding of the Wiener process in stochastic integrals and simulation algorithms can hence be gained on a very elementary level. For example, the trajectories of W_t^2 in Fig. 2.7 were obtained by introducing the time step $\Delta t = 5. \times 10^{-5}$, summing independent Gaussian random numbers with variance Δt, and plotting the squares of the sums. Because of the finite resolution of the plot smaller time steps are clearly unnecessary. The continuity of the trajectories suggests that the points at discrete times are connected by straight lines.

One might ask why it is not sufficient to consider only such discretizations of the Wiener process. The reason is that one would like not only to introduce discrete approximations to stochastic integrals but rather to look at the limit of vanishing time step width as well. The information about arbitrarily small time steps is contained in the continuous-time Wiener process.

The trajectories of the Wiener process can also be understood as limits of random walks. Consider random walks on a one-dimensional lattice with spacing σ starting at zero. We assume that each independent step to the left or right, which is taken with probability $1/2$, corresponds to a time step Δt, and we choose $\sigma = \sqrt{\Delta t}$. As $\Delta t \to 0$, the random walk seems to proceed continuously in space and time, and one obtains trajectories of the Wiener process. In a more general situation, a precise formulation of this idea is worked out in the following exercise (where j corresponds to $1/\Delta t$).

Exercise 2.80 By doing this exercise, the reader can develop a better understanding of the rather abstract Wiener process by means of a relationship to a simple game of dice. The basic idea is to think of diffusion as the effect of many independent random kicks and, accordingly, to approximate the Wiener process by the sum of properly normalized independent random variables.

Let $(X_n)_{n\in\mathbb{N}}$ be a sequence of independent, identically distributed, real-valued random variables on the same probability space. Define a stochastic process by $Y_t := \sum_{n\leq t}(X_n - a)/\sigma$ for $t \in [0, \infty[$, where $a = \langle X_n \rangle$, $\sigma^2 = \mathrm{Var}(X_n)$. What are the values of a and σ if X_n is the outcome in a game of dice? Define a sequence of rescaled stochastic processes by

$$X_t^{(j)} = Y_{jt}/\sqrt{j} \quad j = 1, 2, \ldots \quad t \geq 0 .$$

Show that for any t the sequence of random variables $X_t^{(j)}$ for $j \to \infty$ converges in distribution to a Gaussian random variable with zero mean and variance t. Similarly, for any $t > t'$ the differences $X_t^{(j)} - X_{t'}^{(j)}$ converge in distribution to a Gaussian random variable with zero mean and variance $t - t'$. Show that the independence of the random variables X_n implies the independence of any two increments of the processes $X^{(j)}$ over disjoint intervals, and hence all finite-dimensional marginal distributions of $X_t^{(j)}$ converge as $j \to \infty$ to those of the Wiener process. In short, the Wiener process is the limit in distribution of the processes $X^{(j)}$.

Exercise 2.81 Show that the functions $\alpha_t = 0$ and $\Theta_{tt'} = \mathrm{e}^{-|t-t'|} - \mathrm{e}^{-(t+t')}$ define a Gaussian process. Is there a version of that process with continuous trajectories?

Vector-valued Gaussian processes can be defined in an analogous manner. They are characterized by a vector-valued function $\boldsymbol{\alpha}_t$ and a matrix-valued function $\boldsymbol{\Theta}_{tt'}$ which, for a d-component process, satisfies the following generalizations of the conditions (2.76) and (2.77),

$$\boldsymbol{\Theta}_{tt'} = \boldsymbol{\Theta}_{t't}^T \quad \text{for all } t, t' \in \mathbb{T}, \tag{2.81}$$

and

$$\sum_{j,k=1}^{n} \boldsymbol{x}_j \cdot \boldsymbol{\Theta}_{t_j t_k} \cdot \boldsymbol{x}_k \geq 0 \quad \text{for all } n \in \mathbb{N};\ t_1, \ldots, t_n \in \mathbb{T};\ \boldsymbol{x}_1, \ldots, \boldsymbol{x}_n \in \mathbb{R}^d . \tag{2.82}$$

The d-dimensional Wiener process is defined by

$$\boldsymbol{\alpha}_t = \boldsymbol{0}, \quad \boldsymbol{\Theta}_{tt'} = \min(t, t')\, \boldsymbol{\delta} \quad \text{for all } t, t' \in \mathbb{T}, \tag{2.83}$$

where the unit matrix $\boldsymbol{\delta}$ occurring in the second part of (2.83) implies that the d components of a d-dimensional Wiener process are independent one-dimensional Wiener processes.

2.3.3 Markov Processes

Another important class of stochastic processes that can be constructed and handled in a convenient manner are Markov processes. Loosely speaking, a stochastic process or a stochastic dynamical system satisfies the *Markov property* if the future state of the system at any time $t' > t$ is independent of the past behavior of the system at times $t'' < t$, given the present state at time t. In other words, the further evolution of the process depends on its history only through the current state of the process. This property, formulated by the Russian mathematician Andrei Andreievich Markov (1856–1922) in 1906, plays an important role in the theory of stochastic differential equations for which, like for ordinary differential equations, the future time evolution can be expressed in terms of the current state, independent of previous states of the system. Even though the above formulation of the Markov property sounds rather vague, the careful introduction of σ-algebras (representing information content) and the precise definition of conditional expectations and probabilities (relating random variables with different information content) immediately allow us to define Markov processes rigorously in the above-mentioned spirit.

Definition 2.82 *A real-valued stochastic process $X = (X_t)_{t\in\mathbb{T}}$ on $(\Omega, \mathcal{A}, P)$ is called a* Markov process *if and only if for all $n \in \mathbb{N}$, for all $t_1, t_2, \ldots, t_n, t \in \mathbb{T}$ with $t_1 < t_2 < \ldots < t_n < t$, and for all $B \in \mathcal{B}$*

$$P(X_t \in B | X_{t_1} = x_1, X_{t_2} = x_2, \ldots, X_{t_n} = x_n) = P(X_t \in B | X_{t_n} = x_n) . \quad (2.84)$$

Notice that the equality of the functions of $x_1, \ldots, x_n$ in (2.84) holds only with probability unity, that is, for all $(x_1, \ldots, x_n)$ contained in an event $A \in \mathcal{B}^n$ with $P^{(X_{t_1},\ldots,X_{t_n})}(A) = 1$; this must be so because conditional probabilities are unique only with the exception of a null set. The formulation (2.84) of the Markov property does not require that the distribution of the process X is continuous or discrete; it even applies to processes with general mixed distributions.

The rigorous formulation of the Markov property (2.84) suggests that the conditional probabilities $P(X_t \in B | X_{t'} = x')$ for $t' < t$ play a crucial role in specifying the marginal distributions of Markov processes. Such conditional probabilities may be regarded as *transition probabilities* from a state x' at t' into a set of states B at a later time t. The Markov property (2.84) states that such transitions are independent of the history of the process before t'. Joint distributions for any finite number of times may then be constructed by considering a sequence of transitions governed by the transition probabilities. We can obtain the joint probability that the Markov process X assumes values in $B_1, \ldots, B_n$ at times $t_1 < \ldots < t_n$ by considering the (initial) probability of finding a value of X_{t_1} in B_1, and by then accounting for the transition probabilities from any of the states in B_1 into B_2 between t_1 and t_2, and so forth. We next employ these ideas, which are the key to a simple description of Markov processes, in order to obtain a rigorous construction of Markov processes with continuous distributions by means of the Kolmogorov extension theorem.

Assume that a family of functions $p_{tt'}(x|x')$ for $t, t' \in \mathbb{T}$ with $t' < t$ is given, where each $p_{tt'}$ is a nonnegative function of x and $x' \in \mathbb{R}$ with $\int p_{tt'}(x|x')\, dx = 1$ (that is, $p_{tt'}(\cdot|x')$ is a probability density labeled by the three parameters x', t', t).

For $t_1, \ldots, t_n \in \mathbb{T}$ with $0 = t_0 < t_1 < \ldots < t_n$ and a given probability density $p(x_0)$, the function

$$\overline{p}_{t_0 t_1 \ldots t_n}(x_0, x_1, \ldots, x_n) = p_{t_n t_{n-1}}(x_n|x_{n-1})\, p_{t_{n-1} t_{n-2}}(x_{n-1}|x_{n-2}) \times \ldots p_{t_2 t_1}(x_2|x_1)\, p_{t_1 0}(x_1|x_0)\, p(x_0)\,, \tag{2.85}$$

is nonnegative with $\int \ldots \int \overline{p}_{t_0 t_1 \ldots t_n}(x_0, x_1, \ldots, x_n)\, \mathrm{d}x_0\, \mathrm{d}x_1 \ldots \mathrm{d}x_n = 1$. Hence a family of continuous finite-dimensional measures is defined. The occurrence of $t_0 = 0$ in the list of time arguments can be avoided by integrating over x_0 to obtain the probability density $\overline{p}_{t_1 \ldots t_n}$ with $t_1 > 0$. The compatibility condition (2.73) for the above family of finite-dimensional measures is satisfied when

$$\int p_{tt'}(x|x')\, p_{t't''}(x'|x'')\, \mathrm{d}x' = p_{tt''}(x|x'')\,, \tag{2.86}$$

for all $t, t', t'' \in \mathbb{T}$ with $t'' < t' < t$ and for all $x, x'' \in \mathbb{R}$. This compatibility condition for the functions $p_{tt'}$ is known as the *Chapman-Kolmogorov equation.*

When a family of functions $p_{tt'}(x|x')$ satisfies the Chapman-Kolmogorov equation (2.86), the Kolmogorov extension theorem states that there exists a stochastic process X whose finite-dimensional marginal distributions are continuous probability measures with densities given by (2.85). Equation (2.59) for the conditional probability densities gives

$$\begin{aligned} p(X_t = x|X_{t_1} = x_1, \ldots, X_{t_n} = x_n) &= \frac{\overline{p}_{t_1 \ldots t_n t}(x_1, \ldots, x_n, x)}{\overline{p}_{t_1 \ldots t_n}(x_1, \ldots, x_n)} \\ = p_{tt_n}(x|x_n) = \frac{\overline{p}_{t_n t}(x_n, x)}{\overline{p}_{t_n}(x_n)} &= p(X_t = x|X_{t_n} = x_n)\,, \end{aligned} \tag{2.87}$$

and, therefore, X is indeed a Markov process. The second line of (2.87) implies that $p_{tt'}(x|x')$ is the transition probability density $p(X_t = x|X_{t'} = x')$ for $t' < t$. With the interpretation of the functions $p_{tt'}$ as transition probability densities, the Chapman-Kolmogorov equation (2.86) states that one-step transition probability densities can be expressed in terms of two-step transition probabilities involving an arbitrary intermediate time at which one has to integrate over all possible states of the stochastic dynamical system. According to (2.85), the function $p(x_0)$ is the probability density for the distribution of X_0 at time $t_0 = 0$.

A similar construction can be performed for Markov processes with discrete or arbitrary mixed distributions instead of the continuous ones considered here. In the discrete case, one has to deal with transition probabilities rather than probability densities, and the integral in (2.86) is replaced with a sum; in that case, the structure of the Chapman-Kolmogorov equation suggests a matrix notation for the transition probabilities.

The considerations of this subsection show that the distribution of a Markov process is much easier to construct than the distribution of a general stochastic process. Instead of the family $\overline{P}_{t_1 \ldots t_n}$ of all marginal distributions for any finite number of parameters $t_1 < \ldots < t_n$ with all the compatibility conditions

(2.73), one has to specify only the two-parameter family of transition probability densities $p(X_t = x|X_{t'} = x')$ for $t' < t$ and the initial distribution P^{X_0}. This two-parameter family of transition probability densities still contains a large amount of redundant information, because each time step can be decomposed into a sequence of small time steps. By going to the limit of infinitesimal time steps one can obtain a one-parameter description of Markov processes and, at the same time, eliminate the remaining compatibility conditions (2.86) by avoiding redundant information. A very useful concept in studying infinitesimally small time steps is the *infinitesimal generator* which, for general Markov processes, can be defined by

$$\mathcal{L}_t g(x) := \lim_{\Delta t \downarrow 0} \frac{1}{\Delta t} \Big[E\Big(g(X_{t+\Delta t}) \Big| X_t = x \Big) - g(x) \Big] . \tag{2.88}$$

The infinitesimal generator $\mathcal{L}_t$ is a linear operator whose domain is the set of measurable functions $g\colon \mathbb{R} \to \mathbb{R}$ for which the limit on the right side of (2.88) exists (see Sect. 2.4 of [4]). Of course, the domain must be sufficiently rich, for example, dense in the class of bounded functions (cf. Sect. XIII.9 of Vol. 2 of [1]). Equation (2.56) shows that the infinitesimal generator can be determined from the transition probabilities. When operating on a function g, the infinitesimal generator describes the rate of change of g under the condition that the process starts at a given position at time t.

The one-parameter family of linear operators $\mathcal{L}_t$, $t \in \mathbb{T}$, together with the initial distribution P^{X_0}, yields the most compact characterization of a general Markov process. It remains to be shown that the transition probabilities can be reconstructed from the infinitesimal generator. We show here how to obtain the transition probabilities from the infinitesimal generator of a Markov process with continuous distribution. Since the following arguments lead to the famous Fokker-Planck equation, which plays a central role in the kinetic theory of polymeric liquids, we give all steps of the calculation in detail.

From (2.56), together with Example 2.56, and from (2.88) we obtain

$$\mathcal{L}_t g(x) = \lim_{\Delta t \downarrow 0} \frac{1}{\Delta t} \Big[\int g(y)\, p_{t+\Delta t\, t}(y|x)\, \mathrm{d}y - g(x) \Big] . \tag{2.89}$$

After multiplying (2.89) by $p_{tt'}(x|x')$ where $t' < t$ and integrating over x we obtain by means of the Chapman-Kolmogorov equation

$$\int \big[\mathcal{L}_t g(x)\big]\, p_{tt'}(x|x')\, \mathrm{d}x = \lim_{\Delta t \downarrow 0} \frac{1}{\Delta t} \Big[\int g(y)\, p_{t+\Delta t\, t'}(y|x')\, \mathrm{d}y - \int g(x)\, p_{tt'}(x|x')\, \mathrm{d}x \Big] .$$

When the dummy integration variable y in the first term on the right side is replaced with x, we find after taking the limit $\Delta t \downarrow 0$

$$\int \big[\mathcal{L}_t g(x)\big]\, p_{tt'}(x|x')\, \mathrm{d}x = \int g(x)\, \frac{\partial}{\partial t}\, p_{tt'}(x|x')\, \mathrm{d}x . \tag{2.90}$$

If $\mathcal{L}_t^\dagger$ denotes the adjoint of the operator $\mathcal{L}_t$,[9] so that

[9] Acting with $\mathcal{L}_t$ on one factor in a scalar product of two arbitrary functions is equivalent to acting with $\mathcal{L}_t^\dagger$ on the other factor.

$$\int \left[\mathcal{L}_t g(x)\right] p_{tt'}(x|x')\,\mathrm{d}x = \int g(x)\left[\mathcal{L}_t^\dagger p_{tt'}(x|x')\right]\mathrm{d}x\,, \tag{2.91}$$

then we obtain from the fact that (2.90) holds for arbitrary g (or, at least, for a sufficiently large class of functions g),

$$\boxed{\frac{\partial}{\partial t}\, p_{tt'}(x|x') = \mathcal{L}_t^\dagger\, p_{tt'}(x|x')\,.} \tag{2.92}$$

Equation (2.92) is known as *Kolmogorov's forward equation* or, in cases where $\mathcal{L}_t^\dagger$ is a differential operator containing only first and second order derivatives with respect to x, as the *Fokker-Planck equation* or *diffusion equation*. For given x', t' and the initial condition $p_{tt'}(x|x') = \delta(x-x')$ for $t = t'$ (see Example 2.16), one can integrate (2.92) in time to reconstruct all transition probability densities when the infinitesimal generator of a Markov process with continuous distribution is given. The prominent role of the time-evolution equation (2.92) in the theory of Markov processes is underlined by the many different names associated with that type of equation in various areas of science.

Exercise 2.83 Derive the time-evolution equation $\partial\overline{p}_t(x)/\partial t = \mathcal{L}_t^\dagger\, \overline{p}_t(x)$ for the probability density of a Markov process with continuous distribution. By using suitable initial conditions, all transition probability densities can be reconstructed from this time-evolution equation for the probability density.

Example 2.84 Markov Property of the Wiener Process
Consider the Gaussian transition probability densities

$$p_{tt'}(x|x') = p_{x'\,t-t'}(x) = \frac{1}{\sqrt{2\pi(t-t')}}\exp\left\{-\frac{(x-x')^2}{2(t-t')}\right\}, \tag{2.93}$$

depending on the three parameters x', t', t. The transition probability density is a Gaussian distribution around the starting point x' with a width σ given by $\sigma^2 = t - t'$. The Chapman-Kolmogorov equation can be verified by rearranging quadratic terms in exponentials,

$$\int p_{tt'}(x|x')\, p_{t't''}(x'|x'')\,\mathrm{d}x' = p_{tt''}(x|x'') \int p_{\alpha\Theta}(x')\,\mathrm{d}x'\,, \tag{2.94}$$

with

$$\alpha = \frac{t-t'}{t-t''}\,x'' + \frac{t'-t''}{t-t''}\,x\,, \qquad \Theta = \frac{(t-t')(t'-t'')}{t-t''}\,,$$

so that the integral on the right side of (2.94), being the integral of a probability density, is equal to unity. With the initial probability density $p(x_0) = \delta(x_0)$, the distribution of a Markov process X is uniquely defined through the Gaussian finite-dimensional marginal probability densities (2.85). With the corresponding one- and two-dimensional marginal densities, $\overline{p}_{t_1}(x_1) = p_{0\,t_1}(x_1)$ and $\overline{p}_{t_1t_2}(x_1,x_2) = p_{x_1\,t_2-t_1}(x_2)\,p_{0\,t_1}(x_1)$, we obtain $\langle X_t\rangle = 0$ and $\langle X_{t_1}X_{t_2}\rangle = t_1$ for $t_1 \le t_2$. In other words, the Gaussian Markov process X defined here coincides with the Wiener process of Example 2.79. More precisely, since in both cases only the distributions of the processes are defined, these two processes have identical distributions. The Wiener process thus satisfies the Markov property. The approach to the Wiener process based on the theory of Markov processes in this example is equivalent to the approach based on the theory of Gaussian processes (Example 2.79).

In evaluating the infinitesimal generator of the Wiener process by means of (2.89), one can use the Taylor expansion of $g(y)$ around x in order to obtain

$$\mathcal{L}_t g(x) = \sum_{n=1}^{\infty} \frac{1}{n!} \frac{\partial^n}{\partial x^n} g(x) \lim_{\Delta t \downarrow 0} \frac{1}{\Delta t} \int (y-x)^n p_{0\Delta t}(y-x)\, dy \,. \tag{2.95}$$

Due to symmetry, all terms with odd n vanish. After using $y - x$ as a new integration variable one finds by expanding the characteristic function $\tilde{p}_{0\Delta t}(q) = \exp\{-\frac{1}{2}\Delta t\, q^2\}$ in q^2 (cf. Theorem 2.45) that for even n the integral on the right side of (2.95) is proportional to $(\Delta t)^{n/2}$, so that only the term for $n = 2$ survives after taking the limit $\Delta t \downarrow 0$. We thus obtain

$$\mathcal{L}_t g(x) = \frac{1}{2} \frac{\partial^2}{\partial x^2} g(x) \,, \tag{2.96}$$

and, since integrating by parts twice gives $\mathcal{L}_t^\dagger = \mathcal{L}_t$, the Fokker-Planck equation (2.92) reads

$$\frac{\partial}{\partial t} p_{tt'}(x|x') = \frac{1}{2} \frac{\partial^2}{\partial x^2} p_{tt'}(x|x') \,. \tag{2.97}$$

This partial differential equation can be verified explicitly for the transition probability densities (2.93), thus confirming that the differential operator $(1/2)\, \partial^2/\partial x^2$ is indeed the infinitesimal generator for the Markov process with transition probability densities (2.93). This second-order differential operator describes how a function g is "smeared out" on average when its argument exhibits the diffusion-type time evolution modeled by the Wiener process (in the same way that a concentration profile is "smeared out" due to the diffusive motion of individual, noninteracting particles). □

The Wiener process is an example of a Markov process that is *homogeneous* with respect to time, that is, a Markov process for which $P(X_t \in B | X_{t'} = x')$ depends only on the difference $t - t'$ rather than on t and t' separately (of course, there is an additional dependence on x' and B). For a homogeneous Markov process, the infinitesimal generator $\mathcal{L}$ is independent of time. The distribution of such a process is fully characterized by the initial distribution and a single operator $\mathcal{L}$. For example, one can define the Wiener process as the Markov process with $P^{X_0} = P_0$ (the probability measure concentrated at 0) and $\mathcal{L} = (1/2)\, \partial^2/\partial x^2$ rather than as the Gaussian process with first and second moments given in (2.78). All stationary Markov processes are homogeneous. The Wiener process is homogeneous but not stationary.

The extension of the definition of Markov processes to vector-valued processes is obvious. In general, the components of vector-valued Markov processes are not real-valued Markov processes because the further evolution of each component depends on the current state of all the components. Consideration of the full set of variables is hence crucial for recognizing a Markov process. Given a certain phenomenon, "the art of the physicist is to find those variables that are needed to make the description (approximately) Markovian" (p. 77 of [9]). In many cases, one can introduce some time derivatives of the variable of interest in order to eliminate memory effects and to obtain a Markovian description.

Exercise 2.85 Let the functions α_t and $\Theta_{tt'}$ be given. Assume that the symmetry condition $\Theta_{t't} = \Theta_{tt'}$ and the positivity conditions $\Theta_{tt} \geq 0$ and $\Theta_{t't'}\Theta_{tt} - \Theta_{tt'}^2 \geq 0$ hold for all $t', t \in \mathbb{T}$. Show that these functions define a Gaussian Markov process if and only if

$$\Theta_{tt'}\,\Theta_{t't''} = \Theta_{tt''}\,\Theta_{t't'} \qquad \text{for } t'' < t' < t\,.$$

Determine the corresponding infinitesimal generator for sufficiently smooth functions α_t and $\Theta_{tt'}$.

Exercise 2.86 Verify that both the Wiener process and the Gaussian process defined in Exercise 2.81 satisfy all the conditions of Exercise 2.85, so that both of these Gaussian processes are Markov processes. What is the infinitesimal generator for the process defined in Exercise 2.81?

2.3.4 Martingales

In Example 2.79, it was mentioned that under quite general conditions stochastic processes with independent increments are Gaussian. It is the purpose of this subsection to introduce a class of stochastic processes which is (i) more general than processes with independent increments, (ii) relevant to many applications, particularly in the theory of stochastic integrals presented in the next chapter, and (iii) tractable in a rather convenient manner. Moreover, the theory of these processes provides a unified method for dealing with various limit theorems of probability theory. The class of stochastic processes discussed here is known under the name *martingales*, which emphasizes the strong coupling between the random variables of such processes at different times.[10]

It might be more suggestive to think of a martingale as the analog of a courtroom process in which the truth is exposed in the course of time. At any given time during such a process, one has only a coarse-grained version of the truth given the current state of available information. In analogy with such a courtroom process, a martingale is (essentially) a process that can be represented as $X_t = E(X|\mathcal{A}_t)$ where $\mathcal{A}_t$ is the smallest σ-algebra such that all

[10]The name *martingale* was introduced by the French mathematician Jean Ville in 1939. The word is derived from the name of the French community Martigues in the Bouches-du-Rhône departement. The usual meaning of the word martingale is that part of a bridle which prevents a horse from tossing its head. For that purpose, a harness attached to the mouthpiece by two rings goes between the horse's front legs and under its belly where it is connected to the saddle belt.

$X_{t'}$ for $t' \leq t$ are measurable, that is, the σ-algebra induced by the history of X up to time t. The random variable X represents "the truth," and the *increasing family of σ-algebras* $\mathcal{A}_t$ ($\mathcal{A}_{t'} \subset \mathcal{A}_t$ for $t' \leq t$) represents the increasing "knowledge about the truth" accumulated up to time t. For example, the three functions in Fig. 2.5 may be regarded as members of a martingale for which, in the course of time, the third-order-polynomial character of the underlying random variable X is revealed.

Definition 2.87 *Let $(\Omega, \mathcal{A}, P)$ be a probability space and $(\mathcal{A}_t)_{t\in\mathbb{T}}$ an increasing family of σ-algebras with $\mathcal{A}_t \subset \mathcal{A}$ for all $t \in \mathbb{T}$. A stochastic process $(X_t)_{t\in\mathbb{T}}$ is a* martingale *with respect to $(\mathcal{A}_t)_{t\in\mathbb{T}}$ if and only if*

(i) *each X_t is measurable with respect to $\mathcal{A}_t$;*
(ii) *each X_t has finite expectation $\langle X_t \rangle$;*
(iii) $X_{t'} = E(X_t|\mathcal{A}_{t'})$ *for all $t, t' \in \mathbb{T}$ with $t' < t$.*

As the underlying increasing family of σ-algebras we usually take the σ-algebras $\mathcal{A}_t$ induced by all the random variables $X_{t'}$ for $t' \leq t$. While being a Gaussian or Markov process is a matter of the full distribution of a process, being a martingale is only a matter of conditional expectations.

Martingales are sometimes referred to as "fair game" processes; if X_t represents a gambler's fortune at time t, the game is fair if her or his expected fortune at a time t given the game history up to some previous time t' is precisely the fortune at time t'.

Equation (2.50) and the martingale property (iii) imply that for martingales $\langle X_{t'} \rangle = \langle X_t \rangle$, and the expectation $\langle X_t \rangle$ is thus independent of t. By exploiting the linearity property (2.51) and the fact that $X_{t'}$ is measurable with respect to $\mathcal{A}_{t'}$ we obtain an equivalent formulation of the martingale property (iii),

$$E(X_t - X_{t'}|\mathcal{A}_{t'}) = 0 \qquad \text{for all } t, t' \in \mathbb{T} \text{ with } t' < t. \tag{2.98}$$

In this formulation, the idea of generalizing the class of processes with independent increments can be best recognized. The formulation (2.98) will make it very obvious why martingales constitute an important tool in the theory of stochastic integrals: the martingale property of an integrator, such as the Wiener process, is immediately inherited by stochastic integrals.

Example 2.88 Processes with Independent Increments
Let $X = (X_t)_{t\in\mathbb{T}}$ be a stochastic process with $X_0 = 0$, for which all increments over disjoint time intervals are independent random variables with zero expectations. Then, for $t' < t$, the σ-algebra induced by $X_t - X_{t'}$ is by assumption independent of $\mathcal{A}_{t'}$ which is the σ-algebra induced by all previous increments of X. Equation (2.53) then gives

$$E(X_t - X_{t'}|\mathcal{A}_{t'}) = \langle X_t - X_{t'} \rangle = 0 \,,$$

which according to (2.98) implies that the stochastic process X is a martingale. Since the Wiener process satisfies the assumptions of this example, it is a martingale. The Wiener process thus belongs to all three classes of "tractable" processes considered in this section: Gaussian processes, Markov processes, and martingales. □

It follows from (2.98) and the properties of conditional expectations that for any function Y that is bounded and measurable with respect to $\mathcal{A}_{t'}$,

$$\langle Y(X_t - X_{t'})\rangle = \langle E(Y(X_t - X_{t'})|\mathcal{A}_{t'})\rangle = \langle Y\, E(X_t - X_{t'}|\mathcal{A}_{t'})\rangle = 0\,,$$

and thus

$$\mathrm{Cov}(Y, X_t - X_{t'}) = \langle Y(X_t - X_{t'})\rangle - \langle Y\rangle \langle X_t - X_{t'}\rangle = 0\,. \tag{2.99}$$

Conversely, if Y is chosen as an arbitrary indicator function then, according to the definition of conditional expectations, $\langle Y(X_t - X_{t'})\rangle = 0$ implies that (2.98) holds. Theorem 2.31 suggests that we think of any random variable Y that is measurable with respect to $\mathcal{A}_{t'}$ as a functional of the history of the process X up to time t' or, in particular, as an arbitrary function of any number of increments prior to t'. On the one hand, the martingale property (2.99) is weaker than the independence of $X_t - X_{t'}$ and the history of X prior to t'; for guaranteeing independence one would have to admit arbitrary functions of $X_t - X_{t'}$ rather than only $X_t - X_{t'}$ itself in the factorization of expectations (2.99) (see Theorem 2.49 and the remarks following that theorem). On the other hand, the condition (2.99) is stronger than having uncorrelated increments because Y is allowed to be an arbitrary function of previous increments rather than only a previous increment itself. In summary, processes with independent increments are martingales (see Example 2.88), and martingales have uncorrelated increments (see (2.99)). The class of martingales lies between the class of processes with independent increments and the class with uncorrelated increments. For Gaussian processes, the latter two classes coincide.

Exercise 2.89 Is the Gaussian process defined in Exercise 2.81 a martingale?

Exercise 2.90 Is the process $(W_t^2 - t)_{t\in[0,\infty[}$ a martingale with respect to the increasing family of σ-algebras induced by the Wiener process?

Exercise 2.91 Is the process $(\exp\{W_t - (t/2)\})_{t\in[0,\infty[}$ a martingale with respect to the increasing family of σ-algebras induced by the Wiener process?

The definition of a martingale requires that the random variables at different times differ only due to the increasing information content. In the spirit of the courtroom process analogy, our knowledge of some case increases with time, but the existence of an "ultimate truth" is not clear. In stochastics, this is established by the following theorem (even if this "ultimate truth" is not revealed in a finite time).

Theorem 2.92 *Let $(X_t)_{t\in\mathbb{T}}$ be a martingale with respect to $(\mathcal{A}_t)_{t\in\mathbb{T}}$. If, for some random variable Y with finite expectation, $|X_t| \leq Y$ holds almost surely for any $t \in \mathbb{T}$, then there exists a random variable X with finite expectation such that $X_t = E(X|\mathcal{A}_t)$ for all $t \in \mathbb{T}$.*

For $\mathbb{T} = [0, t_{\max}]$, this is trivial since one may choose $X = X_{t_{\max}}$; for $\mathbb{T} = [0, \infty[$, the reader is referred to the proof of Theorem 11.4.5 in [3] for the construction of an appropriate limiting random variable X.

Among the many results of martingale theory, inequalities play an important role (see Chap. VII of [6]). Just to illustrate the motivation behind the name martingale (see footnote in the beginning of this subsection), we quote an inequality which shows how the martingale property keeps the random variables at all times from "tossing their heads." More precisely, the probability for assuming large values ("high heads") is small and under explicit control.

Theorem 2.93 *If $(X_t)_{t\in\mathbb{T}}$ is a martingale with right continuous trajectories, and if $c > 0$ is a real number, then*

$$P(\{\omega \mid \sup_{t'\leq t} X_{t'}(\omega) \geq c\}) \leq \frac{\langle |X_t| \rangle}{c}.$$

A proof of this martingale inequality can be found in Chap. VII of [6].

References

1. Feller W (1950, 1966) *An Introduction to Probability Theory and Its Applications*, 2 Vols. Wiley, New York
2. Loève M (1977) *Probability Theory*, 2 Vols, 4th Edn. Springer, New York Heidelberg Berlin
3. Bauer H (1981) *Probability Theory and Elements of Measure Theory*, 2nd Edn. Academic, London New York Toronto Sydney San Francisco
4. Arnold L (1974) *Stochastic Differential Equations: Theory and Applications*. Wiley-Interscience, New York London Sydney Toronto
5. Kolmogoroff A (1933) *Grundbegriffe der Wahrscheinlichkeitsrechnung*. Springer, Berlin (Ergebnisse der Mathematik und ihrer Grenzgebiete, Bd 2)
6. Doob JL (1990) *Stochastic Processes*, Wiley Classics Library Edn. Wiley-Interscience, New York Chichester Brisbane Toronto Singapore
7. Kallianpur G (1980) *Stochastic Filtering Theory*. Springer, New York Heidelberg Berlin (Applications of Mathematics, Vol 13)
8. Rogers LCG, Williams D (1994) *Diffusions, Markov Processes, and Martingales, Vol 1, Foundations*, 2nd Edn. Wiley, Chichester New York Brisbane Toronto Singapore (Wiley Series in Probability and Mathematical Statistics)
9. van Kampen NG (1992) *Stochastic Processes in Physics and Chemistry*, 2nd Edn. North-Holland, Amsterdam London New York Tokyo

3. Stochastic Calculus

The theory of stochastic processes provides the framework for describing stochastic systems evolving in time. Our next goal is to characterize the dynamics of such stochastic systems, that is, to formulate equations of motion for stochastic processes.

In the description of the motion of interacting particles in classical deterministic mechanics, differential equations play a most important role. A deep understanding of the laws of motion is impossible without the mathematical background of calculus. It was certainly not by accident that Isaac Newton was one of the pioneers both of *classical mechanics* and of *calculus*, and probably many readers first learned about derivatives from their physics teachers in connection with the definition of the instantaneous velocity and acceleration of a body in nonuniform motion.

If one studies movements that are affected by *random noise*, for example Brownian motion, the particle trajectories can be so irregular that ordinary calculus clearly fails (according to Example 2.79, the trajectories of the Wiener process modeling Brownian motion are nowhere differentiable; in more fashionable words, the trajectories of the Wiener process can be regarded as fractal objects with Hausdorff-Besicovich dimension 3/2). A more general theory is hence required, namely *stochastic calculus*. Since the irregular Brownian forces caused by rapid thermal fluctuations are ubiquitous in the kinetic theory equations of motion for polymer molecules, stochastic calculus constitutes a very important tool for studying the dynamics of polymeric liquids.

Many applied scientists working in various fields of science used the ideas and rules of deterministic calculus in studying equations of motion subject to random noise in a very naive manner; total confusion and displeasing controversies were the only definite results. The subtleties and pitfalls of stochastic calculus, which can easily be mastered once one is fully aware of them, are the motivation for a careful exposition of stochastic calculus in this chapter before investigating kinetic theory models of polymeric fluids. This chapter culminates in the theory of stochastic differential equations and, in particular, in the discussion of numerical integration schemes.

3.1 Motivation

As a motivation for developing the theory of stochastic differential equations and the underlying stochastic calculus we consider the motion of a large, or Brownian, particle in a surrounding fluid consisting of much smaller particles. The historical prototype of that kind of system is pollen in water for which in 1828 the British botanist Robert Brown (1773–1858) published his observation of a very irregular and unpredictable motion of particles in the absence of external forces (other researchers had observed this "Brownian motion" more than a century before Brown, but he was the first to establish Brownian motion as an important phenomenon and to investigate it in more detail[1]). The origin of this motion was much later discovered to lie in the exceedingly frequent collisions between the Brownian particle and the many surrounding fluid particles which are in incessant thermal motion. In this section, we first formulate and solve a (heuristic) equation of motion for the Brownian particle, and we then discuss the problems associated with differential equations involving rapidly fluctuating Brownian forces represented by a stochastic process with vanishing correlation time. The Gaussian stochastic process $(V_t)_{t\in\mathbb{T}}$ with $\mathbb{T} = [0, \infty[$ which is considered in this section is known as the *Ornstein-Uhlenbeck process*.

3.1.1 Naive Approach to Stochastic Differential Equations

If we consider a Brownian particle of mass M in one space dimension we can try to determine its time-dependent velocity V_t from the equation of motion,

$$M\frac{\mathrm{d}V_t}{\mathrm{d}t} = -\zeta V_t + F_t^{\mathrm{B}}\,, \tag{3.1}$$

where we assume that the total force on the Brownian particle is the sum of a frictional force and a Brownian force. The frictional force is assumed to be proportional to the particle velocity where the proportionality factor is the friction coefficient ζ, and this force is opposed to the motion. The Brownian force describes the rate of momentum transfer resulting from the frequent impacts of small fluid particles on the large Brownian particle, and we try to model its very irregular nature by a stochastic process F_t^{B}. Hence, (3.1) is an example of a stochastic differential equation from which the stochastic process $(V_t)_{t\in\mathbb{T}}$ is to be determined. The naive approach described in this section is a natural extension of the pioneering work of the French physicist Paul Langevin (1872–1946) [3]. In the physics literature, stochastic differential equations are often referred to as *Langevin equations*.

As a convenient initial condition we assume that the Brownian particle is at rest at the initial time $t = 0$. If we solve (3.1) just like a deterministic ordinary differential equation its obvious solution for the initial condition $V_0 = 0$ is

[1] An interesting survey of the history of Brownian motion can be found in §§2–4 of a monograph by Edward Nelson [1]. The early history of the stochastic description of Brownian motion has been reviewed by Subrahmanyan Chandrasekhar [2].

$$V_t = \frac{1}{M} \int_0^t \mathrm{e}^{-\zeta(t-t')/M} \, F_{t'}^{\mathrm{B}} \, \mathrm{d}t' \,. \tag{3.2}$$

Since for the random impacts on the Brownian particle there is no preferred direction of momentum transfer from the isotropically distributed fluid particles, we expect that the average of the Brownian force vanishes at all times (the average effect because of the velocity of the Brownian particle relative to the fluid is taken into account through the frictional force),

$$\langle F_t^{\mathrm{B}} \rangle = 0 \,. \tag{3.3}$$

Equation (3.3), together with (3.2) and the linearity of expectations, immediately implies that the Brownian particle does not acquire an average velocity because of the random impacts of fluid particles, $\langle V_t \rangle = 0$. The second moment of V_t is a more interesting quantity,

$$\langle V_t^2 \rangle = \frac{1}{M^2} \int_0^t \mathrm{d}t' \int_0^t \mathrm{d}t'' \; \mathrm{e}^{-\zeta(2t-t'-t'')/M} \, \langle F_{t'}^{\mathrm{B}} F_{t''}^{\mathrm{B}} \rangle \,. \tag{3.4}$$

From the idea that the Brownian force F_t^{B} results from the frequent collisions between the heavy Brownian particle and the small fluid particles it is natural to assume that the correlation time for F_t^{B} is very short compared to the time scale on which we observe the motion of the Brownian particle. As an idealization, we assume that this collision time scale is infinitesimally small. In order to obtain a nonvanishing result for the second moment (3.4) in the limit of zero correlation time we need to assume that the second moments of the Brownian forces are given by a δ-function,

$$\langle F_{t'}^{\mathrm{B}} F_{t''}^{\mathrm{B}} \rangle = \alpha_{\mathrm{B}} \, \delta(t' - t'') \,, \tag{3.5}$$

where the strength or amplitude of the Brownian forces, α_{B}, remains to be determined. After inserting (3.5) into (3.4) and performing the integrations one obtains

$$\frac{1}{2} M \, \langle V_t^2 \rangle = \frac{\alpha_{\mathrm{B}}}{4\zeta} \left(1 - \mathrm{e}^{-2\zeta t/M} \right) . \tag{3.6}$$

Equation (3.6) shows that the kinetic energy of the Brownian particle, acquired through the incessant collisions with tireless fluid particles, on the time scale M/ζ increases from the given initial value 0 to the equilibrium value $\alpha_{\mathrm{B}}/(4\zeta)$. According to the *principle of equipartition of energy*, the equilibrium value of the kinetic energy in (3.6) should be $\frac{1}{2} k_{\mathrm{B}} T$ where k_{B} is Boltzmann's constant and T is the absolute temperature, and we thus obtain

$$\alpha_{\mathrm{B}} = 2 \, k_{\mathrm{B}} T \, \zeta \,. \tag{3.7}$$

The occurrence of ζ both in the frictional force and in the second moments of the Brownian force indicates that both forces have the same origin, namely collisions between the Brownian and the fluid particles. The relationship (3.7) between the parameters ζ and α_B associated with the frictional and Brownian (random) forces, respectively, is a special case of the *fluctuation-dissipation theorem of the second kind.*[2] The above arguments show that the fluctuation-dissipation theorem and the equipartition principle are intimately related. In its general formulation, the fluctuation-dissipation theorem additionally implies that only the limit of vanishing correlation time assumed in (3.5) is consistent with a frictional force that depends only on the instantaneous velocity and not on the velocity at previous times. A δ-correlated force such as F_t^B (see (3.5)) is often referred to as *white noise*. This term reflects the fact that, after Fourier transformation of the second moments of F_t^B, the Fourier components are independent of frequency—just as the contribution of all colors (or frequencies) to the intensity of white light is almost independent of frequency (at least within the visible range).

Equations (3.3), (3.5), and (3.7) determine the first and second moments of the Brownian forces F_t^B. Since the Brownian forces F_t^B represent the result of many independent collisions, according to the central limit theorem discussed in Example 2.71, it is natural to assume that $(F_t^B)_{t\in\mathbb{T}}$ is a Gaussian process. Under this assumption, the first and second moments determined in this section would completely characterize the stochastic properties of the Brownian forces occurring in the stochastic differential equation (3.1) and, furthermore, the velocity process should be fully determined by the explicit integral expression (3.2). In particular, since the Ornstein-Uhlenbeck process V_t is given as the linear transformation (3.2) of the Gaussian process $(F_t^B)_{t\in\mathbb{T}}$, it is itself found to be Gaussian.

Since the velocity process is Gaussian, (3.6) implies that, for times larger than M/ζ, the Maxwellian velocity distribution is approached exponentially fast. It should be noted that the time scale M/ζ for velocity correlations is typically very short compared to the natural observation time for Brownian motion (even for large pollen grains in water, M/ζ is of the order of 10^{-2}s and, for smaller pollen grains, this time scale can become as small as 10^{-6}s). The existence of these extremely rapid velocity fluctuations explains why the trajectories of a Brownian particle are so irregular on time scales large compared to M/ζ. In the limit where the time scale M/ζ vanishes, the position X_t of the Brownian particle at time t is governed by a stochastic differential equation obtained from (3.1),

$$\frac{dX_t}{dt} = \frac{1}{\zeta} F_t^B, \tag{3.8}$$

for which we obtain the explicit solution in the next subsection. The effect of M/ζ on a trajectory of a Brownian particle is illustrated in Fig. 3.1.

---

[2]This type of fluctuation-dissipation theorem, which will repeatedly be mentioned in the subsequent chapters, can be formulated for generalized Langevin equations in which the frictional forces depend on the velocities at previous times. The fluctuation-dissipation theorem then states that the time-dependent friction coefficient, or memory function, is related to the second moments of the Brownian forces, or random forces, by $\langle F_{t'}^B F_{t''}^B \rangle = k_B T\, \zeta(|t'-t''|)$ (see Sects. 1.6 and 2.9 of [4]).

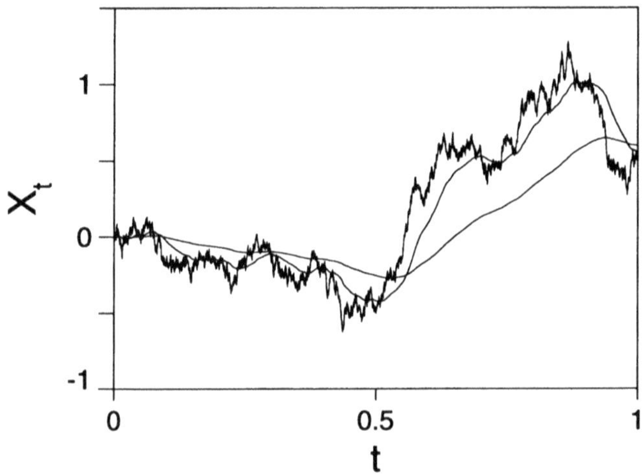

Fig. 3.1. Effect of the time scale M/ζ on a trajectory of the Ornstein-Uhlenbeck process for $\alpha_B = \zeta = 1$. The three trajectories correspond to $M = 0.0001$, $M = 0.04$, and $M = 0.2$, where increasing M leads to smoother trajectories. Within the resolution of this figure, the trajectory with $M = 0.0001$ is indistinguishable from a continuous but nowhere differentiable trajectory of the Wiener process.

Exercise 3.1 Evaluate the averages $\langle V_t V_{t'} \rangle$ for $t, t' > 0$.

3.1.2 Criticism of the Naive Approach

The formulation and solution of the stochastic differential equation (3.1) governing Brownian motion as discussed in Sect. 3.1.1 raises several problems and questions: Is it justified to treat $(F_t^{\mathrm{B}})_{t \in \mathbb{T}}$ as if it were a well-defined Gaussian process? Because of the δ-function occurring in the second moments (3.5), the theory of Gaussian processes developed in Sect. 2.3.2 is not directly applicable. Even if a well-defined meaning could be given to the process $(F_t^{\mathrm{B}})_{t \in \mathbb{T}}$, is it allowable to treat the stochastic differential equation (3.1) like a deterministic differential equation, in particular, in view of the singular nature of F_t^{B}? How can integrals of very irregular stochastic processes such as the one in (3.2) be defined?

In order to further discuss the problems caused by the heuristic approach to stochastic differential equations we formally introduce a stochastic process defined by the integral

$$W_t = \frac{1}{\sqrt{2 k_{\mathrm{B}} T \zeta}} \int_0^t F_{t'}^{\mathrm{B}} \, \mathrm{d}t' \,, \tag{3.9}$$

which is of the type occurring in (3.2) but simpler. The motivation for this further formal manipulation of the merely heuristically introduced Brownian forces is the hope that the integration removes the singular character of the process $(F_t^{\mathrm{B}})_{t\in\mathbb{T}}$ associated with the δ-function in (3.5). As a linear transformation of a Gaussian process, $W = (W_t)_{t\in\mathbb{T}}$ should itself be a Gaussian process, and (3.3) implies $\langle W_t \rangle = 0$. From (3.5), (3.7), and (3.9) we obtain the second moments

$$\langle W_{t_1} W_{t_2} \rangle = \int_0^{t_1} \mathrm{d}t' \int_0^{t_2} \mathrm{d}t''\, \delta(t'-t'') = \min(t_1, t_2)\,. \tag{3.10}$$

Equation (3.10) suggests that the Gaussian process W formally introduced in (3.9) should be interpreted as the Wiener process of Example 2.79; W is thus a well-defined Gaussian stochastic process and, for the first time in this section, we are on safe ground.

According to (3.9), the process $(F_t^{\mathrm{B}})_{t\in\mathbb{T}}$ has the formal representation

$$F_t^{\mathrm{B}} = \sqrt{2k_{\mathrm{B}}T\zeta}\,\frac{\mathrm{d}W_t}{\mathrm{d}t}\,. \tag{3.11}$$

By comparing the equation of motion (3.8) for vanishing M/ζ with (3.11) we immediately obtain the result

$$X_t = \sqrt{2k_{\mathrm{B}}T/\zeta}\, W_t \tag{3.12}$$

for the random trajectories of a Brownian particle which is initially at the origin. Equation (3.12) shows that the Wiener process is an appropriate model for Brownian motion on time scales large compared to M/ζ, and it gives an explicit expression for the diffusion coefficient of the Brownian particle, $D = k_{\mathrm{B}}T/\zeta$, which is to be read off from the equation $\langle X_t^2 \rangle = 2D\,t$ (typical trajectories of the square displacement X_t^2 are shown in Fig. 2.7). Notice that the Nernst-Einstein relation between D and ζ occurs as another manifestation of the fluctuation-dissipation theorem.

In view of the fact that the trajectories of the Wiener process are not differentiable, the representation (3.11) of the Brownian forces clearly highlights the problems associated with the definition and solution of the stochastic differential equation describing Brownian motion which was motivated in Sect. 3.1.1. In order to avoid the necessity of giving a direct interpretation of the forces (3.11), we introduce (3.11) into the solution (3.2) of the stochastic differential equation (3.1) in order to obtain the representation

$$V_t = \frac{\sqrt{2k_{\mathrm{B}}T\zeta}}{M} \int_0^t \mathrm{e}^{-\zeta(t-t')/M}\,\mathrm{d}W_{t'}\,. \tag{3.13}$$

The right-hand side of (3.13) is a simple example of a *stochastic integral*. Even though the non-existing time derivatives of the Wiener process are now avoided, the unbounded variation of the trajectories of W in any arbitrarily small time interval observed in Example 2.79 still precludes a pathwise definition of the stochastic integral in (3.13). A rigorous approach to more general stochastic integrals with respect to the Wiener process is described in the following section. The rigorous theory of stochastic integration and stochastic differential equations will demonstrate that the naive approach to the linear stochastic differential equation (3.1) is correct. In general, however, treating stochastic differential equations like deterministic differential equations is found to lead to incorrect results. This is so because stochastic calculus is governed by quite unusual differentiation rules! It is hence obvious that a careful introduction of stochastic integrals is very important before formulating and investigating stochastic differential equations.

3.2 Stochastic Integration

In this section, we outline a rigorous procedure for defining stochastic integrals with respect to the Wiener process, and we discuss the properties of such stochastic integrals. Throughout this section we assume that the Wiener process $W = (W_t)_{t \in \mathbb{T}}$ on a probability space $(\Omega, \mathcal{A}, P)$ is given where the range of time $\mathbb{T}$ is either $[0, t_{\max}]$ or $[0, \infty[$. The theory of stochastic integrals outlined here was initiated and developed by the Japanese mathematician Kiyosi Itô (1915–) in the early 1940s. The theory was independently established by the Ukrainian mathematician Iosif I. Gikhman (1918–1985). In order to render possible a clear and intuitive exposition we skip most of the (usually quite lengthy) proofs; nevertheless, we feel that it is very important to present all results and definitions with mathematical rigor. Elaborations of the mathematical details and complete proofs can be found, for example, in the textbooks by T. C. Gard [5], L. Arnold [6], or R. S. Liptser and A. N. Shiryayev [7]. Readers who feel that the approach followed here is still too (or even unnecessarily) formal are referred to the excellent intuitive introductions to stochastic integrals given by the physicists C. W. Gardiner [8], N. G. van Kampen [9], and J. Honerkamp [10].

3.2.1 Definition of the Itô Integral

Before introducing the stochastic integral of a real-valued stochastic process $X = (X_t)_{t \in \mathbb{T}}$ with respect to the Wiener process W we need to characterize those processes for which stochastic integrals of the form $\int_0^t X_{t'} \mathrm{d}W_{t'}$ can be defined in a meaningful and consistent way. Of course, the process X must be defined on the same, given probability space $(\Omega, \mathcal{A}, P)$ as W. As we will see below, a further, crucial requirement is that the process X be *nonanticipating*; in other words, the history of the process X up to some time t is required to be independent of the future evolution of the Wiener process W after the time t.

Since the concept of nonanticipating processes plays such an important role in the procedure of introducing stochastic integrals a brief discussion of the rigorous definition of this term is in place. As always, when the independence of sets of random variables is to be established we look at induced σ-algebras. A formal mathematical definition of nonanticipating processes can be based on the family of σ-algebras $(\mathcal{A}_t)_{t\in\mathbb{T}}$, where the σ-algebra $\mathcal{A}_t \subset \mathcal{A}$ may be defined as the smallest σ-algebra for which all $X_{t'}$ and $W_{t'}$ with $t' \leq t$ are measurable (for technical reasons, one usually considers the completions of $\mathcal{A}$ and $\mathcal{A}_t$, as defined in the footnote on p. 67). As in the theory of martingales (see Sect. 2.3.4), the increasing family of σ-algebras $(\mathcal{A}_t)_{t\in\mathbb{T}}$ represents the accumulated information contained in the history of the processes X and W up to the time t. Also increasing families of larger σ-algebras can be admitted (for example, when considering several nonanticipating processes at the same time); important is only that X_t and W_t must be measurable with respect to the "accumulated information" $\mathcal{A}_t$. We furthermore introduce the σ-algebras $\bar{\mathcal{A}}_t$ induced by the future of the Wiener process W as the smallest σ-algebra for which all random variables $W_{t'} - W_t$ for $t' > t$ are measurable. One can then define a stochastic process X to be nonanticipating with respect to the given Wiener process W if, and only if, for all $t \in \mathbb{T}$, the two σ-algebras $\mathcal{A}_t$ and $\bar{\mathcal{A}}_t$ are independent. Assuming the independence of $\mathcal{A}_t$ and $\bar{\mathcal{A}}_t$ corresponds to assuming that W is a martingale with respect to the underlying σ-algebras $(\mathcal{A}_t)_{t\in\mathbb{T}}$, which are larger than the σ-algebras induced by the history of W (see Sect. 2.3 of [11]). As a further technical requirement, the function $\mathbb{T}\times\Omega \to \mathbb{R}$, $(t,\omega) \mapsto X_t(\omega)$ is assumed to be measurable when $\mathbb{T}\times\Omega$ is equipped with the product of the Borel σ-algebra on $\mathbb{T}$ and the σ-algebra $\mathcal{A}$ on Ω. In several steps, we next construct the stochastic integral for an increasingly larger class of such nonanticipating processes X. The procedure is very similar to the introduction of expectations in Sect. 2.2.2.

Step 1: We first consider *simple* processes which, by definition, are nonanticipating processes of the form

$$X_t(\omega) = \sum_{j=1}^{n} \widetilde{X}_{j-1}(\omega)\, \chi_{[t_{j-1},t_j[}(t)\,, \tag{3.14}$$

with suitable random variables $\widetilde{X}_{j-1}$ for which the second moments $\langle \widetilde{X}_{j-1}^2 \rangle$ for $j = 1, \dots n$ are assumed to be finite. In (3.14), the times t_j define a partition of the time range $\mathbb{T} = [0, t_{\max}]$ of the process X with $0 = t_0 < t_1 \cdots < t_{n-1} < t_n = t_{\max}$ (the case $\mathbb{T} = [0, \infty[$ is considered in Step 4 below). The indicator functions in (3.14) imply that all trajectories of a simple process are constant except for jumps at the given times t_j (for that reason, simple processes are occasionally also referred to as random step functions). In the Itô approach, the stochastic integral of a simple nonanticipating process is defined as

$$\int_0^{t_{\max}} X_t\, \mathrm{d}W_t := \sum_{j=1}^{n} \widetilde{X}_{j-1}(W_{t_j} - W_{t_{j-1}})\,. \tag{3.15}$$

The integral defined in (3.15), which is independent of the particular representation of a given simple process in the form (3.14), is a random variable which is measurable with respect to $\mathcal{A}_{t_{\max}}$. Since X was assumed to be nonanticipating, the two factors $\widetilde{X}_{j-1} = X_{t_{j-1}}$ and $(W_{t_j} - W_{t_{j-1}})$ in each term of the sum in (3.15) are independent random variables. This independence of $\widetilde{X}_{j-1}$ and $(W_{t_j} - W_{t_{j-1}})$ is a very important feature of the Itô approach to stochastic integration, and it has a crucial influence on the properties of the resulting stochastic calculus, in particular, on the integration and differentiation rules of Itô's calculus. Using $\widetilde{X}_j$ or $(\widetilde{X}_{j-1} + \widetilde{X}_j)/2$ instead of $\widetilde{X}_{j-1}$ as the prefactor of the increment $(W_{t_j} - W_{t_{j-1}})$ in (3.15) would lead to a very different theory of stochastic integration (see Sect. 3.3.6)! The equation

$$\left\langle \int_0^{t_{\max}} X_t \, \mathrm{d}W_t \right\rangle = 0 \,, \tag{3.16}$$

is an immediate but far-reaching consequence of the independence of $\widetilde{X}_{j-1}$ and $(W_{t_j} - W_{t_{j-1}})$ and of the fact that the increments of the Wiener process have zero expectation (here, the relevance of the assumption that X be nonanticipating becomes obvious). Equation (3.16) is an important characteristic of Itô's calculus.

By using the independence of $\widetilde{X}_{j-1}$ and $(W_{t_j} - W_{t_{j-1}})$ together with the fact that the increments of the Wiener process over disjoint time intervals are also independent random variables with zero expectation one obtains the equation

$$\left\langle \left(\int_0^{t_{\max}} X_t \, \mathrm{d}W_t \right)^2 \right\rangle = \sum_{j=1}^{n} \langle \widetilde{X}_{j-1}^2 \rangle \underbrace{\langle (W_{t_j} - W_{t_{j-1}})^2 \rangle}_{t_j - t_{j-1}} = \int_0^{t_{\max}} \langle X_t^2 \rangle \, \mathrm{d}t \,. \tag{3.17}$$

The identity (3.17) relates the second moment of the stochastic integral and the usual integral of the second moment of the process X. This identity is another characteristic of Itô's calculus, and it is very helpful in the subsequent steps of extending stochastic integrals to larger classes of nonanticipating processes. When (3.17) is applied to the difference of two simple processes, which is itself a simple process, it implies that the stochastic integrals are approximately the same, in a mean-square sense, when the two simple processes are close, in a time-averaged mean-square sense.

Step 2: The first stage of generalizing the definition (3.15) is based on the following lemma.

Lemma 3.2 *For every nonanticipating process X with $\int_0^{t_{\max}} \langle X_t^2 \rangle \mathrm{d}t < \infty$, there exists a sequence of simple processes $(X^{(n)})_{n \in \mathbb{N}}$ such that*

$$\lim_{n \to \infty} \int_0^{t_{\max}} \langle (X_t - X_t^{(n)})^2 \rangle \mathrm{d}t = 0 \,. \tag{3.18}$$

A proof of this lemma can be obtained by suitably discretizing the process X. Rather than proving this lemma in its general form we give a typical example of the straightforward construction of an approximating sequence below in Example 3.5.

We are now in a position to extend the definition of stochastic integrals. For a nonanticipating process X with $\int_0^{t_{\max}} \langle X_t^2 \rangle dt < \infty$, we choose a sequence $(X^{(n)})_{n \in \mathbb{N}}$ of simple processes according to the above lemma, and we define

$$\int_0^{t_{\max}} X_t \, dW_t := \text{ms-}\lim_{n\to\infty} \int_0^{t_{\max}} X_t^{(n)} \, dW_t . \tag{3.19}$$

Since each simple process $X^{(n)}$ is of the form (3.14), the integrals on the right side of (3.19) are defined in (3.15) for any given value of n in the superscript.

The existence of the mean-square limit in (3.19) and its independence of the choice of the approximating sequence of simple processes can be established by means of the identity (3.17) and the property (3.18) of approximating sequences. The convergence expressed in (3.18) implies that, in the limit of large n and l, the integral $\int_0^{t_{\max}} \langle (X_t^{(n)} - X_t^{(l)})^2 \rangle dt$ goes to zero. The identity (3.17), applied to the simple process $X_t^{(n)} - X_t^{(l)}$, then leads to the conclusion that $\text{ms-lim}_{n,l\to\infty} \left| \int_0^{t_{\max}} X_t^{(n)} dW_t - \int_0^{t_{\max}} X_t^{(l)} dW_t \right| = 0$ holds. According to the Cauchy Criterion 2.66, this guarantees the existence of the mean-square limit in (3.19). The independence of the stochastic integral (3.19) of the approximating sequence can be demonstrated by similar arguments.

While (3.15) may be considered as the definition of the stochastic integral for each trajectory of a simple process X, the limit in (3.19) has no well-defined meaning for an individual $\omega \in \Omega$. The mean-square limit is only almost surely unique. The remaining freedom of choosing the mean-square limit in (3.19) on a set of zero measure is often used to select versions of stochastic integrals for which all trajectories of the stochastic process obtained by varying the upper limit of integration have particularly nice properties.

Step 3: The extension of the stochastic integral to an even more general class of nonanticipating processes is based on the following lemma (see Lemma 4.5 in [7]).

Lemma 3.3 *For every nonanticipating process X with*

$$\int_0^{t_{\max}} X_t(\omega)^2 \, dt < \infty \qquad \text{for almost all } \omega \in \Omega , \tag{3.20}$$

there exists a sequence of simple processes $(X^{(n)})_{n \in \mathbb{N}}$ such that

$$\text{st-}\lim_{n\to\infty} \int_0^{t_{\max}} (X_t - X_t^{(n)})^2 dt = 0 . \tag{3.21}$$

Exercise 3.4 Show that the processes satisfying (3.20) are more general than those admitted in Step 2. Accordingly, the mode of approximation of the sequence of simple processes in (3.21) is weaker than in (3.18).

The existence of an approximating sequence, in the sense of the stochastic limit (3.21) guaranteed by this lemma, immediately suggests the definition

$$\int_0^{t_{\max}} X_t \,\mathrm{d}W_t := \text{st-}\lim_{n\to\infty} \int_0^{t_{\max}} X_t^{(n)} \,\mathrm{d}W_t \,, \tag{3.22}$$

for an arbitrary nonanticipating process X satisfying (3.20). The proof of the existence of the limit in (3.22) and of its independence of the choice of the approximating sequence of simple processes can be found in any mathematical textbook on stochastic integrals.

Step 4: For any nonanticipating process X satisfying (3.20) and for every set $B \subset [0, t_{\max}]$ contained in the Borel σ-algebra we can define

$$\int_B X_t \,\mathrm{d}W_t := \int_0^{t_{\max}} \chi_B(t)\, X_t \,\mathrm{d}W_t \,. \tag{3.23}$$

In particular, if we choose $B = [a, b[$, the stochastic integral from a to b is defined. By estimating the contribution to the integral from large times and by applying the Cauchy criterion for stochastic convergence, it is finally possible to extend the upper limit of integration to infinity when $\mathbb{T} = [0, \infty[$. By applying a shift in time, it is also possible to use arbitrary negative values in the lower limit of integration. This step completes the construction of the most general stochastic integrals considered in this book.

Example 3.5 Premiere of Stochastic Integration

What the harmonic oscillator is to quantum mechanics, what the Rouse-Zimm model is to polymer kinetic theory, what the *Drosophila* is to genetics (and what the bee is to sex education), such is the integral $\int_0^t W_{t'} \,\mathrm{d}W_{t'}$ to Itô's calculus. It is quite instructive to evaluate this unavoidable example of a stochastic integral in some detail.

Because $\int_0^t \langle W_{t'}^2 \rangle \mathrm{d}t' = \int_0^t t' \,\mathrm{d}t' = \frac{1}{2}\, t^2 < \infty$ and because of the independence of the increments of the Wiener process, $(W_t)_{t\in\mathbb{T}}$ is a nonanticipating process of the class considered in Step 2 of the above construction of stochastic integrals. In order to construct an approximating sequence of simple processes for W, we partition the interval $[0, t]$ by introducing $t_j^{(n)} = jt/n$ for $j = 0, \ldots n$, we define $\widetilde{W}_j^{(n)} = W_{t_j^{(n)}}$, and we obtain the approximating simple processes by straightforward discretization,

$$W_{t'}^{(n)}(\omega) = \sum_{j=1}^{n} \widetilde{W}_{j-1}^{(n)}(\omega)\, \chi_{[t_{j-1}^{(n)}, t_j^{(n)}[}(t') \,. \tag{3.24}$$

One then has

$$\int_0^t \left\langle \left(W_{t'} - W_{t'}^{(n)}\right)^2 \right\rangle \mathrm{d}t' = \sum_{j=1}^{n} \int_{t_{j-1}^{(n)}}^{t_j^{(n)}} (t' - t_{j-1}^{(n)})\,\mathrm{d}t' = \frac{t^2}{2n},$$

and, since these integrals vanish for $n \to \infty$, the processes $W^{(n)}$ indeed constitute an approximating sequence for W. From the definition of the stochastic integral for simple processes we obtain

$$\int_0^t W_{t'}^{(n)}\,\mathrm{d}W_{t'} = \sum_{j=1}^{n} \widetilde{W}_{j-1}^{(n)}(\widetilde{W}_j^{(n)} - \widetilde{W}_{j-1}^{(n)}) = \frac{1}{2}\Big[\widetilde{W}_n^{(n)\,2} - \sum_{j=1}^{n}(\widetilde{W}_j^{(n)} - \widetilde{W}_{j-1}^{(n)})^2\Big]. \quad (3.25)$$

The last sum in (3.25) can be regarded as $1/n$ times the sum of n independent random variables with identical distribution, where the expectation of each of those n random variables is $n\,(t/n) = t$. The strong law of large numbers (see Example 2.67) then implies that the almost sure limit and hence also the mean-square limit of this arithmetic average of independent, identically distributed, real-valued random variables is equal to their expectation, t. The mean-square convergence of the last sum in (3.25) to the limit t can also be verified directly by using the formula

$$\left\langle \Big[t - \sum_{j=1}^{n}(\widetilde{W}_j^{(n)} - \widetilde{W}_{j-1}^{(n)})^2\Big]^2 \right\rangle = \frac{2t^2}{n},$$

which can be derived by reducing the average on the left side to second and fourth moments of increments of the Wiener process given in Example 2.79. We thus obtain the desired stochastic integral as the mean-square limit of (3.25),

$$\int_0^t W_{t'}\,\mathrm{d}W_{t'} = \frac{1}{2}\,(W_t^2 - t)\,. \quad (3.26)$$

By naive application of deterministic integration rules we would have obtained only the first term on the right-hand side of (3.26). The additional term in (3.26) is the first manifestation of the peculiarities of Itô's calculus pointed out at the beginning of this section. This additional term is crucial for obtaining a vanishing expectation of the stochastic integral $\int_0^t W_{t'}\,\mathrm{d}W_{t'}$. □

Example 3.6 Stochastic Integrals of Deterministic Functions
The stochastic integral is also defined in the special case of a deterministic integrand, that is, for $X_t(\omega) = g_t$ with a real function g_t independent of ω. Such a deterministic process is certainly nonanticipating, and we assume $\int_0^t g_{t'}^2 \mathrm{d}t' < \infty$. Following the standard procedure, g_t can be approximated by a sequence of deterministic step functions $g_t^{(n)}$ such that

$$\lim_{n\to\infty} \int_0^t (g_{t'} - g_{t'}^{(n)})^2 \mathrm{d}t' = 0\,. \quad (3.27)$$

For every step function $g_t^{(n)}$, the stochastic integral $\int_0^t g_{t'}^{(n)}\, dW_{t'}$ is a finite sum of Gaussian random variables and hence itself a Gaussian random variable; this property survives in the limit $n \to \infty$ (this follows from the convergence of the corresponding characteristic functions). The first moment of the stochastic integral vanishes, and the second moment of this Gaussian random variable is, according to (3.17) and (3.27), given by $\int_0^t g_{t'}^2 dt'$. Therefore, the integral (3.13) discussed in the motivation of stochastic integrals in Sect. 3.1 now has a rigorous meaning, and (3.6) for the second moment of (3.13) can be derived rigorously without making any use of δ-correlated noise or other singular objects. □

Exercise 3.7 Prove the identity $\int_0^t (W_{t'}^2 - t')\, dW_{t'} = \frac{1}{3} W_t^3 - tW_t$ by means of the definition of stochastic integrals (Step 2).

3.2.2 Properties of the Itô Integral

Among the properties one would certainly expect from any reasonable definition of an integral are linearity and additivity. These properties are guaranteed by the following theorem.

Theorem 3.8 *Let X and Y be nonanticipating stochastic processes satisfying the condition (3.20).*
For arbitrary $c_1, c_2 \in \mathbb{R}$ and $[a,b] \subset \mathbb{T}$, one has the linearity property

$$\int_a^b (c_1 X_t + c_2 Y_t)\, dW_t = c_1 \int_a^b X_t\, dW_t + c_2 \int_a^b Y_t\, dW_t\,. \tag{3.28}$$

For arbitrary $a, b, c \in \mathbb{T}$ with $a \le b \le c$, one has the additivity property

$$\int_a^c X_t\, dW_t = \int_a^b X_t\, dW_t + \int_b^c X_t\, dW_t\,. \tag{3.29}$$

In order to prove the linearity (3.28), one first considers the case of simple nonanticipating processes, for which this property is trivially obtained from the corresponding property of sums. In the general case, the linearity of stochastic integrals is obtained by limiting procedures which follow the steps of the construction of stochastic integrals. This stepwise procedure is very natural for proving properties of stochastic integrals, and one typically has to employ convergence theorems. The additivity property (3.29) is a simple consequence of the definition (3.23) and the linearity property (3.28).

We next extend the properties (3.16) and (3.17) to more general stochastic integrals.

Theorem 3.9 *For any $[a,b[\subset \mathbb{T}$ and for any nonanticipating process X with $\int_a^b \langle X_t^2 \rangle dt < \infty$, one has*

$$\left\langle \int_a^b X_t \, dW_t \right\rangle = 0 , \tag{3.30}$$

and

$$\left\langle \left(\int_a^b X_t \, dW_t \right)^2 \right\rangle = \int_a^b \langle X_t^2 \rangle \, dt . \tag{3.31}$$

Note that this theorem is not generally true for the larger class of processes satisfying (3.20) which was considered in Step 3 of the construction of stochastic integrals. If one considers two processes X and Y satisfying the assumptions of Theorem 3.9, then the property (3.31) can easily be generalized to

$$\boxed{\left\langle \left(\int_{B_1} X_t \, dW_t \right) \left(\int_{B_2} Y_t \, dW_t \right) \right\rangle = \int_{B_1 \cap B_2} \langle X_t Y_t \rangle \, dt ,} \tag{3.32}$$

where B_1 and B_2 are measurable subsets of $\mathbb{T}$.

Exercise 3.10 Rederive the result of Exercise 3.1 in a rigorous manner by applying (3.32) to the stochastic integral (3.13). Notice that with (3.32) one can avoid the handling of δ-functions.

We next consider the properties of stochastic integrals as functions of the upper limit of integration. The general properties are first illustrated by considering a particular example.

Example 3.11 A Day in the Life of a Pollen Grain
In Sect. 3.1, we realized that $X_t = (2k_\mathrm{B}T/\zeta)^{1/2} W_t$ is an adequate model of the time-dependent position of a diffusing pollen grain (or any other Brownian particle). We here consider the case where the prefactor $B_t := (2k_\mathrm{B}T/\zeta)^{1/2}$ in the corresponding equation for the stochastic differential of X_t is itself a stochastic process on the same probability space, and we hence describe the increments in the position X_t by the equation $dX_t = B_t \, dW_t$. The average of B_t is assumed to change in the course of time, say as a consequence of systematic daily temperature variations, and B_t is of course also subject to stochastic fluctuations due to notoriously unpredictable weather conditions (we assume that the current weather and the future collisions of the pollen grain are independent processes so that B_t is nonanticipating). When assuming that the temperature may be considered approximately constant over periods of one hour, the process $B = (B_t)_{t \in \mathbb{T}}$ constitutes a simple process. It then makes sense to look at individual trajectories of the position process $X_t = \int_0^t B_{t'} \, dW_{t'}$.

The step function in Fig. 3.2 is a typical trajectory of B for one day in a geographical location with extreme daily temperature variations (let us consider a lonely pollen grain in the middle of the Sahara). The unit of time is 1 hour, and the unit of length is to be chosen such that a realistic diffusion coefficient is obtained for the Brownian particle. The thick and thin lines in Fig. 3.2 represent trajectories of the Wiener process and of the stochastic integral $X_t = \int_0^t B_{t'} \, dW_{t'}$,

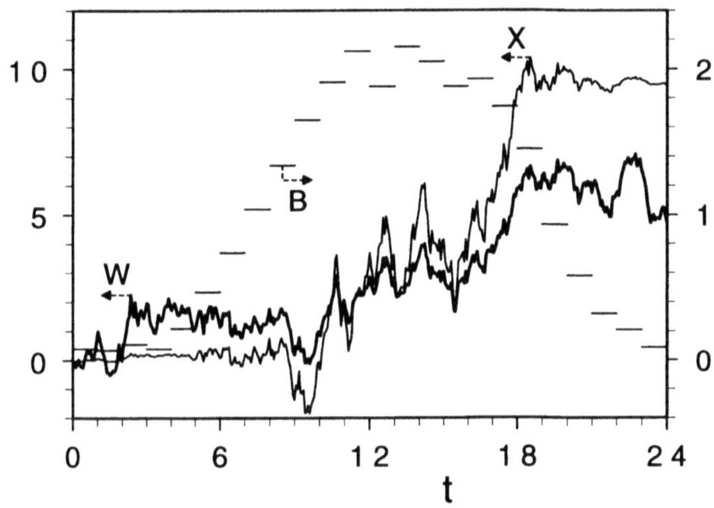

Fig. 3.2. The *thick line* represents a trajectory of the Wiener process W. The *thin line* represents the corresponding trajectory of the process X which is the stochastic integral of the random step function B with respect to the Wiener process W.

respectively. The phases of different activity of the pollen grain are clearly seen in the trajectory of X. In the first four or five chilly hours of the day, the diffusion coefficient is very small, and the trajectory of X shows only minor fluctuations (compared to those of the Wiener process W). Between $t = 6$ and $t = 9$, the prefactor B is of the order of 1, and the trajectories of W and X show almost identical variations and stay "parallel" during that time interval. Around noon, the fluctuations of W are strongly amplified in the stochastic integral X. The total displacement described by X is eventually larger than for the Wiener process W. With the onset of dusk, the pollen grain slows down, and the fluctuations of the Wiener process are clearly damped in the trajectory of X.

The trajectories of the pollen grain in the course of a day are continuous and, at any time, we have no information for making predictions about the expected future increments of X (that is, X is a martingale). The martingale property of the process X, and moreover of stochastic integrals in general, is inherited from the martingale property of the integrator process W, as can be understood by carefully going through all the steps for the following special case (which definitions or which of the properties of conditional expectations compiled in Sect. 2.2.4 are used in each step?),

$$\begin{aligned}
E\Big(X_{9.3} - X_{8.1}\Big|\mathcal{A}_{8.1}\Big) &= E\left(\int_{8.1}^{9.3} B_{t'}\, \mathrm{d}W_{t'}\Big|\mathcal{A}_{8.1}\right) \\
&= E\Big(B_{8.1}(W_9 - W_{8.1}) + B_9(W_{9.3} - W_9)\Big|\mathcal{A}_{8.1}\Big)
\end{aligned}$$

$$= E\Big(B_{8.1}(W_9 - W_{8.1})\Big|\mathcal{A}_{8.1}\Big) + E\Big(E\Big(B_9(W_{9.3} - W_9)\Big|\mathcal{A}_9\Big)\Big|\mathcal{A}_{8.1}\Big)$$
$$= B_{8.1}\, E\Big(W_9 - W_{8.1}\Big|\mathcal{A}_{8.1}\Big) + E\Big(B_9\, E\Big(W_{9.3} - W_9\Big|\mathcal{A}_9\Big)\Big|\mathcal{A}_{8.1}\Big) = 0\,,$$

where, for clarity, no use was made of the fact that $B_{8.1} = B_8$. The fact that stochastic integrals inherit the martingale property from the integrators is the deeper reason for the importance of martingales in stochastic calculus.

In general, the increments of stochastic integrals are uncorrelated but not independent; the martingale concept allows a finer distinction of where stochastic integrals lie between the rather large class of processes with uncorrelated increments and the rather restricted class of processes with independent increments. Notice that the assumption of nonanticipating integrands is crucial for proving the martingale property of stochastic integrals.

Finally, the process X is clearly nonanticipating. The stochastic integral X shares all these properties with the Wiener process W. In the remainder of this subsection, we compile the theorems which show that these properties are typical for all stochastic integrals defined in the preceding subsection and are not just valid for the particular example considered here. □

For formulating the properties of stochastic integrals as a function of the upper limit we define the process

$$Y_t := \int_0^t X_{t'}\, \mathrm{d}W_{t'} \qquad \text{for} \quad 0 \le t \le t_{\max}\,. \tag{3.33}$$

Theorem 3.12 *For any nonanticipating process X satisfying (3.20), the stochastic integral $(Y_t)_{t\in\mathbb{T}}$ defined in (3.33) is a nonanticipating process. There exists a version of Y for which all trajectories $Y_t(\omega)$ are continuous, and for which $X_{t'}(\omega) = 0$ for $t' \le t$ implies $Y_t(\omega) = 0$.*

If $\int_0^{t_{\max}} \langle X_t^2 \rangle \mathrm{d}t < \infty$ then, according to Theorem 3.9, we also know that $\int_0^{t_{\max}} \langle Y_t^2 \rangle \mathrm{d}t < \infty$. The stochastic integral Y can then itself be used as an integrand in a stochastic integral so that iterated stochastic integrals can be defined.

Theorem 3.13 *For any nonanticipating process X with $\int_0^{t_{\max}} \langle X_t^2 \rangle\, \mathrm{d}t < \infty$, the stochastic integral $Y = (Y_t)_{t\in\mathbb{T}}$ defined in (3.33) for $\mathbb{T} = [0, t_{\max}]$ is a martingale with respect to the underlying increasing family of σ-algebras $(\mathcal{A}_t)_{t\in\mathbb{T}}$ (cf. the remarks in the beginning of Sect. 3.2.1).*

Theorem 3.13 is a weaker version of the stochastic independence of the increments of the stochastic integral Y that is obtained when X is a deterministic function, for example, $X_t = 1$ for the Wiener process.

Exercise 3.14 Verify the martingale property for the explicit expression for the stochastic integral $\int_0^t W_{t'}\,\mathrm{d}W_{t'}$.

3.2.3 Itô's Formula

Evaluation of stochastic integrals by means of the definition is, as in the deterministic case, extremely cumbersome. We need to generalize the well-known integration and differentiation rules of deterministic calculus to the stochastic case in order to have tools for an efficient computation of stochastic integrals. In order to have a handy notation, we first introduce the concept of stochastic differentials.

Definition 3.15 *A continuous real-valued stochastic process $(X_t)_{t\in\mathbb{T}}$ is called an* Itô process *if there exist two nonanticipating processes $(A_t)_{t\in\mathbb{T}}$ and $(B_t)_{t\in\mathbb{T}}$ such that the following equations hold for almost all $\omega \in \Omega$:*

$$\int_{\mathbb{T}} |A_t(\omega)|\,\mathrm{d}t < \infty\,, \tag{3.34}$$

$$\int_{\mathbb{T}} B_t(\omega)^2\,\mathrm{d}t < \infty\,, \tag{3.35}$$

$$X_t(\omega) = X_0(\omega) + \int_0^t A_{t'}(\omega)\,\mathrm{d}t' + \Big(\int_0^t B_{t'}\,\mathrm{d}W_{t'}\Big)(\omega) \quad \text{for all } t \in \mathbb{T}. \tag{3.36}$$

The process X is then said to possess the stochastic differential

$$\mathrm{d}X_t := A_t\,\mathrm{d}t + B_t\,\mathrm{d}W_t\,. \tag{3.37}$$

The A-integral in (3.36) is an ordinary integral which, as a consequence of the assumption (3.34), can be performed for each trajectory individually; the B-integral in (3.36) is the random variable obtained by stochastic integration, evaluated at $\omega \in \Omega$. Stochastic differentials are thus defined in terms of the corresponding stochastic integrals. The linearity (3.28) of stochastic integrals is hence inherited by stochastic differentials.

For a justification of the notation introduced in (3.37), it is necessary to verify that the Wiener process W is an Itô process with differential $\mathrm{d}W_t$. Indeed, the Wiener process W can be written in the form of (3.36) with $A_t = 0$ and $B_t = 1$ (this follows immediately from the definition (3.15)). For $B_t = 0$, we recover deterministic differentials. In deterministic calculus, there are various differentiation rules which allow an efficient handling of differentials. The central formula is the chain rule which is next generalized to the case of stochastic differentials.

Theorem 3.16 (Itô's Formula). *Suppose that $(X_t)_{t\in\mathbb{T}}$ is an Itô process with the stochastic differential $\mathrm{d}X_t = A_t\,\mathrm{d}t + B_t\,\mathrm{d}W_t$. If $g(t,x)$ is a real-valued* *deterministic function defined for all $t \in \mathbb{T}$, $x \in \mathbb{R}$ with continuous partial derivatives $\partial g/\partial t$, $\partial g/\partial x$, and $\partial^2 g/\partial x^2$, then the stochastic process $(Y_t)_{t\in\mathbb{T}}$ with $Y_t = g(t, X_t)$ is an Itô process with the stochastic differential*

$$\boxed{\mathrm{d}Y_t = \left[\frac{\partial g}{\partial t}(t,X_t) + \frac{\partial g}{\partial x}(t,X_t)\,A_t + \frac{1}{2}\,\frac{\partial^2 g}{\partial x^2}(t,X_t)\,B_t^2\right]\mathrm{d}t + \frac{\partial g}{\partial x}(t,X_t)\,B_t\,\mathrm{d}W_t\,.} \tag{3.38}$$

The derivation of (3.38) is based on a Taylor expansion, where a second-order expansion is required because the second moments of the increments of the Wiener process over a short time interval are given by the length of the time interval (loosely speaking, $(\mathrm{d}W_t)^2$ behaves like $\mathrm{d}t$ in a mean-square sense, or $\mathrm{d}W_t \sim \sqrt{\mathrm{d}t}$). A rigorous derivation of Itô's formula requires several pages of lengthy estimates and various proofs of the existence of certain limits (see, e.g., Sect. 2.4 of [5]). Rather than proving (3.38) we here try to develop some intuition for its meaning and implications.

Itô's formula can be regarded as the rigorous justification of the following procedure: in the formal second-order Taylor expansion

$$\mathrm{d}Y_t = \frac{\partial g}{\partial t}(t,X_t)\,\mathrm{d}t + \frac{\partial g}{\partial x}(t,X_t)\,\mathrm{d}X_t + \frac{1}{2}\,\frac{\partial^2 g}{\partial x^2}(t,X_t)\,(\mathrm{d}X_t)^2\,, \tag{3.39}$$

one introduces the explicit form of the stochastic differential $\mathrm{d}X_t$ and simplifies the result by treating all differentials like real numbers or deterministic differentials. The partial derivative with respect to t and the linear terms in $\mathrm{d}t$ and $\mathrm{d}W_t$ resulting from the first-order term in $\mathrm{d}X_t$ are retained in (3.38). Of the quadratic terms resulting from $(\mathrm{d}X_t)^2$, only the square of $B_t\,\mathrm{d}W_t$ is kept, and $(\mathrm{d}W_t)^2$ is replaced by the first-order differential $\mathrm{d}t$, which is exactly the mean-square increment of the Wiener process in a time step of length $\mathrm{d}t$ (the other terms resulting from $(\mathrm{d}X_t)^2$ are at least of order $(\mathrm{d}t)^{3/2}$ and may hence be neglected).

Itô's formula (3.38) is the fundamental differentiation rule of stochastic (Itô's) calculus; it is most important for evaluating nontrivial stochastic differentials. As long as $B_t = 0$, we recover the familiar chain rule of deterministic calculus. When stochastic differentials are included ($B_t \neq 0$), there is an additional term involving second-order partial derivatives for which there is no counterpart in deterministic calculus. The unusual differentiation rule (3.38) is the source of all problems resulting from a naive approach to stochastic calculus!

Example 3.17 Rerun of a Classic

Consider the Wiener process W, for which the stochastic differential is $\mathrm{d}W_t$. For $g(t,x) = x^2$, Itô's formula gives

$$\mathrm{d}W_t^2 = 2\,W_t\,\mathrm{d}W_t + \mathrm{d}t\,. \tag{3.40}$$

Equation (3.38) thus allows a very simple rederivation of the result (3.26) of Example 3.5 in the corresponding differential version. No discretization or limiting processes are required when the Itô formula is employed. □

Example 3.18 A Stochastic Exponential
If we apply Itô's formula in order to evaluate the stochastic differential of $Y_t = \exp\{W_t - (t/2)\}$ in terms of the known differential of $X = W$, we obtain for $g(t,x) = \exp\{x - (t/2)\}$

$$\mathrm{d}Y_t = -\frac{1}{2}\,Y_t\,\mathrm{d}t + Y_t\,\mathrm{d}W_t + \frac{1}{2}\,Y_t\,\mathrm{d}t = Y_t\,\mathrm{d}W_t\,. \tag{3.41}$$

Thus, the explicit time dependence of Y_t compensates for the effect of the second-order derivative term in Itô's formula. The stochastic differential equation $\mathrm{d}Y_t = Y_t\,\mathrm{d}W_t$ (or, heuristically, $\mathrm{d}Y_t/\mathrm{d}W_t = Y_t$) is hence not solved by the simple exponential $\exp\{W_t\}$, which one would expect according to the familiar rules of ordinary calculus, but rather by the explicitly time-dependent Itô process $Y_t = \exp\{W_t - (t/2)\}$. This is the severe warning that, even in solving stochastic differential equations which are linear in Y_t, one cannot naively employ the rules of ordinary calculus. □

Exercise 3.19 Evaluate the stochastic differential of $\frac{1}{3}\,W_t^3 - tW_t$ by means of Itô's formula, and verify the result of Exercise 3.7.

In view of the applications in the subsequent chapters (polymer systems involving many degrees of freedom), it is very useful to generalize all the above results to multivariable systems. We introduce the stochastic differential of a process $(\boldsymbol{X}_t)_{t\in\mathbb{T}}$ with values in $\mathbb{R}^d$ by

$$\mathrm{d}\boldsymbol{X}_t = \boldsymbol{A}_t\,\mathrm{d}t + \mathbf{B}_t \cdot \mathrm{d}\boldsymbol{W}_t\,, \tag{3.42}$$

where the d' components of the d'-dimensional Wiener process $(\boldsymbol{W}_t)_{t\in\mathbb{T}}$ are independent one-dimensional Wiener processes (see end of Sect. 2.3.2), and the components of the d-dimensional column vector $(\boldsymbol{A}_t)_{t\in\mathbb{T}}$ and the $d \times d'$-matrix $(\mathbf{B}_t)_{t\in\mathbb{T}}$ are nonanticipating processes (all component processes of $\boldsymbol{A}$, $\mathbf{B}$, $\boldsymbol{W}$, and $\boldsymbol{X}$ are defined on the same probability space $(\Omega, \mathcal{A}, P)$). All the stochastic integrals implied by the matrix notation in (3.42) can be reduced to one-dimensional stochastic integrals for the component processes.

Following McKean [12], for a sufficiently smooth function $\boldsymbol{g}(t,\boldsymbol{x}) : \mathbb{T}\times\mathbb{R}^d \to \mathbb{R}^{d''}$ we can then generalize Itô's formula (3.38) by simply stating that in the formal second-order Taylor expansion

$$\begin{aligned}\mathrm{d}\boldsymbol{g}(t,\boldsymbol{X}_t) = \frac{\partial \boldsymbol{g}}{\partial t}(t,\boldsymbol{X}_t)\,\mathrm{d}t \;&+\; \sum_{j=1}^{d} \frac{\partial \boldsymbol{g}}{\partial x_j}(t,\boldsymbol{X}_t)\,\mathrm{d}X_j(t) \\ &+\; \frac{1}{2}\sum_{j,k=1}^{d} \frac{\partial^2 \boldsymbol{g}}{\partial x_j \partial x_k}(t,\boldsymbol{X}_t)\,\mathrm{d}X_j(t)\,\mathrm{d}X_k(t)\,,\end{aligned} \tag{3.43}$$

the products $\mathrm{d}X_j(t)\,\mathrm{d}X_k(t)$ have to be developed to terms involving $(\mathrm{d}t)^2$, $\mathrm{d}t\,\mathrm{d}W_j(t)$, $\mathrm{d}W_j(t)\,\mathrm{d}W_k(t)$, and then to be reduced according to Table 3.1 in order to obtain the correct stochastic differential of the transformed process.

Table 3.1. Multiplication table for reducing products of stochastic differentials in the multivariate version of Itô's formula

$\times$	dW_1	dW_2	$\cdots$	$dW_{d'}$	dt
dW_1	dt	0	$\cdots$	0	0
dW_2	0	dt	$\cdots$	0	0
$\vdots$	$\vdots$	$\vdots$	$\ddots$	$\vdots$	$\vdots$
$dW_{d'}$	0	0	$\cdots$	dt	0
dt	0	0	$\cdots$	0	0

For reasons of clarity, the time arguments of the components of $\boldsymbol{X}_t$ in (3.43) are given in parentheses rather than as subscripts; in Table 3.1, the time argument t is suppressed in all the differentials $dW_j(t)$.

For the product of two real-valued Itô processes, (3.43) for $d = 2$, $d'' = 1$, and $\boldsymbol{g}(t, x_1, x_2) = x_1 x_2$ implies the product rule

$$d[X_1(t)\, X_2(t)] = X_2(t)\, dX_1(t) + X_1(t)\, dX_2(t) + dX_1(t)\, dX_2(t)\,. \tag{3.44}$$

There are two interesting special cases for stochastic differentials of the form $dX_1(t) = A_1(t)\, dt + B_1(t)\, dW(t)$ and $dX_2(t) = A_2(t)\, dt + B_2(t)\, dW'(t)$. If the Wiener processes W and W' are the independent components of a two-dimensional Wiener process (this situation corresponds to a diagonal 2×2-matrix $\mathbf{B}$ in (3.42)), one recovers the familiar product rule

$$d[X_1(t)\, X_2(t)] = X_2(t)\, dX_1(t) + X_1(t)\, dX_2(t)\,. \tag{3.45}$$

If $W' = W$ (this situation corresponds to a 2×1-matrix $\mathbf{B}$ in (3.42)), one has the unusual product rule

$$d[X_1(t)\, X_2(t)] = X_2(t)\, dX_1(t) + X_1(t)\, dX_2(t) + B_1(t)\, B_2(t)\, dt\,. \tag{3.46}$$

Note that the familiar product rule (3.45) is also obtained whenever the stochastic differential of at least one of the factors contains only the dt-contribution, because only products of two differentials $dW_j(t)$ can lead to unusual terms. In other words, if at least one of the factors in a product of two Itô processes has a deterministic differential, then the familiar product rule is applicable.

In concluding this section on stochastic integration, several generalizations of stochastic integrals considered in the mathematical literature should be mentioned briefly. For example, instead of the Wiener process, more general stochastic processes have been used as integrators. Using the Itô processes introduced in Definition 3.15 as integrators requires only minor generalizations of the definitions given in this section, whereas use of more general square integrable martingales or continuous local martingales introduces considerable complications into the corresponding theory of stochastic integration [7, 11]. The occurrence of martingales, which are a generalization of processes with independent increments, as integrators in more general stochastic integrals is quite natural because the integrals inherit the martingale property from the integrators (see Example 3.11). For all generalizations, it is important to find the most general class of processes that may be used as integrands in order to retain a proper definition of stochastic integrals. In characterizing the most general classes of both integrands and integrators semimartingales play an important role; roughly speaking, a semimartingale is the sum of a martingale and a process with trajectories of bounded variation in every finite time interval (such trajectories can be treated with the tools of ordinary calculus). If the classes of integrals and integrators can be made to coincide, a particularly elegant formulation of stochastic calculus is possible (for example, allowing for very general integrations by parts).

Exercise 3.20 Let $\boldsymbol{W}$ be a d'-dimensional Wiener process. Evaluate the stochastic differential of $\boldsymbol{W}^2 = \boldsymbol{W} \cdot \boldsymbol{W}$.

3.3 Stochastic Differential Equations

Our original motivation for considering stochastic calculus was the desire to study equations of motion in which there occur random forces with very short correlation times. After defining stochastic integrals and differentials in a rigorous manner, we are now in a position to develop the theory of stochastic differential equations. For simplicity, some results of this section, in particular of the first subsection, are formulated for scalar stochastic differential equations. It is understood that, with certain obvious modifications, the results extend to systems of scalar equations. We consider only equations containing first-order differentials; higher-order differential equations should be reduced to first-order equations by introducing derivatives as additional variables. As in the previous section, we assume that the Wiener process $W = (W_t)_{t \in \mathbb{T}}$ on a probability space $(\Omega, \mathcal{A}, P)$ is given where the range of time $\mathbb{T}$ is henceforth $[0, t_{\max}]$.

We first introduce the basic notions and we briefly summarize some fundamental results concerning the existence and uniqueness of solutions of stochastic differential equations. Linear stochastic differential equations are the most important class of equations that can be solved in closed form, and we develop the procedures for obtaining their solutions. In the general case, we discuss the relationship between stochastic differential equations for stochastic processes and Fokker-Planck equations or diffusion equations for their transition probability densities; this relationship between stochastic ordinary differential equations and deterministic partial differential equations is the corner-stone for the second part of this book. A generalization of this relationship to stochastic differential equations with mean field interactions (that is, coefficient functions depending on expectations) is described briefly. The relationship between stochastic differential equations and Fokker-Planck equations is supplemented with a discussion of corresponding boundary conditions. In a final subsection, we discuss the consequences of using an alternative approach to stochastic calculus, namely Stratonovich's approach instead of Itô's approach.

3.3.1 Definitions and Basic Theorems

In this section, we are concerned with stochastic differential equations of the form

$$\mathrm{d}X_t = A(t, X_t)\,\mathrm{d}t + B(t, X_t)\,\mathrm{d}W_t\,. \tag{3.47}$$

In solving the stochastic differential equation (3.47) we look for an Itô process with a stochastic differential given by the right side of (3.47), where the coefficient processes A and B in the differential depend on the solution X of the stochastic differential equation. The solution X and the Wiener process W are defined on the same given probability space $(\Omega, \mathcal{A}, P)$; in the mathematical literature, such a solution is also referred to as a *strong solution*. A strong solution is called *unique* if, and only if, for any two solutions almost all trajectories are identical (that is, any two solutions are indistinguishable). Since every Itô process is nonanticipating, it is sufficient to assume that $A(t,x)$ and $B(t,x)$ are measurable real-valued functions on $\mathbb{T} \times \mathbb{R}$ in order to guarantee that $A(t, X_t)$ and $B(t, X_t)$ are nonanticipating processes. (As a reminder: $\mathbb{R}$ and the subset $\mathbb{T} \times \mathbb{R}$ of $\mathbb{R}^2$ are always considered together with the corresponding Borel σ-algebras.)

For $B = 0$, (3.47) is a deterministic ordinary differential equation. The systematic or deterministic term $A(t, X_t)\,\mathrm{d}t$ in (3.47) is usually referred to as the *drift term*. Because of the interpretation of the Wiener process as a model of Brownian motion or diffusion (cf. Example 3.11), $B(t, X_t)\,\mathrm{d}W_t$ is usually referred to as the *diffusion term*. Depending on whether the prefactor B in the noise term $B(t, X_t)\,\mathrm{d}W_t$ depends on the configuration X_t or not, one distinguishes between *multiplicative* and *additive noise*. This distinction is helpful because in the case of additive noise the theory of stochastic differential equations is considerably less subtle.

We next collect some fundamental results concerning the existence and uniqueness of solutions to stochastic differential equations.

Theorem 3.21 *Suppose that the measurable functions A and B for all $t \in \mathbb{T}$ and for all $x, y \in \mathbb{R}$ satisfy the* Lipschitz conditions

$$|A(t,x) - A(t,y)| \;\leq\; c|x-y|\,, \tag{3.48}$$

$$|B(t,x) - B(t,y)| \;\leq\; c|x-y|\,, \tag{3.49}$$

and the linear growth conditions

$$|A(t,x)| \leq c(1+|x|)\,, \tag{3.50}$$
$$|B(t,x)| \leq c(1+|x|)\,, \tag{3.51}$$

for some constant $c \in \mathbb{R}$. If the real-valued random variable X_0 is independent of the Wiener process W, there exists a unique solution of the stochastic differential equation (3.47) with initial condition X_0 on the entire time interval $\mathbb{T}$.

As in the case of deterministic differential equations, the above theorem can be proved by considering a sequence of successive approximations which converge to the unique solution (Picard iteration proof). A rigorous proof takes of the order of five pages [5–7]. The conclusion of Theorem 3.21 is valid even when the linear growth conditions are removed (see Sect. 3.1 of [5]). Keeping the linear growth conditions in Theorem 3.21 has the advantage that the Lipschitz conditions can be replaced by weaker (local Lipschitz) conditions which, for example, hold whenever the derivatives of $A(t,x)$ and $B(t,x)$ with respect to x are continuous. If A and B are general continuous functions of t and x, existence of a solution can still be established, however, the solution is in general not unique (see Sect. 3.1 of [5]).

A standard result usually given in connection with Theorem 3.21 guarantees that all moments of the solution X_t stay finite on $\mathbb{T}$ when the moments of X_0 are finite (the moments can grow at most exponentially in time) [5, 7]. A number of very general results concerning the boundedness and stability of solutions to stochastic differential equations has been discussed by Gard (see Chap. 5 of [5]). Such results can be derived by adapting the qualitative theory of deterministic differential equations (Gard follows the Lyapunov-type approach).

Are the basic existence and uniqueness results summarized above of any use in practical applications, say in the kinetic theory applications considered in the second part of this book? Surprisingly, the answer is positive. All coefficient functions, A and B, considered in the applications are continuous, so that the existence of solutions is never a problem. Furthermore, the growth conditions guaranteeing uniqueness of the solutions are not only fulfilled for the important class of linear stochastic differential equations but also for certain nonlinear equations such as those resulting from hydrodynamic interactions in dilute polymer solutions.

The basic existence and uniqueness theorem (Theorem 3.21) is easily generalized to the case of the multicomponent stochastic differential equation

$$\mathrm{d}\boldsymbol{X}_t = \boldsymbol{A}(t,\boldsymbol{X}_t)\,\mathrm{d}t + \mathbf{B}(t,\boldsymbol{X}_t)\cdot\mathrm{d}\boldsymbol{W}_t\,. \tag{3.52}$$

The components of the d-dimensional column vector $\boldsymbol{A}$ and of the $d \times d'$-matrix $\mathbf{B}$ are assumed to be measurable functions on $\mathbb{T} \times \mathbb{R}^d$, and the processes $\boldsymbol{X}$ and $\boldsymbol{W}$ were previously introduced in connection with (3.42). It is sufficient to assume that each component of $\boldsymbol{A}$ and $\mathbf{B}$ satisfies the Lipschitz and growth conditions with $|\boldsymbol{x}-\boldsymbol{y}|$ and $|\boldsymbol{x}|$ now representing the length of a d-component column vector.

The components of the process $\boldsymbol{X}$ considered in (3.52) are labelled by a discrete parameter taking the values $1 \ldots d$. One can also consider the problem in which the components of $\boldsymbol{X}$ are labelled by a continuous parameter $y \in \mathbb{D}$, where $\mathbb{D}$ is a subset of $\mathbb{R}$ (or some higher dimensional continuous space). The values of the random variable $\boldsymbol{X}_t$ at any time t are thus functions defined on the domain $\mathbb{D}$ of the label y, and they are hence also referred to as time-dependent random fields. This generalization causes considerable mathematical difficulties; for example, if each component of $\boldsymbol{X}_t$ is modified on a null set N_y depending on the component y (say, in order to have trajectories that are all continuous in t) the uncountable union of the null sets for all components of $\boldsymbol{X}_t$, that is, $\bigcup_{y\in\mathbb{D}} N_y$, may not be a null set. It is hence very subtle to construct modifications of time-dependent random fields for which all components have certain properties (like continuity in t for all $\omega \in \Omega$ and $y \in \mathbb{D}$). It is even more subtle to establish the existence of versions of time-dependent random fields for which all $\boldsymbol{X}_t(\omega)$ are continuous or differentiable functions of the continuous label y. A rigorous theory of *stochastic partial differential equations*, for which the time evolution of the random field $\boldsymbol{X}_t$ may also depend on the (so-called spatial) derivative(s) of $\boldsymbol{X}_t$ with respect to the continuous variable y labelling the components of $\boldsymbol{X}_t$, can be found in the textbook by H. Kunita [13]. A more elementary approach to stochastic partial differential equations is developed in Chap. 8 of Gardiner's book [8]. The prototype of a stochastic partial differential equation arises in fluctuating hydrodynamics where, at each position in space, a suitable Gaussian random noise is added to the Navier-Stokes equation for the velocity field of an incompressible Newtonian fluid. Such stochastic partial differential equations are beyond the scope of this book.

As a further generalization of Theorem 3.21 for processes with a finite number of components, the existence of a unique solution can be established when the processes $\boldsymbol{A}$ and $\mathbf{B}$ in (3.52) are functionals of the entire history of the process $\boldsymbol{X}$ up to the time t and, possibly, also of the history of a further nonanticipating process modeling external random perturbations. The stochastic analog of integrodifferential equations falls into this category of generalizations. On the right sides of the Lipschitz and growth conditions, the lengths $|\boldsymbol{x} - \boldsymbol{y}|$ and $|\boldsymbol{x}|$ are then to be replaced by the suprema of $|\boldsymbol{x}_t - \boldsymbol{y}_t|$ and $|\boldsymbol{x}_t|$ taken for the entire trajectories $\boldsymbol{x}_t$ and $\boldsymbol{y}_t$ to be inserted into the functionals $\boldsymbol{A}$ and $\mathbf{B}$ (furthermore, some additional growth of $\boldsymbol{A}$ and $\mathbf{B}$ is allowed for growing external perturbations). The rigorous theory for this more general situation can be found in Sect. 5.1 of the book by G. Kallianpur [11].

3.3.2 Linear Stochastic Differential Equations

In the theory of deterministic differential equations, linear equations are particularly simple and can, in most cases, be solved in a closed form. The situation is fully analogous for stochastic differential equations. Moreover, linear equations are practically the only interesting stochastic differential equations that can be solved analytically. This section may thus be considered as a guide to distinguish "simple" (i.e., solvable) stochastic differential equations from "difficult" (i.e., unsolvable) ones. After reading this section one should be able not only to recognize and classify "simple" stochastic differential equations, but also to solve them. In the second part of the book, the theory developed in this section is applied to solving the Rouse and Zimm models of polymers in dilute solution.

We here consider the most general case of a system of equations (3.52) where both $\boldsymbol{A}$ and $\mathbf{B}$ are linear in $\boldsymbol{X}$,

$$\mathrm{d}\boldsymbol{X}_t = [\boldsymbol{a}(t) + \mathbf{A}(t)\cdot\boldsymbol{X}_t]\,\mathrm{d}t + \mathbf{B}(t)\cdot\mathrm{d}\boldsymbol{W}_t + \sum_{j=1}^{d'}\mathbf{B}_j(t)\cdot\boldsymbol{X}_t\,\mathrm{d}W_j(t)\,; \qquad (3.53)$$

in this equation, $\boldsymbol{a}(t)$ is a time-dependent d-dimensional column vector, $\mathbf{A}(t)$ and $\mathbf{B}_j(t)$ for $j = 1, 2, \ldots d'$ are time-dependent $d \times d$-matrices, and $\mathbf{B}(t)$ is a time-dependent $d \times d'$-matrix. As long as the components of $\boldsymbol{a}(t)$, $\mathbf{A}(t)$, $\mathbf{B}(t)$, and $\mathbf{B}_j(t)$ are bounded over the time range of interest, $\mathbb{T}$, the Lipschitz and linear growth conditions can easily be verified for the general linear stochastic differential equation (3.53). Theorem 3.21 thus guarantees the existence of a unique solution of (3.53).

We first consider the simpler case of *narrow-sense* linear equations for which, by definition, the matrices $\mathbf{B}_j(t)$ vanish identically,

$$\mathrm{d}\boldsymbol{X}_t = [\boldsymbol{a}(t) + \mathbf{A}(t)\cdot\boldsymbol{X}_t]\,\mathrm{d}t + \mathbf{B}(t)\cdot\mathrm{d}\boldsymbol{W}_t\,. \qquad (3.54)$$

Then, the noise is additive whereas, in general, the noise in (3.53) is multiplicative. Narrow-sense linear stochastic differential equations can be solved by the following procedure: one formally divides (3.54) by $\mathrm{d}t$ and solves the resulting equation like a deterministic linear ordinary differential equation with an inhomogeneous term $\boldsymbol{a}(t) + \mathbf{B}(t)\cdot\mathrm{d}\boldsymbol{W}_t/\mathrm{d}t$ (cf. the naive approach to stochastic differential equations in Sect. 3.1.1). In other words, one first solves the homogeneous narrow-sense linear equation (in terms of time-ordered exponentials) and then employs the method of variation of constants for solving the inhomogeneous equation. The time integral of $\mathrm{d}\boldsymbol{W}_t/\mathrm{d}t$ occurring in the resulting formal solution is then interpreted as a stochastic integral. The following theorem states that the explicit solution obtained by this heuristic procedure is correct.

Theorem 3.22 *The solution of the system of narrow-sense linear stochastic differential equations (3.54) can be written as*

$$\boldsymbol{X}_t = \boldsymbol{\Phi}_t\cdot\left[\boldsymbol{X}_0 + \int_0^t \boldsymbol{\Phi}_{t'}^{-1}\cdot\boldsymbol{a}(t')\,\mathrm{d}t' + \int_0^t \boldsymbol{\Phi}_{t'}^{-1}\cdot\mathbf{B}(t')\cdot\mathrm{d}\boldsymbol{W}_{t'}\right], \qquad (3.55)$$

where the (nonsingular) time-dependent $d \times d$-matrix $\boldsymbol{\Phi}_t$ is given by

$$\begin{aligned}\boldsymbol{\Phi}_t &= \mathcal{T}\text{-}\exp\left\{\int_0^t \mathbf{A}(t')\,\mathrm{d}t'\right\} \qquad (3.56)\\ &:= \boldsymbol{\delta} + \sum_{n=1}^{\infty}\int_0^t \mathrm{d}t_1\int_0^{t_1}\mathrm{d}t_2\cdots\int_0^{t_{n-1}}\mathrm{d}t_n\,\mathbf{A}(t_1)\cdot\mathbf{A}(t_2)\cdots\mathbf{A}(t_n)\,.\end{aligned}$$

Each of the column vectors of the *fundamental matrix* $\boldsymbol{\Phi}_t$, defined as a time-ordered exponential in (3.56) (note that $t_1 \geq t_2 \geq \ldots \geq t_n$ for the nth term in the expansion of the exponential), is a solution of the deterministic homogeneous equation associated with (3.54), where each initial condition has only one nonvanishing component.

Exercise 3.23 Derive the expression

$$\boldsymbol{\Phi}_t \cdot \boldsymbol{\Phi}_{t'}^{-1} = \mathcal{T}\text{-}\exp\left\{\int_{t'}^{t} \mathbf{A}(t'')\,\mathrm{d}t''\right\}.$$

Exercise 3.24 Prove Theorem 3.22.

From the explicit expression (3.55) and Example 3.6 it is clear that the solution of a narrow-sense linear stochastic differential equation is Gaussian provided, of course, that the initial random variable $\boldsymbol{X}_0$ is Gaussian. In order to completely determine the distribution of the process $\boldsymbol{X}$ it is hence sufficient to give the expectations and covariances of $\boldsymbol{X}$, which can be obtained directly from the explicit solution (3.55),

$$\boxed{\langle \boldsymbol{X}_t \rangle = \boldsymbol{\Phi}_t \cdot \left[\langle \boldsymbol{X}_0 \rangle + \int_0^t \boldsymbol{\Phi}_{t'}^{-1} \cdot \boldsymbol{a}(t')\,\mathrm{d}t'\right],} \tag{3.57}$$

$$\boxed{\begin{aligned}\langle \boldsymbol{X}_t \boldsymbol{X}_{t'} \rangle - \langle \boldsymbol{X}_t \rangle \langle \boldsymbol{X}_{t'} \rangle &= \boldsymbol{\Phi}_t \cdot \Bigg[\langle \boldsymbol{X}_0 \boldsymbol{X}_0 \rangle - \langle \boldsymbol{X}_0 \rangle \langle \boldsymbol{X}_0 \rangle \\ &+ \int_0^{\min(t,t')} \boldsymbol{\Phi}_{t''}^{-1} \cdot \mathbf{B}(t'') \cdot \mathbf{B}^T(t'') \cdot \boldsymbol{\Phi}_{t''}^{-1\,T}\,\mathrm{d}t''\Bigg] \cdot \boldsymbol{\Phi}_{t'}^T .\end{aligned}} \tag{3.58}$$

In deriving the covariances (3.58), a straightforward multivariable generalization of (3.32) has been used, and the initial condition and Wiener process have been assumed to be independent. Application of these expressions requires only the calculation of ordinary integrals in order to obtain a complete characterization of the Gaussian solutions of narrow-sense linear stochastic differential equations.

Example 3.25 Checking Our Motivation

The example discussed as a motivation of stochastic differential equations and stochastic calculus in Sect. 3.1 can now be analyzed in the framework of the rigorous theory. The stochastic differential equation (3.1) for the Ornstein-Uhlenbeck process $(V_t)_{t\in\mathbb{T}}$ which models the velocity of a Brownian particle can, according to (3.11), be written as

$$dV_t = -\frac{\zeta}{M} V_t \, dt + \frac{\sqrt{2k_B T \zeta}}{M} \, dW_t . \qquad (3.59)$$

Since this equation is of the narrow-sense linear type, the naive solution obtained by following the procedures of deterministic calculus in Sect. 3.1 is correct. The function Φ_t occurring in the explicit solution (3.55) of the single-variable equation (3.59) is given by $\Phi_t = e^{-\zeta t/M}$. From (3.57) and (3.58) we see that for $\langle V_0 \rangle = 0$ the Gaussian process $(V_t)_{t \in \mathbb{T}}$ is characterized by the expectation $\langle V_t \rangle = 0$ and the covariances

$$\langle V_t V_{t'} \rangle = \langle V_0^2 \rangle e^{-\zeta(t+t')/M} + \frac{k_B T}{M} \left(e^{-\zeta |t-t'|/M} - e^{-\zeta(t+t')/M} \right) . \qquad (3.60)$$

The special case of this result for $t = t'$ and $\langle V_0^2 \rangle = 0$ was previously obtained from the naive approach (see (3.6)). Also the more general result (3.60) can be derived easily by the heuristic arguments of Sect. 3.1 (see Exercise 3.1). □

Exercise 3.26 Consider the process V_t of Example 3.25, and define X_t as the solution of $dX_t = V_t \, dt$ with $X_0 = 0$. Derive a stochastic differential equation for $Y_t = 2X_t V_t$, and show that

$$\frac{d}{dt} \langle Y_t \rangle = -\frac{\zeta}{M} \langle Y_t \rangle + 2 \langle V_t^2 \rangle .$$

Remark: In this exercise, we follow in P. Langevin's footsteps [3]. We obtain the stationary result $\langle Y_t \rangle = (2M/\zeta) \langle V_t^2 \rangle = 2k_B T/\zeta$ and, from $Y_t = dX_t^2/dt$, we find the diffusion coefficient $D = k_B T/\zeta$.

For the most general linear stochastic differential equations (3.53), the situation is more complicated than for the narrow-sense linear equation (3.54). The fundamental matrix $\mathbf{\Phi}_t$ is now to be determined from the homogeneous *stochastic* differential equation

$$d\mathbf{\Phi}_t = \mathbf{A}(t) \cdot \mathbf{\Phi}_t \, dt + \sum_{j=1}^{d'} \mathbf{B}_j(t) \cdot \mathbf{\Phi}_t \, dW_j(t) , \qquad (3.61)$$

with the initial condition $\mathbf{\Phi}_0 = \boldsymbol{\delta}$. For $\mathbf{B}_j(t) \neq 0$, the components of the fundamental matrix $\mathbf{\Phi}_t$ are thus Itô processes. One now has to be very careful both with solving the homogeneous equation (3.61) (cf. Example 3.18) and with the variation of constants – both steps are affected by the peculiarities of stochastic calculus! Formally, the solution of (3.61) can still be written as a time-ordered exponential,

$$\mathbf{\Phi}_t = \mathcal{T}\text{-}\exp\left\{ \int_0^t \mathbf{A}(t') \, dt' + \int_0^t \sum_{j=1}^{d'} \mathbf{B}_j(t') \, dW_j(t') \right\} , \qquad (3.62)$$

where this time-ordered exponential is defined as the natural generalization of the expansion (3.56) in terms of nested integrals. The evaluation of time-ordered exponentials for non-commuting matrices is nontrivial even in the deterministic case; for stochastic integrals, the evaluation of time-ordered exponentials is extremely tedious (see Example 3.27 below, where the time-ordered exponential of the simplest stochastic integral is considered explicitly). When all the matrices $\mathbf{A}$ and $\mathbf{B}_j$ commute with each other at all times, a more explicit expression for the stochastic fundamental matrix $\mathbf{\Phi}_t$ can be given,

$$\mathbf{\Phi}_t = \exp\left\{\int_0^t \left[\mathbf{A}(t') - \frac{1}{2}\sum_{j=1}^{d'} \mathbf{B}_j(t')^2\right] dt' + \int_0^t \sum_{j=1}^{d'} \mathbf{B}_j(t')\, dW_j(t')\right\}. \tag{3.63}$$

The validity of (3.63) can be demonstrated by applying Itô's formula in order to evaluate $d\mathbf{\Phi}_t$ which is then found to satisfy the homogeneous stochastic differential equation (3.61), and by verifying the proper initial condition. The fact that, even when all matrices commute, the time-ordered exponential is not simply obtained by replacing "$\mathcal{T}$-exp" with "exp" is a consequence of stochastic calculus, as the following example illustrates.

Example 3.27 Time-Ordered Exponential of a Stochastic Integral
In solving the homogeneous linear stochastic differential equation $dY_t = Y_t\, dW_t$ of Example 3.18 the time-ordered exponential

$$\mathcal{T}\text{-}\exp\left\{\int_0^t dW_{t'}\right\} = 1 + \sum_{n=1}^{\infty} I_t^{(n)}, \tag{3.64}$$

with

$$I_t^{(n)} := \int_0^t \left\{\int_0^{t_1}\left[\cdots \int_0^{t_{n-2}}\left(\int_0^{t_{n-1}} dW_{t_n}\right) dW_{t_{n-1}} \cdots\right] dW_{t_2}\right\} dW_{t_1}, \tag{3.65}$$

has to be evaluated. Since we here consider a one-dimensional Wiener process W, no problems with rearranging non-commuting factors in the n-fold integrals occur. However, for stochastic integrals, the n-fold nested integrals in (3.65) are not equal to $(1/n!)$ times the nth power of the full integral from 0 to t; this is the source of the additional term in the exponential in (3.63) when carrying out the time-ordering requested in (3.62) for the case of commuting matrices. The problem may heuristically be thought of as arising from nonvanishing contributions to stochastic integrals for equal time arguments because the product $dW_{t_j}\, dW_{t_k}$ may be replaced with the first-order differential dt_j on the hypersurface with $t_j = t_k$. In a rigorous approach, the iterated stochastic integrals $I_t^{(n)}$ can be evaluated by means of the recursion relation

$$I_t^{(n)} = \int_0^t I_{t'}^{(n-1)}\, dW_{t'}, \qquad I_t^{(0)} = 1. \tag{3.66}$$

Such iterated stochastic integrals are of great importance because they occur, among other stochastic integrals, in the systematic derivation of higher-order numerical integration schemes for stochastic differential equations (see Sect. 3.4.2 below).

The explicit expressions for the first few $I_t^{(n)}$ are

$$I_t^{(1)} = W_t\,, \qquad I_t^{(2)} = \frac{1}{2}\,W_t^2 - \frac{1}{2}\,t\,, \qquad I_t^{(3)} = \frac{1}{6}\,W_t^3 - \frac{1}{2}\,t\,W_t\,. \tag{3.67}$$

The integral $I_t^{(1)}$ is trivial, $I_t^{(2)}$ was derived in Example 3.5, and the expression for $I_t^{(3)}$ was established in Exercise 3.7 (see also Exercise 3.19). As a consequence of the peculiarities of stochastic calculus, the stochastic integrals $I_t^{(n)}$ are not given by $W_t^n/n!$. The integral $I_t^{(n)}$ is of nth order in W_t where, formally, a factor of t has to be considered to be of the order of W_t^2. The book-keeping of the coefficients can be handled in a very elegant manner. After introducing Hermite polynomials by the recursion relation

$$H_{n+1}(t,x) = x\,H_n(t,x) - \frac{nt}{2}\,H_{n-1}(t,x)\,, \tag{3.68}$$

together with $H_0(t,x) = 1$, $H_1(t,x) = x$ (a more standard form of Hermite polynomials is obtained as $2^n\,H_n(1,x)$), we may write the stochastic integrals $I_t^{(n)}$ as (see Sect. 2.7 of [12])

$$I_t^{(n)} = \frac{1}{n!}\,H_n(2t, W(t))\,. \tag{3.69}$$

The result (3.69) can be verified by means of the Itô formula (3.38) and the properties $\partial H_n(t,x)/\partial x = n\,H_{n-1}(t,x)$ and $\partial^2 H_n(t,x)/\partial x^2 = -4\,\partial H_n(t,x)/\partial t$ which follow from the recursion relation (3.68).

With the result (3.69), the time-ordered differential in (3.64) can be summed by means of the generating function for Hermite polynomials (see Sect. 2.7 of [12]),

$$\mathcal{T}\text{-}\exp\left\{\int_0^t \mathrm{d}W_{t'}\right\} = \exp\left\{W_t - \frac{1}{2}\,t\right\}. \tag{3.70}$$

The reader should compare the lowest-order terms on the right side of (3.70) to the sum of the iterated integrals listed in (3.67). This example clearly illustrates how useful (3.63) is for commuting matrices. □

After discussing the solution of the general system of homogeneous linear equations, we are now in a position to give the solution of the general inhomogeneous stochastic differential equation (3.53).

Theorem 3.28 *The solution of the most general system of linear stochastic differential equations (3.53) can be written as*

$$\begin{aligned}\boldsymbol{X}_t = \boldsymbol{\Phi}_t \cdot \Bigg\{ \boldsymbol{X}_0 \; &+ \int_0^t \boldsymbol{\Phi}_{t'}^{-1} \cdot \left[\boldsymbol{a}(t') - \sum_{j=1}^{d'} \mathbf{B}_j(t') \cdot \boldsymbol{b}_j(t')\right] \mathrm{d}t' \\ &+ \int_0^t \boldsymbol{\Phi}_{t'}^{-1} \cdot \mathbf{B}(t') \cdot \mathrm{d}\boldsymbol{W}_{t'} \Bigg\}\,,\end{aligned} \tag{3.71}$$

where $\mathbf{b}_j(t')$ *is the* jth *column vector of the matrix* $\mathbf{B}(t')$, *and* $\mathbf{\Phi}_t$ *is the fundamental matrix (3.62).*

The solution given in this theorem can be verified in a straightforward manner by applying Itô's formula.

Exercise 3.29 Find the explicit solution of the most general single-variable linear stochastic differential equation,

$$\mathrm{d}X_t = [a(t) + A(t)\,X_t]\,\mathrm{d}t + [B(t) + B_1(t)\,X_t]\,\mathrm{d}W_t\,. \tag{3.72}$$

Determine the process Y_t with $\mathrm{d}Y_t = Y_t\,\mathrm{d}W_t$ previously considered in Examples 3.18 and 3.27.

Readers who are interested in making money from what they learned about the theory of stochastic integration and linear stochastic differential equations are referred to an example from the financial market discussed in Sect. 10.5 of [14]. For the less lucrative examples from polymer kinetic theory considered in the subsequent chapters we need to introduce several further ideas.

3.3.3 Fokker-Planck Equations

The time evolution of the solution X_t of the stochastic differential equation (3.47) is given in terms of X_t and of the independent increment $\mathrm{d}W_t$ of the Wiener process. Except for X_t, no further information about the history of the process X is required. We hence expect the process X to be Markovian. Furthermore, the transition probabilities $P(X_t \in B | X_{t_0} = x_0)$ for $t > t_0$ should be obtainable as the probability distribution of the random variable

$$X_t = x_0 + \int_{t_0}^{t} A(t', X_{t'})\,\mathrm{d}t' + \int_{t_0}^{t} B(t', X_{t'})\,\mathrm{d}W_{t'}\,, \tag{3.73}$$

which is the solution of the stochastic differential equation (3.47) with initial condition $X_{t_0} = x_0$. From (3.73) and the Itô formula (3.38), we obtain for every real function $g(x)$ with continuous second-order derivatives

$$\begin{aligned} g(X_t) = g(x_0) \quad &+ \int_{t_0}^{t} \left[A(t', X_{t'})\,\frac{\partial g}{\partial x}(X_{t'}) + \frac{1}{2}\,B(t', X_{t'})^2\,\frac{\partial^2 g}{\partial x^2}(X_{t'}) \right] \mathrm{d}t' \\ &+ \int_{t_0}^{t} B(t', X_{t'})\,\frac{\partial g}{\partial x}(X_{t'})\,\mathrm{d}W_{t'}\,. \end{aligned} \tag{3.74}$$

After taking the average of this equation, letting t go to t_0, exploiting the fact that the Itô process X has continuous trajectories, and assuming that the functions A and B are continuous in both arguments, we obtain the following expression for the infinitesimal generator (2.88)

$$\mathcal{L}_t = A(t,x)\,\frac{\partial}{\partial x} + \frac{1}{2}\,B(t,x)^2\,\frac{\partial^2}{\partial x^2}\,. \tag{3.75}$$

For $A = 0$ and $B = 1$, X is the Wiener process, and we can verify (3.75) by comparing it to the previously derived result (2.96) for the Wiener process. Like the Wiener process, the solutions of stochastic differential equations generally have continuous probability distributions.

For the multivariable stochastic differential equation (3.52), the infinitesimal generator can be obtained by a completely analogous procedure,

$$\mathcal{L}_t = \boldsymbol{A}(t,\boldsymbol{x})\cdot\frac{\partial}{\partial \boldsymbol{x}} + \frac{1}{2}\,\mathbf{D}(t,\boldsymbol{x}) : \frac{\partial}{\partial \boldsymbol{x}}\,\frac{\partial}{\partial \boldsymbol{x}}\,, \tag{3.76}$$

where the so-called diffusion matrix,

$$\mathbf{D}(t,\boldsymbol{x}) = \mathbf{B}(t,\boldsymbol{x})\cdot\mathbf{B}^T(t,\boldsymbol{x})\,, \tag{3.77}$$

is a positive-semidefinite symmetric $d \times d$-matrix.

The above heuristic arguments motivate the following theorem which can be proved rigorously (see Sects. 9.2 and 9.4 of [6] or Sects. 5.4 and 5.5 of [11]).

Theorem 3.30 *If the functions* $\boldsymbol{A}$ *and* $\mathbf{B}$ *on* $\mathbb{T}\times\mathbb{R}^d$ *satisfy the Lipschitz and linear growth conditions (3.48)–(3.51), the unique solution of the multivariable stochastic differential equation*

$$\boxed{\mathrm{d}\boldsymbol{X}_t = \boldsymbol{A}(t,\boldsymbol{X}_t)\,\mathrm{d}t + \mathbf{B}(t,\boldsymbol{X}_t)\cdot\mathrm{d}\boldsymbol{W}_t\,,} \tag{3.78}$$

is a Markov process. If $\boldsymbol{A}$ *and* $\mathbf{B}$ *are continuous functions, the infinitesimal generator of this Markov process is given by the second-order differential operator (3.76), (3.77).*

The domain of the infinitesimal generator $\mathcal{L}_t$ consists of the real-valued functions of $\boldsymbol{x} \in \mathbb{R}^d$ which are continuous and bounded, together with their first- and second-order partial derivatives. If the functions $\boldsymbol{A}$ and $\mathbf{B}$ do not explicitly depend on time, the solution of (3.78) is a homogeneous Markov process.

After determining the operator adjoint to $\mathcal{L}_t$ through several integrations by parts, Kolmogorov's forward equation (2.92) implies that the evolution of the probability density $p(t,\boldsymbol{x})$, which characterizes the continuous distribution of $\boldsymbol{X}_t$, is governed by the partial differential equation (see Exercise 2.83)

$$\boxed{\frac{\partial}{\partial t}\,p(t,\boldsymbol{x}) = -\frac{\partial}{\partial \boldsymbol{x}}\cdot\left[\boldsymbol{A}(t,\boldsymbol{x})\,p(t,\boldsymbol{x})\right] + \frac{1}{2}\,\frac{\partial}{\partial \boldsymbol{x}}\,\frac{\partial}{\partial \boldsymbol{x}} : \left[\mathbf{D}(t,\boldsymbol{x})\,p(t,\boldsymbol{x})\right]\,.} \tag{3.79}$$

By using suitable initial conditions, all transition probability densities can be obtained from the differential equation (3.79), which is more commonly known as the Fokker-Planck equation associated with the solution of the stochastic differential equation (3.78).

The transition probability densities for the solution of the stochastic differential equation (3.78) are governed by the Fokker-Planck equation (3.79). If, in turn, a Fokker-Planck equation (3.79) with a symmetric positive-semidefinite diffusion matrix $\mathbf{D}$ is given, we can decompose the matrix $\mathbf{D}$ according to (3.77) in order to construct a stochastic differential equation that characterizes a stochastic process with transition probability densities equal to those obtained from the Fokker-Planck equation (at least for functions $\boldsymbol{A}$ and $\mathbf{B}$ satisfying the assumptions of Theorem 3.30, this equivalence is mathematically rigorous).

The decomposition of a positive-semidefinite symmetric matrix $\mathbf{D}$ in the form (3.77) is always possible but, in general, not unique. Since $\mathbf{D}$ is a symmetric matrix, the matrix equation (3.77) stands for $d(d+1)/2$ independent real equations for the dd' unknown entries of $\mathbf{B}$. For example, one can always choose $\mathbf{B}$ to be a symmetric $d \times d$-matrix (the square root of the positive-semidefinite matrix $\mathbf{D}$) or a lower triangular $d \times d$-matrix (the so-called Cholesky decomposition), the latter choice being particularly convenient for the successive numerical calculation of the $d(d+1)/2$ nonvanishing entries of $\mathbf{B}$. If $\mathbf{D}$ is a singular matrix of rank $d' < d$, one can find a $d \times d'$-matrix $\mathbf{B}$ with $\mathbf{D} = \mathbf{B} \cdot \mathbf{B}^T$ (this can be shown by diagonalizing $\mathbf{D}$); it is then sufficient to consider a Wiener process having d' independent components only. For different choices of $\mathbf{B}$, the trajectories corresponding to a given $\omega \in \Omega$ are different; however, all transition probabilities and hence all averages are identical (and this is all one is interested in when working with Fokker-Planck equations).

Even when a decomposition $\mathbf{D} = \mathbf{B} \cdot \mathbf{B}^T$ is specified, the correspondence between Fokker-Planck equations and stochastic differential equations is not one-to-one. A Fokker-Planck equation specifies only the distribution of the stochastic process, whereas a stochastic differential equation determines the actual trajectories. Given a Fokker-Planck equation for the transition probabilities one has the freedom of choosing a probability space, a Wiener process, and a random variable with the desired initial distribution before the trajectories of the process can be determined by solving the stochastic differential equation in the strong sense introduced in Sect. 3.3.1. When the choice of a suitable probability space, Wiener process, and initial condition is part of solving a given stochastic differential equation, we speak of a *weak solution* [11, 14]. When working with weak solutions only the distribution of the process is of interest, just like in the Fokker-Planck approach. In that sense, weak stochastic differential equations and Fokker-Planck equations are equivalent.

The equivalence between stochastic differential equations and Fokker-Planck equations expressed in Theorem 3.30 is *the corner-stone of this book*. This theorem establishes the link between the theory of stochastic processes on the one hand and polymer kinetic theory on the other. Fokker-Planck equations are ubiquitous in kinetic theory: in that context, the process $\boldsymbol{X}$ for which one constructs trajectories and the dummy variable $\boldsymbol{x}$ occurring in the probability density characterize the configurations of polymer molecules. The theory of stochastic processes and stochastic differential equations developed in the first part of this book helps us to understand the stochastic dynamics behind any Fokker-Planck equation of kinetic theory.

In most kinetic theory applications only certain expectations such as the stress tensor are of interest, and hence only the distribution of the solution of a stochastic differential equation matters. In other words, one is mainly interested in weak solutions of stochastic differential equations. However, one can also think of situations in which strong solutions might actually be very useful in polymer kinetic theory. For example, in order to understand the effect of a certain parameter or of an approximation on the stochastic dynamics of a system it can be very illuminating and much more efficient to compare particular trajectories on a given probability space rather than just distributions or averages. The illustration of the effect of mass in Fig. 3.1 and of the influence of hydrodynamic interaction in Fig. 4.9 below are examples of such situations.

The fact that in the literature the Fokker-Planck equation is also referred to as Kolmogorov's forward equation (in the theory of Markov processes), as the diffusion equation (e.g., in polymer kinetic theory), as the Klein-Kramers equation (in the theory of Brownian motion of particles in an external field), or as the Smoluchowski equation (in the theory of Brownian motion of particles in an external field when only position variables, but no momentum variables, are considered) underlines two important issues: (i) the Fokker-Planck equation arises independently in different fields of science and hence clearly represents a fundamental and crucial concept; (ii) the natural occurrence of stochastic differential equations has not generally been recognized as the common source of all the phenomena usually described by Fokker-Planck equations. An extensive discussion of the derivation of the Fokker-Planck equation, the methods of solving it, and some of its applications can be found in a book by H. Risken [15]. It is the purpose of the present book to illustrate the multitude of advantages gained by reformulating the diffusion equations of polymer kinetic theory in terms of stochastic differential equations. In the subsequent chapters, we show that it is much more intuitive and very illuminating to work directly with stochastic processes rather than with deterministic auxiliary equations for probability distributions. In particular, one can easily obtain computer simulation algorithms by developing numerical integration procedures for stochastic differential equations.

The different approaches to kinetic theory can be compared to different ways of finding an answer to the question "What is the proof of the pudding?" On the one hand, a diffusion equation is like a recipe—one knows all the ingredients, how to process them and, in principle, one can predict all the properties of the finished pudding. Integration of a stochastic differential equation, on the other hand, is like eating the pudding. It is the purpose of this book to sketch many appetizing pictures of delicious puddings and to show where to get the cheapest and the best pudding.

Exercise 3.31 Rederive the infinitesimal generator for the Gaussian process introduced in Exercise 2.81 (cf. Exercise 2.86).

Exercise 3.32 Find decompositions of

$$\mathbf{D} = \begin{pmatrix} 81 & 27 & 18 \\ 27 & 45 & 0 \\ 18 & 0 & 54 \end{pmatrix}$$

in the form $\mathbf{D} = \mathbf{B} \cdot \mathbf{B}^T$ with lower triangular and symmetric 3×3-matrices $\mathbf{B}$.

3.3.4 Mean Field Interactions

The Fokker-Planck equation corresponding to the stochastic differential equation (3.78) is linear in the probability density. In polymer kinetic theory, one is sometimes faced with the situation that the diffusion equation is nonlinear in the probability density $p(t, \boldsymbol{x})$ because the functions $\boldsymbol{A}$ and $\mathbf{D}$ in (3.79) depend on averages evaluated with $p(t, \boldsymbol{x})$. For example, this situation occurs for the consistent averaging method and the Gaussian approximation for hydrodynamic interaction (see Sect. 4.2.4), the FENE-P model (see Sect. 4.3.2), the Doi model for liquid crystal polymers (see Sect. 5.2.2), and the Doi-Edwards model without independent alignment (see Sect. 6.3.2). In the mathematical literature, such processes are referred to as *nonlinear diffusions* or *processes with mean field interactions*.

We here consider the situation in which the coefficients $\boldsymbol{A}(t, \boldsymbol{x})$ and $\mathbf{B}(t, \boldsymbol{x})$ in the stochastic differential equation (3.78) are replaced by $\boldsymbol{A}(\boldsymbol{x}, \langle \boldsymbol{g}_A(\boldsymbol{x}, \boldsymbol{X}_t)\rangle)$ and $\mathbf{B}(\boldsymbol{x}, \langle \boldsymbol{g}_B(\boldsymbol{x}, \boldsymbol{X}_t)\rangle)$, that is, they depend on expectations of one or several real-valued functions. Provided that the functions $\boldsymbol{A}$, $\mathbf{B}$, $\boldsymbol{g}_A$, and $\boldsymbol{g}_B$ satisfy Lipschitz conditions and suitable growth conditions [16], the existence of a unique solution of the modified stochastic differential equation can be established. For the FENE-P model, which satisfies neither Lipschitz nor growth conditions, it has been shown that the mean field approximation strongly affects the physical predictions by introducing multiple stationary solutions [17].

We next describe a law of large numbers that is of great importance for the numerical simulation of processes with mean field interactions. Consider the system of n stochastic differential equations,

$$\begin{aligned} d\boldsymbol{X}_t^{(j)} &= \boldsymbol{A}\left(\boldsymbol{X}_t^{(j)}, \frac{1}{n}\sum_{k=1}^{n} \boldsymbol{g}_A(\boldsymbol{X}_t^{(j)}, \boldsymbol{X}_t^{(k)})\right) dt \\ &+ \mathbf{B}\left(\boldsymbol{X}_t^{(j)}, \frac{1}{n}\sum_{k=1}^{n} \boldsymbol{g}_B(\boldsymbol{X}_t^{(j)}, \boldsymbol{X}_t^{(k)})\right) \cdot d\boldsymbol{W}_t^{(j)}, \end{aligned} \tag{3.80}$$

for $j = 1, 2, \ldots, n$, where the processes $\boldsymbol{W}^{(j)}$ are independent d'-dimensional Wiener processes. In other words, the expectations in the drift and diffusion terms are replaced by ensemble averages. We furthermore assume that the initial conditions $\boldsymbol{X}_0^{(j)}$ are independent, identically distributed random variables.

If $\boldsymbol{X}_t$ is the solution of the stochastic differential equation with mean field interactions, where the distribution of the initial condition $\boldsymbol{X}_0$ coincides with the distribution of the random variables $\boldsymbol{X}_0^{(j)}$ and the coefficient functions satisfy Lipschitz and suitable growth conditions, then

$$\text{ms-}\lim_{n\to\infty} \frac{1}{n} \sum_{j=1}^{n} g(\boldsymbol{X}_t^{(j)}) = \langle g(\boldsymbol{X}_t) \rangle \;, \tag{3.81}$$

for all bounded continuous functions $g: \mathbb{R}^d \to \mathbb{R}$ [16]. Convergence theorems under weaker conditions for the coefficient functions have been described in [18]. Typical fluctuations for finite n are expected to be of the order of $n^{-1/2}$; a very detailed discussion of these fluctuations can be found in [19]. In a pioneering paper on nonlinear diffusions, McKean [20] has shown that, under suitable smoothness conditions, one has the stronger mode of almost sure convergence in (3.81), so that one has a strong law of large numbers.

The processes $\boldsymbol{X}^{(j)}$ are often referred to as *weakly interacting stochastic processes*. The weakness of the interaction between any two of these processes is obvious from the factor $1/n$ in front of the interaction terms in (3.80). For large n, any finite number of processes becomes essentially independent; this asymptotic independence property is what is sometimes referred to as *propagation of chaos* [18, 20, 21]. The convergence of weakly interacting processes to processes with mean field interactions is very important also for recoil calculations and for the CONNFFESSIT approach outlined in Sect. 1.2.3: the stochastic differential equations for the polymer configurations contain velocity gradients that are computed from the stresses in a flowing liquid; the polymer contribution to these stresses is taken from ensemble averages, so that a weak coupling arises.

In numerical simulations of processes with mean field interactions we need a law of large numbers similar to (3.81) for the time-discretized version of the underlying stochastic differential equation in the stronger mode of almost sure convergence. Then, the simulation of a single ensemble of weakly interacting processes is sufficient to obtain averages for the discretized problem.

3.3.5 Boundary Conditions

In the discussion of stochastic differential equations, we have so far assumed that the range of the solutions is $\mathbb{R}$ or $\mathbb{R}^d$ and that there has been no need to consider the boundary conditions at infinity. For example, for the Wiener process (Example 2.79) or the Ornstein-Uhlenbeck process (Example 3.25), the probability density p is Gaussian; both p and its derivatives vanish very rapidly at infinity. We expect this to be a generic case because, for reasonably well-behaved p, the normalization implies that p vanishes at infinity and, if its derivatives do not oscillate rapidly near infinity, we also expect the derivatives of p to vanish at infinity.

If the range of a stochastic process is some region $\mathbb{D} \subset \mathbb{R}^d$ with a boundary $\mathbb{B}$, the situation is more complicated, and boundary conditions deserve more than just the marginal attention usually paid to them in the literature on stochastic differential equations. An important concept in formulating and analyzing various boundary conditions are *Markov times*. For the Markovian solution $\boldsymbol{X} = (\boldsymbol{X}_t)_{t\in\mathbb{T}}$ of the stochastic differential equation (3.78), we can apply the following definition of a *first passage time*, which is the prototype of a Markov time. If $\mathbb{D}, \mathbb{B} \in \mathcal{B}^d$, for example when $\mathbb{D}$ is a closed set, and a Markov process with continuous trajectories starts in $\mathbb{D}$, that is $P^{X_0}(\mathbb{D}) = 1$, we can ask when does $\boldsymbol{X}$ pass the boundary $\mathbb{B}$ of $\mathbb{D}$ for the first time. We hence define the first passage time,

$$T^{\mathrm{FP}}(\omega) := \inf\{t \in \mathbb{T} \mid \boldsymbol{X}_t(\omega) \in \mathbb{D}^C\}\,, \tag{3.82}$$

and we set $T^{\mathrm{FP}}(\omega) = \infty$ when the set over which the infimum has to be taken is empty (when a trajectory does not pass the boundary $\mathbb{B}$ within the time range $\mathbb{T}$). It can be shown that the function $T^{\mathrm{FP}} : \Omega \to \mathbb{R} \cup \{\infty\}$ is a random variable in the measurable space $(\mathbb{R} \cup \{\infty\}, \mathcal{B} \cup \{B \cup \{\infty\} \mid B \in \mathcal{B}\})$. Moreover, one finds that the event $\{\omega \mid T^{\mathrm{FP}}(\omega) \leq t\}$ for $t \in \mathbb{T}$ is contained in the σ-algebra $\mathcal{A}_t$ induced by the random variables $\boldsymbol{X}_{t'}$ for $t' \leq t$; that is, the occurrence of a passage before t can be recognized with the information available at time t.[3]

On the level of stochastic differential equations, the effect of an absorbing boundary can be obtained by terminating a trajectory as soon as it reaches the boundary for the first time. The first passage time is thus an important concept in the formulation of this type of boundary condition. For the probability density p in the Fokker-Planck equation, such an *absorbing boundary condition* means that p vanishes at the boundary. Then, the probability of $\Omega_t = \{\omega \mid \boldsymbol{X}_{t'}(\omega) \in \mathbb{D} - \mathbb{B}$ for $t' \leq t\}$, that is the probability of the unterminated trajectories which always remained inside the range $\mathbb{D}$, decreases with increasing time t (the fact that Ω_t is measurable, which is closely related to the fact that first passage times are random variables, is nontrivial, and its proof depends on the continuity of the trajectories of $\boldsymbol{X}$).

If the process $\boldsymbol{X}$ includes some angular variables it is natural to impose *periodic boundary conditions*. On the level of Fokker-Planck equations, this means that the probability density is unchanged if its period is added to an angular variable. On the level of stochastic differential equations, values of an angular variable may be identified if they differ only by a full period.

More subtle are *reflecting boundary conditions*. In order to understand reflecting boundary conditions on the Fokker-Planck level, we note that the Fokker-Planck equation (3.79) can be regarded as a continuity equation expressing the conservation of probability when we introduce the *probability current*

$$\boldsymbol{J}(t,\boldsymbol{x}) := \boldsymbol{A}(t,\boldsymbol{x})\,p(t,\boldsymbol{x}) - \frac{1}{2}\frac{\partial}{\partial \boldsymbol{x}} \cdot [\mathbf{D}(t,\boldsymbol{x})\,p(t,\boldsymbol{x})]\,. \tag{3.83}$$

If we consider the situation where the process is confined to some region $\mathbb{D}$ with a sufficiently smooth boundary $\mathbb{B}$, there is no flow of probability across the boundary, and the component of the probability current $\boldsymbol{J}$ normal to the boundary must hence vanish on $\mathbb{B}$. Since there is no probability current through

[3] This is the defining property of Markov times.

the boundary $\mathbb{B}$ the trajectories of the process must be reflected there (in case they can reach the boundary). In other words, reflecting boundary conditions mean that the probability currents to and away from the boundary sum to zero. The trajectories of the corresponding Markov process are reflected back into the range $\mathbb{D}$ whenever they hit the boundary. In general, a rigorous construction of such trajectories is subtle. We first discuss an example for which the reflecting boundaries can be handled in a simple way, and then we describe an alternative procedure that is applicable in more general situations.

Example 3.33 Wiener Process with Reflecting Boundaries
In order to illustrate the construction of a process with reflecting boundaries we consider the solution of the one-dimensional stochastic differential equation

$$dS_t = \sqrt{\frac{2}{\lambda}}\, dW_t \tag{3.84}$$

in the interval $[0,1]$, where $\lambda > 0$. The distribution of the initial random variable S_0 is assumed to be uniform in $[0,1]$. For this example, with vanishing drift and constant diffusion, a rigorous treatment of the reflecting boundaries is very simple.

The trajectories of the solution of (3.84) in the presence of reflecting boundaries at 0 and 1 can be constructed as follows. Starting from the value $x = S_0(\omega) + \sqrt{2/\lambda}\, W_t(\omega)$ of the exact solution of (3.84) in the absence of the boundaries we can easily introduce the effect of reflections as follows. If $x < 0$, we replace x with $-x > 0$. Then, for $x \in [n, n+1[$ with $n = 0, 1, 2 \ldots$, we do the following replacement,

$$x \to S_t(\omega) = \begin{cases} x - n & \text{for even } n \\ n + 1 - x & \text{for odd } n \end{cases},$$

in order to obtain the value of the trajectory in the presence of reflecting boundaries. The trajectories of the process S look locally like those of the Wiener process (continuous but nowhere differentiable and of unbounded variation over finite time intervals), however, they are confined to the interval $[0,1]$. A typical trajectory starting at 0 is shown in Fig. 3.3 for $\lambda = 2$. The very irregular nature of the paths implies that, whenever one reflection occurs, there will be a countably infinite number of reflections. This wild banging at the boundary is the source of the problems with a rigorous treatment of reflecting boundary conditions in more general situations.

The solution of the corresponding Fokker-Planck equation,

$$\frac{\partial p(t,s)}{\partial t} = \frac{1}{\lambda} \frac{\partial^2 p(t,s)}{\partial s^2}, \tag{3.85}$$

is $p(t,s) = 1$, which satisfies (i) the initial condition, (ii) the Fokker-Planck equation (3.85) in $]0,1[$, and (iii) the condition of a vanishing probability current (not only at the boundaries but everywhere in $[0,1]$ one has $\partial p(t,s)/\partial s = 0$). □

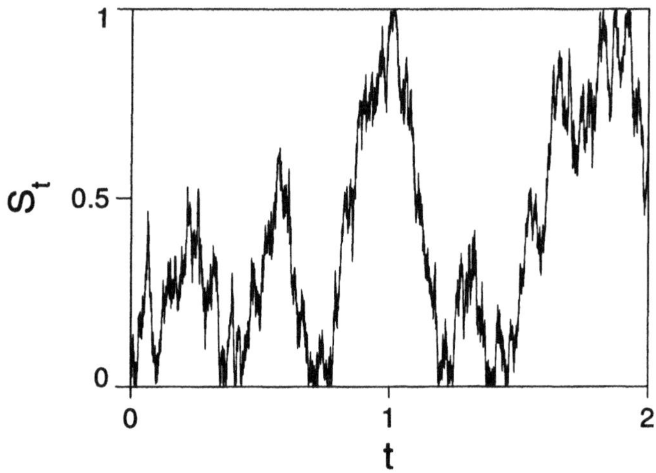

Fig. 3.3. A trajectory of the Wiener process in the presence of reflecting boundaries at 0 and 1.

The construction of the preceding example suffers from a serious disadvantage. For any piece of the trajectory of S for which the corresponding trajectory in the absence of boundaries would be contained in the interval $[n, n+1[$ with odd n, one actually finds $dS_t = -\sqrt{2/\lambda}\, dW_t$ due to the reflections. In view of the symmetric distribution of the increments of the Wiener process this change in sign may seem irrelevant. However, in the presence of a drift term, the sign of the drift may not be changed in carrying out the reflections, and hence one needs a more general approach.

Rigorous treatments of reflecting boundary conditions on the level of stochastic differential equations are very scarce in the literature; here we follow Sect. IV.7 of [22]. In the language of [22], reflecting boundary conditions correspond to a *boundary operator of the Wentzell type* which is given by the normal component of the gradient operator (directed into the domain). For a given stochastic differential equation in $\mathbb{D}$, reflections are introduced by adding a term of the form $\boldsymbol{n}(\boldsymbol{X}_t)\, d\phi_t$ to the differential equation, where $\boldsymbol{n}(\boldsymbol{x})$ is the unit normal vector in $\boldsymbol{x} \in \mathbb{B}$ (directed into the domain) and $(\phi_t)_{t\in\mathbb{T}}$ is an increasing process with continuous trajectories, which increases only when $\boldsymbol{X}_t$ is at the boundary $\mathbb{B}$; more precisely $\int_0^t \chi_{\mathbb{B}}(\boldsymbol{X}_{t'})\, d\phi_{t'} = \phi_t$ and $\int_0^t \chi_{\mathbb{B}}(\boldsymbol{X}_{t'})\, dt' = 0$ for all t (for an increasing process, by definition, all trajectories are non-decreasing functions of time; integrands with respect to $d\phi_t$ can be handled by using deterministic calculus for each trajectory). Loosely speaking, the process is shifted back into the region $\mathbb{D}$ in the normal direction whenever it attempts to cross the boundary $\mathbb{B}$.

In order to illustrate the explicit construction of the increasing process $(\phi_t)_{t\in\mathbb{T}}$ contained in the proof of Theorem 7.2 of [22], we consider the stochastic differential equation

$$\mathrm{d}X_t = A(X_t)\,\mathrm{d}t + B(X_t)\,\mathrm{d}W_t\,, \tag{3.86}$$

for the range $[0,\infty[$ and a reflecting boundary at 0. The solution with reflecting boundary is characterized by the following pair of equations,

$$X_t = X_0 + \int_0^t A(X_{t'})\,\mathrm{d}t' + \int_0^t B(X_{t'})\,\mathrm{d}W_{t'} + \phi_t\,, \tag{3.87}$$

$$\phi_t = -\min_{t'\le t}\left\{X_0 + \int_0^{t'} A(X_{t''})\,\mathrm{d}t'' + \int_0^{t'} B(X_{t''})\,\mathrm{d}W_{t''}\right\}. \tag{3.88}$$

As long as (3.88) leads to a negative value of ϕ_t, this value has to be replaced with zero (this corresponds to the situation where the boundary has not yet been reached). Obviously, the pair of equations (3.87) and (3.88) implies that $X_t \ge 0$ for all t. The condition $X_t = 0$ holds if and only if the minimum in (3.88) is assumed at the final time t and, due to the continuity of Itô processes, only then ϕ_t can increase (i.e., the minimum in (3.88) can decrease further). This is in agreement with the general idea that ϕ_t can increase only when X_t is at the boundary. Away from the boundary, X_t satisfies (3.86).

The above pair of equations suggests how reflecting boundary conditions can be treated in simulations. In a discretized version of (3.87), (3.88), when the minimum is taken over the values at the discrete times, the construction comes down to resetting $X_t = 0$ whenever the boundary at 0 is crossed. If a time step Δt is used, the true minimum is of the order $(\Delta t)^{1/2}$ smaller than the one obtained at the discrete times, and the naive discretization hence leads to a systematic underestimation of X_t of the order $(\Delta t)^{1/2}$. In order to reduce the order of the error one needs to analyze the increments of the stochastic integrals and of the minimum in (3.87), (3.88) for each time step in more detail. Figure 3.4 shows a typical trajectory for $A = -1$, $B = 1$, $X_0 = 0$.

Exercise 3.34 Consider the stochastic differential equation $\mathrm{d}X_t = -\mathrm{d}t + \mathrm{d}W_t$ on $[0,\infty[$ with a reflecting boundary at 0 (see Fig. 3.4). Calculate the stationary probability density reached for large t and the expectation of X_t in the stationary state.

Example 3.35 Wiener Process with Absorbing Boundaries

In this example, we discuss the solution of (3.85) with absorbing boundary conditions at 0 and 1, that is, $p(t,0) = p(t,1) = 0$. As an initial condition we impose $p(0,s) = \delta(s - s_0)$ with $s_0 \in]0,1[$.

The solution can be given in terms of eigenfunctions,

$$p(t,s) = 2\sum_{n=1}^{\infty} \sin(n\pi s_0)\,\sin(n\pi s)\,\mathrm{e}^{-n^2\pi^2 t/\lambda}\,. \tag{3.89}$$

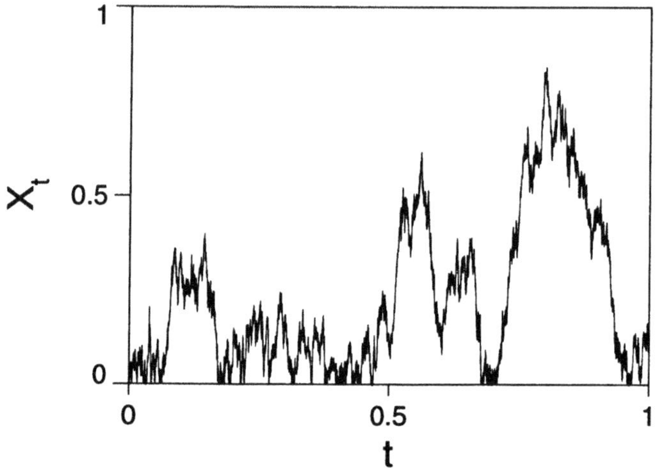

Fig. 3.4. A trajectory of the solution of (3.86) for $A = -1$, $B = 1$, $X_0 = 0$ in the presence of a reflecting boundary at 0. The constant drift makes the trajectory always return to 0 where it is then reflected.

The Fokker-Planck equation and the boundary conditions can be verified for each term in the sum, and the initial condition follows from the completeness of the sine functions. If the initial distribution is uniform we can integrate (3.89) over s_0 from 0 to 1 in order to obtain

$$p(t,s) = \frac{4}{\pi} \sum_{\substack{n=1 \\ n \text{ odd}}}^{\infty} \frac{1}{n} \sin(n\pi s)\, \mathrm{e}^{-n^2\pi^2 t/\lambda} . \tag{3.90}$$

For the uniform initial condition, the probability of the set of unterminated trajectories of the process with absorbing boundaries, Ω_t, is obtained by further integrating (3.90) over s,

$$P(\Omega_t) = \frac{8}{\pi^2} \sum_{\substack{n=1 \\ n \text{ odd}}}^{\infty} \frac{1}{n^2} \mathrm{e}^{-n^2\pi^2 t/\lambda} .$$

The probability density of the first passage or absorption time is obtained from the rate of decrease of $P(\Omega_t)$,

$$p^{T^{\mathrm{FP}}}(t) = -\frac{\mathrm{d}P(\Omega_t)}{\mathrm{d}t} = \frac{8}{\lambda} \sum_{\substack{n=1 \\ n \text{ odd}}}^{\infty} \mathrm{e}^{-n^2\pi^2 t/\lambda} =: \overline{\mu}(t) . \tag{3.91}$$

The integral of $p^{T^{\mathrm{FP}}}$ from 0 to ∞ is equal to unity; this result indicates that all trajectories are absorbed in a finite time. The expectation of the first passage time calculated from its probability distribution is $\lambda/12$. Within this time, the mean-square displacement of the solution of (3.84) without boundary effects is 1/6, which is twice the mean-square distance of the initial position from the nearest boundary.

The same procedure can be repeated if the process starts at s_0. One then obtains

$$p(T^{\mathrm{FP}} = t|S_0 = s_0) = \frac{4\pi}{\lambda} \sum_{\substack{n=1 \\ n \text{ odd}}}^{\infty} n \sin(n\pi s_0)\, \mathrm{e}^{-n^2\pi^2 t/\lambda} =: \overline{\mu}(t)\, \overline{\pi}(t, s_0)\,, \tag{3.92}$$

where the second part of (3.92) is the definition of the function $\overline{\pi}(t, s_0)$. From (3.92) we find that, if the process starts at s_0, the mean first passage time is $\lambda s_0(1 - s_0)/2$. The construction of first passage time distributions by means of absorbing boundaries described here is a standard technique in the theory of first passage times. □

Exercise 3.36 Consider the stochastic differential equation $\mathrm{d}X_t = -\mathrm{d}t + \mathrm{d}W_t$ of Exercise 3.34 on $[0, \infty[$. Let the deterministic initial position be x_0, and let 0 be an absorbing boundary. Calculate the probability density of X_t and the first passage or absorption time. Are all trajectories absorbed in a finite time? What is the expectation of the first passage time?

Example 3.37 Last Reflection Times

In this example, we consider a Markov process with continuous trajectories in some closed region $\mathbb{D}$ with boundary $\mathbb{B}$. For a given time t, we define a random variable that is measurable with respect to $\mathcal{A}_t$, namely the *last reflection time* before t,

$$T_t^{\mathrm{LR}}(\omega) := \sup\{t' \leq t \mid \boldsymbol{X}_{t'}(\omega) \in \mathbb{B}\}\,, \tag{3.93}$$

and we take $T_t^{\mathrm{LR}}(\omega) = -\infty$ when the set over which the supremum has to be taken is empty (when a trajectory was not at the boundary $\mathbb{B}$ before t).

As a concrete example we consider the Wiener process with reflecting boundaries of Example 3.33, and we determine the distribution of T_t^{LR} under the condition that the process at time t is at the position $s_0 \in]0, 1[$, $p(T_t^{\mathrm{LR}} = t'|S_t = s_0)$. This example is relevant to the discussion of reptation models in Chap. 6.

Since the distributions of the increments of the Wiener process are symmetric, trajectories evolving forward and backward in time have equal probabilities. If we let run time backwards from t we obtain

$$p(T_t^{\mathrm{LR}} = t'|S_t = s_0) = p(T^{\mathrm{FP}} = t - t'|S_0 = s_0)\,, \tag{3.94}$$

where the latter conditional probability density can be taken over from Example 3.35. By integrating over s_0 we furthermore obtain the distribution of the last reflection time irrespective of the current position,

$$p^{T_t^{\mathrm{LR}}}(t') = \overline{\mu}(t - t')\,. \tag{3.95}$$

Finally, we determine the conditional probability $p(S_t = s|T_t^{\mathrm{LR}} = t')$. From the definition (2.59) and the preceding results we obtain

$$p(S_t = s | T_t^{\rm LR} = t') = p^{S_t}(s) \, \frac{p(T_t^{\rm LR} = t' | S_t = s)}{p^{T_t^{\rm LR}}(t')} = \frac{p(T^{\rm FP} = t - t' | S_0 = s)}{\overline{\mu}(t - t')} \, .$$

With the result (3.92) of Example 3.35 we obtain

$$p(S_t = s | T_t^{\rm LR} = t') = \overline{\pi}(t - t', s) \, . \tag{3.96}$$

The results (3.95) and (3.96) will be used in the discussion of reptation models in Chap. 6. □

3.3.6 Stratonovich's Stochastic Calculus

The discussion of stochastic differential equations presented so far is based on the concept of stochastic integration developed by Kiyosi Itô in 1944. In concluding this section, we should point out that there are alternative theories of stochastic integrals, the most important of which is associated with the name of Rouslan L. Stratonovich. After using the properties of this new type of integrals for several years, Stratonovich explicitly stated their definition and properties in 1963 (see remarks in [23]).

In the applied literature, there is a never-ending discussion as to whether Itô or Stratonovich integrals are preferable, and most authors seem to be convinced that one must have a clear preference. The intensity of the discussion is even more surprising in view of the fact that the whole issue may be regarded as a matter of taste: in virtually all applications, there is a straightforward conversion procedure for the two types of integrals. In the mathematical literature, on the other hand, many authors jump back and forth between the two different types of integrals whenever it is convenient. The mere fact that Stratonovich integrals are introduced in rigorous textbooks on mathematical stochastics should discredit associations like "imprecise mathematical definition" (see Sect. 1.4.4 of [8]) in connection with Stratonovich integrals.

Mathematicians have a certain preference for Itô integrals because of their powerful martingale properties which are very helpful in deriving many profound results. Some excellent theoretical physicists strongly dislike Itô integrals because of the unusual form of the resulting differentiation and integration rules. For example, even though he admits the mathematical consistency of both types of integrals and the rigorous relationship between them, van Kampen makes the following strong statement (p. 186 of [24]): "The final conclusion is that a physicist cannot go wrong by regarding the Itô interpretation as one of those vagaries of the mathematical mind that are of no concern to him." By developing Itô calculus in detail, we obviously express more sympathy with the mathematicians' point of view. The reason for this is that we wish to develop and apply efficient numerical integration schemes for stochastic differential equations, and this requires profound mathematical results. Although it is rather clear that one should listen to mathematicians when doing numerical mathematics, we here wish to attempt a fair comparison of the different approaches to stochastic calculus.

In coming to an unemotional conclusion as to whether Itô's or Stratonovich's calculus is preferable *in a given situation* one should first try to find reasonable criteria for the selection of one or the other calculus. Among the possible considerations are: (i) Which one can be defined more generally? (ii) Which one is more convenient for deriving theorems? (iii) Which one can be justified in formulating stochastic differential equations in certain classes of applications? (iv) Which one is more convenient in analytical calculations for a given application? (v) Which one is more convenient in numerical calculations? As we develop the theory of Stratonovich stochastic integrals in this subsection, we keep coming back to these questions, and we comment on all these considerations—as unemotionally as possible.

We here introduce Stratonovich's stochastic integral for integrands which are Itô processes (see Definition 3.15). Given an Itô process X with the stochastic differential $\mathrm{d}X_t = A_t\,\mathrm{d}t + B_t\,\mathrm{d}W_t$, we define for $a, b \in \mathbb{T}$ with $a \leq b$,

$$\int_a^b X_t \circ \mathrm{d}W_t := \int_a^b X_t\,\mathrm{d}W_t + \frac{1}{2}\int_a^b B_t\,\mathrm{d}t\,. \tag{3.97}$$

The integral $\int_a^b X_t \circ \mathrm{d}W_t$ is referred to as the *Stratonovich integral*, and the notation "$\circ$" is occasionally called *Itô's circle*. The general mathematical theory of Stratonovich's integral can, for example, be found in the stochastics textbooks by H. Kunita [13] and by P. Protter [25]. The definition (3.97) remains valid in the more general case where one adds a nonanticipating process of bounded variation for which jumps in the trajectories are allowed to the Itô process X (see Sect. V.5 of [25]).

In order to define the Stratonovich integral we have made extensive use of the Itô integral, both in the definition (3.97) and in the characterization of the processes X for which Stratonovich integrals are defined. Even though the Stratonovich integral can be defined without falling back on the Itô integral, the Itô integral may be considered as more fundamental because it can be defined for a larger class of integrands. On the other hand, Stratonovich integrals can be defined for processes on more general manifolds (see remarks on p. 107 of [26]). For our purposes, the idea of having a class of integrands as large as possible is not an important criterion because we are eventually interested in stochastic differential equations, the solutions of which are Itô processes anyway. In other words, the criterion (i) is of no relevance to our selection because for all applications in this book both Itô and Stratonovich integrals are general enough.

The Stratonovich integral clearly inherits the linearity and additivity properties expected for any kind of integral from the corresponding properties (3.28) and (3.29) of the Itô integral and of ordinary integrals. However, the simple expressions for the first and second moments of Itô integrals in Theorem 3.9 have to be replaced with more complicated expressions involving the process B; in particular, the average of Stratonovich integrals does not vanish in general. This feature is one that makes the Itô integral more attractive in certain analytical calculations (see the criterion (iv) listed above). This advantage of Itô's calculus is related to the fact that, in the perturbation expansion of solutions of stochastic differential equations, many terms vanish immediately, whereas in Stratonovich's calculus they cancel only after lengthy calculations (see Sect. 4.2.6 of [8]).

While Itô integrals are martingales (Theorem 3.13), this is not the case for Stratonovich integrals. Since there exists a vast literature on martingales, providing many profound results, Itô integrals are often preferable in rigorous mathematical developments. According to the criterion (ii) listed above, Itô's calculus is hence widely accepted as preferable. However, this consideration is not particularly relevant for the purpose of this book.

A direct calculation of Stratonovich integrals without any reference to Itô integrals is made possible by means of the identity (3.98) established in the following theorem (see Sect. 2.3 of [13] or Sect. V.5 of [25]).

Theorem 3.38 *Let $(X_t)_{t\in\mathbb{T}}$ be a nonanticipating process for which the Stratonovich integral is defined. If, for the sequence of partitions $a = t_0^{(n)} < t_1^{(n)} \cdots < t_{n-1}^{(n)} < t_n^{(n)} = b$ of the interval $[a, b]$ (with $a, b \in \mathbb{T}$), the maximum time step $\max_{j\leq n}(t_j^{(n)} - t_{j-1}^{(n)})$ converges to zero as $n \to \infty$, one has*

$$\int_a^b X_t \circ \mathrm{d}W_t = \text{st-}\lim_{n\to\infty} \sum_{j=1}^{n} \frac{1}{2}(X_{t_{j-1}^{(n)}} + X_{t_j^{(n)}})(W_{t_j^{(n)}} - W_{t_{j-1}^{(n)}}) \,. \tag{3.98}$$

For a simple process X, the sum in (3.98) should be compared to the corresponding definition (3.15) of the Itô integral: instead of multiplying the increments of the Wiener process over a small time step by the initial value of X we now multiply by the arithmetic mean of the initial and final values of X for the corresponding time step. In connection with the definition (3.15) we already pointed out that this difference has important consequences, and that the Itô choice leads to certain simplifications. Equation (3.97) quantifies the difference between the corresponding stochastic integrals. According to (3.98), for Stratonovich integrals it is obviously important to have information about the joint distribution of $X_{t_j^{(n)}}$ and $W_{t_j^{(n)}} - W_{t_{j-1}^{(n)}}$ because these are not independent random variables. This information enters via B_t in (3.97).

Rather than proving (3.98), we next verify it for an important special case.

Example 3.39 The Ultimate Rerun of a Classic

Since it was very instructive to calculate the Itô integral $\int_0^t W_{t'}\,\mathrm{d}W_{t'}$ we now evaluate the analogous Stratonovich integral $\int_0^t W_{t'} \circ \mathrm{d}W_{t'}$ by means of (3.98). In the notation of Example 3.5 we have

$$\sum_{j=1}^{n} \frac{1}{2}(\widetilde{W}_{j-1}^{(n)} + \widetilde{W}_j^{(n)})(\widetilde{W}_j^{(n)} - \widetilde{W}_{j-1}^{(n)}) = \frac{1}{2}\sum_{j=1}^{n}(\widetilde{W}_j^{(n)2} - \widetilde{W}_{j-1}^{(n)2}) = \frac{1}{2}\,\widetilde{W}_n^{(n)2} \,,$$

and we hence obtain the result expected according to deterministic calculus,

$$\int_0^t W_{t'} \circ \mathrm{d}W_{t'} = \frac{1}{2} W_t^2 .$$

The same result follows immediately from (3.26) and (3.97) because $B_t = 1$ for the Wiener process. We have thus verified the equivalence of (3.97) and (3.98) for the special case $X = W$. □

Exercise 3.40 Calculate the Stratonovich integral $\int_0^t W_{t'}^2 \circ \mathrm{d}W_{t'}$.

Example 3.43 below shows that the use of the rules of deterministic calculus is not restricted to the particular example $\int_0^t W_{t'} \circ \mathrm{d}W_{t'}$ but is generally allowed in transforming Stratonovich stochastic differential equations. The fact that the rules of deterministic calculus can be applied in evaluating Stratonovich integrals is the most important advantage of this alternative approach to stochastic calculus (in view of the criterion (iv) listed in the beginning of this subsection). However, even if deterministic differentiation rules can be used in Stratonovich's calculus, this should not be used as an excuse to ignore the subtleties of stochastic calculus: the difference between the Itô and Stratonovich approaches is only one manifestation of these subtleties, and serious difficulties in developing higher-order numerical integration schemes are another one; even when using Stratonovich's calculus one cannot expect that naively adapted deterministic integration schemes yield the familiar order of convergence! Before proving the applicability of deterministic differentiation rules in Stratonovich's calculus in Example 3.43, we first establish a relationship between Itô and Stratonovich stochastic differential equations.

Example 3.41 Equivalent Itô and Stratonovich Equations
We again consider the solution $\boldsymbol{X}$ of the stochastic differential equation

$$\mathrm{d}\boldsymbol{X}_t = \boldsymbol{A}(t, \boldsymbol{X}_t)\, \mathrm{d}t + \mathbf{B}(t, \boldsymbol{X}_t) \cdot \mathrm{d}\boldsymbol{W}_t . \qquad (3.99)$$

If $\mathbf{B}$ is sufficiently smooth, we can apply Itô's formula (3.43) to the column vectors of $\mathbf{B}$ in order to show that

$$\mathrm{d}\mathbf{B}(t, \boldsymbol{X}_t) = (\ldots)\, \mathrm{d}t + \sum_{j=1}^{d} \frac{\partial \mathbf{B}}{\partial x_j}(t, \boldsymbol{X}_t)\, \mathrm{d}X_j(t) . \qquad (3.100)$$

The multicomponent generalization of the definition (3.97) obtained by applying (3.97) to the corresponding component processes implies after substituting $\mathrm{d}\boldsymbol{X}_t = (\ldots)\, \mathrm{d}t + \mathbf{B}(t, \boldsymbol{X}_t) \cdot \mathrm{d}\boldsymbol{W}_t$ in (3.100)

$$\begin{aligned} \int_0^t \mathbf{B}(t', \boldsymbol{X}_{t'}) \odot \mathrm{d}\boldsymbol{W}_{t'} &= \int_0^t \mathbf{B}(t', \boldsymbol{X}_{t'}) \cdot \mathrm{d}\boldsymbol{W}_{t'} \\ &+ \frac{1}{2} \int_0^t \left[\mathbf{B}^T(t', \boldsymbol{X}_{t'}) \cdot \frac{\partial}{\partial \boldsymbol{x}} \right] \cdot \mathbf{B}^T(t', \boldsymbol{X}_{t'})\, \mathrm{d}t' , \end{aligned}$$

where the symbol "$\odot$" stands for Itô's circle when a matrix multiplication is implied. Here, and in the subsequent equations of this section, it is understood that the partial derivatives with respect to $\boldsymbol{x}$ are to be taken before $\boldsymbol{X}_{t'}$ or $\boldsymbol{X}_t$ is inserted into the respective functions of $\boldsymbol{x}$. The solution of the Itô stochastic differential equation (3.99) hence satisfies the Stratonovich stochastic differential equation

$$\mathrm{d}\boldsymbol{X}_t = \left\{\boldsymbol{A}(t,\boldsymbol{X}_t) - \frac{1}{2}\left[\mathbf{B}^T(t,\boldsymbol{X}_t)\cdot\frac{\partial}{\partial\boldsymbol{x}}\right]\cdot\mathbf{B}^T(t,\boldsymbol{X}_t)\right\}\mathrm{d}t + \mathbf{B}(t,\boldsymbol{X}_t)\odot\mathrm{d}\boldsymbol{W}_t\,. \tag{3.101}$$

Conversely, the Stratonovich stochastic differential equation

$$\boxed{\mathrm{d}\boldsymbol{X}_t = \boldsymbol{A}(t,\boldsymbol{X}_t)\,\mathrm{d}t + \mathbf{B}(t,\boldsymbol{X}_t)\odot\mathrm{d}\boldsymbol{W}_t\,.} \tag{3.102}$$

and the Itô stochastic differential equation

$$\boxed{\mathrm{d}\boldsymbol{X}_t = \left\{\boldsymbol{A}(t,\boldsymbol{X}_t) + \frac{1}{2}\left[\mathbf{B}^T(t,\boldsymbol{X}_t)\cdot\frac{\partial}{\partial\boldsymbol{x}}\right]\cdot\mathbf{B}^T(t,\boldsymbol{X}_t)\right\}\mathrm{d}t + \mathbf{B}(t,\boldsymbol{X}_t)\cdot\mathrm{d}\boldsymbol{W}_t\,,} \tag{3.103}$$

have the same solution $\boldsymbol{X}$. In summary, the pair (3.99) and (3.101) (or, alternatively, of (3.102) and (3.103)) of Itô and Stratonovich differential equations is solved by the same Itô process. Note that these conversion rules imply that, in the case of additive noise, there is no need to distinguish between the Itô and Stratonovich definitions in formulating a stochastic differential equation. □

Exercise 3.42 Transform the Itô stochastic differential equation $\mathrm{d}X_t = X_t\,\mathrm{d}W_t$ with $X_0 = 1$ into a Stratonovich stochastic differential equation. Verify that the correct solution of Example 3.18 is obtained when the resulting Stratonovich stochastic differential equation is solved by applying the rules of deterministic calculus.

Example 3.43 Transformation of Stratonovich Equations
We are now in a position to investigate the transformation behavior of Stratonovich stochastic differential equations. Let $\boldsymbol{g} : \mathbb{R}^d \to \mathbb{R}^d$ be a one-to-one mapping with continuous second-order partial derivatives, and let $\boldsymbol{X}$ be the solution of the Stratonovich differential equation (3.102). We next determine the Stratonovich stochastic differential equation satisfied by $\boldsymbol{Y}_t = \boldsymbol{g}(\boldsymbol{X}_t)$ by finding the corresponding Itô differential equation for $\boldsymbol{X}$, applying Itô's formula, and translating the Itô equation for the transformed process back into the transformed Stratonovich stochastic differential equation. As a useful auxiliary quantity we introduce the $d \times d$-matrix $\mathbf{T}(\boldsymbol{x}) = [(\partial/\partial\boldsymbol{x})\boldsymbol{g}(\boldsymbol{x})]^T$ occurring in the familiar chain rule of deterministic calculus. By applying Itô's formula (3.43) to the equivalent Itô differential (3.103) we obtain

$$\mathrm{d}\boldsymbol{Y}_t = \mathbf{T}(\boldsymbol{X}_t)\cdot\mathrm{d}\boldsymbol{X}_t + \frac{1}{2}\mathbf{D}(t,\boldsymbol{X}_t) : \frac{\partial}{\partial\boldsymbol{x}}\frac{\partial}{\partial\boldsymbol{x}}\boldsymbol{g}(\boldsymbol{X}_t)\,\mathrm{d}t\,, \tag{3.104}$$

where $\mathrm{d}\boldsymbol{X}_t$ stands for the Itô differential on the right side of (3.103), and $\mathbf{D} = \mathbf{B} \cdot \mathbf{B}^T$. More explicitly, one has after combining the "unusual" terms containing explicit derivatives in (3.103) and (3.104),

$$\begin{aligned} \mathrm{d}\boldsymbol{Y}_t &= \mathbf{T}(\boldsymbol{X}_t) \cdot [\boldsymbol{A}(t, \boldsymbol{X}_t)\, \mathrm{d}t + \mathbf{B}(t, \boldsymbol{X}_t) \cdot \mathrm{d}\boldsymbol{W}_t] \\ &\quad + \frac{1}{2} \left[\mathbf{B}^T(t, \boldsymbol{X}_t) \cdot \frac{\partial}{\partial \boldsymbol{x}} \right] \cdot [\mathbf{T}(\boldsymbol{X}_t) \cdot \mathbf{B}(t, \boldsymbol{X}_t)]^T \, \mathrm{d}t \,. \end{aligned} \tag{3.105}$$

When $\boldsymbol{X}_t = \boldsymbol{g}^{-1}(\boldsymbol{Y}_t)$ is introduced into (3.105), we arrive at an Itô stochastic differential equation for $\boldsymbol{Y}_t$. According to the rules for going back and forth between Itô and Stratonovich differential equations obtained in Example 3.41, (3.105) can be rewritten in the form of the equivalent Stratonovich stochastic differential equation

$$\mathrm{d}\boldsymbol{Y}_t = \mathbf{T}(\boldsymbol{X}_t) \cdot [\boldsymbol{A}(t, \boldsymbol{X}_t)\, \mathrm{d}t + \mathbf{B}(t, \boldsymbol{X}_t) \odot \mathrm{d}\boldsymbol{W}_t] \,. \tag{3.106}$$

In order to realize that the term in the second line of (3.105) is exactly the modification of the drift term required to obtain the simple Stratonovich stochastic differential equation (3.106) one should keep in mind (i) that the role of $\mathbf{B}$ is now taken over by $\mathbf{T} \cdot \mathbf{B}$ (see first line of (3.105)), and (ii) that $(\partial / \partial \boldsymbol{x})$ can be written as $(\partial / \partial \boldsymbol{x}) = \mathbf{T}^T(\boldsymbol{x}) \cdot (\partial / \partial \boldsymbol{y})$.

In (3.106), $\boldsymbol{X}_t = \boldsymbol{g}^{-1}(\boldsymbol{Y}_t)$ has to be introduced in order to obtain the final Stratonovich stochastic differential equation. The simple transformation between the Stratonovich equations (3.102) and (3.106) is obtained by multiplying with the matrix of first-order partial derivatives $\mathbf{T}(\boldsymbol{X}_t)$, that is, by the rules of ordinary calculus. It is worth underlining the fact that the derivation of this simple transformation rule for Stratonovich stochastic differential equations was based on two essential ingredients: Itô's formula (3.43) and the definition (3.97) of Stratonovich integrals in terms of Itô integrals. □

We are now in a position to comment on the question as to which calculus is more suitable for the numerical integration of stochastic differential equations (see criterion (v) listed in the beginning of this subsection). If we compare (3.15) and (3.98) we notice that the Itô and Stratonovich discretizations for differential equations correspond, respectively, to explicit and implicit time integration schemes. Since, for most kinetic theory applications, the models can be chosen such that there are no problems with numerical stability, the less expensive explicit schemes are preferable (there is no system of equations to be solved in updating the configuration in each time step). Then, Itô's calculus is preferable. The desirable conversion of Stratonovich into Itô differential equations is in general problematic because it requires the evaluation of derivatives of $\mathbf{B}$ which is a very expensive step when the derivatives need to be evaluated numerically (e.g., when the decomposition $\mathbf{D} = \mathbf{B} \cdot \mathbf{B}^T$ has to be done numerically). A more detailed discussion of numerical integration schemes is postponed to the next section.

Throughout this subsection we have commented on all our initially formulated criteria for selecting Itô's or Stratonovich's calculus, except for probably the most controversial one: which calculus can be justified in formulating stochastic differential equations for certain classes of applications? This consideration deserves some more detailed comments.

If the δ-correlated noise in a stochastic differential equation is an idealization of a noise with a very short but nonzero correlation time, a standard argument leads to the conclusion that Stratonovich's calculus should be applied. That is, since the trajectories of a noise with nonzero correlation time can be treated by deterministic calculus, the familiar form of differentiation rules should be applicable no matter how short the correlation time is. Indeed, there is a theorem by Wong and Zakai (see Theorem 6.1 of [5]), which under quite general conditions guarantees that if $W_t = \text{as-}\lim_{n\to\infty} W_t^{(n)}$ (uniformly in t) for a sequence of processes $W^{(n)}$ with continuous trajectories of bounded variation with piecewise continuous derivatives, then the solution of a Stratonovich differential equation is the as-limit of the solutions to the differential equations obtained when replacing W by $W^{(n)}$ (note that each trajectory of the latter differential equations can be obtained by deterministic calculus). There are even more general results which can be used to justify Stratonovich's calculus based on the idea of convergence in distribution. For example, in a theorem by Papanicolaou and Kohler (see Theorem 6.2 in [5] or Sect. 6.5 of [8]), the irregular behavior of the Wiener process is approximated in distribution by rescaling time and amplitude of a given smoothly correlated process (for increasing rescaling factors, the trajectories become more and more irregular).

While the theorem of Wong and Zakai offers a strong argument in favor of Stratonovich's stochastic calculus, it should be noted that in many applications the underlying idealizations are not based on rigorous limiting procedures but on certain more or less justified approximations. For example, if a stochastic differential equation for the slow variables of a deterministic system is derived by separating the dynamics of slow and fast variables by means of the projection operator formalism, what is then the appropriate calculus? One might be tempted to argue that Stratonovich's calculus is the proper choice because the starting point equations are deterministic (although the theorem of Wong and Zakai is clearly not applicable). However, the fact that a noise derived by means of the projection operator formalism should have a vanishing expectation implies that, in the multiplicative noise case, Itô's calculus must be used [27] (only in Itô's calculus does one have $\langle \mathbf{B}(t, \boldsymbol{X}_t) \cdot \mathrm{d}\boldsymbol{W}_t \rangle = \boldsymbol{0}$, which is a loosely stated version of (3.30)). Admittedly, "projection operator techniques are purely formal and cannot therefore provide a justification for the use of stochastic methods" (p. 186 of [24]); however, if it is assumed that a system may be described by a stochastic differential equation and if its precise form is obtained from projection operator techniques, then one is forced to use the Itô approach.

There are other situations in which Itô's calculus can be justified in a rigorous manner. In approximating jump processes by means of stochastic differential equations, Itô's calculus arises very naturally (see Sect. 6.1 of [5] or Sect. 7.2.1 of [8]). Therefore, Itô's stochastic calculus is usually applicable when stochastic differential equations are used to describe chemical reactions or population dynamics. For a model of diffusion for spatially varying temperature, van Kampen has derived the Itô version of the Fokker-Planck equation [28]. In summary, the selection of the Itô or Stratonovich interpretation for a given stochastic differential equation should be based on a very careful analysis of the underlying problem or of the derivation of the equation. For certain classes of problems or situations we have indicated the proper choice of stochastic calculus.

In concluding this subsection on Stratonovich's stochastic calculus we want to obtain the Fokker-Planck equation corresponding to a given Stratonovich stochastic differential equation (3.102). Knowing the equation governing the time-evolution of the probability density and the transition probabilities for the solution of the equivalent Itô equation (3.103), we immediately obtain for the desired Fokker-Planck equation corresponding to (3.102),

$$\boxed{\frac{\partial}{\partial t}\,p(t,\boldsymbol{x}) = -\frac{\partial}{\partial \boldsymbol{x}}\cdot[\boldsymbol{A}(t,\boldsymbol{x})\,p(t,\boldsymbol{x})] + \frac{1}{2}\,\frac{\partial}{\partial \boldsymbol{x}}\cdot\left\{\mathbf{B}(t,\boldsymbol{x})\,\frac{\partial}{\partial \boldsymbol{x}}:[\mathbf{B}(t,\boldsymbol{x})\,p(t,\boldsymbol{x})]\right\}.} \tag{3.107}$$

In component form, the second-order derivative term in (3.107) reads $(1/2)(\partial/\partial x_j)B_{jn}(\partial/\partial x_k)B_{kn}\,p$ where it is understood that the repeated letter indices j and k are to be summed over from 1 to d, that n is to be summed over from 1 to d', and that the partial derivatives act on all terms to the right of them.

In the Itô version of the Fokker-Planck equation, only the drift vector $\boldsymbol{A}$ and the diffusion matrix $\mathbf{D}$ occur. The separation of drift and diffusion is unambiguous, and the quantities $\boldsymbol{A}$ and $\mathbf{D}$ are fundamental concepts in the general theory of *diffusion processes* (see, e.g., Sect. 1.8 of [5]). A decomposition of the diffusion matrix $\mathbf{D}$ in the form $\mathbf{D} = \mathbf{B}\cdot\mathbf{B}^T$ is not required. For the Stratonovich version of the Fokker-Planck equation, the situation is completely different. Different matrices $\mathbf{B}$ in the decomposition of a given $\mathbf{D} = \mathbf{B}\cdot\mathbf{B}^T$ lead to different Fokker-Planck equations (3.107) when $\boldsymbol{A}$ is kept constant. In order not to change the solution of the Fokker-Planck equation when using different matrices $\mathbf{B}$ one has to modify the drift term appropriately. In other words, there is no unique or obvious distinction between drift and diffusion effects in Stratonovich's calculus. The contribution to the drift term depending on the particular decomposition of $\mathbf{D}$ is known as the "spurious drift" or the "noise-induced drift." The most general structure of spurious drifts has been analyzed in [29, 30]. For example, for $d \geq 3$ and $\det\mathbf{D} > 0$, any drift term can be compensated by the spurious drift for a suitable decomposition of $\mathbf{D}$; the drift is completely arbitrary.

Exercise 3.44 For

$$\mathbf{B}(\boldsymbol{x}) = \begin{pmatrix} \cos g(\boldsymbol{x}) & -\sin g(\boldsymbol{x}) \\ \sin g(\boldsymbol{x}) & \cos g(\boldsymbol{x}) \end{pmatrix},$$

where $g : \mathbb{R}^2 \to \mathbb{R}$ is a differentiable mapping, the solution of the Itô stochastic differential equation $\mathrm{d}\boldsymbol{X}_t = \mathbf{B}(\boldsymbol{X}_t) \cdot \mathrm{d}\boldsymbol{W}_t$ with $\boldsymbol{X}_0 = \boldsymbol{0}$ is a two-dimensional Wiener process (where $\boldsymbol{W}$ is also a two-dimensional Wiener process). Show that the corresponding Stratonovich stochastic differential equation contains a spurious drift term which depends on the mapping g and vanishes only if g is constant. Notice that the divergence of the spurious drift for this two-dimensional problem vanishes for all sufficiently smooth g [29].

3.4 Numerical Integration Schemes

We are now in a position to formulate stochastic differential equations rigorously and, in particular, we have established a close relationship between stochastic differential equations and Fokker-Planck equations. For example, we can rewrite the polymer dynamics that are usually described by diffusion or Fokker-Planck equations in terms of stochastic differential equations of motion. What can we gain by such a reformulation? Is the reformulated problem any more tractable than the original one?

Like deterministic differential equations their stochastic counterparts can be solved only in very special cases. Solvable stochastic differential equations are essentially the linear ones discussed in Sect. 3.3.2 and those which can be reduced to linear equations by suitable nonlinear transformations. The stochastic differential equations associated with all the interesting nonlinear models of polymer kinetic theory cannot be solved analytically. However, powerful numerical integration schemes have been developed for solving stochastic differential equations. Since random numbers are involved, such integration schemes may be regarded as simulation algorithms. The significance of the relationship between Fokker-Planck equations and stochastic differential equations can hence be described as follows: for a high-dimensional configuration space, it is hopeless to solve a Fokker-Planck equation by the usual numerical methods for treating partial differential equations; the associated stochastic differential equations, however, can be integrated numerically even for a large number of degrees of freedom, and averages or other properties of the distribution of the solution can be evaluated from an ensemble of trajectories.

The advantages of using stochastic differential equations rather than Fokker-Planck equations seem to violate an empirical conservation law which states that solving a given problem by any reasonable approach requires roughly the same amount of effort. This paradox can be resolved by noting that solution of the Fokker-Planck equation for a high-dimensional problem yields the full information about the distribution. The numerical integration of a stochastic differential equation yields an ensemble of trajectories from which one can evaluate averages, but one cannot obtain the distribution of the solution in a high-dimensional space. One might try to obtain information about the distribution by dividing the configuration space into cells and counting the number of trajectories which assume values in each cell. In other words, one could try to construct a histogram approximating the distribution. However, in a high-dimensional configuration space, for any reasonable approximation to the distribution function the number of cells would be so large that most cells would be unoccupied. In conclusion, high-dimensional distribution functions can be obtained neither from Fokker-Planck equations nor from stochastic differential equations; however, averages or contracted distribution functions can be obtained directly by numerical integration of stochastic differential equations, whereas the Fokker-Planck equation approach fails when the full distribution cannot be determined.

In this section, we provide the mathematical background for the numerical integration of stochastic differential equations. Since the mid-1970s many papers on sophisticated numerical integration schemes for stochastic differential equations appeared in the mathematical literature, and 20 years later this is still a field of very active research. A comprehensive description of efficient numerical integration methods can be found in a recent book by P. E. Kloeden and E. Platen [31]. A number of rigorous results on the convergence and stability of various numerical integration methods can be found in Chap. 7 of Gard's book [5]. Motivated by kinetic theory equations for polymeric liquids, a pioneering study of sophisticated integration methods for stochastic differential equations was performed by A. Greiner, W. Strittmatter and J. Honerkamp [32]. Earlier work on higher-order integration schemes for equations with additive noise by E. Helfand was also done to provide methods for treating complex physical systems, such as polymer solutions [33]. The methods described in the following subsections constitute the background for the simulation algorithms developed in the second part of the book, and plenty of examples can be found in all the subsequent chapters on various types of kinetic theory models. Since a considerable fraction of the remainder of this book may be regarded as a collection of examples and exercises illustrating the usefulness and the implementation of numerical integration schemes we refrain from giving numerical exercises in this section.

3.4.1 Euler's Method

Assume that we are interested in the solution $\boldsymbol{X} = (\boldsymbol{X}_t)_{t\in\mathbb{T}}$, $\mathbb{T} = [0, t_{\max}]$, of the multicomponent Itô stochastic differential equation

$$d\boldsymbol{X}_t = \boldsymbol{A}(t, \boldsymbol{X}_t)\, dt + \mathbf{B}(t, \boldsymbol{X}_t) \cdot d\boldsymbol{W}_t\,, \tag{3.108}$$

with a d-component initial condition $\boldsymbol{X}_0$ which is independent of the given d'-dimensional Wiener process $\boldsymbol{W} = (\boldsymbol{W}_t)_{t\in\mathbb{T}}$. The solution $\boldsymbol{X}$ is defined on the same given probability space $(\Omega, \mathcal{A}, P)$ on which $\boldsymbol{X}_0$ and $\boldsymbol{W}$ are defined. The components of the d-dimensional column vector $\boldsymbol{A}$ and of the $d \times d'$-matrix $\mathbf{B}$ are assumed to be measurable functions on $\mathbb{T} \times \mathbb{R}^d$. The simplest discretization of (3.108) is the Euler scheme, which is sometimes also called the Euler-Maruyama scheme (the mean-square convergence of this scheme was first established by G. Maruyama in 1955). For a given partition $0 = t_0 < t_1 \cdots < t_{n-1} < t_n = t_{\max}$ of the time range $\mathbb{T} = [0, t_{\max}]$, the Euler scheme is given by

$$\boxed{\widetilde{\boldsymbol{Y}}_{j+1} = \widetilde{\boldsymbol{Y}}_j + \boldsymbol{A}(t_j, \widetilde{\boldsymbol{Y}}_j)\,(t_{j+1} - t_j) + \mathbf{B}(t_j, \widetilde{\boldsymbol{Y}}_j) \cdot (\boldsymbol{W}_{t_{j+1}} - \boldsymbol{W}_{t_j})\,,} \tag{3.109}$$

for $j = 0, 1, \ldots, n-1$, where the initial value is $\widetilde{\boldsymbol{Y}}_0 = \boldsymbol{X}_0$. We certainly expect that, for any j, the vector-valued random variable $\widetilde{\boldsymbol{Y}}_j$ generated by this iterative scheme constitutes an approximation to $\boldsymbol{X}_{t_j}$. Equation (3.109) is often referred to as the stochastic difference equation associated with the differential equation (3.108). Whenever it is convenient or important to have an approximate solution of a continuous time argument rather than only at discrete times one needs an interpolation rule; the most commonly used rule is the linear interpolation $\boldsymbol{Y} = (\boldsymbol{Y}_t)_{t\in\mathbb{T}}$,

$$\boldsymbol{Y}_t = \widetilde{\boldsymbol{Y}}_j + \frac{t - t_j}{t_{j+1} - t_j}\,(\widetilde{\boldsymbol{Y}}_{j+1} - \widetilde{\boldsymbol{Y}}_j) \qquad \text{for } t \in [t_j, t_{j+1}[\,, \tag{3.110}$$

but for higher-order integration methods more sophisticated interpolation rules in terms of iterated stochastic integrals would be more appropriate (see Sect. 10.6 of [31]).

A recursion relation like (3.109) for calculating the values $\widetilde{\boldsymbol{Y}}_j$ at given discrete times together with an interpolation rule like (3.110) for constructing the continuous-time process $(\boldsymbol{Y})_{t\in\mathbb{T}}$ is referred to as an *approximation scheme* or a *numerical integration scheme*. Whenever we are interested only in the solution at the given time $t_{\max}$, we do not need to specify the interpolation rule (we then refer to a recursion relation as an approximation or numerical integration scheme). The Euler scheme is motivated by the definition of the stochastic integral in (3.15); as it has been discussed in connection with that definition, it is crucial that the matrix $\mathbf{B}$ be evaluated at $\widetilde{\boldsymbol{Y}}_j$ and not at $\widetilde{\boldsymbol{Y}}_{j+1}$ or any other linear combination of $\widetilde{\boldsymbol{Y}}_j$ and $\widetilde{\boldsymbol{Y}}_{j+1}$. The matrix $\mathbf{B}(t_j, \widetilde{\boldsymbol{Y}}_j)$ and the increment $\boldsymbol{W}_{t_{j+1}} - \boldsymbol{W}_{t_j}$ are hence independent, which is the most important characteristic of Itô's calculus.

In order to formulate statements about the quality of approximation schemes for solving stochastic differential equations, we introduce the concept of the *order of strong convergence*. Towards that end we consider different partitions of the time range $\mathbb{T}$ and the processes $\boldsymbol{Y}^{\Delta t}$ generated by a given approximation scheme where Δt indicates the maximum time step, $\max_{j\leq n}(t_j - t_{j-1})$. We say that an approximation scheme converges strongly with order $v > 0$ at time $t_{\max}$ if there exists a positive constant c, which does not depend on Δt, such that for sufficiently small Δt

$$\left\langle |\boldsymbol{X}_{t_{\max}} - \boldsymbol{Y}^{\Delta t}_{t_{\max}}|^2 \right\rangle^{1/2} \leq c\,(\Delta t)^v\,. \tag{3.111}$$

This definition is a generalization of the usual deterministic order of convergence because it reduces to the latter when the diffusion coefficient vanishes and the initial condition is deterministic (under these conditions the angular brackets in (3.111) can be omitted). The generalization of the deterministic global truncation error used here is based on an approximation of the solution of a given stochastic differential equation according to the mean-square difference. For an approximation scheme with order of strong convergence $\upsilon > 0$, the mean-square limit of any sequence of approximations for which the maximum time step goes to zero is given by the solution of the underlying stochastic differential equation. According to Sect. 2.2.6, the mean-square limit is unique outside a null set so that a strongly convergent scheme gives approximations to the individual trajectories. The concept of a strongly convergent scheme is related to strong solutions to stochastic differential equations because it ensures the convergence of trajectories for given probability space, Wiener process, and initial condition. Such approximation schemes are relevant when trajectories for a given setup need to be compared or when the characteristic features of trajectories need to be faithfully resolved (e.g., in studying or optimizing the influence of certain parameters, in understanding the consequences of certain interactions, in testing approximation schemes, or in preparing movies to illustrate or better understand some stochastic motions; see, e.g., Figs. 3.1 and 4.9). Furthermore, strong approximation schemes are the starting point for constructing the weak schemes considered in Sect. 3.4.3.

For deterministic differential equations, the Euler scheme is known to converge with order 1. The modified result for stochastic differential equations is given in the following theorem [5, 31].

Theorem 3.45 *Consider the stochastic differential equation (3.108), where* $\left\langle \boldsymbol{X}_0^2 \right\rangle$ *is finite, the coefficient functions* $\boldsymbol{A}$ *and* $\mathbf{B}$ *satisfy the Lipschitz and linear growth conditions (3.48)–(3.51), and*

$$\begin{aligned} |\boldsymbol{A}(t,\boldsymbol{x}) - \boldsymbol{A}(t',\boldsymbol{x})| &\le c\,(1+|\boldsymbol{x}|)\,|t-t'|^{1/2}\,, \\ |\mathbf{B}(t,\boldsymbol{x}) - \mathbf{B}(t',\boldsymbol{x})| &\le c\,(1+|\boldsymbol{x}|)\,|t-t'|^{1/2}\,, \end{aligned}$$

for all $t, t' \in \mathbb{T}$ *and* $\boldsymbol{x} \in \mathbb{R}^d$*, where* c *is a constant. Then, the Euler scheme for solving (3.108) converges strongly with order* $\upsilon = 1/2$.

Instead of giving a proof of this theorem (see, e.g., [5, 31]) we heuristically elucidate the reasons for the low order of strong convergence of the stochastic Euler scheme ($\upsilon = 1/2$ instead of 1 for the deterministic case). In writing down the Euler scheme (3.109) one assumes that the coefficient matrix $\mathbf{B}$ throughout the interval $[t_j, t_{j+1}]$ may be evaluated at the initial configuration $\widetilde{\boldsymbol{Y}}_j$ instead of a time-dependent configuration. During a time interval of size Δt, these time-dependent configurations differ from the initial ones by stochastic terms of order $(\Delta t)^{1/2}$, and so does then the coefficient $\mathbf{B}$. This is the source of the order 1/2 deviations for nonzero, configuration-dependent coefficients $\mathbf{B}$. These arguments show that, in the case of additive noise, a higher order of convergence can be expected.

In view of this low order of strong convergence of the Euler scheme it is natural to look for higher-order schemes. From the experience with deterministic differential equations, a natural approach would be to take into account higher-order terms in the time steps $t_{j+1} - t_j$ and in the increments of the Wiener process $\boldsymbol{W}_{t_{j+1}} - \boldsymbol{W}_{t_j}$. However, it can be shown that, in general, this strategy does not lead to higher-order schemes (see Chap. 7 of [5]). It is the purpose of the following subsection to illustrate why more detailed information about the time evolution of the Wiener process in each time step is required, and how higher-order schemes can be developed.

3.4.2 Mil'shtein's Method

Usually the Euler scheme gives good numerical results when the drift and diffusion coefficients are nearly constant. This scheme is still most widely used in Brownian dynamics simulations of polymer molecules. In general, however, it is not particular satisfactory and higher-order schemes should be developed. We here discuss a scheme proposed by G. N. Mil'shtein in 1974 which is of strong order 1. The strong order $v = 1$ of the Mil'shtein scheme corresponds to that of the simpler Euler scheme in the deterministic case without any noise, that is when $\mathbf{B} = \mathbf{0}$.

In order to explain the Mil'shtein scheme we start from the integral version of the stochastic differential equation (3.108) for the interval $[t_j, t_{j+1}]$,

$$\boldsymbol{X}_{t_{j+1}} = \boldsymbol{X}_{t_j} + \int_{t_j}^{t_{j+1}} \boldsymbol{A}(t, \boldsymbol{X}_t)\, \mathrm{d}t + \int_{t_j}^{t_{j+1}} \mathbf{B}(t, \boldsymbol{X}_t) \cdot \mathrm{d}\boldsymbol{W}_t \,. \tag{3.112}$$

The Euler scheme is obtained when the integrands in (3.112) are evaluated at the initial time t_j. While the error introduced by replacing t with t_j is of order $t - t_j$, the deviation between $\boldsymbol{X}_t$ and $\boldsymbol{X}_{t_j}$ is of order $(t-t_j)^{1/2}$, where the precise form of these leading-order corrections is

$$\boldsymbol{X}_t \approx \boldsymbol{X}_{t_j} + \mathbf{B}(t_j, \boldsymbol{X}_{t_j}) \cdot (\boldsymbol{W}_t - \boldsymbol{W}_{t_j}) \,.$$

The corresponding leading-order corrections in $\mathbf{B}$ can be obtained from the Itô formula,

$$\mathbf{B}(t, \boldsymbol{X}_t) \approx \mathbf{B}(t_j, \boldsymbol{X}_{t_j}) + (\boldsymbol{W}_t - \boldsymbol{W}_{t_j}) \cdot \mathbf{B}^T(t_j, \boldsymbol{X}_{t_j}) \cdot \frac{\partial}{\partial \boldsymbol{x}} \mathbf{B}(t_j, \boldsymbol{X}_{t_j}) \,, \tag{3.113}$$

and these corrections of order $(t - t_j)^{1/2}$ should be taken into account in approximating the stochastic integral in (3.112) in order to achieve an improved order of strong convergence, namely $v = 1$. When this improved approximation for $\mathbf{B}(t, \boldsymbol{X}_t)$ is introduced into (3.112) we obtain the Mil'shtein scheme,

$$\begin{aligned} \widetilde{\boldsymbol{Y}}_{j+1} &= \widetilde{\boldsymbol{Y}}_j + \boldsymbol{A}(t_j, \widetilde{\boldsymbol{Y}}_j)\,(t_{j+1} - t_j) + \mathbf{B}(t_j, \widetilde{\boldsymbol{Y}}_j) \cdot (\boldsymbol{W}_{t_{j+1}} - \boldsymbol{W}_{t_j}) \\ &\quad + \mathbf{I}^{(2)}_{t_j t_{j+1}} : \left[\mathbf{B}^T(t_j, \widetilde{\boldsymbol{Y}}_j) \cdot \frac{\partial}{\partial \boldsymbol{x}}\right] \mathbf{B}^T(t_j, \widetilde{\boldsymbol{Y}}_j) \,, \end{aligned} \tag{3.114}$$

where the symbol $\mathbf{I}^{(2)}_{t_j t_{j+1}}$ stands for the following stochastic integral,

$$\mathbf{I}^{(2)}_{t_j t_{j+1}} := \left[\int_{t_j}^{t_{j+1}} (\boldsymbol{W}_t - \boldsymbol{W}_{t_j})\, \mathrm{d}\boldsymbol{W}_t \right]^T \,. \tag{3.115}$$

The Mil'shtein scheme differs from the Euler scheme by the correction term in the second line of (3.114). In this correction term, the derivative of $\mathbf{B}(t, \boldsymbol{x})$ with respect to $\boldsymbol{x}$ is to be taken before $\widetilde{\boldsymbol{Y}}_j$ is inserted into $\mathbf{B}(t, \boldsymbol{x})$. The analytic calculation of the derivatives of $\mathbf{B}$ can be avoided by replacing them with the corresponding finite differences [31] (at the expense of evaluating the diffusion term $d' + 1$ times in each time step; see also Sect. 3.4.4). In order to illustrate the meaning of the double contraction in the correction term in the second line of (3.114) we give the k-component of that term in component form, $I^{(2)}_{n'n} \, B_{ln} \, (\partial/\partial x_l) \, B_{kn'}$, where all arguments have been suppressed.

Under the conditions of Theorem 3.45, augmented by similar Lipschitz and growth conditions for certain first and second spatial derivatives of the coefficient functions $\boldsymbol{A}$ and $\mathbf{B}$ (see Theorem 10.3.5 of [31]), the order of strong convergence $v = 1$ of the Mil'shtein scheme can be established in a rigorous manner (see also Sect. 7.5 of [5]). If there is additive noise, the correction term in the second line of (3.114) is absent, and the result for the Mil'shtein scheme implies that the Euler scheme already has the order of strong convergence $v = 1$.

The symmetric part of the stochastic integral $\mathbf{I}^{(2)}_{t_j t_{j+1}}$ can easily be expressed in terms of the increments of the Wiener process,

$$\mathbf{I}^{(2)}_{t_j t_{j+1}} + \left[\mathbf{I}^{(2)}_{t_j t_{j+1}}\right]^T = (\boldsymbol{W}_{t_{j+1}} - \boldsymbol{W}_{t_j})(\boldsymbol{W}_{t_{j+1}} - \boldsymbol{W}_{t_j}) - (t_{j+1} - t_j)\, \boldsymbol{\delta} \,, \qquad (3.116)$$

as can be verified by evaluating its differential by means of the Itô formula. The integral $\mathbf{I}^{(2)}_{t_j t_{j+1}}$ itself, however, cannot be expressed in terms of the increments of the Wiener process and is quite expensive to simulate (see Sects. 5.8 or 10.3 of [31]). More detailed information about the time evolution of the Wiener process between t_j and t_{j+1} is required in order to simulate the antisymmetric part of $\mathbf{I}^{(2)}_{t_j t_{j+1}}$. The Mil'shtein scheme is hence particularly useful when only the symmetric part of $\mathbf{I}^{(2)}_{t_j t_{j+1}}$ matters, and we next discuss special cases of this type.

Only the symmetric part of the stochastic integrals $\mathbf{I}^{(2)}_{t_j t_{j+1}}$ is relevant whenever the following condition holds for all $k = 1, \ldots, d$, for all $n, n' = 1, \ldots, d'$, $t \in \mathbb{T}$, and $\boldsymbol{x} \in \mathbb{R}^d$,

$$B_{ln}(t, \boldsymbol{x}) \, \frac{\partial B_{kn'}(t, \boldsymbol{x})}{\partial x_l} = B_{ln'}(t, \boldsymbol{x}) \, \frac{\partial B_{kn}(t, \boldsymbol{x})}{\partial x_l} \,. \qquad (3.117)$$

In the case of additive noise, that is, when $\mathbf{B}$ is a function of t only, the condition (3.117) is trivially fulfilled; $\mathbf{I}^{(2)}_{t_j t_{j+1}}$ is not needed at all. An important special case in which (3.117) is fulfilled and simulation of the symmetrized stochastic integrals (3.116) is sufficient for applying the Mil'shtein scheme occurs when $\mathbf{B}$ is diagonal and its kth diagonal element depends only on x_k and t. Finally, the Mil'shtein scheme requires only the knowledge of the increments of the Wiener process when $d' = 1$, that is, when only one real-valued Wiener process is involved in a stochastic differential equation.

Higher-order schemes can be developed in a way that is very similar to the procedure for the Mil'shtein scheme. They are based on better approximations for the coefficients $\boldsymbol{A}(t, \boldsymbol{X}_t)$ and $\mathbf{B}(t, \boldsymbol{X}_t)$ in (3.112), which can be constructed by an iterated application of the Itô formula (by such an iterative procedure one can improve approximations like (3.113)). Since the increments of the Wiener process are of the order of the square root of the corresponding time step, a general scheme with order v of strong convergence involves up to $2v$-fold nested stochastic integrals. A straightforward, systematic procedure for constructing higher-order approximations is most conveniently based on the Itô-Taylor expansion (see Chap. 5 of [31]). The rather complicated explicit recursion relations for schemes converging strongly with order $v = 3/2$ and 2, both involving derivatives of the coefficient functions or avoiding them by means of finite differences, are given in Chaps. 10 and 11 of [31].

The occurrence of stochastic integrals such as (3.115), which in general cannot be expressed in terms of increments of the Wiener process over the given time intervals, is the source of the biggest problems in adapting numerical integration techniques for deterministic differential equations to stochastic equations. A very clear and comprehensive discussion of the limitations of approximation schemes that work only with increments of the underlying Wiener process is contained in Chap. 7 of [5]. Evaluation of integrals such as (3.115) requires more detailed information about the time evolution of the Wiener process in each time step and is quite expensive in computer simulations. The simplest simulation of $\mathbf{I}^{(2)}_{t_j t_{j+1}}$ is obtained by further dividing each time step into n subintervals and discretizing the integrals (3.115). In order to achieve the theoretical order of strong convergence of the Mil'shtein scheme, n has to increase as $1/\Delta t$ for decreasing Δt. A more efficient simulation scheme for $\mathbf{I}^{(2)}_{t_j t_{j+1}}$ can be based on the Fourier representation of the trajectories of the Wiener process (see Sects. 5.8 or 10.3 of [31], where a truncation parameter p ten times smaller than n is sufficient to obtain roughly the same accuracy).

3.4.3 Weak Approximation Schemes

In the previous sections, we studied the strong convergence of certain approximation schemes for solving stochastic differential equations. Often, one is not interested in constructing the individual trajectories of the solutions on a given probability space but only in certain moments or averages. In particular, this is the case when a stochastic differential equation has been constructed from a Fokker-Planck equation because the latter equation contains only information about the distribution of a stochastic process and not about its trajectories; in other words, one is then interested in a weak solution of the stochastic differential equation.

We say that an approximation scheme converges weakly with order $v > 0$ at time $t_{\max}$ if, for all sufficiently smooth functions $g : \mathbb{R}^d \to \mathbb{R}$ with polynomial growth, there exists a positive constant c_g, which does not depend on Δt, such that for sufficiently small Δt

$$\left|\left\langle g(\boldsymbol{X}_{t_{\max}})\right\rangle - \left\langle g(\boldsymbol{Y}^{\Delta t}_{t_{\max}})\right\rangle\right| \leq c_g\, (\Delta t)^v . \tag{3.118}$$

More precisely, the functions g must be $2(v+1)$ times continuously differentiable, and all partial derivatives up to order $2(v+1)$, say $\tilde{g}$, must satisfy the polynomial growth condition $|\tilde{g}(\boldsymbol{x})| \leq c\,(1+|\boldsymbol{x}|^n)$, where c and n may depend only on $\tilde{g}$ (cf. the linear growth condition of Theorem 3.21). In the deterministic case, we again recover the deterministic order of convergence by choosing $g(\boldsymbol{x}) = x_j$ for $j = 1, 2, \ldots, d$, so that the orders of strong and weak convergence coincide. For an approximation scheme with order of weak convergence $v > 0$, the convergence of expectations for any sequence of approximations for which the maximum time step goes to zero implies that the limit in distribution is given by the solution of the underlying stochastic differential equation (see Sect. 2.2.6). The concept of a weakly convergent scheme is related to weak solutions to stochastic differential equations because only the distribution of the solution is of interest.

It is much simpler to construct schemes of a given weak rather than strong order of convergence. For a heuristic explanation, consider the Mil'shtein scheme (3.114). In that scheme, it follows from the definition (3.115) that the stochastic integrals $\mathbf{I}^{(2)}_{t_j t_{j+1}}$ are random variables of order Δt with vanishing average (see also (3.116)). In order to avoid errors of order Δt in the trajectories, one has to account for the effects of the stochastic integrals $\mathbf{I}^{(2)}_{t_j t_{j+1}}$, whereas in averages these terms of order Δt do not contribute and may thus be neglected. In summary, we expect that the Euler scheme (3.109), which we rewrite in an obvious notation as

$$\widetilde{\boldsymbol{Y}}_{j+1} = \widetilde{\boldsymbol{Y}}_j + \boldsymbol{A}(t_j, \widetilde{\boldsymbol{Y}}_j)\,\Delta t_j + \mathbf{B}(t_j, \widetilde{\boldsymbol{Y}}_j) \cdot \Delta \boldsymbol{W}_j\,, \tag{3.119}$$

converges weakly with order $v = 1$. Under suitable smoothness and growth conditions on the drift and diffusion coefficients, the order of weak convergence $v = 1$ can be rigorously established for the Euler scheme (see Theorem 14.5.1 of [31]).

Not only is the Euler scheme sufficient for first-order weak convergence but this first-order scheme can further be simplified by replacing the Gaussian increments $\Delta \boldsymbol{W}_j$ by other random variables with similar moments. The only requirement is that the probability distribution of the solution be approximated to the desired order of Δt. For the first-order Euler scheme, deviations should occur only in second order of the time step and all the components of the $\Delta \boldsymbol{W}_j$ thus have to be independent real-valued random variables X satisfying the following three moment conditions for some constant c,

$$\begin{aligned} |\langle X\rangle| &\leq c\,(\Delta t_j)^2\,, \\ \left|\left\langle X^2\right\rangle - \Delta t_j\right| &\leq c\,(\Delta t_j)^2\,, \\ \left|\left\langle X^3\right\rangle\right| &\leq c\,(\Delta t_j)^2\,. \end{aligned} \tag{3.120}$$

Of course, just as for the increments of the Wiener process, the random variables used in different time steps j also have to be independent. Simple possible choices are the two-point distributed random variable X taking the values $\pm(\Delta t_j)^{1/2}$ with probability $\frac{1}{2}$, or $X = (12\Delta t_j)^{1/2}\,(Y - \frac{1}{2})$ where the distribution of the random variable Y is uniform in $[0, 1]$.

Since it is so simple to obtain an approximation scheme of weak order $v = 1$ there is good reason for hoping that higher-order weak schemes might also be tractable. Such schemes can, for example, be developed in a systematic way by simplifying higher-order strong schemes resulting from the procedure described in Sect. 3.4.2. Just as for the weak first-order scheme, multiple stochastic integrals above a certain order do not contribute,[4] and the increments of the Wiener process and the contributing multiple integrals can be replaced with simpler random variables satisfying suitable moment conditions. In this way one obtains the following simplified approximation scheme of weak order $v = 2$ (see Sect. 14.2 of [31]),

$$\begin{aligned}
\widetilde{\boldsymbol{Y}}_{j+1} &= \widetilde{\boldsymbol{Y}}_j + \boldsymbol{A}(t_j, \widetilde{\boldsymbol{Y}}_j)\, \Delta t_j + \frac{1}{2}\left(\frac{\partial}{\partial t} + \mathcal{L}_t\right) \boldsymbol{A}(t_j, \widetilde{\boldsymbol{Y}}_j)\, (\Delta t_j)^2 \\
&+ \mathbf{B}(t_j, \widetilde{\boldsymbol{Y}}_j) \cdot \Delta \boldsymbol{W}_j + \frac{1}{2}\left(\frac{\partial}{\partial t} + \mathcal{L}_t\right) \mathbf{B}(t_j, \widetilde{\boldsymbol{Y}}_j) \cdot \Delta \boldsymbol{W}_j\, \Delta t_j \\
&+ \frac{1}{2} \Delta t_j\, \Delta \boldsymbol{W}_j \cdot \mathbf{B}^T(t_j, \widetilde{\boldsymbol{Y}}_j) \cdot \frac{\partial}{\partial \boldsymbol{x}} \boldsymbol{A}(t_j, \widetilde{\boldsymbol{Y}}_j) \\
&+ \frac{1}{2} (\Delta \boldsymbol{W}_j \Delta \boldsymbol{W}_j - \Delta t_j\, \boldsymbol{\delta} + \Delta \mathbf{V}_j) : \left[\mathbf{B}^T(t_j, \widetilde{\boldsymbol{Y}}_j) \cdot \frac{\partial}{\partial \boldsymbol{x}}\right] \mathbf{B}^T(t_j, \widetilde{\boldsymbol{Y}}_j) .
\end{aligned} \tag{3.121}$$

Here the increments of the Wiener process $\Delta \boldsymbol{W}_j$ may be replaced with independent random variables with independent components X satisfying the following moment conditions (saying that only third-order deviations from the corresponding Gaussian moments are allowed in a second-order scheme),

$$\begin{aligned}
|\langle X \rangle| &\leq c\,(\Delta t_j)^3 , \\
\left|\left\langle X^2 \right\rangle - \Delta t_j\right| &\leq c\,(\Delta t_j)^3 , \\
\left|\left\langle X^3 \right\rangle\right| &\leq c\,(\Delta t_j)^3 , \\
\left|\left\langle X^4 \right\rangle - 3(\Delta t_j)^2\right| &\leq c\,(\Delta t_j)^3 , \\
\left|\left\langle X^5 \right\rangle\right| &\leq c\,(\Delta t_j)^3 .
\end{aligned} \tag{3.122}$$

The diagonal components of the $d' \times d'$-matrix $\Delta \mathbf{V}_j$ occurring in (3.121) vanish, the components above the diagonal are independent two-point distributed random variables assuming the values $\pm \Delta t_j$ with probability $\frac{1}{2}$ or any other random variables satisfying the moment conditions,

$$\begin{aligned}
|\langle X \rangle| &\leq c\,(\Delta t_j)^3 , \\
\left|\left\langle X^2 \right\rangle - (\Delta t_j)^2\right| &\leq c\,(\Delta t_j)^3 ,
\end{aligned} \tag{3.123}$$

[4]More precisely, for a scheme with order v of weak convergence, the moments of the stochastic integrals are such that all terms involving more than v nested integrations do not contribute to averages (see Sect. 5.7 of [31]); therefore, all integrals of the scheme with order v of strong convergence that involve more than v-fold integrations can be omitted immediately.

and the components below the diagonal are chosen such that $\Delta \mathbf{V}_j$ is antisymmetric. The infinitesimal generator $\mathcal{L}_t$ was defined in (3.76), and it is understood that all derivatives with respect to $\boldsymbol{x}$ occurring in (3.121) are performed before $\boldsymbol{x}$ is replaced with $\widetilde{\boldsymbol{Y}}_j$.

Even though the second-order scheme (3.121) looks rather complicated, one can easily understand all the occurring terms. The first two lines contain the terms of the Euler scheme and the effects of systematic variations of the coefficients $\boldsymbol{A}$ and $\mathbf{B}$ during a time step; the systematic changes due to variation of $\widetilde{\boldsymbol{Y}}$ are contained in the differential operator $\mathcal{L}_t$ which accounts both for the deterministic drift of $\widetilde{\boldsymbol{Y}}$ (first-order derivatives) and the average mean-square displacement due to random forces (second-order derivatives). Notice that the combination $\partial/\partial t + \mathcal{L}_t$ is exactly what occurs in the systematic term of the Itô formula. The third line accounts for stochastic variations in the coefficient $\boldsymbol{A}$, and the corresponding term for the coefficient $\mathbf{B}$ is contained in the fourth line. The fourth line is closely related to the correction term in the Mil'shtein scheme (3.114) where, for a weak approximation, the antisymmetric part of $\mathbf{I}^{(2)}_{t_j t_{j+1}}$ is replaced with $\frac{1}{2}\Delta \mathbf{V}_j$. In the case of additive noise, the second-order scheme (3.121) is considerably simplified.

The moment conditions (3.122) can be realized in various simple ways. The simplest discrete representation is a three-point distributed random variable X assuming the values $\pm(3\,\Delta t_j)^{1/2}$ with probability $\frac{1}{6}$ and the value 0 with probability $\frac{2}{3}$. Alternatively, one can construct random variables X that have a continuous distribution and satisfy the moment conditions (3.122) from random variables Y with uniform distribution in $[0, 1]$ according to

$$X = \sqrt{\Delta t_j}\left(Y - \frac{1}{2}\right)\left[c_1\left|Y - \frac{1}{2}\right| + c_2\right], \tag{3.124}$$

or

$$X = \sqrt{\Delta t_j}\left(Y - \frac{1}{2}\right)\left[c'_1\left(Y - \frac{1}{2}\right)^2 + c'_2\right], \tag{3.125}$$

where the constants c_1, c_2, c'_1, and c'_2 can easily be determined to any desired precision by solving a pair of equations, one of which is quadratic and the other quartic. For most purposes, the following values should be (more than) sufficient,

$$c_1 = 10.72458949\,, \qquad c_2 = -0.71691890\,,$$
$$c'_1 = 14.14855378\,, \qquad c'_2 = 1.21569221\,.$$

The distribution of the random variable (3.124) was considered in Example 2.38, which showed that the nonmonotonic transformation from Y to X resulting from the opposite signs of c_1 and c_2 leads to an integrable singularity in the probability density of X.

Exercise 3.46 Determine the smooth probability density of the random variable (3.125).

3.4.4 More Sophisticated Methods

When applying the above approximation schemes to the solution of certain stochastic differential equations, for example in polymer kinetic theory, one may run into various problems. This section addresses several fundamental problems and shows how to avoid or overcome these. Moreover, we discuss various possibilities for making numerical approximation schemes for stochastic differential equations more efficient, so that this section may be regarded as a *checklist of ideas* to consider before starting a major simulation. Even though this checklist is certainly not complete and the improvement of numerical integration techniques for stochastic differential equations is a field of very active research, the following comments on more sophisticated methods may be helpful in many applications. Of course, the various ideas discussed separately in this section can and should be combined whenever it is appropriate.

For deterministic differential equations, Runge-Kutta and predictor-corrector methods are classical examples of higher-order schemes. The Runge-Kutta method, in particular when used with adaptive time-step control, is still considered as a useful workhorse when convenience is more important than high precision, and Runge-Kutta succeeds virtually always [34]. On the other hand, the usefulness of predictor-corrector methods now seems to be very limited; they are recommended only when high-precision solutions of very smooth equations involving complicated functions are required. Otherwise, if high-precision results are required, the method of choice should be based on extrapolations to zero time-step width. Both polynomial and rational extrapolation methods are used. The Bulirsch-Stoer method based on this idea has replaced predictor-corrector methods in most applications requiring high precision [34]. In the context of molecular dynamics simulations, that is, in integrating Newton's equations of motion for a set of molecules, the schemes of Gear and Verlet are very popular [35]. While the Gear algorithm is based on the predictor-corrector idea, the Verlet algorithm is an explicit two-step scheme especially designed for second-order differential equations. If the stability of numerical solutions is a problem in solving deterministic differential equations, in particular for stiff problems involving a wide range of time scales, implicit methods are a valuable alternative. In this section, we consider the generalization of all these ideas to stochastic differential equations, and we mention the additional possibility of variance reduction methods.

For stochastic differential equations, one of the most obvious problems is that higher-order weak or strong approximation schemes like (3.114) or (3.121) require derivatives of the coefficient functions $\boldsymbol{A}$ and $\mathbf{B}$, and these may be difficult or impossible to evaluate. In particular when the matrix $\mathbf{B}(t_j, \widetilde{\boldsymbol{Y}}_j)$ has to be constructed by numerical Cholesky decomposition of the diffusion matrix at a particular value $\widetilde{\boldsymbol{Y}}_j$, it is clearly impossible to obtain the spatial derivatives of $\mathbf{B}$ analytically. Another important problem is the numerical stability of the proposed integration schemes. Extrapolation and variance reduction methods for minimizing the effects of time-step width and the statistical error bars are further issues of this section.

We first discuss the problem caused by derivatives of the coefficient functions occurring in higher-order schemes, and we present several *derivative-free approximation schemes*. An obvious idea for avoiding such derivatives is to calculate the coefficient functions at several supporting values and to replace derivatives with appropriate differences. For example in the Mil'shtein scheme (3.114), we can use the replacement

$$B_{ln}(t_j, \widetilde{\boldsymbol{Y}}_j) \frac{\partial B_{kn'}(t_j, \widetilde{\boldsymbol{Y}}_j)}{\partial x_l} \to \frac{1}{\sqrt{\Delta t_j}} \left[B_{kn'}(t_j, \boldsymbol{\Upsilon}_j^n) - B_{kn'}(t_j, \widetilde{\boldsymbol{Y}}_j) \right] , \qquad (3.126)$$

where the supporting values $\boldsymbol{\Upsilon}_j^n$ for $n = 1, \ldots, d'$ are given by

$$\boldsymbol{\Upsilon}_j^n = \widetilde{\boldsymbol{Y}}_j + \boldsymbol{b}_n(t_j, \widetilde{\boldsymbol{Y}}_j) \sqrt{\Delta t_j} \qquad (3.127)$$

(as in Sect. 3.3.2, $\boldsymbol{b}_n$ is the nth column vector of the matrix $\mathbf{B}$). In the replacement (3.126) it is understood that the decomposition of the diffusion matrix is constructed such that the existence of the spatial derivatives of the diffusion matrix implies the existence of the derivatives of $\mathbf{B}$. For the second-order scheme (3.121), the same term requires a more symmetric and hence more accurate discretization of the derivatives,

$$B_{ln}(t_j, \widetilde{\boldsymbol{Y}}_j) \frac{\partial B_{kn'}(t_j, \widetilde{\boldsymbol{Y}}_j)}{\partial x_l} \to \frac{1}{2\sqrt{\Delta t_j}} \left[B_{kn'}(t_j, \boldsymbol{\Upsilon}_j^n) - B_{kn'}(t_j, \overline{\boldsymbol{\Upsilon}}_j^n) \right] , \qquad (3.128)$$

where, in addition to the $\boldsymbol{\Upsilon}_j^n$, the following d' supporting values have been introduced,

$$\overline{\boldsymbol{\Upsilon}}_j^n = \widetilde{\boldsymbol{Y}}_j - \boldsymbol{b}_n(t_j, \widetilde{\boldsymbol{Y}}_j) \sqrt{\Delta t_j} \, . \qquad (3.129)$$

The second-order spatial derivatives of $\mathbf{B}$ implied by $\mathcal{L}_t$ in (3.121) can be avoided by the replacement

$$\frac{1}{2} D_{nn'}(t_j, \widetilde{\boldsymbol{Y}}_j) \frac{\partial^2 B_{kl}(t_j, \widetilde{\boldsymbol{Y}}_j)}{\partial x_n \partial x_{n'}} \to \frac{1}{2\Delta t_j} \sum_{n=1}^{d'} \left[B_{kl}(t_j, \boldsymbol{\Upsilon}_j^n) + B_{kl}(t_j, \overline{\boldsymbol{\Upsilon}}_j^n) \right.$$
$$\left. - 2 \, B_{kl}(t_j, \widetilde{\boldsymbol{Y}}_j) \right] . \qquad (3.130)$$

The replacement (3.130) can be used to avoid not only the second-order-derivative contribution to $\mathcal{L}_t$ but also the first-order-derivative contribution; the latter contribution can be included by adding $\boldsymbol{A} \, \Delta t_j$ to one particular pair of supporting values $\boldsymbol{\Upsilon}_j^n$, $\overline{\boldsymbol{\Upsilon}}_j^n$, or by adding $\boldsymbol{A} \, \Delta t_j / d'$ to each of the d' pairs of supporting values. It does no harm if these shifted supporting values are also used in (3.128) in order to keep the number of evaluations of $\mathbf{B}$ as small as possible. It may even be worthwhile to add $\boldsymbol{A} \, \Delta t_j$ to $\boldsymbol{\Upsilon}_j^n$ in (3.126); this does not affect the order of strong convergence of the derivative-free Mil'shtein scheme, but it is an inexpensive way of reducing the coefficient of the leading-order corrections by accounting for some of the higher-order terms.

For many polymer problems, the numerical evaluation of **B** is the most expensive step. When this step has to be performed $2d'+1$ times for each time step, where d' typically is a large number, derivative-free second-order schemes are prohibitively expensive.

The approximation schemes obtained after eliminating derivatives of the coefficient functions by means of supporting values are referred to as *Runge-Kutta schemes*. It must be emphasized that these schemes are not straightforward generalizations of deterministic Runge-Kutta schemes to stochastic differential equations; the common name only refers to the common strategy for avoiding derivatives. As mentioned above, a major drawback of Runge-Kutta schemes is the fact that the coefficient functions must be evaluated many times. Derivative-free higher-order schemes requiring fewer function evaluations can sometimes be achieved by *multistep methods* in which $\widetilde{\boldsymbol{Y}}_{j+1}$ is expressed in terms of several previous values $\widetilde{\boldsymbol{Y}}_j$, $\widetilde{\boldsymbol{Y}}_{j-1}$ ... and the coefficient functions evaluated at several previous times (see Sect. 11.4 of [31]).

Multistep methods can sometimes lead to improved numerical stability. Improving the stability of integration schemes is also the idea behind using *implicit approximation schemes*. These implicit schemes are hence of crucial importance in solving stiff stochastic differential equations, that is, equations describing dynamical behavior with vastly different time scales. Under sufficient smoothness and regularity conditions for the coefficient functions all schemes discussed in this book are stable for sufficiently small time steps (for the Euler scheme, this can be proved under the conditions of Theorem 3.45). In practice, one is of course interested in keeping stability for time steps as large as possible, and this is why implicit methods are used.

Implicit methods can be applied in both strong and weak approximation schemes. For example, the implicit versions of the Euler and Mil'shtein schemes are obtained by the following replacements in (3.109) and (3.114),

$$\boldsymbol{A}(t_j, \widetilde{\boldsymbol{Y}}_j) \to c\,\boldsymbol{A}(t_{j+1}, \widetilde{\boldsymbol{Y}}_{j+1}) + (1-c)\,\boldsymbol{A}(t_j, \widetilde{\boldsymbol{Y}}_j), \tag{3.131}$$

where $c \in [0,1]$ is the degree of implicitness, and $c = \frac{1}{2}$ is a particularly popular choice. In general, $\widetilde{\boldsymbol{Y}}_{j+1}$ then has to be determined by solving a nonlinear equation. Notice that only the drift term, but not the diffusion term, can be made implicit for strong approximation schemes. Even for very small time steps, there is a nonzero probability for large increments $\Delta \boldsymbol{W}_j$ which cause the solution $\widetilde{\boldsymbol{Y}}_{j+1}$ of the implicit equation to become singular so that $\left\langle |\widetilde{\boldsymbol{Y}}_{j+1}| \right\rangle$ is infinite and the concept of strong convergence becomes meaningless. Since implicit schemes modify the drift term whereas Runge-Kutta schemes are primarily intended for avoiding derivatives in the diffusion term, there is no problem with combining these two ideas.

For weak approximation schemes, the diffusion term can also be treated implicitly when the Gaussian increments are replaced with bounded random variables satisfying proper moment conditions. However, the Itô-Stratonovich problem indicates that it is necessary to be careful when the diffusion term is not evaluated with the configuration at the beginning of a time step. For example, the fully implicit weak Euler scheme is

$$\begin{aligned}\widetilde{\boldsymbol{Y}}_{j+1} = \widetilde{\boldsymbol{Y}}_j \; &+ \; \left[c\,\overline{\boldsymbol{A}}(t_{j+1}, \widetilde{\boldsymbol{Y}}_{j+1}) + (1-c)\,\overline{\boldsymbol{A}}(t_j, \widetilde{\boldsymbol{Y}}_j)\right] \Delta t_j \\ &+ \; \left[c'\,\mathbf{B}(t_{j+1}, \widetilde{\boldsymbol{Y}}_{j+1}) + (1-c')\,\mathbf{B}(t_j, \widetilde{\boldsymbol{Y}}_j)\right] \cdot \Delta \boldsymbol{W}_j\,, \qquad (3.132)\end{aligned}$$

where

$$\overline{\boldsymbol{A}}(t, \boldsymbol{x}) = \boldsymbol{A}(t, \boldsymbol{x}) - c' \left[\mathbf{B}^T(t, \boldsymbol{x}) \cdot \frac{\partial}{\partial \boldsymbol{x}}\right] \cdot \mathbf{B}^T(t, \boldsymbol{x})\,,$$

and $c, c' \in [0,1]$ determine the degree of implicitness. For $c' = 0$, (3.132) corresponds to the Itô approach, whereas $c' = \frac{1}{2}$ corresponds to the equivalent Stratonovich equation (see Sect. 3.3.6). If the diffusion matrix cannot be decomposed in a closed form then the calculation of $\overline{\boldsymbol{A}}$ requires $d' + 1$ evaluations of $\mathbf{B}$; this makes implicit schemes for problems involving many degrees of freedom prohibitively expensive.

Implicit methods are very time-consuming also because they require the solution of an equation or a system of equations for $\widetilde{\boldsymbol{Y}}_{j+1}$ which, in general, are nonlinear. We next discuss an idea for avoiding this situation without losing too much of the improved stability of implicit methods. This idea, which was introduced as the *semi-implicit method* [36, 37], can be applied when the drift term is the sum of a linear and a nonlinear contribution or when it can be decomposed into a dominating linear contribution and a nonlinear correction. In that situation, the replacement (3.131) is done only for the linear contribution to the drift term. The resulting equation for $\widetilde{\boldsymbol{Y}}_{j+1}$ is then linear and can be solved very efficiently. Such semi-implicit methods can lead to drastically improved stability [36]. In Sect. 4.3.2, we use a semi-implicit scheme for avoiding stability problems for an equation with a singular drift term; in doing so, only the nonlinear term is treated implicitly, while the linear term is treated by the predictor-corrector scheme described below.

In concluding the discussion of integration schemes with improved stability we would like to point out that the occurrence of a vast range of time scales is very typical for polymer problems. The local motions of monomers or small segments of a polymer chain are very fast and, for long chains, the global motion of a chain can be slower by many orders of magnitude. This seems like a typical situation in which implicit schemes would be useful. However, if one wants to resolve all the details of very local motions one has to use correspondingly small time steps. In this situation the possibility offered by implicit schemes of using larger time steps without losing stability is of no help since the fast processes of interest would not be faithfully reproduced. If one does not need to resolve the rapid local motions then a correspondingly coarser polymer model should be used, thus eliminating the shorter time scales and the need to use an implicit method. In summary, if the coarsest polymer model that resolves all the physically relevant time scales in a given problem is chosen, implicit methods do not offer major advantages. It is a particularly fortunate situation in polymer kinetic theory that the range of time scales, or the stiffness of a problem, can be easily controlled by the number of degrees of freedom used to model the polymer molecules.

Another route to constructing more efficient (in particular, higher-order) integration methods for stochastic differential equations is inspired by classical *predictor-corrector schemes*. One can think of such schemes as the first steps in an iterative solution of the implicit equations. For example, the predictor-corrector scheme related to the implicit Euler scheme (3.132) can be obtained as follows. In a first step, one evaluates the right side of (3.132) when $\widetilde{\boldsymbol{Y}}_{j+1}$ is replaced with the zeroth approximation $\widetilde{\boldsymbol{Y}}_j$ (corresponding to $c = c' = 0$),

$$\boxed{\overline{\boldsymbol{Y}}_{j+1} = \widetilde{\boldsymbol{Y}}_j + \boldsymbol{A}(t_j, \widetilde{\boldsymbol{Y}}_j)\, \Delta t_j + \mathbf{B}(t_j, \widetilde{\boldsymbol{Y}}_j) \cdot \Delta \boldsymbol{W}_j\,,} \tag{3.133}$$

in order to obtain the better first approximation $\overline{\boldsymbol{Y}}_{j+1}$, which is the prediction for $\widetilde{\boldsymbol{Y}}_{j+1}$. Then, as the next step in the iterative solution of the implicit equation (3.132), we use $\overline{\boldsymbol{Y}}_{j+1}$ instead of $\widetilde{\boldsymbol{Y}}_{j+1}$ in evaluating the right side of (3.132) in order to obtain the corrector

$$\boxed{\begin{aligned} \widetilde{\boldsymbol{Y}}_{j+1} = \widetilde{\boldsymbol{Y}}_j \; &+ \; \left[c\, \overline{\boldsymbol{A}}(t_{j+1}, \overline{\boldsymbol{Y}}_{j+1}) + (1-c)\, \overline{\boldsymbol{A}}(t_j, \widetilde{\boldsymbol{Y}}_j)\right] \Delta t_j \\ &+ \; \left[c'\, \mathbf{B}(t_{j+1}, \overline{\boldsymbol{Y}}_{j+1}) + (1-c')\, \mathbf{B}(t_j, \widetilde{\boldsymbol{Y}}_j)\right] \cdot \Delta \boldsymbol{W}_j\,. \end{aligned}} \tag{3.134}$$

In the above first-order predictor-corrector scheme, the Euler scheme is used as a predictor, and the corrector is obtained from the implicit Euler scheme by replacing $\widetilde{\boldsymbol{Y}}_{j+1}$ with the predicted value $\overline{\boldsymbol{Y}}_{j+1}$ on the right-hand side of the implicit scheme. The same idea can be used with higher-order methods. For example, the second-order scheme (3.121) can be used as a predictor, and any implicit version of the same scheme can be turned into a corrector by replacing $\widetilde{\boldsymbol{Y}}_{j+1}$ with the predicted value $\overline{\boldsymbol{Y}}_{j+1}$. Predictor-corrector methods have an important advantage: the difference between predictor and corrector provides us with information about the local error in each time step, which we can use for on-line improvement of simulations, for example, by controlling a variable time-step width.

Predictor-corrector schemes may also lead to a higher order of weak convergence (without evaluating derivatives of coefficient functions). For an important class of stochastic differential equations, the second-order convergence of the predictor-corrector scheme based on the Euler algorithm with $c = c' = 1/2$ is established in the following exercise.

Exercise 3.47 For additive noise, that is, for a diffusion term $\mathbf{B}$ depending only on t, show that the predictor-corrector scheme (3.133) and (3.134) with $c = c' = 1/2$ yields an integration scheme of weak order $\upsilon = 2$. Why, in the case of multiplicative noise, is the weak order of convergence reduced to $\upsilon = 1$?

In the remainder of this section, we discuss two further possibilities for improving the efficiency of simulations when weak solutions of stochastic differential equations are required: time-step width extrapolation and variance reduction methods.

For any approximation scheme, the order of convergence is known in advance. If one performs simulations for several different time steps, one can plot the simulation result for some averaged quantity versus time-step width and check whether the expected order of convergence is achieved. For higher-order schemes, this is a very simple and helpful means of testing a code. Moreover, *time-step width extrapolation* is a powerful means for increasing the efficiency of simulations. Instead of choosing the time step Δt so small that the deviation from the exact result (corresponding to $\Delta t = 0$) is acceptable, one can perform simulations at several larger time steps and extrapolate to $\Delta t = 0$ (of course, the extrapolation procedure depends on the order of the algorithm). Often reliable extrapolation is possible with quite large time steps, and the quality of the extrapolation fit indicates the appropriateness of the selected range of time steps. A careful simulation should always be done with different time steps in order to check the code, to select the proper range of time steps, and to improve the efficiency by extrapolation.

Time-step extrapolation can be regarded also as a method for increasing the order of convergence. For example, if the first-order Euler scheme is used for two different time steps and the results are extrapolated linearly, the extrapolated result is of second order in the time-step width. Even higher-order convergence can be achieved by working with three (or more) different time steps and using quadratic (or higher-order) extrapolation. Of course, one can also improve the order of convergence for higher-order schemes, such as (3.121), by extrapolation methods. Time-step extrapolation is therefore a very simple and valuable tool for improving simulations, and it should always be employed when a weak approximation scheme is used for integrating a stochastic differential equation.

There is another means of increasing the efficiency of weak numerical integration schemes which is, compared to time-step extrapolation, of much more limited applicability but can lead to dramatic improvements: *variance reduction*. The idea of variance reduction methods is to transform a given stochastic differential equation so that the transformed equation yields the same averages but with a smaller variance for a certain quantity of interest. Then, when a given number of trajectories is simulated, a correspondingly smaller statistical error bar is obtained. The possibility of finding a miraculous transformation reducing the error bars in a simulation seems to be too good to be true. Indeed, there is a catch: the construction of the transformation requires an understanding of the simulated system. Typically, an approximate solution is needed in order to construct a variance reduced integration scheme, and the better the approximation is, the more the variance is reduced. The possibility of constructing considerably more efficient integration schemes can thus be a motivation for developing good approximations. How this very general and abstract idea works in detail can best be seen by applying it to a particular example (see Sect. 4.1.5). In spite of the fact that, in polymer kinetic theory, many good approximation schemes have been known for a long time, Sect. 4.1.5 resulted from the first attempts to exploit the potential power of variance reduction methods in this field.

Finally, we would like to point out that it would be an interesting and important mathematical problem to generalize the rigorous convergence results for various numerical integration schemes to the processes with mean field interactions considered in Sect. 3.3.4. Simulations of processes with mean field interactions in kinetic theory so far must rely on intuition.

Let us now summarize the various considerations that are required, or at least helpful, before a simulation program for a given problem is developed. The first step is to decide whether the underlying stochastic differential equation is of the Itô or Stratonovich type. There is a straightforward procedure for going back and forth between Itô and Stratonovich equations, and if one starts from a Fokker-Planck equation or diffusion equation an unambiguous formulation is possible. If the starting point is a stochastic differential equation or Langevin equation, such an equation should also be accompanied by an interpretation rule, which has to be found from a deeper understanding of the physical system. The second step is to decide whether a strong or weak solution is needed. In the weak case, it is much simpler to construct good approximation schemes, and the efficiency can be further improved by replacing Gaussian random variables with more readily generated ones, or by using shifted random numbers (see the remarks at the end of Sect. 4.1.4). Finally, the nature and the complexity of the problem determine the order of strong or weak convergence of the approximation scheme to be selected, the necessity of eliminating derivatives of the coefficient functions (Runge-Kutta schemes, multistep schemes, predictor-corrector schemes), and the requirements for obtaining numerical stability (implicit schemes, semi-implicit schemes). Further possibilities for improving the efficiency of weak integration schemes are offered by time-step width extrapolation and variance reduction methods.

References

1. Nelson E (1967) *Dynamical Theories of Brownian Motion*. Princeton University Press, Princeton
2. Chandrasekhar S (1943) Rev. Mod. Phys. 15: 1
3. Langevin P (1908) Comptes rendus 146: 530
4. Kubo R, Toda M, Hashitsume N (1985) *Statistical Physics II, Nonequilibrium Statistical Mechanics*. Springer, Berlin Heidelberg New York Tokyo (Springer Series in Solid-State Sciences, Vol 31)
5. Gard TC (1988) *Introduction to Stochastic Differential Equations*. Marcel Dekker, New York Basel (Monographs and Textbooks in Pure and Applied Mathematics, Vol 114)
6. Arnold L (1974) *Stochastic Differential Equations: Theory and Applications*. Wiley-Interscience, New York London Sydney Toronto

7. Liptser RS, Shiryayev AN (1977) *Statistics of Random Processes, Vol 1, General Theory*. Springer, New York Heidelberg Berlin (Applications of Mathematics, Vol 5)
8. Gardiner CW (1990) *Handbook of Stochastic Methods for Physics, Chemistry and the Natural Sciences*, 2nd Edn. Springer, Berlin Heidelberg New York London Paris Tokyo Hong Kong (Springer Series in Synergetics, Vol 13)
9. van Kampen NG (1992) *Stochastic Processes in Physics and Chemistry*, 2nd Edn. North-Holland, Amsterdam London New York Tokyo
10. Honerkamp J (1993) *Stochastic Dynamical Systems*. VCH Publishers, New York Weinheim Cambridge
11. Kallianpur G (1980) *Stochastic Filtering Theory*. Springer, New York Heidelberg Berlin (Applications of Mathematics, Vol 13)
12. McKean HP (1969) *Stochastic Integrals*. Academic Press, New York London (Probability and Mathematical Statistics, Vol 5)
13. Kunita H (1990) *Stochastic Flows and Stochastic Differential Equations*. Cambridge University Press, Cambridge (Cambridge Studies in Advanced Mathematics, Vol 24)
14. Chung KL, Williams RJ (1990) *Introduction to Stochastic Integration*, 2nd Edn. Birkhäuser, Boston Basel Berlin (Probability and Its Applications)
15. Risken H (1989) *The Fokker-Planck Equation*, 2nd Edn. Springer, Berlin Heidelberg New York London Paris Tokyo Hong Kong Barcelona Budapest (Springer Series in Synergetics, Vol 18)
16. Oelschläger K (1984) Ann. Probab. 12: 458
17. Fan XJ, Bird RB, Renardy M (1985) J. Non-Newtonian Fluid Mech. 18: 255
18. Gärtner J (1988) Math. Nachr. 137: 197
19. Sznitman AS (1985) in: Albeverio S (ed) *Infinite Dimensional Analysis and Stochastic Processes*. Pitman, Boston London Melbourne, p 145 (Research Notes in Mathematics, Vol 124)
20. McKean HP (1967) in: Lecture Series in Differential Equations 7: 41. Catholic University, Washington D.C.
21. Sznitman AS (1991) in: Hennequin PL (ed) *École d'Été de Probabilités de Saint-Flour*. Springer, Berlin Heidelberg New York, p 165 (Lecture Notes in Mathematics, Vol 1464)
22. Ikeda N, Watanabe S (1989) *Stochastic Differential Equations and Diffusion Processes*, 2nd Edn. North-Holland, Amsterdam Oxford New York (North-Holland Mathematical Library, Vol 24)
23. Stratonovich RL (1966) J. SIAM Control 4: 362
24. van Kampen NG (1981) J. Stat. Phys. 24: 175
25. Protter P (1990) *Stochastic Integration and Differential Equations*. Springer, Heidelberg Berlin New York (Applications of Mathematics, Vol 21)
26. Rogers LCG, Williams D (1987) *Diffusions, Markov Processes, and Martingales, Vol 2, Itô Calculus*. Wiley, Chichester New York Brisbane Toronto Singapore (Wiley Series in Probability and Mathematical Statistics)
27. Español P, Öttinger HC (1993) Z. Phys. B 90: 377
28. van Kampen NG (1988) J. Math. Phys. 29: 1220
29. Ryter D, Deker U (1980) J. Math. Phys. 21: 2662
30. Deker U, Ryter D (1980) J. Math. Phys. 21: 2666

31. Kloeden PE, Platen E (1992) *Numerical Solution of Stochastic Differential Equations*. Springer, Berlin Heidelberg New York London Paris Tokyo Hong Kong Barcelona Budapest (Applications of Mathematics, Vol 23)
32. Greiner A, Strittmatter W, Honerkamp J (1988) J. Stat. Phys. 51: 95
33. Helfand E (1979) Bell System Tech. J. 58: 2289
34. Press WH, Teukolsky SA, Vetterling WT, Flannery BP (1992) *Numerical Recipes in FORTRAN. The Art of Scientific Computing*, 2nd Edn. Cambridge University Press, Cambridge
35. Allen MP, Tildesley DJ (1987) *Computer Simulation of Liquids*. Clarendon Press, Oxford
36. Petersen WP (1990) *Stability and Accuracy of Simulations for Stochastic Differential Equations*. IPS Research Report No. 90-02, ETH Zürich (unpublished)
37. Petersen WP (1994) J. Comput. Phys. 113: 75

Part II

Polymer Dynamics

The analysis of polymer dynamics in the second part of this book is based on the reformulation of Fokker-Planck or diffusion equations, which are ubiquitous in polymer kinetic theory, as stochastic differential equations. The emphasis in Part II is on the method of rewriting well-known models in terms of stochastic processes, and on how such a reformulation can be exploited to obtain analytical results or numerical solutions (computer simulations). Several advantages of working with ensembles of trajectories instead of probability densities are elaborated in the course of the subsequent chapters. It is shown that the stochastic equations of motion for the polymer configurations allow a more *direct understanding of models* than the deterministic equations for the configurational distribution functions, it is illustrated how solvable models can be recognized and how *analytical solutions* can be obtained, and it is emphasized how numerical integration schemes for stochastic differential equations can be used to construct *efficient simulation algorithms* for the dynamics of polymers.

The more direct understanding of the physical content of kinetic theory models and the availability of powerful simulation techniques are important keys to developing new, more realistic models. It should be emphasized that the *purpose of the subsequent chapters* is the presentation and illustration of the ideas and methods of stochastics introduced in Part I in connection with well-established polymer models—a complete survey of all the existing models and their properties or a systematic derivation of existing or new models is clearly not attempted here. A comprehensive discussion of bead-spring models without and with constraints is contained in Chapters 13–16 of [1], and the derivation of the diffusion equations, which we use here as starting points for formulating stochastic differential equations of motion, is presented in great detail in Chapters 17–19 of the same standard textbook reference.

> "An actual polymer molecule is an extremely complex mechanical system with an enormous number of degrees of freedom. To study the detailed motions of this complicated system and their relations

> to the nonequilibrium properties would be prohibitively difficult. As a result it has been customary for polymer scientists to resort to mechanical models to simulate the mechanical behavior of the macromolecule."

These sentences from Ref. 1 (p. 1) underline that the use of *coarse-grained mechanical models* is very crucial in polymer kinetic theory. Only after eliminating the very fast processes associated with local motions in favor of stochastic noise does it become possible to investigate the polymer dynamics on the longer time scales that are responsible for many physical properties of polymeric fluids (such as their viscoelastic behavior which is very important in polymer processing). The necessity of eliminating fast processes in favor of noise explains why stochastic processes are so important in the theory of polymeric fluids.

Bead-rod-spring models are the prototypes of the coarse-grained models investigated in polymer kinetic theory. In spite of the coarse-graining, bead-rod-spring models can be used to work out the universal physical properties of high-molecular-weight polymers—provided that the number of beads N is large enough. The *universality* of many static and dynamic properties of polymers can actually be used to justify the simple mechanical models of polymer kinetic theory: in the limit of large N, most of the details of a mechanical model on the bead scale become irrelevant, and only the proper inclusion of fundamental effects like hydrodynamic interaction, excluded volume, or anisotropic mobility is important. However, there is also interest in mechanical models for small values of N, both from chemical engineers and from theoretical physicists, and not just for pedagogical convenience.

For the chemical engineer, small values of N reduce the calculations to an extent that allows the treatment of problems of practical interest, and they may still capture the qualitative behavior of actual polymeric fluids resulting from the orientibility and stretchability of polymer molecules. For the theoretical physicist, small values of N must be used in any perturbation theory for certain nonlinear interactions because low-order perturbation theory can account for only a few interactions. The effects of most of the necessarily large number of interactions in a long chain can be retained only by a proper choice of the parameters characterizing the beads and their interactions. This "proper choice of the parameters" can be found by means of the renormalization group theory which relies on a successive coarse-graining in many small steps such that each small step can be handled by perturbation theory. Even though this is usually hidden in the questionable reliability of formal expansions in space dimensionality, when extrapolating to three dimensions, a renormalization group refinement of perturbation theory can hardly capture more of the polymer dynamics than a chain with very few beads.

4. Bead-Spring Models for Dilute Solutions

The idea of establishing a relationship between diffusion equations and stochastic differential equations has a long tradition in the kinetic theory of dilute polymer solutions. In 1969, R. Zwanzig [2] described the rigorous equivalence between the two approaches for bead-spring chains with hydrodynamic interaction, and he pointed out the fundamental importance of this equivalence for computer simulations. In particular, Zwanzig rearranged the diffusion tensor in the Fokker-Planck equation so that he obtained the Stratonovich-version of the stochastic differential equation. In 1978, M. Fixman [3] showed in two landmark papers how simulations of polymer dynamics can be constructed and applied (*Brownian dynamics simulations*). Fixman discussed the relationship between Fokker-Planck equations and stochastic differential equations for bead-spring models with hydrodynamic interaction, even in the presence of constraints. Moreover, he presented numerical integration schemes for stochastic differential equations, in particular, the explicit Euler scheme and the corresponding implicit and predictor-corrector treatments of the diffusion term (which requires an intuitive understanding of the difference between the Itô- and Stratonovich approaches). Fixman also pointed out that the usual integration schemes may not be useful for stochastic differential equations. In the same year, D. L. Ermak and J. A. McCammon carried out a computer simulation for short chains in order to study the effect of hydrodynamic interaction on dynamic equilibrium properties. Their paper [4], which contains a very clear presentation of their explicit Euler scheme for integrating the Itô-version of the stochastic equations of motion (in the mathematical literature, the validity of this most basic integration scheme has been known since 1955) and an explicit procedure for carrying out the Cholesky decomposition of the diffusion tensor, has most frequently been cited in papers on Brownian dynamics. We here show how the rigorous theory of stochastic differential equations can be exploited to obtain more efficient simulation algorithms, even in more general cases where a formulation of the stochastic equations of motion by direct physical arguments is not so straightforward.

We first describe the Rouse model in which the stochastic equations of motion for the polymer configurations are linear so that closed-form solutions can be obtained. Even though it is not very realistic, the Rouse model is quite important for illustrating the kinetic theory approach to polymeric liquids, for understanding some basic features of dilute solutions, and as a starting point for developing more realistic nonlinear models. We then incorporate various nonlinear effects into the equations for the Rouse model: hydrodynamic interaction, excluded volume, and finite polymer extensibility. For these nonlinear models, we discuss both how approximation schemes and how simulation algorithms can be developed.

4.1 Rouse Model

In the Rouse model [5], the polymers are represented by linear chains of N identical, spherical "beads" connected by $N-1$ Hookean "springs" (see Fig. 4.1). The solvent is modeled as an incompressible, Newtonian fluid which is completely characterized by its viscosity η_s. Since the Rouse model was developed for dilute polymer solutions no interactions between different chains are considered. The "beads" do not represent individual monomers but rather chain segments of twenty or more monomers: the beads need to be large compared to the solvent molecules in order to justify the continuum description of the solvent in which the beads are immersed. The "springs" do not represent molecular interactions but rather entropic effects due to the elimination of more local degrees of freedom: if a spring represents the difference vector between the ends of a subchain consisting of many monomers or independent flexible units then the central limit theorem implies that the distribution of end-to-end vectors is Gaussian; for these purely entropic effects, the Helmholtz free energy obtained as the logarithm of the Gaussian probability density is quadratic, and this corresponds to a linear spring force.

The Hookean bead-spring chain model, which was developed by P. E. Rouse in 1953 [5], has been widely used by polymer chemists for interpreting linear viscoelastic measurements, and it has had considerable impact on the direction of experimental programs. The dumbbell version of the model ($N = 2$) was previously investigated by J. J. Hermans [6].[1] The Rouse model is certainly not generally capable of giving quantitative results, but nevertheless the qualitative understanding that it provides in flows with small velocity gradients has proven to be quite valuable. Furthermore, the Rouse model is the starting point for the development of many more realistic models.

We first present the elegant rigorous solution of the Rouse model based on the theory of linear stochastic differential equations developed in Sect. 3.3.2. From that solution we then derive a constitutive equation for the stress tensor and the predictions for various material functions in shear and extensional flows. Finally, we use the exactly solvable Rouse model as an example for illustrating the general ideas behind both conventional and variance reduced Brownian dynamics simulations.

[1] A model that is mathematically equivalent to a Hookean dumbbell model in two dimensions was considered by W. Kuhn and H. Kuhn [7]. These authors began with a thread model and considered only the end-to-end vector.

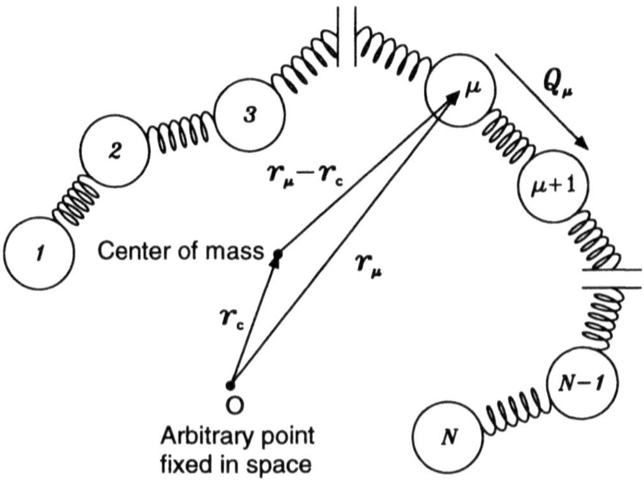

Fig. 4.1. The freely jointed bead-spring chain model formed from N "beads" and $N-1$ "springs." The internal configuration of the chain is given by the connector vectors $\boldsymbol{Q}_\mu := \boldsymbol{r}_{\mu+1} - \boldsymbol{r}_\mu$ $(\mu = 1, 2, \ldots, N-1)$.

4.1.1 Analytical Solution for the Equations of Motion

The notation used for describing the configurations of bead-spring chains is introduced in Fig. 4.1. In the conventional approach to polymer kinetic theory, the dynamics of the bead spring chains in a given homogeneous solvent flow field $\boldsymbol{v}(\boldsymbol{r}, t) = \boldsymbol{v}_0(t) + \boldsymbol{\kappa}(t) \cdot \boldsymbol{r}$ is described by a diffusion equation for the probability density in configuration space, $p = p(t, \boldsymbol{r}_1, \boldsymbol{r}_2, \ldots, \boldsymbol{r}_N)$, which is often referred to as *configurational distribution function*. The famous diffusion equation for bead-spring chain models can be found in any textbook on polymer kinetic theory (see, e.g., (15.1-1) and (15.1-4) of [1] or (3.121) and (4.1) of [8]),

$$\frac{\partial p}{\partial t} = -\sum_{\mu=1}^{N} \frac{\partial}{\partial \boldsymbol{r}_\mu} \cdot \left\{ \left(\boldsymbol{v}_0 + \boldsymbol{\kappa} \cdot \boldsymbol{r}_\mu + \frac{1}{\zeta} \boldsymbol{F}_\mu \right) p \right\} + \frac{k_B T}{\zeta} \sum_{\mu=1}^{N} \frac{\partial}{\partial \boldsymbol{r}_\mu} \cdot \frac{\partial}{\partial \boldsymbol{r}_\mu} p \,, \tag{4.1}$$

where ζ is the bead friction coefficient and $\boldsymbol{F}_\mu$ represents the potential force on bead μ. For the Rouse model one has in the absence of external forces,

$$\boldsymbol{F}_\mu = \begin{cases} H(\boldsymbol{r}_2 - \boldsymbol{r}_1) & \text{if } \mu = 1 \\ H(\boldsymbol{r}_{\mu+1} - \boldsymbol{r}_\mu) - H(\boldsymbol{r}_\mu - \boldsymbol{r}_{\mu-1}) & \text{if } 1 < \mu < N \\ -H(\boldsymbol{r}_N - \boldsymbol{r}_{N-1}) & \text{if } \mu = N \,, \end{cases} \tag{4.2}$$

where H is the Hookean spring constant.

The diffusion equation (4.1) is of the form of the Fokker-Planck equation (3.79). A more detailed comparison shows: the configuration vector $\boldsymbol{x}$ of (3.79) corresponds to a $3N$-dimensional column vector in which all the spatial components of the N three-dimensional vectors $\boldsymbol{r}_\mu$, $\mu = 1, 2, \ldots, N$, are listed. Similarly, the drift vector in (3.79) corresponds to a $3N$-dimensional column vector in which all the components of the N three-dimensional vectors $\boldsymbol{v}_0 + \boldsymbol{\kappa} \cdot \boldsymbol{r}_\mu + \frac{1}{\zeta}\boldsymbol{F}_\mu$, $\mu = 1, 2, \ldots, N$, are compiled. Finally, the diffusion matrix in (3.79) is given by $2k_{\rm B}T/\zeta$ times the $3N \times 3N$-unit matrix. After establishing these correspondences, we can immediately write down the corresponding system of stochastic differential equations of motion for the bead positions,

$$\boxed{\mathrm{d}\boldsymbol{r}_\mu(t) = \left[\boldsymbol{v}_0(t) + \boldsymbol{\kappa}(t) \cdot \boldsymbol{r}_\mu(t) + \frac{1}{\zeta}\boldsymbol{F}_\mu(t)\right] \mathrm{d}t + \sqrt{\frac{2k_{\rm B}T}{\zeta}}\, \mathrm{d}\boldsymbol{W}_\mu(t)\,.} \tag{4.3}$$

In (4.3) each $\boldsymbol{r}_\mu(t)$, $\mu = 1, 2, \ldots, N$, represents a time-dependent random variable, whereas the $\boldsymbol{r}_\mu$ in the diffusion equation (4.1) is a dummy variable (in Sect. 3.3.3 we used capital letters for the random variables and the corresponding lower-case letters for the dummy variables; here, we regard the occurrence of the time argument, which we include in parentheses because $\boldsymbol{r}_\mu$ already carries a subscript, as sufficient for a clear conceptual distinction). Each of the N independent processes $\boldsymbol{W}_\mu(t)$ in (4.3) is a three-dimensional Wiener process, so that in total we deal with a collection of $3N$ independent one-dimensional Wiener processes.

The stochastic differential equations of motion (4.3) resulting from the diffusion equation (4.1) allow a direct physical interpretation of bead-spring models. It is particularly illuminating to compare (4.3) to the stochastic equation of motion for Brownian particles given in (3.8) and (3.11), which is $\mathrm{d}X_t = (2k_{\rm B}T/\zeta)^{1/2}\,\mathrm{d}W_t$. Both equations of motion result from a force balance from which the effects of bead inertia have been eliminated. In Sect. 3.1, only frictional and Brownian forces enter the force balance (3.1), and neglect of the bead mass M leads to (3.8). In the stochastic equations of motion for bead-spring models (4.3), also the interaction forces between the beads are included into the force balance, and the frictional forces are now proportional to the differences between bead and solvent velocities (in Sect. 3.1, the latter velocity was assumed to vanish); otherwise, the arguments in formulating the equations of motion are exactly the same. In summary, our previous discussion of the motion of Brownian particles helps us to understand all the physics behind the equations of motion (4.3); in particular, we see that the factor $(2k_{\rm B}T/\zeta)^{1/2}$ in front of the increments of the Wiener process is dictated by the fluctuation-dissipation theorem of the second kind, as discussed in Sect. 3.1.1. The formulation of the equations of motion (4.3) for bead-spring models is thus rather straightforward. A detailed discussion of the effects of bead inertia [9] shows that bead inertia in polymer models may indeed be neglected for all practical purposes.

The system of stochastic differential equations (4.3) for the Rouse model is of the narrow-sense linear type considered in Sect. 3.3.2 (with $d = d' = 3N$). Since the noise is additive, the Itô and Stratonovich interpretations of those equations of motion coincide. Section 3.3.2 contains the complete solution of the stochastic differential equation (4.3). In particular, we know that for Gaussian initial conditions the solution is a Gaussian stochastic process, which is fully characterized by the first and second moments (3.57) and (3.58). The main problem in obtaining the explicit solution is the evaluation of a time-ordered exponential of a $3N \times 3N$-matrix. In order to avoid this difficulty we decouple the system of equations by two successive linear transformations into 3×3-problems before solving them. This procedure is motivated by the methods for solving systems of deterministic linear differential equations which, in Sect. 3.3.2, we found to be applicable for narrow-sense linear stochastic differential equations.

We first transform to the internal connector vectors $\boldsymbol{Q}_j(t)$, for $j = 1, 2, \ldots, N-1$, and the center of mass position $\boldsymbol{r}_c(t)$ introduced in Fig. 4.1. Since, for linear transformations $\boldsymbol{g}$, the Itô formula (3.43) coincides with the deterministic transformation rules (because the second-order derivatives of $\boldsymbol{g}$ vanish), we can simply add all the equations in (4.3) and also take differences between equations with successive indices μ in order to obtain the equations,

$$\mathrm{d}\boldsymbol{Q}_j(t) = \left[\boldsymbol{\kappa}(t) \cdot \boldsymbol{Q}_j(t) - \frac{1}{\zeta} \sum_{k=1}^{N-1} A_{jk}^{\mathrm{R}} \, \boldsymbol{F}_k^{\mathrm{c}}(t)\right] \mathrm{d}t + \sqrt{\frac{2k_{\mathrm{B}}T}{\zeta}} \left[\mathrm{d}\boldsymbol{W}_{j+1}(t) - \mathrm{d}\boldsymbol{W}_j(t)\right], \tag{4.4}$$

and

$$\mathrm{d}\boldsymbol{r}_{\mathrm{c}}(t) = [\boldsymbol{v}_0(t) + \boldsymbol{\kappa}(t) \cdot \boldsymbol{r}_{\mathrm{c}}(t)] \, \mathrm{d}t + \sqrt{\frac{2k_{\mathrm{B}}T}{\zeta}} \frac{1}{N} \sum_{\mu=1}^{N} \mathrm{d}\boldsymbol{W}_\mu(t), \tag{4.5}$$

where $\boldsymbol{F}_k^{\mathrm{c}}(t)$ is the force in the kth spring, that is $\boldsymbol{F}_k^{\mathrm{c}}(t) := H \, \boldsymbol{Q}_k(t)$ for the Rouse model, and the quantities A_{jk}^{R} are the elements of the $(N-1) \times (N-1)$ Rouse matrix,

$$A_{jk}^{\mathrm{R}} := \begin{cases} 2 & \text{if } |j-k| = 0 \\ -1 & \text{if } |j-k| = 1 \\ 0 & \text{otherwise.} \end{cases} \tag{4.6}$$

Here and throughout this chapter, the convention that Greek indices are used for numbering beads (from 1 to N) and Latin indices are used for numbering connectors (from 1 to $N-1$) is adopted from [1]. In the following, we hence need not specify the range of summation when a sum extends over the natural range of a Greek or Latin index.

After decoupling center of mass and internal motions, we further decouple the system (4.4) by introducing normal modes, that is by diagonalizing the Rouse matrix (4.6). In order to do so, we introduce the symmetric orthogonal $(N-1) \times (N-1)$-matrix with elements

$$\Omega_{jk}^{\mathrm{R}} := \sqrt{\frac{2}{N}} \sin \frac{jk\pi}{N}, \tag{4.7}$$

which diagonalizes the Rouse matrix (see §15.3 of [1]),

$$\sum_{l,n} \Omega_{lj}^{\mathrm{R}} \, A_{ln}^{\mathrm{R}} \, \Omega_{nk}^{\mathrm{R}} = a_j^{\mathrm{R}} \, \delta_{jk}, \quad a_j^{\mathrm{R}} = 4 \sin^2 \frac{j\pi}{2N}. \tag{4.8}$$

Exercise 4.1 Show that the matrix introduced in (4.7) is orthogonal, and verify that it diagonalizes the Rouse matrix where the eigenvalues $a_j^{\rm R}$ are given in (4.8).

After performing the linear transformation $\boldsymbol{Q}'_k(t) := \sum_j \Omega^{\rm R}_{jk}\, \boldsymbol{Q}_j(t)$ we obtain from (4.4)

$$\mathrm{d}\boldsymbol{Q}'_k(t) = \left[\boldsymbol{\kappa}(t)\cdot\boldsymbol{Q}'_k(t) - \frac{H}{\zeta}\, a_k^{\rm R}\, \boldsymbol{Q}'_k(t)\right]\mathrm{d}t + \sqrt{\frac{2k_{\rm B}T a_k^{\rm R}}{\zeta}}\,\mathrm{d}\boldsymbol{W}'_k(t)\,, \tag{4.9}$$

where

$$\boldsymbol{W}'_k(t) := \frac{1}{\sqrt{a_k^{\rm R}}} \sum_j \Omega^{\rm R}_{jk}\, \left[\boldsymbol{W}_{j+1}(t) - \boldsymbol{W}_j(t)\right]\,, \tag{4.10}$$

for $k = 1, 2, \ldots, N-1$. If we further introduce

$$\boldsymbol{W}'_N(t) := \frac{1}{\sqrt{N}} \sum_\mu \boldsymbol{W}_\mu(t)\,, \tag{4.11}$$

we can write

$$\mathrm{d}\boldsymbol{r}_{\rm c}(t) = \left[\boldsymbol{v}_0(t) + \boldsymbol{\kappa}(t)\cdot\boldsymbol{r}_{\rm c}(t)\right]\mathrm{d}t + \sqrt{\frac{2k_{\rm B}T}{\zeta N}}\,\mathrm{d}\boldsymbol{W}'_N(t)\,. \tag{4.12}$$

Exercise 4.2 Show that the processes $\boldsymbol{W}'_\mu(t)$ for $\mu = 1, 2, \ldots, N$ defined in (4.10) and (4.11) are independent three-dimensional Wiener processes.

The transformation to center of mass motion and internal normal modes not only decouples all the equations of motion but, according to Exercise 4.2, the resulting inhomogeneous terms moreover happen to involve independent Wiener processes $\boldsymbol{W}'_\mu(t)$. For independent initial conditions, for example for equilibrium initial conditions, and assuming no complications due to boundary conditions, the center of mass motion and the internal modes are all independent stochastic processes. The solution of (4.9) is obtained from Theorem 3.22,

$$\boldsymbol{Q}'_k(t) = \mathrm{e}^{-H a_k^{\rm R} t/\zeta}\, \mathbf{E}(t,0)\cdot\boldsymbol{Q}'_k(0) + \sqrt{\frac{2k_{\rm B}T a_k^{\rm R}}{\zeta}} \int_0^t \mathrm{e}^{-H a_k^{\rm R}(t-t')/\zeta}\, \mathbf{E}(t,t')\cdot\mathrm{d}\boldsymbol{W}'_k(t')\,, \tag{4.13}$$

where

$$\mathbf{E}(t,t') = \mathcal{T}\text{-}\exp\left\{\int_{t'}^{t} \boldsymbol{\kappa}(t'')\,\mathrm{d}t''\right\}. \tag{4.14}$$

We similarly obtain the solution of (4.12),

$$\boldsymbol{r}_{\rm c}(t) = \mathbf{E}(t,0)\cdot\boldsymbol{r}_{\rm c}(0) + \int_0^t \mathbf{E}(t,t')\cdot\boldsymbol{v}_0(t')\,\mathrm{d}t' + \sqrt{\frac{2k_{\rm B}T}{\zeta N}} \int_0^t \mathbf{E}(t,t')\cdot\mathrm{d}\boldsymbol{W}'_N(t')\,. \tag{4.15}$$

After transforming back to the original variables, (4.13) and (4.15) give the complete explicit solutions for the stochastic process modeling polymer configurations in the Rouse model,

$$\begin{aligned}
\boldsymbol{Q}_j(t) &= \sum_{k=1}^{N-1} \Omega^{\mathrm{R}}_{jk}\, \boldsymbol{Q}'_k(t)\,, \\
\boldsymbol{r}_1(t) &= \boldsymbol{r}_{\mathrm{c}}(t) - \sum_{j=1}^{N-1} \frac{N-j}{N}\, \boldsymbol{Q}_j(t)\,, \\
\boldsymbol{r}_\nu(t) &= \boldsymbol{r}_1(t) + \sum_{j=1}^{\nu-1} \boldsymbol{Q}_j(t)\,, \qquad \text{for } \nu = 2,3,\ldots,N\,.
\end{aligned} \tag{4.16}$$

Exercise 4.3 Verify the transformations (4.16).

The time-ordered exponential in (4.14) still needs to be evaluated in order to obtain the tensor $\mathbf{E}(t,t')$ for a given flow field. Note that this tensor depends only on the flow field and has nothing to do with the molecular model. This deformation tensor is well known in continuum mechanics, and the tensor

$$\mathbf{C}^{-1}(t,t') = \mathbf{E}(t,t') \cdot \mathbf{E}^T(t,t') \tag{4.17}$$

is the famous *Finger strain tensor* of continuum mechanics, which is the inverse of the equally famous *Cauchy strain tensor* $\mathbf{C}(t,t')$ (see §8.1 of [10] for a systematic discussion of various finite strain tensors). The molecular model naturally leads to these deformation tensors. The following exercise breathes life into these rather formal definitions of deformation tensors.

Exercise 4.4 Calculate the deformation tensors $\mathbf{E}(t,t')$, $\mathbf{C}^{-1}(t,t')$, and $\mathbf{C}(t,t')$ for the time-dependent shear flows and general extensional flows defined in (1.4) and (1.17), respectively.

The formulation of the stochastic differential equations of motion for the Rouse model and the analysis of their solutions given in this subsection are mathematically rigorous. The normal-mode analysis is particularly simple when carried out on the level of stochastic differential equations. If we treat the Gaussian solution via its first and second moments, the quantities $\langle \boldsymbol{r}_\mu(t) \boldsymbol{r}_\nu(t) \rangle$ are to be determined from $(3N)^2$ linear deterministic ordinary differential equations. It is then not so obvious that the normal-mode analysis requires only the diagonalization of a $N \times N$-matrix, but the diagonalization of the $(3N)^2 \times (3N)^2$-matrix in the linear differential equations for $\langle \boldsymbol{r}_\mu(t) \boldsymbol{r}_\nu(t) \rangle$ can be achieved by using the results of this section (the solution of these equations in terms of the exponential of a $3N \times 3N$-matrix can be obtained from (3.58)). Even more delicate is the normal-mode analysis on the level of probability densities and Fokker-Planck equations [11]. The probability density can be written as a product of Gaussians for the individual modes only because the Wiener processes $\boldsymbol{W}'_\mu(t)$ happen to be independent (see Exercise 4.2); the decoupling achieved on the level of stochastic differential equations would not require the independence of the processes $\boldsymbol{W}'_\mu(t)$. The linearity of the equations of motion is the key to the solvability of the Rouse model. The crucial role of this linearity in the configurations is not always realized in work on the Fokker-Planck level because all Fokker-Planck equations are linear in the probability density p.

4.1.2 Stress Tensor

From the above solution of the Rouse model, all kinds of static and dynamic quantities can be determined by evaluating certain expectations of the configurations. For example, (4.12) describes the center of mass motion with respect to the flowing solvent, and we immediately obtain the Rouse prediction for the center of mass diffusion coefficient, $D_c = k_B T/(\zeta N)$. The predicted decrease of the diffusion coefficient with the molecular weight of the polymer molecules (the molecular weight is proportional to the number of beads N) is much stronger than observed experimentally, and this discrepancy is a first reminder of the drastic simplifications in the Rouse model.

For the investigation of the flow behavior of dilute polymer solutions the calculation of the stress tensor is of special interest. The polymer contribution to the stress tensor, which describes the momentum flux through, or the force exerted across, an oriented surface in a fluid, can be expressed in terms of bead-spring chain configurations as follows (see §15.2 of [1]),

$$\boxed{\boldsymbol{\tau}^{\mathrm{p}}(t) = (N-1)\, n_{\mathrm{p}}\, k_{\mathrm{B}} T\, \boldsymbol{\delta} - n_{\mathrm{p}} \sum_j \left\langle \boldsymbol{Q}_j(t) \boldsymbol{F}^{\mathrm{c}}_j(t) \right\rangle ,} \tag{4.18}$$

where n_{p} is the number density of polymers, which is assumed to be independent of position (homogeneous system). Equation (4.18) is known as the *Kramers expression* for the stress tensor. The isotropic first term results from the internal motions of the beads in a polymer molecule; whenever a bead moves through a surface the bead momentum is exchanged through that surface. This is the well-known isotropic pressure associated with internal kinetic energy. The more interesting second term resulting from the spring forces can be understood with the help of Fig. 4.2.

Let $\boldsymbol{Q}_j(t)$, $j = 1, 2, \ldots, N-1$, be the configuration of the jth spring of a given chain in the polymer solution at time t. The average force exerted by all springs of the given chain from the left of the oriented plane of area S_V in Fig. 4.2 on the right of that plane is

$$-\sum_j \left\langle \mathrm{sgn}\big(\boldsymbol{n} \cdot \boldsymbol{Q}_j(t)\big)\, \boldsymbol{F}^{\mathrm{c}}_j(t)\, \chi_{\{\omega \mid \boldsymbol{Q}_j(t,\omega) \text{ intersects plane}\}} \right\rangle ,$$

where the sign-function (sgn) takes care of the proper direction of the connector force. The event $\{\omega \mid \boldsymbol{Q}_j(t,\omega) \text{ intersects plane}\}$ occurs if and only if $\boldsymbol{r}_j(t)$ is contained in the box of volume $S_V\, |\boldsymbol{n} \cdot \boldsymbol{Q}_j(t)|$ shown in Fig. 4.2. Since, for

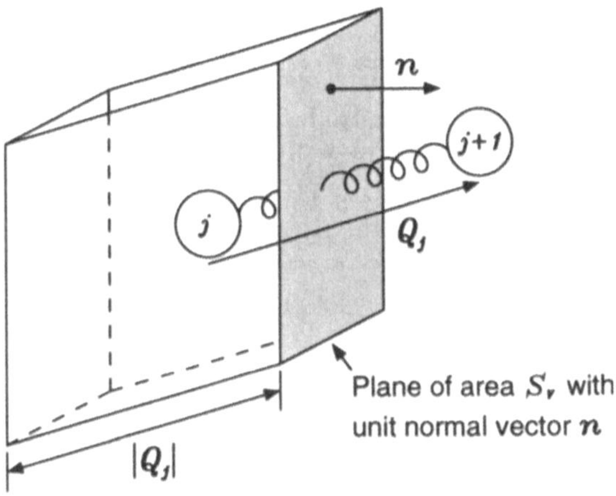

Fig. 4.2. The volume that may be occupied by bead "j" when the jth connector in the configuration $\boldsymbol{Q}_j$ intersects the shaded plane.

the assumed homogeneous system, $\boldsymbol{Q}_j(t)$ and $\boldsymbol{r}_j(t)$ are independent random variables, we can first integrate over $\boldsymbol{r}_j$ in order to obtain the force exerted by all springs through the given plane,

$$-\frac{N_\mathrm{p}}{V}\, S_V\, \boldsymbol{n}\cdot\sum_j \left\langle \boldsymbol{Q}_j(t)\boldsymbol{F}_j^\mathrm{c}(t)\right\rangle .$$

Here, a factor N_p has been introduced in order to account for all N_p polymer chains in the system, and the volume of the box divided by the volume of the system, $S_V\,|\boldsymbol{n}\cdot\boldsymbol{Q}_j(t)|/V$, has been inserted as the probability for finding $\boldsymbol{r}_j(t)$ in the box in Fig. 4.2. When $n_\mathrm{p} = N_\mathrm{p}/V$ is used, the force exerted by all springs of all chains through a plane of area S_V in the direction of $\boldsymbol{n}$ is found to correspond to the second contribution to the stress tensor in (4.18).

The average in (4.18) can be evaluated by means of the explicit solution for the Rouse model. Since the transformation to normal modes is orthogonal we obtain

$$\sum_j \left\langle \boldsymbol{Q}_j(t)\boldsymbol{F}_j^\mathrm{c}(t)\right\rangle = H\sum_{j,k,l} \Omega_{jk}^\mathrm{R}\,\Omega_{jl}^\mathrm{R}\,\langle \boldsymbol{Q}'_k(t)\boldsymbol{Q}'_l(t)\rangle = H\sum_k \langle \boldsymbol{Q}'_k(t)\boldsymbol{Q}'_k(t)\rangle .$$

By inserting the explicit solution (4.13) we find

$$\begin{aligned}\boldsymbol{\tau}^\mathrm{p}(t) &= (N-1)\,n_\mathrm{p}\,k_\mathrm{B}T\,\boldsymbol{\delta} - n_\mathrm{p}\,k_\mathrm{B}T\sum_j \mathrm{e}^{-t/\lambda_j}\,\mathbf{C}^{-1}(t,0)\\ &\quad - n_\mathrm{p}\,k_\mathrm{B}T\sum_j \frac{1}{\lambda_j}\int_0^t \mathrm{e}^{-(t-t')/\lambda_j}\,\mathbf{C}^{-1}(t,t')\,\mathrm{d}t' ,\end{aligned}$$

where the relaxation times

$$\lambda_j := \frac{\zeta}{2Ha_j^{\rm R}} = \frac{\zeta}{8H}\left(\sin\frac{j\pi}{2N}\right)^{-2}, \tag{4.19}$$

$j = 1, 2, \ldots, N-1$, have been introduced, and equilibrium initial conditions have been assumed, that is $\langle \boldsymbol{Q}'_j(0)\boldsymbol{Q}'_k(0)\rangle = \delta_{jk}\,(k_{\rm B}T/H)\,\boldsymbol{\delta}$. If the system was at equilibrium for all $t' \leq 0$, that is $\mathbf{C}^{-1}(t,t') = \mathbf{C}^{-1}(t,0)$ for $t' \leq 0$, we can rewrite the above expression for the stress tensor in the more compact form

$$\boxed{\boldsymbol{\tau}^{\rm p}(t) = n_{\rm p}\,k_{\rm B}T\sum_j \frac{1}{\lambda_j}\int_{-\infty}^{t} {\rm e}^{-(t-t')/\lambda_j}\left[\boldsymbol{\delta} - \mathbf{C}^{-1}(t,t')\right]{\rm d}t'.} \tag{4.20}$$

Equation (4.20) is the famous constitutive equation for the Rouse model in integral form [11]. The total stress is obtained by adding the Newtonian stress contribution of the unperturbed solvent flow field, $\boldsymbol{\tau}(t) = -\eta_{\rm s}\left[\boldsymbol{\kappa}(t) + \boldsymbol{\kappa}^T(t)\right] + \boldsymbol{\tau}^{\rm p}(t)$. For the dumbbell model, we have only one relaxation time $\lambda_H := \lambda_1 = \zeta/(4H)$, and we recover the convected Jeffreys model or Oldroyd-B model of continuum mechanics. The Rouse chain model is hence a multimode generalization of these famous models of continuum mechanics. Equations (4.8) and (4.19) imply that λ_j decreases with increasing j, and that λ_j divided by the longest relaxation time λ_1 is a function of j and N only. For large N (more precisely, for $j \ll N$), we have $\lambda_j = \lambda_1/j^2$, where the longest relaxation time is proportional to N^2, $\lambda_1 = \zeta N^2/(2H\pi^2)$.

Let us summarize and discuss the parameters occurring in the equations of motion and in the stress tensor for the Rouse model. If we assume that the solvent viscosity $\eta_{\rm s}$ and the velocity gradients $\boldsymbol{\kappa}(t)$ are known in a given experiment, then the three model parameters entering the equation of motion are N, λ_H, and $k_{\rm B}T/H$. The parameters λ_H and $(k_{\rm B}T/H)^{1/2}$ are characteristic time and length scales associated with the Hookean springs. These parameters can be taken to be unity by choosing suitable units of time and length, or they can be used to fit some polymer properties related to time and length (for example, one can fit the longest relaxation time and the root-mean-square radius of gyration of the polymer molecules at equilibrium). The only dimensionless parameter is N. This situation does not change when we consider also the stress tensor. Then, the additional parameter $n_{\rm p}\,k_{\rm B}T$ occurs. This parameter can be taken to be unity by choosing suitable units of mass, or it can be used to fit some polymer property related to mass (for example, one can fit a modulus, or the quantities $n_{\rm p}$ and $k_{\rm B}T$ can be measured directly). In conclusion, the Rouse model can be considered as essentially parameter-free: no mechanical contrivance is able to predict the fundamental time and length scales (a prohibitively complicated quantum mechanical calculation would be required for any given chemical structure of the monomers), and $n_{\rm p}\,k_{\rm B}T$ is a measurable quantity. If N can be taken large enough, as should be possible for long polymer molecules, then all polymer properties which depend on chemical details only through the fundamental time and length scales should be independent of N (see, for example, the result for the longest relaxation times, $\lambda_j = \lambda_1/j^2$).

4.1.3 Material Functions in Shear and Extensional Flows

Since we have derived a closed-form constitutive equation for the Rouse model, the calculation of material functions in various types of flow can be regarded as a problem of continuum mechanics. Many rheological properties of the Rouse model have been discussed repeatedly in the literature (see, for example, §15.3 of [1] for a summary). We here discuss only one example for illustration, namely stress growth at start-up of a steady general extensional flow of the form (1.17) at time $t = 0$.

If we introduce the result of Exercise 4.4,

$$\mathbf{C}^{-1}(t,t') = \begin{pmatrix} \mathrm{e}^{2\dot\epsilon(t-t')} & 0 & 0 \\ 0 & \mathrm{e}^{2m\dot\epsilon(t-t')} & 0 \\ 0 & 0 & \mathrm{e}^{-2(1+m)\dot\epsilon(t-t')} \end{pmatrix}$$

for $t' > 0$, and $\mathbf{C}^{-1}(t,t') = \mathbf{C}^{-1}(t,0)$ for $t' \leq 0$, into the constitutive equation for the Rouse model, and if we perform the integration over t', we obtain for the extensional stress growth coefficients defined in Sect. 1.2.2,

$$\begin{aligned} \mu_1^+(t,\dot\epsilon,m) &= \eta_\mathrm{s} + n_\mathrm{p}\, k_\mathrm{B}T \sum_j \frac{\lambda_j}{(1-2\lambda_j\dot\epsilon)[1+2(1+m)\lambda_j\dot\epsilon]} \\ &- \frac{n_\mathrm{p}\, k_\mathrm{B}T}{2(2+m)\dot\epsilon} \sum_j \left\{ \frac{2\lambda_j\dot\epsilon}{1-2\lambda_j\dot\epsilon}\, \mathrm{e}^{-(1-2\lambda_j\dot\epsilon)\,t/\lambda_j} \right. \\ &\qquad \left. + \frac{2(1+m)\lambda_j\dot\epsilon}{1+2(1+m)\lambda_j\dot\epsilon}\, \mathrm{e}^{-[1+2(1+m)\lambda_j\dot\epsilon]\,t/\lambda_j} \right\}, \end{aligned} \tag{4.21}$$

and

$$\begin{aligned} \mu_2^+(t,\dot\epsilon,m) &= \eta_\mathrm{s} + n_\mathrm{p}\, k_\mathrm{B}T \sum_j \frac{\lambda_j}{(1-2m\lambda_j\dot\epsilon)[1+2(1+m)\lambda_j\dot\epsilon]} \\ &- \frac{n_\mathrm{p}\, k_\mathrm{B}T}{2(1+2m)\dot\epsilon} \sum_j \left\{ \frac{2m\lambda_j\dot\epsilon}{1-2m\lambda_j\dot\epsilon}\, \mathrm{e}^{-(1-2m\lambda_j\dot\epsilon)\,t/\lambda_j} \right. \\ &\qquad \left. + \frac{2(1+m)\lambda_j\dot\epsilon}{1+2(1+m)\lambda_j\dot\epsilon}\, \mathrm{e}^{-[1+2(1+m)\lambda_j\dot\epsilon]\,t/\lambda_j} \right\}. \end{aligned} \tag{4.22}$$

Only for sufficiently small strain rates, $\dot\epsilon < 1/(2\lambda_1)$, are the extensional stress growth coefficients (4.21) and (4.22) bounded functions of t; otherwise, these coefficients diverge as t goes to infinity. For $\dot\epsilon < 1/(2\lambda_1)$, the steady state extensional viscosities $\mu_1(\dot\epsilon,m)$ and $\mu_2(\dot\epsilon,m)$ are given by the first lines of (4.21) and (4.22), respectively.

By introducing the dimensionless parameters $\lambda_1\dot\epsilon$ and t/λ_1 and using the result $\lambda_j = \lambda_1/j^2$ one can reveal the universal character of the material functions for long chains, for example, by writing

$$\frac{\mu_1(\dot\epsilon,m)-\eta_\mathrm{s}}{n_\mathrm{p}\, k_\mathrm{B}T\lambda_1} = \sum_{j=1}^{\infty} \frac{1/j^2}{(1-2\lambda_1\dot\epsilon/j^2)[1+2(1+m)\lambda_1\dot\epsilon/j^2]}\,. \tag{4.23}$$

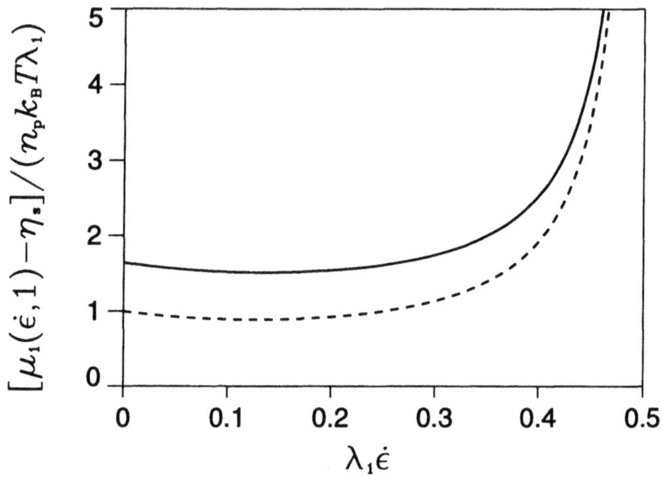

Fig. 4.3. The universal extensional viscosity for exquibiaxial extension for long Rouse chains (*continuous line*). For comparison, the corresponding curve for Hookean dumbbells has been included (*dashed line*).

The dominating contribution to $\mu_1(\dot\epsilon, m)$ stems from terms with small j so that the approximation $\lambda_j = \lambda_1/j^2$ is justified, and the sum can be extended to infinity. The universal function in (4.23) for $m = 1$ is shown in Fig. 4.3.

For the steady state extensional viscosities in the linear viscoelastic limit $\dot\epsilon \to 0$, we obtain

$$\mu_1(0, m) = \mu_2(0, m) = \eta_s + n_p\, k_B T \sum_j \lambda_j\,, \tag{4.24}$$

which, according to the general formalism of continuum mechanics, coincides with the zero-shear-rate viscosity η (the shear viscosity predicted by the Rouse model is actually independent of shear rate).

In steady shear flow, all properties of the Rouse model that depend only on the internal configurations of the chains can most conveniently be obtained from the second moments,

$$\boxed{\left\langle \boldsymbol{Q}_j(t)\boldsymbol{Q}_k(t)\right\rangle = \frac{k_B T}{H}\left[\delta_{jk}\,\boldsymbol{\delta} + 2\lambda_H C^{R}_{jk}(\boldsymbol{\varkappa} + \boldsymbol{\varkappa}^T) + 8\lambda_H^2 (C^{R})^2_{jk}\,\boldsymbol{\varkappa}\cdot\boldsymbol{\varkappa}^T\right],} \tag{4.25}$$

where the quantities C^{R}_{jk} are the elements of the $(N-1)\times(N-1)$ *Kramers matrix*,

$$C^{R}_{jk} := \min(j,k) - \frac{jk}{N}\,. \tag{4.26}$$

Note that, due to the simple structure of C^{R}_{jk}, the square in (4.25) and all higher powers of the Kramers matrix and their traces can be evaluated in closed form (see §11.6 of [1]). The result (4.25) is particularly useful in determining the viscometric functions, in constructing perturbation theories around the Rouse model, and in renormalization group calculations.

Exercise 4.5 Show that the Kramers matrix is the inverse of the Rouse matrix. Use this result in order to prove the relationship (4.25).

Exercise 4.6 Calculate the viscometric functions η, Ψ_1 and Ψ_2 for all shear rates $\dot{\gamma}$, and verify that the results are independent of $\dot{\gamma}$. Discuss the molecular weight dependence for long chains, and express the results for the viscometric functions in terms of the longest relaxation time λ_1.

Exercise 4.7 Calculate the mean-square gyration tensor $\frac{1}{N}\sum_\mu \langle(\boldsymbol{r}_\mu - \boldsymbol{r}_c)(\boldsymbol{r}_\mu - \boldsymbol{r}_c)\rangle$ for all shear rates $\dot{\gamma}$, and express the results for long chains in terms of $\lambda_1\dot{\gamma}$.

The predictions of the Rouse model are not in quantitative agreement with experimental results for dilute solutions. In contrast to the Rouse predictions, experimental results for the viscometric functions η, Ψ_1 and Ψ_2 depend on shear rate, and a nonvanishing second normal stress difference is observed. Also the predicted molecular weight dependences of viscosity, first normal stress coefficient and diffusion coefficient are in conflict with experimental findings, and the divergence of the extensional viscosities at a finite extensional strain rate is unrealistic. On the other hand, the Rouse model is a first step in the direction of a qualitative understanding of dilute polymer solutions. It explains the occurrence of viscoelastic effects, in particular of a first normal stress difference, and it suggests a power law dependence of various polymer properties on molecular weight.

Even if the Rouse model fails in predicting the properties of dilute solutions, it might still be useful in other applications. This statement is related to the fact that the constitutive equation derived from the Hookean dumbbell model formally coincides with the Oldroyd-B model of continuum mechanics and with the constitutive equation for a temporary network model for polymer melts.[2] The Oldroyd-B model has been used in many studies of complex flow problems which need to be solved by numerical methods such as finite elements. In such flow calculations it may be useful to read off the stress tensor from model molecules rather than getting them from closed-form integral expressions or differential equations. Hookean dumbbells thus offer a possibility of treating the Oldroyd-B model in flow calculations by stochastic simulation techniques.

4.1.4 A Primer in Brownian Dynamics Simulations

The fact that the Rouse model can be solved exactly makes it an ideal playground for developing some fundamental insights into the numerical integration

[2] Of course, the interpretation of the parameters is then different.

of stochastic differential equations which leads to Brownian dynamics simulations. For a given partition $0 = t_0 < t_1 \cdots < t_{n-1} < t_n = t_{\max}$ of the time range $\mathbb{T} = [0, t_{\max}]$, the simple Euler scheme for (4.3) is

$$\boldsymbol{r}_\mu(t_{j+1}) = \boldsymbol{r}_\mu(t_j) + \left[\boldsymbol{v}_0(t_j) + \boldsymbol{\varkappa}(t_j) \cdot \boldsymbol{r}_\mu(t_j) + \frac{1}{\zeta}\boldsymbol{F}_\mu(t_j)\right] \Delta t_j + \sqrt{\frac{2k_\mathrm{B}T}{\zeta}}\, \Delta \boldsymbol{W}_{\mu,j}\,, \tag{4.27}$$

with $\Delta t_j := t_{j+1} - t_j$ and $\Delta \boldsymbol{W}_{\mu,j} := \boldsymbol{W}_\mu(t_{j+1}) - \boldsymbol{W}_\mu(t_j)$. Whenever there is no risk of confusion, we here use the same symbols for the solution of the stochastic differential equation (4.3) and the approximation given by the integration scheme (4.27). The general theory of Sect. 3.4 implies that, for the Rouse dynamics with additive noise, the order of both weak and of strong convergence is $\upsilon = 1$.

Various material properties in shear and extensional flows have been derived for the discrete equations of motion with constant time step $\Delta t_j = \Delta t$, and several important features of the results have been worked out [12]. For the evaluation of material functions one is interested in weak solutions. The explicit solution of (4.27) shows that the material functions depend only on the first and second moments of $\Delta \boldsymbol{W}_{\mu,j}$, and that other characteristics of the distribution of the random variables have no effect whatsoever. For fixed Δt, the relative discretization errors have been found to vanish for $N \to \infty$; this should be expected if the relevant parameter determining the size of the corrections is of order $\Delta t/\lambda_1$ (if only large scale properties of the polymer molecules affect the quantity of interest). Surprisingly, it has also been observed that the discretization errors in the shear viscosity and in the extensional viscosity for $m = -1/2$ (simple extension) are only of second order in Δt; of course, for certain moments the discretization error may happen to be smaller than guaranteed by the order of convergence of the numerical integration scheme.

The numerical integration of (4.27) requires a proper initial condition and realizations of the increments $\Delta \boldsymbol{W}_{\mu,j}$. Equilibrium initial conditions can be realized by placing the center of mass position arbitrarily, then choosing all components of all the connector vectors as independent real-valued Gaussian random variables with mean 0 and variance $k_\mathrm{B}T/H$ (see (4.25) for $\boldsymbol{\varkappa} = \boldsymbol{0}$), and finally using the transformations (4.16). The components of the increments $\Delta \boldsymbol{W}_{\mu,j}$ are all independent real-valued random variables with mean 0 and variance Δt_j; only in strong approximation schemes do these random variables need to be Gaussian. In short, the numerical integration or Brownian dynamics simulation can be carried out as soon as we can generate the required random variables on the computer.

The realization of independent random numbers on a computer is a highly developed art (or should one say "Black Art?"). As deterministic machines computers provide only *pseudo-random numbers*. We here restrict ourselves to a few unsophisticated remarks on *random number generators*. The basic problem is to produce a sequence of random numbers from the interval $[0, 1]$ with uniform distribution; other distributions can be produced by suitable transformations (as we will discuss below when we need them; see, for example, Exercise 4.9, Exercise 4.10, and Example 4.17). Strictly speaking, even the precise formulation of what a "random sequence" is, in particular for finite sequences, is a major undertaking (see Sect. 3.5 of [13]). As an excellent reference for the theory of random number generators we recommend a book by D. E. Knuth [13].

Most random number generators used in the literature are based on the linear congruential method. These may not be the best random number generators, but they are the ones which are best understood. With the linear congruential method one generates a sequence of integers between 0 and $n-1$ (a large number) by the recurrence relation

$$n_{j+1} = k\, n_j + l \quad (\mathrm{mod}\; n)\,. \tag{4.28}$$

Here n is called the modulus, and k and l are positive integers called the multiplier and the increment, respectively. An initial "seed" n_0 is required to start the sequence. The returned random numbers between 0 and 1 are n_j/n, $j = 1, 2, \ldots$. The modulus n should be large because it determines the maximum period after which the sequence of random numbers repeats itself; k and l should be chosen such that a period close to the maximum possible period is achieved. The choice $l = 0$ cuts down the length of the period of the sequence (because 0 clearly may not occur in the sequence), but it is still possible to make the period reasonably long. The idea behind the linear congruential method is a folding mechanism which is closely related to mappings used in studying chaotic systems. The constructed pseudo-random numbers appear to be highly unpredictable.

A major problem with the linear congruential method on current 32-bit workstations is period exhaustion. Since the maximum possible period is of the order of $2 \cdot 10^9$, the full period is typically exhausted after about half an hour of computer time. For more sophisticated random number generators, such as `ran2` of [14], the generation of $2 \cdot 10^9$ random numbers takes 1 or 2 hours. Therefore, there exists a rather natural time window (say, 15 minutes to 2 hours) in which efficient simulations are impossible. In the examples of this book, we mostly use several hours of computer time, so that the linear congruential method on 32-bit machines is not good enough. Of course, these remarks reflect the state of computer technology in the year 1995. As soon as faster machines and/or full 64-bit technology become available (including the corresponding compilers), the generation of more than $2 \cdot 10^9$ random numbers by the linear congruential method will be unproblematic.

Even for a careful choice of the integers k, l, n (see Sect. 7.1 of [14] for a table of "good" choices), a random number generator based on the linear congruential method may introduce sequential correlations. Such correlations can be broken up by an additional random shuffling of the random numbers, and period exhaustion can be avoided by combining several random number generators (see the routines `ran1` and `ran2` of [14]). One can then construct random number generators that pass the long list of sensible statistical tests accumulated over several decades (see Sect. 3.3 of [13]). Any good generator ought to pass all these tests; however, there is still no guarantee that it cannot fail in a particular application. Knuth's parenthetical statement that "every random number generator will fail in at least one application" (after 173 pages of theory!) underlines that the art of generating random numbers is somewhat unsatisfactory. From time to time, publications demonstrating the failure of popular random number generators indeed stir a lot of attention and lead to new additions to the list of statistical tests [15]. As a rule, system-supplied random number generators should be used with great care.

Exercise 4.8 By seeking information from the literature, develop a routine for generating random numbers with uniform distribution in $[0,1]$. Your random number generator should provide sequences of more than 10^{10} "good" random numbers, and sequential correlations should be avoided.

Exercise 4.9 Develop a routine for generating Gaussian random numbers with mean zero and variance unity. (Hint: Combine the results of Exercises 2.36 and 4.8; see also p. 117 of [13].)

Exercise 4.10 Develop a routine for generating random unit vectors. (Hint: Combine the results of Exercises 2.37 and 4.8; see also p. 131 of [13].)

In most of the simulations presented in this book, the generation of random numbers contributes significantly or even dominantly to the total simulation time. The random number generator is called extremely frequently, and the simulation time may be reduced considerably by incorporating the corresponding subroutine or function into the main program. This gain in efficiency can most conveniently be obtained by invoking suitable compiler options which activate "inlining." A package of excellent portable random number generators ideally suited for vector computers is described in [16]; all the established statistical tests have been tried for this public-domain package.

Once we have generated appropriate realizations of the stochastic processes $\boldsymbol{r}_\mu(t)$ by means of a suitable random number generator and an integration scheme like (4.27) we would like to evaluate material properties such as the stress tensor. The information about the anisotropic polymer contribution to the stress tensor contained in one particular trajectory $\boldsymbol{r}_\mu(t,\omega)$ is given by the random variable

$$\boldsymbol{\tau}^{\mathrm{rv}}(t,\omega) := -n_{\mathrm{p}}\, H \sum_j \boldsymbol{Q}_j(t,\omega)\, \boldsymbol{Q}_j(t,\omega)\,, \tag{4.29}$$

which describes the stresses in the configuration corresponding to ω at time t. Of course, for an estimation of stresses based on only one trajectory we expect large statistical errors. More precisely, the statistical error for any component in (4.29) can be characterized by the standard deviation of that component (each component of $\boldsymbol{\tau}^{\mathrm{rv}}(t)$ is a real-valued random variable). For example, for steady shear flow we obtain by using Wick's theorem for decomposing fourth moments and from (4.25) for the second moments the following variance for the (j,k)-component of the fluctuating stress tensor (4.29),

$$\begin{aligned}\mathrm{Var}\left(\frac{\tau_{jk}^{\mathrm{rv}}}{n_{\mathrm{p}}\, k_{\mathrm{B}}T}\right) &= 4\lambda_H^2 c_2 \left[(\boldsymbol{\varkappa}+\boldsymbol{\varkappa}^T)_{jk}\right]^2 \\ &+ (1+\delta_{jk})\Big\{(N-1) + 8\lambda_H^2 c_2 \left[(\boldsymbol{\varkappa}\cdot\boldsymbol{\varkappa}^T)_{jj} + (\boldsymbol{\varkappa}\cdot\boldsymbol{\varkappa}^T)_{kk}\right] \\ &\quad +64\lambda_H^4\, c_4\, (\boldsymbol{\varkappa}\cdot\boldsymbol{\varkappa}^T)_{jj}(\boldsymbol{\varkappa}\cdot\boldsymbol{\varkappa}^T)_{kk}\Big\}\,, \end{aligned} \tag{4.30}$$

where c_2 and c_4 are the traces of the second and fourth powers of the Kramers matrix, respectively. For example, the variance of the estimate for the polymer contribution to the viscosity is

$$\mathrm{Var}\left(\frac{\tau_{12}^{\mathrm{rv}}}{\eta_{\mathrm{p}}\dot\gamma}\right) = \frac{45(N-1)+3(N^2-1)(2N^2+7)\lambda_H^2\dot\gamma^2}{5(N^2-1)^2\lambda_H^2\dot\gamma^2}\,, \tag{4.31}$$

where the exact result for the polymer contribution to the viscosity is denoted by $\eta_{\mathrm{p}} := \eta - \eta_{\mathrm{s}}$ (see (4.24)). This expression underlines the problem with simulations of the viscosity at small shear rates, where the statistical error diverges as $1/\dot\gamma$. In the limit of high shear rates or long chains, the statistical errors become independent of shear rate. The long-chain limit value of the right side of (4.31) is 6/5; furthermore note that $N^2\lambda_H = (\pi^2/2)\lambda_1$ for large N so that it is very natural when each $\lambda_H^2\dot\gamma^2$ is accompanied by a factor of order N^4.

The usual procedure for obtaining small statistical errors is to generate a large number N_{T} of independent trajectories. For any tensor component τ_{jk}^{rv}, the values obtained from an ensemble of trajectories at time t may be regarded as realizations of N_{T} independent, identically distributed, real-valued random variables for a given ω, and the strong law of large numbers (see Example 2.67) guarantees that the arithmetic mean of the N_{T} results for $\tau_{jk}^{\mathrm{rv}}(t,\omega)$ converges to $\langle\tau_{jk}^{\mathrm{rv}}(t)\rangle$. The discussion of Example 2.71 shows that the statistical error for the ensemble is

$$\sqrt{\frac{\mathrm{Var}\left(\tau_{jk}^{\mathrm{rv}}(t)\right)}{N_{\mathrm{T}}}}\,.$$

Since $\mathrm{Var}\left(\tau_{jk}^{\mathrm{rv}}(t)\right)$ can be estimated from the N_{T} results for $\tau_{jk}^{\mathrm{rv}}(t,\omega)$ according to Exercise 2.68, one can give a statistical error bar for the simulation without knowing the variance in advance.

Exercise 4.11 Develop a routine for Brownian dynamics simulations of Hookean dumbbells for start-up of steady shear flow based on the Euler scheme

$$\boldsymbol{Q}(t_{j+1}) = \boldsymbol{Q}(t_j) + \left[\boldsymbol{\varkappa}\cdot\boldsymbol{Q}(t_j) - \frac{2H}{\zeta}\boldsymbol{Q}(t_j)\right]\Delta t_j + \sqrt{\frac{4k_{\mathrm{B}}T}{\zeta}}\,\Delta\boldsymbol{W}_j'\,, \tag{4.32}$$

where the equalities $\boldsymbol{Q}(t) := \boldsymbol{Q}_1(t) = \boldsymbol{Q}_1'(t)$ and $a_1^{\mathrm{R}} = 2$ for dumbbells have been used. Use units of time, length and mass such that $\lambda_H = 1$, $k_{\mathrm{B}}T/H = 1$, and $n_{\mathrm{p}}\, k_{\mathrm{B}}T = 1$. For $\lambda_H\dot\gamma = 1$, investigate the dependence of the viscosity and of the first normal stress coefficient at time $t = \lambda_H$ on time-step width, and extrapolate your results to zero time-step width. Estimate the statistical error bars for your simulation results **(it goes without saying that this should always be done)**.

As the order of weak convergence for the Euler scheme is $\upsilon = 1$, it is interesting to note that an efficient second-order algorithm can be developed by means of a predictor-corrector scheme. The dynamics of the Rouse model fits into the class of stochastic differential equations discussed in Exercise 3.47; from the results of that exercise we obtain the following second-order weak scheme for Hookean dumbbells with predictor

$$\overline{\boldsymbol{Q}}(t_{j+1}) = \boldsymbol{Q}(t_j) + \left[\boldsymbol{\kappa}(t_j) \cdot \boldsymbol{Q}(t_j) - \frac{2H}{\zeta}\boldsymbol{Q}(t_j)\right] \Delta t_j + \sqrt{\frac{4k_{\mathrm{B}}T}{\zeta}}\, \Delta \boldsymbol{W}'_j , \qquad (4.33)$$

and corrector

$$\begin{aligned} \boldsymbol{Q}(t_{j+1}) &= \boldsymbol{Q}(t_j) + \frac{1}{2}\left[\boldsymbol{\kappa}(t_{j+1}) \cdot \overline{\boldsymbol{Q}}(t_{j+1}) - \frac{2H}{\zeta}\overline{\boldsymbol{Q}}(t_{j+1})\right. \\ &+ \left. \boldsymbol{\kappa}(t_j) \cdot \boldsymbol{Q}(t_j) - \frac{2H}{\zeta}\boldsymbol{Q}(t_j)\right] \Delta t_j + \sqrt{\frac{4k_{\mathrm{B}}T}{\zeta}}\, \Delta \boldsymbol{W}'_j . \end{aligned} \qquad (4.34)$$

Exercise 4.12 Develop a routine for second-order Brownian dynamics simulations of Hookean dumbbells for start-up of steady shear flow based on the predictor-corrector scheme (4.33), (4.34) by modifying the program of Exercise 4.11. Check the dependence of the viscometric functions at time $t = \lambda_H$ on time-step width. Is the predictor-corrector scheme more efficient than the Euler scheme?

Since the generation of random numbers is the by far most expensive part of the simulations described in this subsection one should try to use these random numbers as efficiently as possible. A very simple idea is to apply a time shift to the random numbers used in the construction of a trajectory, and to use the *shifted random numbers* in constructing another trajectory (see Sects. 3.A.5–3.A.7, 5.B.4, and 5.F of [17] for several variations on this idea). For example, when shifting the random numbers to smaller times, the first few random numbers are discarded, the remaining ones are used for starting the new trajectory, and new random numbers need to be generated for the last few time steps. For sufficiently large shifts, the trajectories will look significantly different, thus providing essentially new information about the system in spite of the lack of stochastic independence of the different trajectories. The statistical error bars for the simulation results can be obtained by performing several independent simulations. In special situations, intermediate computations can be stored and need not be repeated for the shifted random numbers.

We have tested the idea of shifting random numbers for the simulation of Hookean dumbbells for start-up of shear flow based on the Euler scheme (see Exercise 4.11). For the fixed time step $\Delta t = 0.04\,\lambda_H$, the random numbers have been shifted to smaller times by various numbers of time steps. Independent initial conditions have been used. In comparing the efficiency of the corresponding simulations, the decrease in computer time and the increase in the statistical error bars need to be taken into account. The optimum simulation for the selected parameters was obtained for a shift by six time steps, or by $0.24\,\lambda_H$, where the simulation with shifted random numbers is roughly twice as efficient as the standard simulation of Exercise 4.11.

4.1.5 Variance Reduced Simulations

In this subsection, we illustrate how the statistical error bars in a simulation can be reduced without increasing the number of simulated trajectories. The approach described below was inspired by Sect. 16.3 of [18]. Apparently, the variance reduction technique described here had not been used in polymer dynamics before this subsection was written. We believe that the idea of improving the efficiency of simulations by means of approximate solutions is so appealing, and that the approach is so promising that a few remarks on the principles of variance reduction in this book seem to be worthwhile (a summary of the ideas presented in this subsection has first been published in [19]). Maybe the discussion of variance reduction in a textbook on polymer dynamics will even stimulate the development of a potentially powerful tool in this field. Readers who are not interested in the most recent and still somewhat preliminary developments in the area of efficient simulation schemes may skip this subsection, which is considerably more difficult than the primer in Brownian dynamics simulations in the preceding subsection; Sect. 4.1.5 is not required for understanding the rest of the book.

For the illustration of variance reduction we choose the Hookean dumbbell model at start-up of steady shear flow, and we point out what needs to be done if a model cannot be solved in closed form. Suppose that we are interested in the mean-square dumbbell extension at some time which we denote by $t_{\rm max}$. In the usual approach we would generate realizations of the estimator $\boldsymbol{Q}^2(t_{\rm max})$ by means of some numerical integration scheme for $\boldsymbol{Q}(t)$ such as the Euler scheme (4.32). If we set $\Theta = \mathrm{Var}[\boldsymbol{Q}^2(t_{\rm max})]$, the statistical error for an ensemble of $N_{\rm T}$ trajectories is $(\Theta/N_{\rm T})^{1/2}$. In variance reduced simulations, we try to construct a random variable Y with $\langle Y\rangle = \langle \boldsymbol{Q}^2(t_{\rm max})\rangle$ and $\mathrm{Var}(Y) = \Theta_{\rm red} < \Theta$. Since we restrict our attention to averages we are interested only in weak solutions.

The variance reduction method pursued in the following is based on the same ideas as *Monte-Carlo integration* schemes. We first explain the background of Monte-Carlo integration by considering dumbbells at equilibrium. We then develop the general ideas and formulas which eventually are made more and more explicit.

Example 4.13 Importance Sampling at Equilibrium

According to (4.25) we may express the mean-square dumbbell extension at equilibrium (after setting $\boldsymbol{\kappa} = \boldsymbol{0}$) as follows in terms of Gaussian integrals,

$$\frac{H}{k_{\rm B}T}\left\langle \boldsymbol{Q}^2\right\rangle = \int \boldsymbol{Q}^2\, p_{0\delta}(\boldsymbol{Q})\,\mathrm{d}^3Q = \int Q^4\, p_{0\delta}(Q)\,\mathrm{d}Q\,. \tag{4.35}$$

Both the three- and the one-dimensional integral in (4.35) can immediately be evaluated to be equal to three, which follows also directly from (4.25). Imagine that we do not know this value of the integral, and that we want to obtain it by simulation of random variables $\boldsymbol{Q}$ and Q' with the appropriate three- and one-dimensional Gaussian distributions, respectively. The statistical errors for N_{T} independent realizations are proportional to $(N_{\mathrm{T}})^{-1/2}$, where the respective prefactors are given by $[\mathrm{Var}(Y)]^{1/2}$ and $[\mathrm{Var}(Y')]^{1/2}$ for $Y = \boldsymbol{Q}^2$ and $Y' = Q'^4$. If we evaluate these prefactors by means of Wick's theorem then we find $\mathrm{Var}(Y) = 6$ and $\mathrm{Var}(Y') = 96$. In summary, both estimators Y and Y' give the same average, but one needs 16 times as many realizations of Y' as of Y to obtain the desired average with the same statistical error bars.

Why does a simulation based on Y' lead to such large statistical errors? The answer to this question lies in the following observation. The Gaussian distribution is peaked around 0. The majority of the generated configurations Q' is hence around 0, and such configurations make only a very small contribution to the average of $Y' = Q'^4$. The rare configurations with large $|Q'|$ make a large contribution to that average, and this is the source of the large statistical error bars. It would be much better to preferably generate those configurations that contribute most to the integral, and to suitably estimate the required integral from those biased configurations. This is the idea of importance sampling in Monte-Carlo integration, and this is the key to the variance reduced simulations developed in this subsection. For the simulation based on Y, configurations that contribute most to the integral are generated more frequently (even though the factor $\boldsymbol{Q}^2$ is still unfavorable). In the following exercise, an estimator based on an even better adaptation to the last integral in (4.35) is discussed—that exercise leads the way to successful variance reduction. □

Exercise 4.14 Let X be a random variable with a continuous distribution with density $p_{\mathrm{red}}(x)$. We use

$$Y = f_{\mathrm{c}}(X)\, X^4, \qquad f_{\mathrm{c}}(X) := \frac{p_{0\delta}(X)}{p_{\mathrm{red}}(X)}, \tag{4.36}$$

as an estimator for the last integral in (4.35), where the correction factor $f_{\mathrm{c}}(X)$ makes up for the biased distribution that we used. Explain why this estimator produces the correct average. Show that a random variable $Y = \boldsymbol{Q}^2$ where $\boldsymbol{Q}$ is distributed according to the three-dimensional Gaussian $p_{0\delta}(\boldsymbol{Q})$ corresponds to $X = |\boldsymbol{Q}|$ and $p_{\mathrm{red}}(x) = x^2\, p_{0\delta}(x)$.
Show that the function

$$p_{\mathrm{red}}(x) = \frac{1}{3\sqrt{2\pi}}\, x^4 \exp\left\{-\frac{1}{2}\, x^2\right\} = \frac{1}{3}\, x^4\, p_{0\delta}(x)$$

is a probability density on $\mathbb{R}$. This density is particularly well adapted to the last integral in (4.35). What is the variance of Y for this choice of the probability density? How can it be that our variance reduced estimator is so suspiciously successful?

Guided by the ideas of Example 4.13 and Exercise 4.14, we can now develop the basic procedure for the construction of variance reduced simulations for stochastic differential equations. We first describe the general construction and we emphasize which conditional probability densities are required (in general, only approximate expressions are available). We then discuss the role of the approximations in variance reduced simulations. In order to be specific, we consider the size of a Hookean dumbbell at start-up of steady shear flow at a given time which we denote by $t_{\max}$.

Assume that the distribution of the initial random variable $\boldsymbol{Q}(t_0)$ at time t_0 is continuous with probability density $p(\boldsymbol{Q}_0)$. For the optimal importance sampling suggested by Exercise 4.14, we would like to simulate connector vectors with a probability density $p_{\text{red}}(\boldsymbol{Q})$ proportional to

$$\boldsymbol{Q}^2 \, p^Q(\boldsymbol{Q}) \, ,$$

where $p^Q(\boldsymbol{Q})$ denotes the probability density for the distribution of the random variable $\boldsymbol{Q}(t_{\max})$,

$$p^Q(\boldsymbol{Q}) = \int p(\boldsymbol{Q}(t_{\max}) = \boldsymbol{Q} | \boldsymbol{Q}(t_0) = \boldsymbol{Q}_0) \, p(\boldsymbol{Q}_0) \, \mathrm{d}^3 Q_0 \, . \tag{4.37}$$

Since the exact transition probability density for the solution of the underlying stochastic differential equation required in (4.37) is in general unknown, we need an approximation $p_{\text{app}}(\boldsymbol{Q}(t_{\max}) = \boldsymbol{Q} | \boldsymbol{Q}(t_0) = \boldsymbol{Q}_0)$ as the *first basic input for variance reduced simulations*, and we define

$$p^Q_{\text{app}}(\boldsymbol{Q}) := \int p_{\text{app}}(\boldsymbol{Q}(t_{\max}) = \boldsymbol{Q} | \boldsymbol{Q}(t_0) = \boldsymbol{Q}_0) \, p(\boldsymbol{Q}_0) \, \mathrm{d}^3 Q_0 \, . \tag{4.38}$$

Once $p^Q_{\text{app}}(\boldsymbol{Q})$ is available, we choose $p_{\text{red}}(\boldsymbol{Q})$ proportional to

$$\boldsymbol{Q}^2 \, p^Q_{\text{app}}(\boldsymbol{Q}) \, .$$

According to Exercise 4.14, we can then introduce a correction factor

$$f_{\text{c}}(\boldsymbol{Q}) = \frac{p^Q(\boldsymbol{Q})}{p_{\text{red}}(\boldsymbol{Q})} = \frac{p^Q(\boldsymbol{Q})}{p^Q_{\text{app}}(\boldsymbol{Q})} \, \frac{\langle \boldsymbol{Q}^2(t_{\max}) \rangle_{\text{app}}}{\boldsymbol{Q}^2} \, , \tag{4.39}$$

where $\langle \boldsymbol{Q}^2(t_{\max}) \rangle_{\text{app}}$ is the approximate expression for the expectation of $\boldsymbol{Q}^2$ at time $t_{\max}$,

$$\langle \boldsymbol{Q}^2(t_{\max}) \rangle_{\text{app}} := \int \boldsymbol{Q}^2 \, p^Q_{\text{app}}(\boldsymbol{Q}) \, \mathrm{d}^3 Q \, .$$

We still need to obtain the exact distribution of $\boldsymbol{Q}(t_{\max})$ in the correction factor $f_{\text{c}}(\boldsymbol{Q})$. The basic idea is to construct the required transition probability density in (4.37) from a sequence of transition probabilities at small time steps that can eventually be treated by numerical integration schemes. We hence select a partition $t_0 < t_1 \cdots < t_{n-1} < t_n = t_{\max}$ of the time range $\mathbb{T} = [t_0, t_{\max}]$, and we next describe the construction of a variance reduced simulation that consists of the random variables $\widetilde{\boldsymbol{Q}}(t_0)$, $\widetilde{\boldsymbol{Q}}(t_1)$, ..., $\widetilde{\boldsymbol{Q}}(t_n)$. The construction of a trajectory (labeled by ω) proceeds in three steps:

1. Select the final configuration $\widetilde{\boldsymbol{Q}}(t_n, \omega)$ according to a continuous distribution for which the density is proportional to $\boldsymbol{Q}^2 \, p^Q_{\text{app}}(\boldsymbol{Q})$.

2. Then, select the initial configuration $\widetilde{\boldsymbol{Q}}(t_0,\omega)$ according to a continuous distribution with density $p_{\text{app}}(\boldsymbol{Q}(t_0) = \boldsymbol{Q}|\boldsymbol{Q}(t_n) = \widetilde{\boldsymbol{Q}}(t_n,\omega))$, where the function

$$p_{\text{app}}(\boldsymbol{Q}(t_0) = \boldsymbol{Q}|\boldsymbol{Q}(t_n) = \boldsymbol{Q}_n) := p_{\text{app}}(\boldsymbol{Q}(t_n) = \boldsymbol{Q}_n|\boldsymbol{Q}(t_0) = \boldsymbol{Q})\,\frac{p(\boldsymbol{Q})}{p^{\boldsymbol{Q}}_{\text{app}}(\boldsymbol{Q}_n)}$$

is defined in terms of previously introduced probability densities (this definition is motivated by (2.59)).

3. Finally, for $j = 0, 1, \ldots, n-2$, select the intermediate configurations $\widetilde{\boldsymbol{Q}}(t_{j+1},\omega)$ according to a continuous distribution with density

$$p_{\text{app}}(\boldsymbol{Q}(t_{j+1}) = \boldsymbol{Q}|\boldsymbol{Q}(t_j) = \widetilde{\boldsymbol{Q}}(t_j,\omega), \boldsymbol{Q}(t_n) = \widetilde{\boldsymbol{Q}}(t_n,\omega)) .$$

Note that the last step of this construction requires approximate conditional probability densities $p_{\text{app}}(\boldsymbol{Q}(t_{j+1}) = \boldsymbol{Q}|\boldsymbol{Q}(t_j) = \boldsymbol{Q}_j, \boldsymbol{Q}(t_n) = \boldsymbol{Q}_n)$ as the *second basic input for variance reduced simulations.*

The reader should realize that the approximate probability densities required as basic inputs for variance reduced simulations are just sufficient for constructing the random variables $\widetilde{\boldsymbol{Q}}(t_0)$, $\widetilde{\boldsymbol{Q}}(t_1)$, ..., $\widetilde{\boldsymbol{Q}}(t_n)$; hence, no additional consistency conditions of the Chapman-Kolmogorov type need to be verified. The given initial distribution and the transition probability density $p_{\text{app}}(\boldsymbol{Q}(t_n) = \boldsymbol{Q}_n|\boldsymbol{Q}(t_0) = \boldsymbol{Q}_0)$ are necessary and sufficient for the simulation of $\widetilde{\boldsymbol{Q}}(t_0)$ and $\widetilde{\boldsymbol{Q}}(t_n)$. The conditional probability densities $p_{\text{app}}(\boldsymbol{Q}(t_{j+1}) = \boldsymbol{Q}_{j+1}|\boldsymbol{Q}(t_j) = \boldsymbol{Q}_j, \boldsymbol{Q}(t_n) = \boldsymbol{Q}_n)$ for $j = 0, 1, \ldots, n-2$ provide exactly the information needed to iteratively construct the intermediate random variables $\widetilde{\boldsymbol{Q}}(t_1)$, $\widetilde{\boldsymbol{Q}}(t_2)$, ..., $\widetilde{\boldsymbol{Q}}(t_{n-1})$, where we assume that, like for the Markov process to be approximated, $\widetilde{\boldsymbol{Q}}(t_{j+1})$ can depend on previous values only through $\widetilde{\boldsymbol{Q}}(t_j)$.

Once more motivated by Exercise 4.14, we introduce the correction factor

$$f_{\text{c}}(\boldsymbol{Q}_0, \boldsymbol{Q}_1, \ldots, \boldsymbol{Q}_n) = f'_{\text{c}}(\boldsymbol{Q}_0, \boldsymbol{Q}_1, \ldots, \boldsymbol{Q}_n)\,\frac{\langle \boldsymbol{Q}^2(t_{\max})\rangle_{\text{app}}}{\boldsymbol{Q}_n^2}, \tag{4.40}$$

where

$$\begin{aligned} f'_{\text{c}}(\boldsymbol{Q}_0, \boldsymbol{Q}_1, \ldots, \boldsymbol{Q}_n) &= \frac{p(\boldsymbol{Q}(t_n) = \boldsymbol{Q}_n|\boldsymbol{Q}(t_{n-1}) = \boldsymbol{Q}_{n-1})}{p_{\text{app}}(\boldsymbol{Q}(t_n) = \boldsymbol{Q}_n|\boldsymbol{Q}(t_0) = \boldsymbol{Q}_0)} \\ &\times \prod_{j=0}^{n-2} \frac{p(\boldsymbol{Q}(t_{j+1}) = \boldsymbol{Q}_{j+1}|\boldsymbol{Q}(t_j) = \boldsymbol{Q}_j)}{p_{\text{app}}(\boldsymbol{Q}(t_{j+1}) = \boldsymbol{Q}_{j+1}|\boldsymbol{Q}(t_j) = \boldsymbol{Q}_j, \boldsymbol{Q}(t_n) = \boldsymbol{Q}_n)} . \end{aligned} \tag{4.41}$$

The factor f'_{c} is the ratio of the exact to the approximate probability density for finding the values $\boldsymbol{Q}_0, \boldsymbol{Q}_1, \ldots, \boldsymbol{Q}_n$ at the times $t_0, t_1, \ldots, t_n$. If the transition probability densities $p_{\text{app}}(\boldsymbol{Q}(t_{j+1}) = \boldsymbol{Q}_{j+1}|\boldsymbol{Q}(t_j) = \boldsymbol{Q}_j, \boldsymbol{Q}(t_n) = \boldsymbol{Q}_n)$ are given by the exact Markovian results

$$p_{\text{app}}(\boldsymbol{Q}(t_{j+1}) = \boldsymbol{Q}_{j+1}|\boldsymbol{Q}(t_j) = \boldsymbol{Q}_j, \boldsymbol{Q}(t_n) = \boldsymbol{Q}_n) =$$
$$\frac{p(\boldsymbol{Q}(t_{j+1}) = \boldsymbol{Q}_{j+1}|\boldsymbol{Q}(t_j) = \boldsymbol{Q}_j)\, p(\boldsymbol{Q}(t_n) = \boldsymbol{Q}_n|\boldsymbol{Q}(t_{j+1}) = \boldsymbol{Q}_{j+1})}{p(\boldsymbol{Q}(t_n) = \boldsymbol{Q}_n|\boldsymbol{Q}(t_j) = \boldsymbol{Q}_j)},$$

then the factor $f'_{\rm c}$ is equal to

$$p(\boldsymbol{Q}(t_n) = \boldsymbol{Q}_n|\boldsymbol{Q}(t_0) = \boldsymbol{Q}_0)/p_{\text{app}}(\boldsymbol{Q}(t_n) = \boldsymbol{Q}_n|\boldsymbol{Q}(t_0) = \boldsymbol{Q}_0),$$

and the correction factor $f_{\rm c}$ is very similar to the result in (4.39); after averaging the $\boldsymbol{Q}_0$-dependence of $f_{\rm c}$ with the conditional probability density of $\widetilde{\boldsymbol{Q}}(t_0)$ given that $\widetilde{\boldsymbol{Q}}(t_n) = \boldsymbol{Q}_n$, we actually recover (4.39).

Since the correction factor is constructed in the spirit of Exercise 4.14, we obtain by the same arguments

$$\left\langle \boldsymbol{Q}^2(t_n) \right\rangle = \left\langle f_{\rm c}\left(\widetilde{\boldsymbol{Q}}(t_0), \widetilde{\boldsymbol{Q}}(t_1), \ldots, \widetilde{\boldsymbol{Q}}(t_n)\right) \widetilde{\boldsymbol{Q}}^2(t_n) \right\rangle, \tag{4.42}$$

so that $f_{\rm c}\widetilde{\boldsymbol{Q}}^2(t_n)$ is indeed an estimator for the mean-square dumbbell extension at $t_n = t_{\max}$. Equation (4.42) can be rewritten in the alternative form

$$\left\langle \boldsymbol{Q}^2(t_n) \right\rangle = \left\langle f'_{\rm c}\left(\widetilde{\boldsymbol{Q}}(t_0), \widetilde{\boldsymbol{Q}}(t_1), \ldots, \widetilde{\boldsymbol{Q}}(t_n)\right) \right\rangle \left\langle \boldsymbol{Q}^2(t_n) \right\rangle_{\text{app}}. \tag{4.43}$$

The representation (4.43) clearly shows that we achieved our aim: $f'_{\rm c}\left(\widetilde{\boldsymbol{Q}}(t_0), \widetilde{\boldsymbol{Q}}(t_1), \ldots, \widetilde{\boldsymbol{Q}}(t_n)\right)$ $\langle \boldsymbol{Q}^2(t_n) \rangle_{\text{app}}$ constitutes an estimator for the desired average $\langle \boldsymbol{Q}^2(t_n) \rangle$ and, if all approximate transition probabilities coincide with the exact ones, then $f'_{\rm c} = 1$ and the variance of the estimator is reduced to zero.

Since we now have information about the distribution at the intermediate times t_j we obtain the following identity for an arbitrary function g

$$\begin{aligned} &\left\langle g\big(\boldsymbol{Q}(t_0), \boldsymbol{Q}(t_1), \ldots, \boldsymbol{Q}(t_n)\big) \right\rangle = \\ &\left\langle f_{\rm c}\left(\widetilde{\boldsymbol{Q}}(t_0), \widetilde{\boldsymbol{Q}}(t_1), \ldots, \widetilde{\boldsymbol{Q}}(t_n)\right) g\left(\widetilde{\boldsymbol{Q}}(t_0), \widetilde{\boldsymbol{Q}}(t_1), \ldots, \widetilde{\boldsymbol{Q}}(t_n)\right) \right\rangle. \end{aligned} \tag{4.44}$$

Thus we have an alternative estimation for the average of an arbitrary function g. Note that the results (4.42) and (4.44) are exact regardless of the accuracy of the required approximate transition probabilities. However, the variance reduction crucially depends on the quality of the approximation. In order to see this we evaluate the variance of the estimator,

$$\begin{aligned} &\text{Var}\left[f_{\rm c}\left(\widetilde{\boldsymbol{Q}}(t_0), \widetilde{\boldsymbol{Q}}(t_1), \ldots, \widetilde{\boldsymbol{Q}}(t_n)\right) g\left(\widetilde{\boldsymbol{Q}}(t_0), \widetilde{\boldsymbol{Q}}(t_1), \ldots, \widetilde{\boldsymbol{Q}}(t_n)\right)\right] = \\ &\left\langle \boldsymbol{Q}^2(t_{\max}) \right\rangle_{\text{app}} \left\langle f'_{\rm c}\big(\boldsymbol{Q}(t_0), \boldsymbol{Q}(t_1), \ldots, \boldsymbol{Q}(t_n)\big) \frac{g^2\big(\boldsymbol{Q}(t_0), \boldsymbol{Q}(t_1), \ldots, \boldsymbol{Q}(t_n)\big)}{\boldsymbol{Q}^2(t_n)} \right\rangle \\ &\qquad - \left\langle g\big(\boldsymbol{Q}(t_0), \boldsymbol{Q}(t_1), \ldots, \boldsymbol{Q}(t_n)\big) \right\rangle^2. \end{aligned} \tag{4.45}$$

Two requirements are crucial for a successful variance reduction. First, the ratio of the exact and the approximate probability density f_c' introduced in (4.41) should be close to unity. A good approximation to the trajectories of the underlying stochastic differential equation is hence required. In choosing such an approximation one should keep in mind that one needs to be able to simulate the random variables $\widetilde{\boldsymbol{Q}}(t_j)$ based on the corresponding approximate distribution. The second requirement is that the essential configuration dependence of the function g should be like $\boldsymbol{Q}^2(t_n)$ (for an appropriate importance sampling). This is of course the case when g is $\boldsymbol{Q}^2(t_n)$. Otherwise, the part of $\boldsymbol{Q}^2(t_n)$, which we arbitrarily chose for illustration, may be taken by a quantity more closely related to g. In particular, if g depends only on $\boldsymbol{Q}(t_n)$ and g is nonnegative, we may choose g instead of $\boldsymbol{Q}^2(t_n)$. If g also takes on negative values, $|g|$ can be used instead of $\boldsymbol{Q}^2(t_n)$. Even if the exact transition probabilities are known, the variance can be reduced to zero only if g is either solely positive or solely negative. In other situations, one can carry out separate simulations for the positive and negative parts of the quantity of interest [18].

We now have the correction factor f_c in a form which, once the required approximate transition probability densities are given, can be used in simulations. In (4.40) and (4.41), only the exact small-step transition probabilities $p(\boldsymbol{Q}(t_{j+1}) = \boldsymbol{Q}_{j+1}|\boldsymbol{Q}(t_j) = \boldsymbol{Q}_j)$ are required as an additional ingredient. If we choose sufficiently small time steps, any desired accuracy can be achieved for these transition probabilities by evaluating them by means of suitable numerical integration schemes. For example, for the Euler scheme, the transition probability density $p(\boldsymbol{Q}(t_{j+1}) = \boldsymbol{Q}_{j+1}|\boldsymbol{Q}(t_j) = \boldsymbol{Q}_j)$ is a Gaussian probability density with mean and variance given by the drift and diffusion terms, respectively. If the transition probabilities are obtained from an approximation scheme, (4.42) and (4.44) are still rigorous identities provided that $\boldsymbol{Q}(t_j)$ represents the corresponding time-discretized approximate solution of the underlying stochastic differential equation. In summary, we are now in a position where we can try to design simulations which have smaller statistical error bars, although they suffer from the usual systematic errors due to time discretization. According to the remarks of this paragraph, the development of higher-order variance reduced schemes is possible by using more refined expressions for the small-step transition probabilities appearing in the correction factor.

The above considerations are very general; actually, only the notation was borrowed from the kinetic theory of dumbbell models. We next construct a variance reduced simulation for Hookean dumbbells at start-up of steady shear flow on a more explicit level. In order to avoid unnecessarily lengthy equations, all times and distances are measured in units of λ_H and $(k_B T/H)^{1/2}$, respectively, which corresponds to setting $\lambda_H = k_B T/H = 1$; the appropriate factors can be reintroduced by means of dimensional analysis, and we do that whenever it is convenient. The first basic input required is the transition probability density

$$p_{\text{app}}(\boldsymbol{Q}(t) = \boldsymbol{Q}|\boldsymbol{Q}(0) = \boldsymbol{Q}_0) = p_{0\Theta}(\boldsymbol{Q} - \mathrm{e}^{-t/2}\,(\boldsymbol{\delta} + \boldsymbol{\varkappa} t) \cdot \boldsymbol{Q}_0)\,, \qquad (4.46)$$

where the Gaussian probability density $p_{0\Theta}(\boldsymbol{Q})$ is determined by

$$\boldsymbol{\Theta} = (1-\mathrm{e}^{-t})\,\boldsymbol{\delta} + \left[1-(1+t)\mathrm{e}^{-t}\right](\boldsymbol{\varkappa}+\boldsymbol{\varkappa}^T) + 2\left[1-\left(1+t+\frac{1}{2}t^2\right)\mathrm{e}^{-t}\right]\boldsymbol{\varkappa}\cdot\boldsymbol{\varkappa}^T . \tag{4.47}$$

This transition probability follows from the closed-form solution of the linear stochastic differential equation for the Rouse model. For equilibrium initial conditions at $t_0 = 0$, we obtain the probability density required for the construction of $\widetilde{\boldsymbol{Q}}(t_{\max})$ from (4.38),

$$p_{\rm app}^{Q}(\boldsymbol{Q}) = p_{0\Theta_{\rm f}}(\boldsymbol{Q}) , \tag{4.48}$$

with

$$\boldsymbol{\Theta}_{\rm f} = \boldsymbol{\delta} + (1-\mathrm{e}^{-t_{\max}})(\boldsymbol{\varkappa}+\boldsymbol{\varkappa}^T) + 2\left[1-(1+t_{\max})\mathrm{e}^{-t_{\max}}\right]\boldsymbol{\varkappa}\cdot\boldsymbol{\varkappa}^T . \tag{4.49}$$

In other words, the probability density (4.48) allows us to do the first step of the procedure outlined on p. 171. For the second step, we determine the conditional probability density required for the construction of $\widetilde{\boldsymbol{Q}}(0)$ for given $\widetilde{\boldsymbol{Q}}(t_{\max})$, which is the Gaussian distribution

$$p_{\rm app}(\boldsymbol{Q}(0) = \boldsymbol{Q}|\boldsymbol{Q}(t_{\max}) = \boldsymbol{Q}_n) = p_{0\Theta_{\rm i}}(\boldsymbol{Q} - \mathrm{e}^{-t_{\max}/2}\,(\boldsymbol{\delta}+\boldsymbol{\varkappa}^T t_{\max})\cdot\boldsymbol{\Theta}_{\rm f}^{-1}\cdot\boldsymbol{Q}_n) , \tag{4.50}$$

with

$$\boldsymbol{\Theta}_{\rm i} = (\boldsymbol{\delta}+\boldsymbol{\varkappa}^T t_{\max})\cdot\boldsymbol{\Theta}_{\rm f}^{-1}\cdot\boldsymbol{\Theta}\cdot(\boldsymbol{\delta}-\boldsymbol{\varkappa}^T t_{\max}) , \tag{4.51}$$

where the covariance matrix $\boldsymbol{\Theta}$ given in (4.47) is evaluated for $t = t_{\max} = t_n$.

Exercise 4.15 Derive the (transition) probability densities (4.46)–(4.51) for Hookean dumbbells at start-up of steady shear flow.

The distributions (4.48) and (4.50) allow the simulation of first $\widetilde{\boldsymbol{Q}}(t_{\max})$ and then $\widetilde{\boldsymbol{Q}}(0)$. According to Exercise 4.15, these distributions correspond to the exact solution for Hookean dumbbells at start-up of steady shear flow; for small time steps, they deviate slightly from the time-discrete solution which enters the correction factor in (4.41) through the factors $p(\boldsymbol{Q}(t_{j+1}) = \boldsymbol{Q}_{j+1}|\boldsymbol{Q}(t_j) = \boldsymbol{Q}_j)$. The variance will hence not be reduced to zero. In nonlinear problems, however, the increase in the variance of the estimator due to the approximate transition probability $p_{\rm app}(\boldsymbol{Q}(t) = \boldsymbol{Q}|\boldsymbol{Q}(0) = \boldsymbol{Q}_0)$ is certainly much larger. We once more emphasize that the variance reduction crucially depends on the quality of available approximations. Gaussian approximations are particularly convenient in carrying out variance reduced simulations. If the parameters of the Gaussian distribution are calculated by numerical solution of ordinary differential equations, these equations may be discretized with the same time step as used in the simulations.

For the third and final step of the procedure outlined on p. 171 we need to describe an iterative procedure for constructing $\widetilde{\boldsymbol{Q}}(t_{j+1})$ for given values of $\widetilde{\boldsymbol{Q}}(t_j)$ and $\widetilde{\boldsymbol{Q}}(t_n)$ for $j = 0, 1, \ldots, n-2$. Rather than specifying the conditional probability densities $p_{\rm app}(\boldsymbol{Q}(t_{j+1}) = \boldsymbol{Q}_{j+1}|\boldsymbol{Q}(t_j) = \boldsymbol{Q}_j, \boldsymbol{Q}(t_n) = \boldsymbol{Q}_n)$ we prefer to directly handle the random variables $\widetilde{\boldsymbol{Q}}(t_{j+1})$; this is in the spirit of using random variables instead of their distributions, or stochastic differential equations instead of diffusion equations. Loosely speaking, we need to carry out a time step starting at $\widetilde{\boldsymbol{Q}}(t_j)$, and we have to keep in mind that the trajectory should end at $\widetilde{\boldsymbol{Q}}(t_n)$. In the Rouse model we can calculate the exact distribution of $\widetilde{\boldsymbol{Q}}(t_{j+1})$. Guided by the exact solution, we here suggest the more general procedure,

$$\begin{aligned}\widetilde{\boldsymbol{Q}}(t_{j+1}) &= [\boldsymbol{\delta} - \boldsymbol{\varepsilon}(t_j, t_{j+1}, t_n)] \cdot \left\{\widetilde{\boldsymbol{Q}}(t_j) + [\boldsymbol{\kappa} \cdot \widetilde{\boldsymbol{Q}}(t_j) - \frac{1}{2}\widetilde{\boldsymbol{Q}}(t_j)]\, \Delta t_j\right\} \\ &+ \boldsymbol{\varepsilon}(t_j, t_{j+1}, t_n) \cdot \boldsymbol{\Phi}_{\mathrm{app}}(t_{j+1}, t_n) \cdot \widetilde{\boldsymbol{Q}}(t_n) + \sqrt{\boldsymbol{\delta} - \boldsymbol{\varepsilon}(t_j, t_{j+1}, t_n)} \cdot \Delta \boldsymbol{W}_j\,,\end{aligned} \tag{4.52}$$

for $j = 0, 1, \ldots, n-2$, where $\sqrt{\boldsymbol{\delta} - \boldsymbol{\varepsilon}(t_j, t_{j+1}, t_n)}$ represents some matrix $\boldsymbol{\sigma}$ for which $\boldsymbol{\sigma} \cdot \boldsymbol{\sigma}^T = \boldsymbol{\delta} - \boldsymbol{\varepsilon}(t_j, t_{j+1}, t_n)$. This is a generalization of (4.32), which is recovered for $\boldsymbol{\varepsilon}(t_j, t_{j+1}, t_n) = \boldsymbol{0}$. The idea behind introducing $\boldsymbol{\varepsilon}(t_j, t_{j+1}, t_n)$ is to reduce the influence of the drift and diffusion terms when the given final value $\widetilde{\boldsymbol{Q}}(t_n)$ is approached and, at the same time, to switch on the influence of the final value. The quantity $\boldsymbol{\Phi}_{\mathrm{app}}(t_{j+1}, t_n)$ is introduced in order to appropriately guide the trajectory to its final value. Reasonable forms of these terms can be determined as follows. If in a suitable approximation, the expectation at time t_n, denoted by $\boldsymbol{Q}_n$, depends linearly on the initial condition $\boldsymbol{Q}_{j+1}$ at time t_{j+1}, then we read off the matrix $\boldsymbol{\Phi}_{\mathrm{app}}(t_{j+1}, t_n)$ from the linear relationship $\boldsymbol{Q}_{j+1} = \boldsymbol{\Phi}_{\mathrm{app}}(t_{j+1}, t_n) \cdot \boldsymbol{Q}_n$. The quantity $\boldsymbol{\varepsilon}(t_j, t_{j+1}, t_n)$ can be constructed from the covariance matrix of the process at t_n if it starts with a deterministic initial condition at t_{j+1}, denoted by $\boldsymbol{\Theta}(t_{j+1}, t_n)$ (it is assumed that the covariance matrix $\boldsymbol{\Theta}(t_{j+1}, t_n)$ is independent of the initial value, at least to a satisfactory approximation). We then suggest the following choice,

$$\begin{aligned}\boldsymbol{\varepsilon}(t_j, t_{j+1}, t_n) &= (t_{j+1} - t_j)\left[(t_{j+1} - t_j)\,\boldsymbol{\delta}\right. \\ &+ \left.\boldsymbol{\Phi}_{\mathrm{app}}(t_{j+1}, t_n) \cdot \boldsymbol{\Theta}(t_{j+1}, t_n) \cdot \boldsymbol{\Phi}^T_{\mathrm{app}}(t_{j+1}, t_n)\right]^{-1}\,,\end{aligned} \tag{4.53}$$

where, in general, $(t_{j+1} - t_j)$ should be multiplied by the diffusion tensor which here is $4k_{\mathrm{B}}T/\zeta = k_{\mathrm{B}}T/(H\lambda_H) = 1$ for the units chosen. Note that $\boldsymbol{\varepsilon}(t_j, t_{j+1}, t_n)$ is small except for the last few time steps for which the denominator, like the numerator, is of the order of Δt_j. This suggests that for constructing the intermediate positions it may be sufficient to have good approximations for short-time transitions rather than for the more complicated conditional probability densities $p_{\mathrm{app}}(\boldsymbol{Q}(t_{j+1}) = \boldsymbol{Q} | \boldsymbol{Q}(t_j) = \boldsymbol{Q}_j, \boldsymbol{Q}(t_n) = \boldsymbol{Q}_n)$.

In principle, generalizations in which the diffusion tensor depends on $\widetilde{\boldsymbol{Q}}(t_j)$ and the auxiliary quantities $\boldsymbol{\Theta}(t_{j+1}, t_n)$, $\boldsymbol{\Phi}_{\mathrm{app}}(t_{j+1}, t_n)$ depend on $\widetilde{\boldsymbol{Q}}(t_{j+1})$ are possible. However, if these matrices are independent of configuration, all the coefficients in (4.52) can be evaluated before the simulation is started, and the variance reduced simulation is not much more time consuming than a standard Euler integration scheme.

For the Hookean dumbbell problem in shear flow we have

$$\boldsymbol{\Phi}_{\text{app}}(t_{j+1}, t_n) = \mathrm{e}^{(t_n - t_{j+1})/2} \, [\boldsymbol{\delta} - \boldsymbol{\varkappa}(t_n - t_{j+1})] \,, \tag{4.54}$$

and

$$\boldsymbol{\Phi}_{\text{app}}(t_{j+1}, t_n) \cdot \boldsymbol{\Theta}(t_{j+1}, t_n) \cdot \boldsymbol{\Phi}^T_{\text{app}}(t_{j+1}, t_n) =$$

$$\begin{aligned} &\left(\mathrm{e}^{t_n - t_{j+1}} - 1\right) \boldsymbol{\delta} + \left\{\mathrm{e}^{t_n - t_{j+1}}[1 - (t_n - t_{j+1})] - 1\right\} (\boldsymbol{\varkappa} + \boldsymbol{\varkappa}^T) \\ &+ 2\left\{\mathrm{e}^{t_n - t_{j+1}}\left[1 - (t_n - t_{j+1}) + \frac{1}{2}(t_n - t_{j+1})^2\right] - 1\right\} \boldsymbol{\varkappa} \cdot \boldsymbol{\varkappa}^T . \end{aligned} \tag{4.55}$$

We now have a complete iterative scheme for variance reduced simulations.

Exercise 4.16 Derive (4.54) and (4.55) for Hookean dumbbells in steady shear flow.

Example 4.17 Details of a Variance Reduced Simulation

We here describe some practical details and results of a variance reduced simulation for Hookean dumbbells at start-up of steady shear flow based on the ideas of this section. The goal of this example is to show how variance reduced simulations can actually be carried out on a computer and, in particular, how the correction factor can be handled. The remarks of this example may hence be regarded as instructions for writing a computer program. Going through these details should be helpful for designing variance reduced simulations for other problems.

In the first step, $\widetilde{\boldsymbol{Q}}(t_n)$ is sampled according to the distribution

$$\boldsymbol{Q}^2 \, p_{0\Theta_\mathrm{f}}(\boldsymbol{Q}) / \mathrm{Tr}(\boldsymbol{\Theta}_\mathrm{f})$$

obtained from (4.48) and (4.49). Realizations of $\widetilde{\boldsymbol{Q}}(t_n)$ are constructed from five Gaussian random variables with mean zero and variance unity, W^f_j, $j = 1, \ldots, 5$, by a suitable transformation. To that end we diagonalize the covariance matrix $\boldsymbol{\Theta}_\mathrm{f}$,

$$\boldsymbol{\Omega}^T_{\Theta_\mathrm{f}} \cdot \boldsymbol{\Theta}_\mathrm{f} \cdot \boldsymbol{\Omega}_{\Theta_\mathrm{f}} = \begin{pmatrix} \Theta^{(1)}_\mathrm{f} & 0 & 0 \\ 0 & \Theta^{(2)}_\mathrm{f} & 0 \\ 0 & 0 & \Theta^{(3)}_\mathrm{f} \end{pmatrix} ,$$

where $\boldsymbol{\Omega}_{\Theta_\mathrm{f}}$ is an orthogonal matrix. We first construct an auxiliary three-dimensional column vector $\boldsymbol{X}$. With probability $\Theta^{(k)}_\mathrm{f} / \mathrm{Tr}(\boldsymbol{\Theta}_\mathrm{f})$ we select one of the three space components k and define $X_k = \mathrm{sgn}(W^\mathrm{f}_1) \, [(W^\mathrm{f}_1)^2 + (W^\mathrm{f}_2)^2 + (W^\mathrm{f}_3)^2]^{1/2}$, and the remaining two components are set equal to W^f_4 and W^f_5. The reader should verify that

$$\sum_{k=1}^{3} (\Omega_{\Theta_\mathrm{f}})_{jk} \sqrt{\Theta^{(k)}_\mathrm{f}} \, X_k$$

then constitutes a realization of the components of $\widetilde{\boldsymbol{Q}}(t_n)$. The result

$$p^Q_{\rm app}(\widetilde{\boldsymbol{Q}}(t_n,\omega)) = \frac{1}{\sqrt{(2\pi)^3 \det(\boldsymbol{\Theta}_{\rm f})}} \exp\left\{-\frac{1}{2}\sum_{j=1}^{5}\left[W_j^{\rm f}(\omega)\right]^2\right\}, \tag{4.56}$$

for an arbitrary realization labeled by ω will be useful in evaluating the correction factor $f'_{\rm c}$.

According to (4.50), the configuration at $t = 0$ can subsequently be realized as

$$\widetilde{\boldsymbol{Q}}(0) = {\rm e}^{-t_{\max}/2}\,(\boldsymbol{\delta} + \boldsymbol{\varkappa}^T t_{\max}) \cdot \boldsymbol{\Theta}_{\rm f}^{-1} \cdot \widetilde{\boldsymbol{Q}}(t_n) + \sqrt{\boldsymbol{\Theta}_{\rm i}} \cdot \boldsymbol{W}^{\rm i}, \tag{4.57}$$

where the three components of $\boldsymbol{W}^{\rm i}$ are further independent standard Gaussian random variables and $\sqrt{\boldsymbol{\Theta}_{\rm i}}$ represents some matrix $\boldsymbol{\sigma}$ with $\boldsymbol{\sigma}\cdot\boldsymbol{\sigma}^T = \boldsymbol{\Theta}_{\rm i}$. Notice that

$$p_{\rm app}(\boldsymbol{Q}(0) = \widetilde{\boldsymbol{Q}}(0,\omega)|\boldsymbol{Q}(t_n) = \widetilde{\boldsymbol{Q}}(t_n,\omega)) = \frac{1}{\sqrt{(2\pi)^3 \det(\boldsymbol{\Theta}_{\rm i})}} \exp\left\{-\frac{1}{2}\left[\boldsymbol{W}^{\rm i}(\omega)\right]^2\right\}. \tag{4.58}$$

Finally, all the intermediate configurations can be constructed from (4.52) when $\Delta\boldsymbol{W}_j$ is represented by $(\Delta t)^{1/2}\boldsymbol{W}_j$, where the three-dimensional random vectors $\boldsymbol{W}_j$ have further independent standard Gaussian components, and a constant time step Δt is assumed. The use of non-Gaussian random variables in variance reduced simulations, which aim at weak solutions, remains to be explored (the handling of the correction factor will then be considerably different). For these intermediate steps we have

$$p_{\rm app}(\boldsymbol{Q}(t_{j+1}) = \widetilde{\boldsymbol{Q}}(t_{j+1},\omega)|\boldsymbol{Q}(t_j) = \widetilde{\boldsymbol{Q}}(t_j,\omega), \boldsymbol{Q}(t_n) = \widetilde{\boldsymbol{Q}}(t_n,\omega)) = \frac{1}{\sqrt{(2\pi\Delta t)^3 \det[\boldsymbol{\delta} - \boldsymbol{\varepsilon}(t_j,t_{j+1},t_n)]}} \exp\left\{-\frac{1}{2}\boldsymbol{W}_j^2(\omega)\right\}. \tag{4.59}$$

Finally, we use the Euler approximation for the exact transition probabilities,

$$p(\boldsymbol{Q}(t_{j+1}) = \boldsymbol{Q}_{j+1}|\boldsymbol{Q}(t_j) = \boldsymbol{Q}_j) = p_{0\Delta t}\left(\boldsymbol{Q}_{j+1} - \boldsymbol{Q}_j - \left[\boldsymbol{\varkappa}\cdot\boldsymbol{Q}_j - \frac{1}{2}\boldsymbol{Q}_j\right]\Delta t\right). \tag{4.60}$$

By combining the results (4.56), (4.58), (4.59) and (4.60), we obtain for the square of the correction factor

$$\begin{aligned} f'^2_{\rm c}\left(\widetilde{\boldsymbol{Q}}(0),\widetilde{\boldsymbol{Q}}(t_1),\ldots,\widetilde{\boldsymbol{Q}}(t_n)\right) &= \frac{\det(\boldsymbol{\Theta}_{\rm i})\det(\boldsymbol{\Theta}_{\rm f})}{(\Delta t)^3}\prod_{j=0}^{n-2}\det[\boldsymbol{\delta} - \boldsymbol{\varepsilon}(t_j,t_{j+1},t_n)] \\ \times \exp\Bigg\{\sum_{j=1}^{5}\left(W_j^{\rm f}\right)^2 + (\boldsymbol{W}^{\rm i})^2 + \sum_{j=0}^{n-2}\boldsymbol{W}_j^2 - \widetilde{\boldsymbol{Q}}^2(0) \\ -\frac{1}{\Delta t}\sum_{j=0}^{n-1}\left[\widetilde{\boldsymbol{Q}}(t_{j+1}) - \widetilde{\boldsymbol{Q}}(t_j) - [\boldsymbol{\varkappa}\cdot\widetilde{\boldsymbol{Q}}(t_j) - \frac{1}{2}\widetilde{\boldsymbol{Q}}(t_j)]\,\Delta t\right]^2\Bigg\}. \end{aligned} \tag{4.61}$$

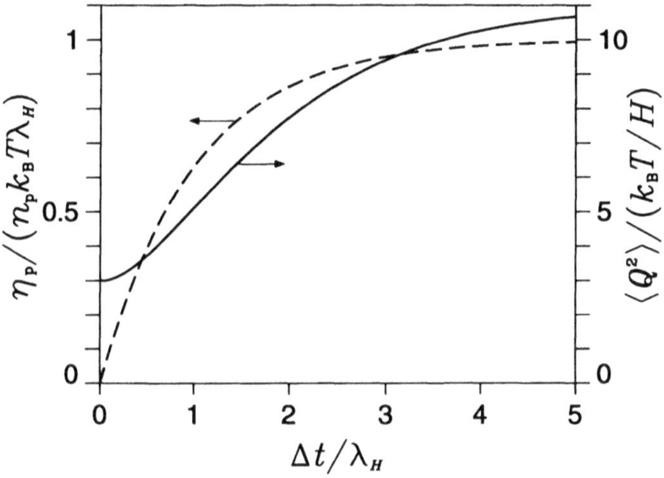

Fig. 4.4. Analytic results for the mean-square size and for the polymer contribution to the viscosity for Hookean dumbbells at start-up of steady shear flow with $\lambda_H \dot{\gamma} = 2$.

In order to check the variance reduction achieved by the ideas presented above, we have performed direct and variance reduced simulations for Hookean dumbbells at start-up of steady shear flow with $\lambda_H \dot{\gamma} = 2$. The maximum time was chosen as $t_{\max} = 5\lambda_H$. In Fig. 4.4, the exact results for the mean-square size and for the polymer contribution to the viscosity are shown. At small t, the time dependence of the viscosity is linear whereas the mean-square size has a quadratic time dependence. The approach to the steady state values is more rapid for the viscosity than for the size. Actually, the exact solution given in [12] for the discrete time step used in the simulations, $\Delta t = 0.01\lambda_H$, was used for checking the simulation results. The agreement was found to be excellent, within the very small statistical error bars.

Figures 4.5 and 4.6 show the results for the statistical error bars in our simulations of 30 000 blocks of 1024 trajectories ($N_T = 30.72 \times 10^6$). Blocks of trajectories were used so that identical operations could be performed on all the trajectories of a block in the innermost loops of the computer program, thus achieving excellent *vectorization* of the programs on a CRAY Y-MP computer. For the variance reduced simulations, only some 10% more CPU-time than for the direct simulations is required.

The dramatic effect of variance reduction for $\langle \boldsymbol{Q}^2(t) \rangle$ near $t_{\max}$, where the variance of the estimator for $\langle \boldsymbol{Q}^2(t_{\max}) \rangle$ was minimized, is obvious in Fig. 4.5. Even in the biased variance reduced simulation, all material properties at all times can be estimated; however, the statistical error bars for short times are larger than in a direct simulation (this is the consequence of an inappropriate bias). For large $t_{\max}$, the variance of an ideal variance reduced estimator ($f'_c \to 1$,

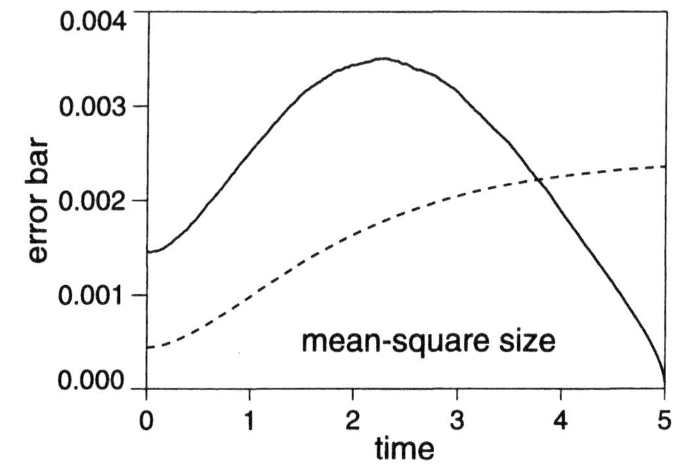

Fig. 4.5. Comparison of the statistical error bars for the mean-square size of a Hookean dumbbell at start-up of steady shear flow with $\lambda_H \dot{\gamma} = 2$ in direct (*dashed curve*) and variance reduced (*continuous curve*) simulations. The units for time and error bar are λ_H and $k_B T/H$, respectively.

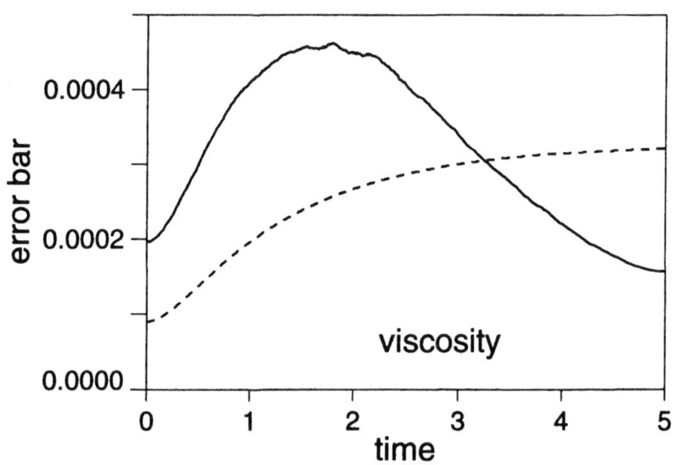

Fig. 4.6. Comparison of the statistical error bars for the polymer contribution to the viscosity for Hookean dumbbells at start-up of steady shear flow with $\lambda_H \dot{\gamma} = 2$ in direct (*dashed curve*) and variance reduced (*continuous curve*) simulations. The units for time and error bar are λ_H and $n_p k_B T \lambda_H$, respectively.

Table 4.1. Detailed comparison of statistical error bars for Hookean dumbbells at start-up of steady shear flow at $t_{\max} = 5\lambda_H$ with $\lambda_H \dot{\gamma} = 2$

	average	error bar in	
		variance reduced /	direct simulation
$\frac{H}{k_B T}\langle \boldsymbol{Q}^2 \rangle$	10.704941	0.000064	0.002361
$\frac{\eta_p}{n_p k_B T \lambda_H}$	0.993134	0.000157	0.000321
$\frac{\Psi_1}{n_p k_B T \lambda_H^2}$	1.924525	0.000144	0.000529
$\frac{\Psi_2}{n_p k_B T \lambda_H^2}$	-0.000039	0.000097	0.000090

which is achieved for $\Delta t \to 0$) for $(H/k_B T)\langle \boldsymbol{Q}^2(0) \rangle$ is

$$15 \left(2c + \frac{1}{c}\right) R_F \left(1, 1 + \frac{\lambda_H \dot{\gamma}}{c}, 1 - \frac{\lambda_H \dot{\gamma}}{c}\right) - 9 \, ,$$

where $c = [1 + (\lambda_H \dot{\gamma})^2]^{1/2}$, and R_F is Carlson's standard elliptic integral of the first kind (see Sect. 6.11 of [14]). For $\lambda_H \dot{\gamma} = 2$, this leads to an increase of the error bars by a factor of about 3.5 at $t = 0$ if the variance is optimized at large $t_{\max}$. The factor observed in simulations with $\Delta t = 0.01\lambda_H$, $t_{\max} = 5\lambda_H$ is 3.3 and, as one can see in Fig. 4.5, this factor decreases with increasing time.

Even though the variance reduced simulation is designed for one particular quantity at one particular time,[3] the statistical error bars for other quantities may also be reduced. Since the stress tensor is quadratic in the connector vector we may expect smaller error bars also for the viscometric functions when the variance of the estimator for $\langle \boldsymbol{Q}^2(t_{\max}) \rangle$ is minimized. A detailed comparison of the error bars for various quantities at $t = t_{\max}$ is given in Table 4.1. Indeed, the error bars for the viscosity and first normal stress coefficient are reduced by factors of 2.0 and 3.7 respectively. The error bar for the second normal stress coefficient is roughly the same in the direct and variance reduced simulations. The statistical error bars for the viscosity at all times are compared in Fig. 4.6. They deviate only by a factor of order two, where the direct simulation is better for short times and the variance reduced simulation is superior near $t_{\max}$.

The variance reduction achieved for $\langle \boldsymbol{Q}^2(t_{\max}) \rangle$ corresponds to a decrease of the statistical error bars by a factor of 37. In order to obtain the same error bar as in a direct simulation, a factor of some 1200 in CPU-time can be saved,

[3] An idea for alleviating these conditions is mentioned in the discussion at the end of this subsection.

where the longer run time for the variance reduced simulation has been taken into account. □

The above construction of variance reduced simulations for Hookean dumbbells uses transition probabilities obtained from the exact solution. In general, approximate expressions for the transition probabilities are required, and Gaussian approximations are particularly convenient for carrying out the simulations. Since there exist excellent approximations for models with hydrodynamic interaction (see Sect. 4.2), for which no exact solution is available, the method is particularly promising for such models. Even with very simple approximations, it has been demonstrated that the ideas presented in this subsection work for Hookean dumbbells with hydrodynamic interaction [20]. In any case, the idea of using approximations in order to improve simulations is very attractive. It makes the development of good approximation schemes even more rewarding. In addition to a good approximation one can obtain the exact results by means of simulations with smaller statistical error bars; in particular, one can better judge the quality of the approximation by means of the more precise simulation results.

The variance reduced simulations presented here in great detail are based on importance sampling. In the most elementary realization of importance sampling for stochastic differential equations, which has been described in this section, a biased final configuration is selected first. Alternatively, one can produce a biased final distribution not by selecting the final configurations but rather by modifying the drift and initial conditions such that the final configurations turn out to be distributed according to the desired probability density. The proper modification of the drift and the determination of the corresponding correction factor can be based on a suitably constructed Girsanov transformation, which is a profound result of Itô calculus (see Sect. 7.1 of [21] for a formal derivation and Sect. 6.4 of [18] for a heuristic justification). Details on this idea, and its successful implementation for Hookean dumbbells with hydrodynamic interaction, can be found in [22].

Very recently, a variance reduction method that is not based on importance sampling has been adapted to polymer kinetic theory [23]. The basic idea is to estimate the fluctuations in the quantity of interest as obtained in an unbiased standard simulation. By subtracting an approximate expression for the fluctuations, one can reduce the fluctuations, that is, the variance, without changing averages. This approach is suited ideally when the fluctuations can be estimated from a Gaussian approximation and, even better, when the quantity of interest is a quadratic function of configuration. Under these conditions, the variance for the viscometric functions and the mean-square size of Hookean dumbbells with hydrodynamic interaction at start-up of shear flow has been reduced by two orders of magnitude, uniformly in time and at little extra costs [23]. If the quantity of interest is not a quadratic function of configuration, the variance can still be reduced by subtracting the results of two simulations, one for the exact stochastic differential equation and another one for an approximate differential equation for which the averages are known. The trajectories of the two simulations must be constructed with the same sequence of random numbers so that the fluctuations are approximately cancelled.

4.2 Hydrodynamic Interaction

Up to this point we have assumed that the beads of a bead-spring chain move through the solvent without disturbing the given homogeneous solvent flow field. In this section we wish to remove this "free-draining" assumption. The perturbation of the solvent flow field then leads to an additional interaction between the beads. We follow here the approach of Kirkwood and Riseman [24], who seem to have been the first to introduce the idea of *hydrodynamic interaction* into polymer kinetic theory. Incorporation of this effect into the Rouse model makes the equations of motion nonlinear so that approximations and simulations play an important role in studying hydrodynamic interactions. The predictions for some material properties are found to become much more realistic when hydrodynamic interaction is accounted for.

After describing the standard approach to incorporating hydrodynamic interaction into bead-spring models and discussing different hydrodynamic-interaction tensors we introduce the preaveraging approximation which leads to the *Zimm model* [25]. The long chain limit of the Zimm model is discussed in detail. A drastically improved description of hydrodynamic-interaction effects is obtained within the Gaussian approximation. Finally, we illustrate the power of Brownian dynamics simulations for Hookean dumbbells with hydrodynamic interaction.

4.2.1 Description of Hydrodynamic Interaction

If the beads of a bead-spring chain move through the solvent they perturb the solvent flow field. Such perturbations propagate through the solvent and influence the motion of the other beads. This hydrodynamic interaction between beads is an extremely complex phenomenon because, in principle, the nonlinear character of the Navier-Stokes equation for the solvent motion, the nonzero size of the beads, and the presence of many beads should be taken into account. In the usual approach, one linearizes the Navier-Stokes equation and assumes that the propagation of solvent flow perturbations is infinitely fast [26]. One then expects that there exists a linear relationship between the force exerted by a bead at a point $\boldsymbol{r}'$ on the solvent, $\boldsymbol{F}(\boldsymbol{r}')$, and the velocity perturbation at some other position $\boldsymbol{r}$, $\Delta\boldsymbol{v}(\boldsymbol{r})$. If we assume that the beads are point particles so that the forces on the solvent are exerted at well-defined positions in space we obtain,

$$\Delta\boldsymbol{v}(\boldsymbol{r}) = \boldsymbol{\Omega}(\boldsymbol{r}-\boldsymbol{r}')\cdot\boldsymbol{F}(\boldsymbol{r}'),$$

where $\boldsymbol{\Omega}(\boldsymbol{r})$ is the Green's function of the time-independent linearized Navier-Stokes equation,

$$\boldsymbol{\Omega}(\boldsymbol{r}) = \frac{1}{8\pi\eta_{\mathrm{s}} r}\left(\boldsymbol{\delta} + \frac{\boldsymbol{r}\boldsymbol{r}}{r^2}\right). \tag{4.62}$$

The tensor function $\boldsymbol{\Omega}(\boldsymbol{r})$ is also known as the *Oseen-Burgers tensor*. In view of the linearization of the Navier-Stokes equation it is natural to assume that the velocity perturbations caused by different beads can be superimposed. Then, the diffusion equation (4.1) for the configurational distribution function $p = p(t, \boldsymbol{r}_1, \boldsymbol{r}_2, \ldots, \boldsymbol{r}_N)$ can be modified as follows in order to incorporate hydrodynamic-interaction effects into an arbitrary bead-spring model (see, e.g., (15.1-4) and (15.4-4) of [1]),

$$\begin{aligned}\frac{\partial p}{\partial t} &= -\sum_{\mu}\frac{\partial}{\partial \boldsymbol{r}_\mu}\cdot\Big\{\Big[\boldsymbol{v}_0 + \boldsymbol{\varkappa}\cdot\boldsymbol{r}_\mu + \frac{1}{\zeta}\boldsymbol{F}_\mu + \sum_\nu \boldsymbol{\Omega}(\boldsymbol{r}_\mu - \boldsymbol{r}_\nu)\cdot\boldsymbol{F}_\nu\Big]p\Big\} \\ &\quad + \frac{k_{\mathrm{B}}T}{\zeta}\sum_{\mu,\nu}\frac{\partial}{\partial \boldsymbol{r}_\mu}\cdot\left[\delta_{\mu\nu}\,\boldsymbol{\delta} + \zeta\boldsymbol{\Omega}(\boldsymbol{r}_\mu - \boldsymbol{r}_\nu)\right]\cdot\frac{\partial}{\partial \boldsymbol{r}_\nu}p,\end{aligned} \tag{4.63}$$

where we define $\boldsymbol{\Omega}(\boldsymbol{0}) := \boldsymbol{0}$ in order to avoid hydrodynamic self-interactions. In this section, we again use the convention that Greek indices are to be summed from 1 to N and Latin indices from 1 to $N-1$. In the Itô approach, the stochastic differential equations of motion equivalent to the Fokker-Planck equation (4.63) are

$$\begin{aligned}\mathrm{d}\boldsymbol{r}_\mu(t) &= \Big[\boldsymbol{v}_0(t) + \boldsymbol{\varkappa}(t)\cdot\boldsymbol{r}_\mu(t) + \frac{1}{\zeta}\boldsymbol{F}_\mu(t) + \sum_\nu \boldsymbol{\Omega}(\boldsymbol{r}_\mu(t) - \boldsymbol{r}_\nu(t))\cdot\boldsymbol{F}_\nu(t)\Big]\,\mathrm{d}t \\ &\quad + k_{\mathrm{B}}T\sum_\nu \boldsymbol{\nabla}\cdot\boldsymbol{\Omega}(\boldsymbol{r}_\nu(t) - \boldsymbol{r}_\mu(t))\,\mathrm{d}t + \sqrt{\frac{2k_{\mathrm{B}}T}{\zeta}}\sum_\nu \mathbf{B}_{\mu\nu}(t)\cdot\mathrm{d}\boldsymbol{W}_\nu(t),\end{aligned} \tag{4.64}$$

where, for all $\mu, \mu' = 1, 2, \ldots, N$,

$$\delta_{\mu\mu'}\,\boldsymbol{\delta} + \zeta\boldsymbol{\Omega}(\boldsymbol{r}_\mu - \boldsymbol{r}_{\mu'}) = \sum_\nu \mathbf{B}_{\mu\nu}\cdot\mathbf{B}^T_{\mu'\nu}. \tag{4.65}$$

Each of the $N \times N$ tensors $\mathbf{B}_{\mu\nu}$ depends on all bead positions, and $\mathbf{B}_{\mu\nu}(t)$ is obtained by inserting the polymer configuration at time t. The terms involving $\boldsymbol{\nabla}\cdot\boldsymbol{\Omega}$ in the second line of (4.64) result from the fact that the second-order derivative term in (4.63) needs to be rearranged in order to obtain the Itô version of the Fokker-Planck equation. For the Oseen-Burgers tensor, the incompressibility of the solvent implies that these terms vanish. Note that the Stratonovich version of the equations of motion would be much more complicated (the partial derivatives of the tensors $\mathbf{B}_{\mu\nu}$ with respect to the bead positions would be required, and no closed-form expressions for these tensors can be given; the numerical evaluation of these derivatives is extremely time-consuming, as was mentioned in connection with implicit methods in Sect. 3.4.4).

The equations of motion (4.64) can be understood as follows. The force exerted on the solvent by bead ν is the negative of the frictional force felt by the polymer. Since we neglect bead inertia in the equations for the bead motion, the sum of frictional force, spring force and Brownian force on each bead vanishes. The force exerted on the solvent by bead ν is hence equal to the sum of the spring and Brownian forces on bead ν. In the first line of (4.64), the velocity perturbations resulting from the spring forces on all beads are added to the given flow field. The contribution of the Brownian forces is much less obvious. This problem is related to the extremely rapid fluctuations of the Brownian forces which cannot be propagated instantaneously, as is assumed in the Oseen-Burgers tensor description. The form of the diffusion term in (4.64) can be obtained from the postulate that, at equilibrium, the solvent acts as a heat bath for the polymer molecules so that the configurational distribution function should be given by the Boltzmann distribution. The second line of (4.63) suggests that it is not the rapidly fluctuating Brownian forces that are propagated through the solvent, but rather the "smoothed Brownian forces" on the beads ν that tend to balance osmotic pressure gradients and that depend on all bead positions,

$$\overline{\boldsymbol{F}}_\nu^{\mathrm{B}} = -k_{\mathrm{B}}T \frac{\partial}{\partial \boldsymbol{r}_\nu} \ln p \,. \tag{4.66}$$

The use of such smoothed Brownian forces is crucial in the usual derivation of kinetic theory models [1].

An alternative derivation of (4.64), which does not use smoothed Brownian forces, can be based on a system of coupled stochastic differential equations for polymers and solvent [26]. The equation for the solvent is taken from the fluctuating-hydrodynamics approach to Newtonian fluids in which a noise term is added to the Navier-Stokes equation. In this approach it can be seen that the occurrence of the Oseen-Burgers tensor in the diffusion term of (4.63) is a direct consequence of the noise in the Navier-Stokes equation of fluctuating hydrodynamics.

Exercise 4.18 Work out the stochastic differential equations of motion for the connector vector and the center of mass position for a dumbbell model with hydrodynamic interaction. Show that the two equations of motion decouple. Write down the corresponding Fokker-Planck equation for the connector vector. Show that, for the Oseen-Burgers tensor, the equation of motion for the connector vector is meaningless for short distances between the beads.

In writing down the equations of motion for bead-spring chains with hydrodynamic interaction we have been very careless: one should always check whether the diffusion term in a Fokker-Planck equation is positive-semidefinite. It is hence not clear whether a decomposition of the type (4.65) is possible. Exercise 4.18 shows how dangerous such carelessness can be; when using the Oseen-Burgers tensor, the diffusion tensor is actually not positive-semidefinite. Of course, such a mathematical problem has a physical origin. On the one hand, in deriving the Oseen-Burgers tensor $\boldsymbol{\Omega}$, we assumed point particles or beads with zero radius. On the other hand, by assuming a bead friction coefficient ζ

for given solvent viscosity η_s, we implicitly introduce a bead radius a_b given by the Stokes law $\zeta = 6\pi\eta_s a_b$ (or, at least, of that order of magnitude). Hence, the Oseen-Burgers tensor yields a realistic description of hydrodynamic interactions only for bead separations large compared to a_b. It is the combination $\boldsymbol{\delta} - \zeta\boldsymbol{\Omega}$ that is not always positive-semidefinite, and the problem occurs at bead separations smaller than $(3/2)a_b$ where two beads of radius a_b already overlap and the assumption of point particles in calculating $\boldsymbol{\Omega}$ is certainly unjustified.

There are two possibilities for restoring a positive-semidefinite diffusion term: we can prevent the beads from overlapping, or we can modify the hydrodynamic-interaction tensor. The first possibility corresponds to introducing *excluded-volume interactions*, another complicated nonlinear phenomenon that changes the physics of the model drastically. Since we are here interested in a separate discussion of hydrodynamic-interaction effects which, for long chains, are dominated by the many hydrodynamic interactions between pairs of beads that are not close in space, we prefer the possibility of modifying the Oseen-Burgers tensor. Various modifications or regularizations have been suggested, the most famous of which is the Rotne-Prager-Yamakawa tensor [27, 28] (see, e.g., [29] for an overview). It should be noted that the necessity of regularizing the Oseen-Burgers tensor is not primarily a consequence of its singularity at short distances, but rather due to the resulting loss of the positive-semidefinite form of the diffusion term. We here choose a hydrodynamic-interaction tensor that smoothly switches off the interactions at short distances (of order a_b) [29],

$$\boldsymbol{\Omega}(\boldsymbol{r}) = \frac{1}{8\pi\eta_s r\,[r^2 + (4/3)a_b^2]^3}\Bigg[\left(r^6 + \frac{14}{3}\,a_b^2 r^4 + 8\,a_b^4 r^2\right)\boldsymbol{\delta} + \left(r^6 + 2\,a_b^2 r^4 - \frac{8}{3}\,a_b^4 r^2\right)\frac{\boldsymbol{rr}}{r^2}\Bigg], \tag{4.67}$$

where a_b is not regarded as a free parameter but is rather assumed to be related to ζ and η_s through the Stokes law $\zeta = 6\pi\eta_s a_b$. Several arguments in favor of the above modification of the Oseen-Burgers tensor can be offered. First of all, $\boldsymbol{\delta} - \zeta\boldsymbol{\Omega}$ is then positive-definite. For dumbbells, this implies that the diffusion matrix in configuration space is positive-definite.[4] Furthermore, the hydrodynamic-interaction tensor (4.67) satisfies the incompressibility condition $\boldsymbol{\nabla}\cdot\boldsymbol{\Omega} = \boldsymbol{0}$ which leads to a simplification of the equations of motion (4.64).[5] For large r, where the nonzero radius of the beads does not matter, we recover

[4] Warning: For longer chains, (4.67) does not generally lead to a positive-semidefinite diffusion matrix. One should then resort to the Rotne-Prager-Yamakawa tensor for which it has been proven that it leads to a positive-semidefinite diffusion matrix for all configurations of chains consisting of any number of beads [27]. The formulation of a hydrodynamic-interaction tensor which leads to a positive-semidefinite diffusion matrix for all chain lengths and, in contrast to the Rotne-Prager-Yamakawa tensor, is continuously differentiable would be desirable when aiming at higher-order numerical integration schemes.

[5] Warning: For the Rotne-Prager-Yamakawa tensor, the condition $\boldsymbol{\nabla}\cdot\boldsymbol{\Omega} = \boldsymbol{0}$ follows from the incompressibility of the solvent. For the full description of the hydrodynamic interaction between two spheres with nonvanishing radii, incompressibility does **not** imply the condition $\boldsymbol{\nabla}\cdot\boldsymbol{\Omega} = \boldsymbol{0}$ [30].

the Oseen-Burgers tensor. Moreover, the $1/r^3$- and $1/r^5$-corrections (where the latter have a vanishing coefficient) correspond to the leading order terms of the mutual mobility tensor for two spheres of radius a_b at large distances [30, 31].[6] A further advantage of the above modification, which is not shared by the Rotne-Prager-Yamakawa tensor, is its smooth dependence on r; this is important for the systematic development of higher-order numerical integration schemes for the stochastic differential equations of motion. Detailed comparisons for dumbbells show that the regularization procedure has little influence on the material functions predicted [29]; in the limit of long chains, only the behavior of the hydrodynamic-interaction tensors at large distances is expected to matter so that all regularizations at short distances should lead to identical results.

Throughout this subsection, there was no need to specify the detailed form of the forces $\boldsymbol{F}_\nu$ so that the formulation of the dynamics in the presence of hydrodynamic interactions is valid for general bead-spring-chain models. We next focus our attention on models with Hookean springs, and we discuss the effect of hydrodynamic interaction on various predicted material properties.

4.2.2 Zimm Model

The nonlinear stochastic differential equations (4.64) cannot be solved in closed from. In order to obtain a tractable approximation, B. H. Zimm [25] replaced the random variables $\boldsymbol{\Omega}(\boldsymbol{r}_\mu(t) - \boldsymbol{r}_\nu(t))$ by their equilibrium averages,[7]

$$\boldsymbol{\Omega}(\boldsymbol{r}_\mu(t) - \boldsymbol{r}_\nu(t)) \to \begin{cases} \dfrac{1}{6\pi\eta_s}\sqrt{\dfrac{2H}{\pi k_B T|\mu-\nu|}}\,\boldsymbol{\delta} & \text{if } \mu \neq \nu \\ \boldsymbol{0} & \text{if } \mu = \nu\,. \end{cases} \tag{4.68}$$

Exercise 4.19 Evaluate the equilibrium average of $\boldsymbol{\Omega}(\boldsymbol{r}_\mu(t) - \boldsymbol{r}_\nu(t))$ for the Oseen-Burgers tensor in order to establish the physical motivation behind the replacement (4.68).

If we introduce an $N \times N$-matrix with elements

$$H_{\mu\nu} := \begin{cases} \sqrt{2}h^* \dfrac{1}{\sqrt{|\mu-\nu|}} & \text{if } \mu \neq \nu \\ 1 & \text{if } \mu = \nu\,, \end{cases} \tag{4.69}$$

with the hydrodynamic-interaction parameter

[6] Additional $1/r^4$-terms arise in systems consisting of three and more spheres. The effect of the presence of one or more other spheres on the mobility of one particular sphere is also of order $1/r^4$ [31].

[7] Zimm worked on the level of configurational distribution functions so that the formulation (4.68) of equilibrium averaging results only after translating to the level of stochastic processes.

$$h^* := \frac{\zeta}{6\pi\eta_s}\sqrt{\frac{H}{\pi k_B T}}, \tag{4.70}$$

then the equations of motion for the Zimm model can be written as

$$\mathrm{d}\boldsymbol{r}_\mu(t) = \Big[\boldsymbol{v}_0(t) + \boldsymbol{\varkappa}(t)\cdot\boldsymbol{r}_\mu(t) + \frac{1}{\zeta}\sum_\nu H_{\mu\nu}\boldsymbol{F}_\nu(t)\Big]\,\mathrm{d}t + \sqrt{\frac{2k_B T}{\zeta}}\sum_\nu B_{\mu\nu}\,\mathrm{d}\boldsymbol{W}_\nu(t), \tag{4.71}$$

where

$$H_{\mu\mu'} = \sum_\nu B_{\mu\nu}\,B_{\mu'\nu}\,. \tag{4.72}$$

Note that the $N \times N$-matrix with elements $B_{\mu\nu}$ has nothing to do with the $N \times (N-1)$-matrix B introduced in §11.6 of [1]. The noise in (4.71) is additive so that there is no need to distinguish between the Itô and Stratonovich interpretations of the stochastic differential equations of motion for the Zimm model. For sufficiently small h^*, the $H_{\mu\nu}$ define a positive-definite matrix so that the unphysical behavior of the Oseen-Burgers tensor at short distances does not matter after equilibrium averaging.

The parameter h^* can be expressed as $h^* = a_b/(\pi k_B T/H)^{1/2}$ which is roughly the bead radius over the root-mean-square distance between two beads connected by a spring at equilibrium. We hence expect that h^* should be smaller than 1/2, but not by more than an order of magnitude. From a theoretical point of view, values of h^* around 0.25 are very attractive because, for the Zimm model, they minimize the effects of chain length [32]. Treating hydrodynamic interaction by first order perturbation theory and refining the results by means of renormalization group theory suggests $h^* \approx 0.18$ [33], and inclusion of excluded volume should reduce this value by a factor of 3/4 [34]. From an experimental point of view, values of h^* between 0.1 and 0.2 are successfully used in fitting rheological data to the predictions for Zimm chains consisting of up to a few hundred beads (see, e.g., the table on p. 171 of [1]). In conclusion, the range of acceptable values for the only additional parameter, h^*, occurring in going from the Rouse to the Zimm model is rather narrow. Moreover, the predictions for long Zimm chains are found to be independent of the precise value of h^*.

The system of stochastic differential equations (4.71) for the Zimm model is of the narrow-sense linear type with additive noise considered in Sect. 3.3.2. Solution of these equations is hence only a matter of linear algebra (decoupling of the equations or exponentiation of a matrix). While *all the steps in solving the Zimm and Rouse models and the structure of the solutions are very similar*, the orthogonal matrix leading to a decoupled system of equations cannot be determined in a closed form for the Zimm model (for ring molecules with equilibrium-averaged hydrodynamic interaction, the cyclic symmetry allows one to find an explicit expression for the corresponding orthogonal matrix [35]).

As in the Rouse model, the time evolution of the connector vectors can be determined by taking differences of the equations of motion for successive beads,

$$\mathrm{d}\boldsymbol{Q}_j(t) = \left[\boldsymbol{\kappa}(t)\cdot\boldsymbol{Q}_j(t) - \frac{1}{\zeta}\sum_k A^{\mathrm{Z}}_{jk}\,\boldsymbol{F}^{\mathrm{c}}_k(t)\right]\mathrm{d}t + \sqrt{\frac{2k_{\mathrm{B}}T}{\zeta}}\sum_\nu (B_{j+1\,\nu} - B_{j\nu})\,\mathrm{d}\boldsymbol{W}_\nu(t)\,, \tag{4.73}$$

where the quantities A^{Z}_{jk} are the elements of the $(N-1)\times(N-1)$ Zimm matrix,

$$A^{\mathrm{Z}}_{jk} := A^{\mathrm{R}}_{jk} + \sqrt{2}h^*\left(\frac{2}{\sqrt{|j-k|}} - \frac{1}{\sqrt{|j-k+1|}} - \frac{1}{\sqrt{|j-k-1|}}\right). \tag{4.74}$$

Here and in the following we use the convention that contributions from values j and k which lead to vanishing denominators must be omitted (exclusion of hydrodynamic self-interactions). Once the Zimm matrix is diagonalized, that is,

$$\sum_{l,n} \Omega^{\mathrm{Z}}_{lj}\,A^{\mathrm{Z}}_{ln}\,\Omega^{\mathrm{Z}}_{nk} = a^{\mathrm{Z}}_j\,\delta_{jk}\,, \tag{4.75}$$

with an orthogonal $(N-1)\times(N-1)$-matrix with elements Ω^{Z}_{jk}, then we obtain decoupled equations of motion for the normal modes $\boldsymbol{Q}'_k(t) := \sum_j \Omega^{\mathrm{Z}}_{jk}\,\boldsymbol{Q}_j(t)$,

$$\mathrm{d}\boldsymbol{Q}'_k(t) = \left[\boldsymbol{\kappa}(t)\cdot\boldsymbol{Q}'_k(t) - \frac{H}{\zeta}\,a^{\mathrm{Z}}_k\,\boldsymbol{Q}'_k(t)\right]\mathrm{d}t + \sqrt{\frac{2k_{\mathrm{B}}Ta^{\mathrm{Z}}_k}{\zeta}}\,\mathrm{d}\boldsymbol{W}'_k(t)\,, \tag{4.76}$$

where

$$\boldsymbol{W}'_k(t) := \frac{1}{\sqrt{a^{\mathrm{Z}}_k}}\sum_{j,\nu}\Omega^{\mathrm{Z}}_{jk}(B_{j+1\,\nu} - B_{j\nu})\,\boldsymbol{W}_\nu(t)\,, \tag{4.77}$$

for $k = 1, 2, \ldots, N-1$.

If we look at the equation for the center of mass motion following from (4.71), all the individual bead positions are found to occur in that equation. For the Zimm model, the dynamics of the connector vectors and the center of mass position are thus correlated. We could first solve the equations of motion for the connector vectors and then calculate the center of mass motion. However, a much more convenient way of describing the motion of the entire molecule in space is obtained by introducing the hydrodynamic center of resistance (see §15.4 of [1]),

$$\boldsymbol{r}_{\mathrm{h}} := \sum_\nu l_\nu\,\boldsymbol{r}_\nu\,, \tag{4.78}$$

where the coefficients l_ν can be expressed in terms of the elements of the inverse of the matrix defined in (4.69),

$$l_\nu := \frac{\sum_\mu H^{-1}_{\mu\nu}}{\sum_{\mu,\mu'} H^{-1}_{\mu\mu'}}\,. \tag{4.79}$$

The motivation behind this definition is immediately clear from the fact that it then follows from (4.71),

$$d\boldsymbol{r}_\mathrm{h}(t) = [\boldsymbol{v}_0(t) + \boldsymbol{\varkappa}(t) \cdot \boldsymbol{r}_\mathrm{h}(t)]\, dt + \sqrt{\frac{2k_\mathrm{B}T}{\zeta}} \sum_{\mu,\nu} l_\mu \, B_{\mu\nu} \, d\boldsymbol{W}_\nu(t) \,. \tag{4.80}$$

If we further introduce

$$\boldsymbol{W}'_N(t) := \sqrt{\sum_{\mu,\nu} H^{-1}_{\mu\nu}} \sum_{\mu,\nu} l_\mu \, B_{\mu\nu} \, \boldsymbol{W}_\nu(t) \,, \tag{4.81}$$

we can write

$$d\boldsymbol{r}_\mathrm{h}(t) = [\boldsymbol{v}_0(t) + \boldsymbol{\varkappa}(t) \cdot \boldsymbol{r}_\mathrm{h}(t)]\, dt + \sqrt{2D_\mathrm{h}}\, d\boldsymbol{W}'_N(t) \,, \tag{4.82}$$

with

$$D_\mathrm{h} = \frac{k_\mathrm{B}T}{\zeta \sum_{\mu,\nu} H^{-1}_{\mu\nu}} \,. \tag{4.83}$$

The independence of the time-evolution of the center of resistance and of the internal degrees of freedom is established in the following exercise.

Exercise 4.20 Show that the processes $\boldsymbol{W}'_\mu(t)$ for $\mu = 1, 2, \ldots, N$ defined in (4.77) and (4.81) are independent three-dimensional Wiener processes.

Exercise 4.21 In steady shear flow, all properties of the Zimm model that depend only on the internal configurations of the chains can most conveniently be obtained from the second moments,

$$\left\langle \boldsymbol{Q}_j(t)\boldsymbol{Q}_k(t) \right\rangle = \frac{k_\mathrm{B}T}{H} \left[\delta_{jk}\, \boldsymbol{\delta} + 2\lambda_H C^\mathrm{Z}_{jk} (\boldsymbol{\varkappa} + \boldsymbol{\varkappa}^T) + 8\lambda_H^2 (C^\mathrm{Z})^2_{jk}\, \boldsymbol{\varkappa} \cdot \boldsymbol{\varkappa}^T \right] .$$

In this equation, the quantities C^Z_{jk} are the elements of the $(N-1) \times (N-1)$ *modified Kramers matrix*, which is the inverse of the Zimm matrix (4.74). Prove this modification of the corresponding Rouse result (4.25).

As for the Rouse model, the transformation to the motion of the center of resistance and internal normal modes not only decouples all the equations of motion but, according to Exercise 4.20, also the resulting inhomogeneous terms happen to involve independent Wiener processes $\boldsymbol{W}'_\mu(t)$. This coincidence is the deeper reason why a closed-form constitutive equation for the Zimm model could be derived by using a product of Gaussian probability densities for the normal modes [11]. The decoupled equations *(4.76) and (4.82) can be solved in exactly the same manner as (4.9) and (4.12)* for the Rouse model. After transforming back to the original variables, we obtain the complete explicit solution for the stochastic process modeling the dynamics of polymer configurations in the Zimm model,

$$\begin{aligned}
\boldsymbol{Q}_j(t) &= \sum_{k=1}^{N-1} \Omega^{\mathrm{Z}}_{jk}\, \boldsymbol{Q}'_k(t)\,, \\
\boldsymbol{r}_1(t) &= \boldsymbol{r}_{\mathrm{h}}(t) - \sum_{j=1}^{N-1}\left(1-\sum_{\mu=1}^{j} l_\mu\right)\boldsymbol{Q}_j(t)\,, \\
\boldsymbol{r}_\nu(t) &= \boldsymbol{r}_1(t) + \sum_{j=1}^{\nu-1}\boldsymbol{Q}_j(t)\,, \qquad \text{for } \nu = 2,3,\ldots,N\,.
\end{aligned} \tag{4.84}$$

Exercise 4.22 Verify the transformations (4.84).

Example 4.23 Velocity Field Inside a Polymer Chain

We here wish to calculate the average perturbation of the given velocity field inside and around a polymer molecule,

$$\Delta\boldsymbol{v}(\boldsymbol{r}) = \sum_\mu \left\langle \boldsymbol{\Omega}(\boldsymbol{r}-\boldsymbol{r}_\mu)\cdot\left(\boldsymbol{F}_\mu + \overline{\boldsymbol{F}}^{\mathrm{B}}_\mu\right)\right\rangle .$$

The Brownian contribution in this equation vanishes. This can be seen by writing the average as an integral, inserting the expression (4.66), and integrating by parts; the result then vanishes due to the incompressibility condition $\boldsymbol{\nabla}\cdot\boldsymbol{\Omega} = \boldsymbol{0}$. In order to facilitate the further evaluation of the velocity perturbations we introduce the Fourier representation of the Oseen-Burgers tensor,

$$\boldsymbol{\Omega}(\boldsymbol{r}) = \frac{1}{(2\pi)^3}\int \mathrm{e}^{-i\boldsymbol{q}\cdot\boldsymbol{r}}\,\frac{1}{\eta_{\mathrm{s}} q^2}\left(\boldsymbol{\delta} - \frac{\boldsymbol{q}\boldsymbol{q}}{q^2}\right)\mathrm{d}^3 q\,. \tag{4.85}$$

This representation makes it more obvious that the Oseen-Burgers tensor is the Green's function for the linearized time-independent Navier-Stokes equation. For a given position of the center of resistance $\boldsymbol{r}_{\mathrm{h}}$, we obtain the following expression for the velocity perturbation around the center of resistance of a molecule,

$$\Delta\boldsymbol{v}(\boldsymbol{r}_{\mathrm{h}}+\boldsymbol{r}) = \frac{1}{(2\pi)^3\eta_{\mathrm{s}}}\sum_\mu \int \mathrm{e}^{-i\boldsymbol{q}\cdot\boldsymbol{r}}\,\frac{1}{q^2}\left(\boldsymbol{\delta}-\frac{\boldsymbol{q}\boldsymbol{q}}{q^2}\right)\cdot\left\langle \boldsymbol{F}_\mu \mathrm{e}^{i\boldsymbol{q}\cdot(\boldsymbol{r}_\mu-\boldsymbol{r}_{\mathrm{h}})}\right\rangle \mathrm{d}^3 q\,.$$

Since both $\boldsymbol{F}_\mu$ and $\boldsymbol{r}_\mu - \boldsymbol{r}_{\mathrm{h}}$ can be expressed in terms of connector vectors (see (4.2) and (4.84)), the remaining average can be expressed in terms of second moments by differentiating the characteristic function of the Gaussian connector vectors,

$$\begin{aligned}
&\left\langle \boldsymbol{F}_\mu \mathrm{e}^{i\boldsymbol{q}\cdot(\boldsymbol{r}_\mu-\boldsymbol{r}_{\mathrm{h}})}\right\rangle = \\
&\qquad \langle \boldsymbol{F}_\mu(\boldsymbol{r}_\mu-\boldsymbol{r}_{\mathrm{h}})\rangle\cdot i\boldsymbol{q}\,\exp\left\{-\frac{1}{2}\boldsymbol{q}\cdot\langle(\boldsymbol{r}_\mu-\boldsymbol{r}_{\mathrm{h}})(\boldsymbol{r}_\mu-\boldsymbol{r}_{\mathrm{h}})\rangle\cdot\boldsymbol{q}\right\}.
\end{aligned}$$

By combining all the above expressions, we obtain the final results for the velocity perturbation and its gradient in the Zimm model. For steady shear

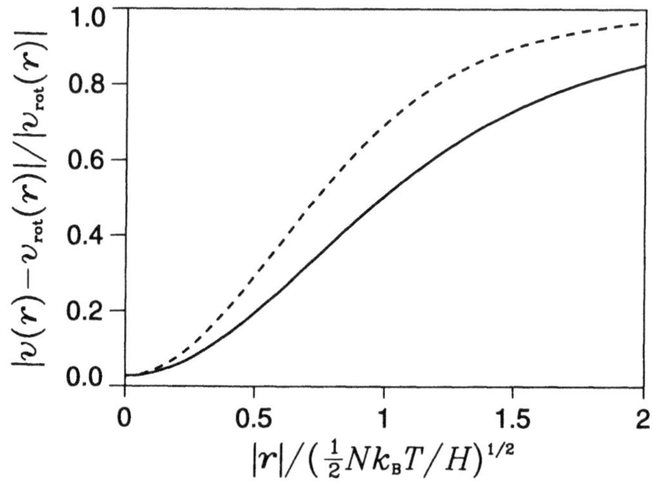

Fig. 4.7. Relative deviation of the perturbed velocity field as a function of the distance from the center of resistance for steady shear flow at small flow rates ($N = 1000$). The *dashed line* is for vectors $\boldsymbol{r}$ parallel to the 1- and 2-directions, and the *continuous line* is for $\boldsymbol{r}$ in the direction of $(\pm 1, \pm 1, 0)$.

flow at small flow rates, the relative deviation of the perturbed velocity field, $\boldsymbol{v}(\boldsymbol{r}) := \boldsymbol{\varkappa} \cdot \boldsymbol{r} + \Delta \boldsymbol{v}(\boldsymbol{r}_\mathrm{h} + \boldsymbol{r})$, from a pure rotation, $\boldsymbol{v}_\mathrm{rot}(\boldsymbol{r}) := (\boldsymbol{\varkappa} - \boldsymbol{\varkappa}^T) \cdot \boldsymbol{r}/2$, around the center of resistance with angular velocity $\dot{\gamma}/2$ is shown in Fig. 4.7. The curves have been calculated for $N = 1000$, and they are indistinguishable from the corresponding curves for $N = 500$. Without hydrodynamic interaction, that is, without any screening of the applied velocity gradients, this ratio is equal to unity. For perfect screening, the polymer molecule rotates like a rigid sphere, and this ratio is zero.

It is obvious from Fig. 4.7 that the velocity screening does not take place in a boundary layer but throughout the polymer molecule. Even at the center of resistance, the screening is not perfect (extrapolation of the results for various N shows that this is not an artifact of using a finite value of N).

The final result for the gradient of the velocity perturbation has the following form,

$$\left[\frac{\partial}{\partial \boldsymbol{r}} \Delta \boldsymbol{v}(\boldsymbol{r}_\mathrm{h} + \boldsymbol{r})\right]^T = \frac{1}{(2\pi)^3 \eta_\mathrm{s}} \sum_\mu \langle \boldsymbol{F}_\mu (\boldsymbol{r}_\mu - \boldsymbol{r}_\mathrm{h}) \rangle : \tag{4.86}$$
$$\int \mathrm{e}^{-i\boldsymbol{q} \cdot \boldsymbol{r}} \frac{\boldsymbol{q}}{q} \left(\boldsymbol{\delta} - \frac{\boldsymbol{q}\boldsymbol{q}}{q^2}\right) \frac{\boldsymbol{q}}{q} \exp\left\{-\frac{1}{2} \boldsymbol{q} \cdot \langle (\boldsymbol{r}_\mu - \boldsymbol{r}_\mathrm{h})(\boldsymbol{r}_\mu - \boldsymbol{r}_\mathrm{h}) \rangle \cdot \boldsymbol{q}\right\} \mathrm{d}^3 q \,.$$

In the spirit of equilibrium averaging of hydrodynamic interactions, one can assume that the average in the exponential may be replaced by its isotropic equilibrium value; we here do not make this simplifying assumption. If $|\boldsymbol{r}|$ is large compared to the polymer size, the gradient of the average velocity perturbation decays as $|\boldsymbol{r}|^{-3}$. In spite of this weak decay,[8] the space integral of (4.86) is finite,

$$\int \left[\frac{\partial}{\partial \boldsymbol{r}} \Delta \boldsymbol{v}(\boldsymbol{r})\right]^T \mathrm{d}^3 r = \frac{1}{5\eta_\mathrm{s}} \sum_\mu \left[\langle \boldsymbol{F}_\mu(\boldsymbol{r}_\mu - \boldsymbol{r}_\mathrm{h})\rangle - \frac{1}{3}\,\boldsymbol{\delta}\,\mathrm{Tr}\,\langle \boldsymbol{F}_\mu(\boldsymbol{r}_\mu - \boldsymbol{r}_\mathrm{h})\rangle\right] .$$

This last result has previously been derived by Schümmer and Otten by means of different arguments [36]. Since the sum of all spring forces $\boldsymbol{F}_\mu$ vanishes, we can replace the center of resistance in this equation by the center of mass. □

Exercise 4.24 Work out the steps involved in the numerical calculation of the velocity screening inside a polymer molecule for steady shear flow as shown in Fig. 4.7. Discuss the velocity field near the center of resistance for large N.

After solving the equations of motion for the Zimm model, we next discuss the expression for the stress tensor. The usual procedure for evaluating stresses is (more or less tacitly) to add the Kramers expression for the polymer contribution to the stress tensor (4.18) to the solvent contribution $-\eta_\mathrm{s}(\boldsymbol{\kappa} + \boldsymbol{\kappa}^T)$, even for models with hydrodynamic interaction. However, Example 4.23 shows that the average velocity gradient in the solvent of a dilute polymer solution with polymer number density n_p does not coincide with the given $\boldsymbol{\kappa}$ but is rather given by [36]

$$\overline{\boldsymbol{\kappa}} = \boldsymbol{\kappa} + \frac{n_\mathrm{p}}{5\eta_\mathrm{s}} \sum_\mu \left[\langle \boldsymbol{F}_\mu(\boldsymbol{r}_\mu - \boldsymbol{r}_\mathrm{c})\rangle - \frac{1}{3}\,\boldsymbol{\delta}\,\mathrm{Tr}\,\langle \boldsymbol{F}_\mu(\boldsymbol{r}_\mu - \boldsymbol{r}_\mathrm{c})\rangle\right] ,$$

so that there is an additional anisotropic solvent contribution to the stress tensor,

$$\Delta \boldsymbol{\tau}^\mathrm{s} = -\frac{2}{5}\, n_\mathrm{p} \sum_\mu \langle \boldsymbol{F}_\mu(\boldsymbol{r}_\mu - \boldsymbol{r}_\mathrm{c})\rangle .$$

This result is at variance with the discussion in Sects. 3.7.3 and 3.7.4 of [8]. According to Table 15.2-1 of [1], this solvent contribution is $-2/5$ of the anisotropic polymer contribution (4.18). When formulated in terms of the given velocity gradient $\boldsymbol{\kappa}$, there is hence an extra solvent contribution to the stress tensor which is of the same form as the polymer contribution. Notice, however, that the velocity gradient $\boldsymbol{\kappa}$ is not observable; the stresses should hence be expressed in terms of the observable velocity gradient $\overline{\boldsymbol{\kappa}}$ so that the total anisotropic solvent contribution to the stress tensor is simply $-\eta_\mathrm{s}(\overline{\boldsymbol{\kappa}} + \overline{\boldsymbol{\kappa}}^T)$. For any given homogeneous flow

[8] The integral of $\Delta \boldsymbol{v}$ over a sphere around $\boldsymbol{r}_\mathrm{h}$ vanishes for symmetry reasons. If the boundaries of the macroscopic domain are considered then, in general, one obtains a nonzero result, and the integral depends on the shape of the domain. Notice, however, that such an approach would be of rather limited value because, at macroscopic distances from the polymer molecule, the neglect of retardation effects in the Navier-Stokes equation, which is necessary to obtain the Oseen-Burgers tensor, cannot be justified [26]. We rather look at averaged velocity gradients which are nonzero, and for which the influence of the boundaries seems to be irrelevant.

field, it is then correct to add the polymer contribution (4.18) to the solvent stress tensor $-\eta_s(\overline{\boldsymbol{\kappa}} + \overline{\boldsymbol{\kappa}}^T)$. In a given flow field, there is no need to distinguish between $\boldsymbol{\kappa}$ and $\overline{\boldsymbol{\kappa}}$ in the polymer contribution to the stress tensor because the difference leads only to an effect of second order in polymer concentration on the stress tensor (which, like other second order effects stemming from interactions between different polymer molecules, may be neglected in the theory of dilute polymer solutions). From now on, we hence no longer distinguish between $\boldsymbol{\kappa}$ and $\overline{\boldsymbol{\kappa}}$ in calculating rheological material functions in given flows. Notice, however, that the ideas discussed in this paragraph should be taken into account in calculating elastic recoil for dilute polymer solutions.

When the expression for the stress tensor of the Zimm model need not be modified, the rheological material functions for this model can be obtained from those of the Rouse model by a very simple substitution. In order to see this one just has to realize that the constitutive equation for the Zimm model is given by (4.20), where the relaxation times λ_j in (4.19) are now formed with the eigenvalues of the Zimm matrix, a_j^{Z}, instead of the corresponding eigenvalues a_j^{R} of the Rouse matrix. These modified relaxation times are to be used in all material functions, for example, in the extensional viscosities (4.21) and (4.22), in the zero-shear-rate viscosity (4.24), or in the other viscometric functions determined in Exercise 4.6. The only problem is that, in general, the eigenvalues of the Zimm matrix have to be determined numerically. For very large N, however, many material properties predicted by the Zimm model can be presented in a more elegant form, as is shown in the next subsection.

Exercise 4.25 Calculate the relaxation times and the polymer contribution to the viscosity for Zimm chains consisting of two and three beads.

4.2.3 Long Chain Limit and Universal Behavior

In the preceding subsection we have seen how various properties of the Zimm model can be expressed in terms of the eigenvalues and inverse of the Zimm matrix and of the inverse of the matrix with elements $H_{\mu\nu}$. Except for very short chains, none of these quantities can be calculated analytically for finite chain length N. In the limit $N \to \infty$, however, some interesting analytical results can be obtained.

For several reasons, the long chain limit of the Zimm model is very important. First of all, it allows us to discuss the properties of the Zimm model without looking at long tables of results for various values of N and h^*. More importantly, in this limit we elaborate the universal properties predicted by the model, the properties that are independent of the details of the mechanical model and hence constitute general consequences of the presence of equilibrium-averaged hydrodynamic interaction in long polymer chains. These universal properties are independent of the strength of the hydrodynamic interaction, h^*; only the presence of hydrodynamic interaction is relevant. Power-law dependences of various material properties on molecular weight with universal exponents are expected (see Sect. 8.2.2.1 of [37]) and, from the prefactors, it should be possible to form universal ratios. The predicted universal exponents and prefactors are ideally suited for a parameter-free test of the Zimm model by means of experimental data for high molecular weight polymer solutions. A similar situation occurs for other models, for example, if better approximations for hydrodynamic interaction or excluded volume are considered. The universal exponents and prefactors for long chains are of great interest for any model and, ideally, they should be obtained by extrapolation from simulation results.

We first consider the diffusion coefficient. Since the process $\boldsymbol{W}'_N$ in the equation of motion for the center of resistance (4.82) is a Wiener process, the diffusion coefficient is given by (4.83). Even in the presence of a flow field, polymer diffusivity is predicted to remain isotropic. This is an artifact of the equilibrium-averaged hydrodynamic interactions in the Zimm model; in general, one expects that polymer diffusivity in flowing solvents has to be described by a diffusion tensor [38]. The asymptotic molecular weight dependence of the Zimm prediction for the diffusion coefficient D_h can be obtained as follows. In the limit of very long chains ($N \to \infty$), the transformed indices

$$x := \frac{2\mu}{N} - 1, \quad y := \frac{2\nu}{N} - 1, \ldots \tag{4.87}$$

can be considered as continuous variables in the interval $[-1, 1]$. With the replacements

$$\sum_{\mu=1}^{N} \quad \to \quad \frac{N}{2} \int_{-1}^{1} \mathrm{d}x \,, \tag{4.88}$$

$$\delta_{\mu\nu} \quad \to \quad \frac{2}{N}\, \delta(x-y) \,, \tag{4.89}$$

one realizes from the definition of the inverse of a matrix and from (4.69) that the matrix elements $H^{-1}_{\mu\nu}$ are of the order $h^{*-1} N^{-3/2}$. After introducing a factor N^2 for the double sum in (4.83) we hence find

$$D_\mathrm{h} = c\, \frac{h^* \, k_\mathrm{B} T}{\zeta \sqrt{N}} \qquad \text{for large } N. \tag{4.90}$$

By introducing hydrodynamic-interaction effects, the molecular weight dependence of the diffusion coefficient changed from N^{-1} to $N^{-1/2}$, which is in excellent agreement with experimental results for Θ conditions[9] (see Table 15.4-2 of

[9] At the Θ temperature, or Flory temperature, the repulsive excluded-volume interactions between different segments of a polymer molecule are compensated for by the polymer-solvent interactions, so that the polymer molecules nearly behave like ideal Hookean chains. In three dimensions, three-body interactions cause corrections to the ideal behavior which decay only very slowly (logarithmically) with chain length. For establishing Θ conditions rather poor solvents are required. Typically, a further small decrease of the temperature causes a collapse transition or phase separation, and a small increase in temperature leads to good solvents with distinct excluded-volume effects. The longer the chains, the more difficult it becomes to fine-tune a polymer solution to the Θ temperature (the required precision in adjusting the temperature is inversely proportional to the square root of the chain length).

[1]). An exact calculation of the numerical coefficient c is tedious but possible by solving an integral equation [39]; the result is $c = \pi^{3/2}\,[\Gamma(3/4)]^{-2} \approx 3.708$. The diffusion coefficient is often used to introduce the hydrodynamic radius of polymer molecules as

$$\overline{R}_{\mathrm{h}} := \frac{k_{\mathrm{B}}T}{6\pi\eta_{\mathrm{s}}\,D_{\mathrm{h}}} = \frac{\sqrt{\pi}}{c}\sqrt{N\,\frac{k_{\mathrm{B}}T}{H}}\,. \tag{4.91}$$

We can compare $\overline{R}_{\mathrm{h}}$ to the root-mean-square radius of gyration $\overline{R}_{\mathrm{g}}$ at equilibrium. Since static equilibrium properties are not affected by hydrodynamic interaction we obtain from Exercise 4.7,

$$U_{\mathrm{RD}} := \frac{\overline{R}_{\mathrm{g}}}{\overline{R}_{\mathrm{h}}} = \frac{\pi}{\sqrt{2}\,[\Gamma(3/4)]^2} \approx 1.47934\,. \tag{4.92}$$

This is a typical example of a universal ratio of prefactors for long chains predicted by the Zimm model.

If the rules (4.87)–(4.89) for long chains are supplemented by the rule that differences should be replaced by derivatives, with a factor $2/N$ for each derivative, then the continuous version of the eigenvalue equation suggests that the eigenvalues a_j^{Z} of the Zimm matrix are of the order $h^*\,N^{-3/2}$. Therefore, the relaxation times are proportional to $N^{3/2}\zeta/(h^*H)$. We hence write

$$\lambda_j = c_j\left(\frac{N}{j}\right)^{3/2}\frac{\zeta}{4h^*H\pi^2}\,, \tag{4.93}$$

where the leading-order dependence of the numerical prefactors on j has been introduced through the idea that the jth relaxation time corresponds to the relaxation of a mode involving N/j beads, and the factor $4\pi^2$ has been introduced for convenience ((4.93) should be compared to the corresponding Rouse result (4.19)). Relaxation times scale as $N^{3/2}$ for the Zimm model, whereas the corresponding chain length dependence for the Rouse model is N^2. The coefficients c_j are numerical constants of order unity (the approximate theory in Sect. 4.2.1 of [8] leads to $c_j = 1$ for all j). The original numerical calculation of the relaxation times corresponds to $c_1 = 1.22$, $c_2 = 1.09$, $c_3 = 1.06$, where the latter two values are already consistent with the asymptotic expression for large j, $c_j = \left(1 - \frac{1}{2\pi j}\right)^{-1}$ [40].

Since a spectrum of relaxation times associated with modes of the chain dynamics is an artifact of models described by linear stochastic differential equations of motion [41], and since even the longest relaxation time is rather difficult to determine by experiments or computer simulations, one usually defines a characteristic time scale in terms of the polymer contribution to the viscosity. Equation (4.24) and the results of Exercise 4.6 suggest that the following definition yields an appropriate characteristic time scale,

$$\boxed{\lambda_\eta := \lim_{n_{\mathrm{p}}\to 0}\frac{\eta_{\mathrm{p}}}{n_{\mathrm{p}}\,k_{\mathrm{B}}T}}\,. \tag{4.94}$$

For sufficiently dilute solutions, η_p is proportional to n_p so that a well-defined limit exists. If, in general, the viscosity depends on the shear rate then the zero-shear-rate viscosity is usually taken in (4.94). The time scale λ_η is often used by experimentalists to introduce dimensionless strain rates such as the reduced shear rate. For long chains, a relationship between different time scales is established by the universal ratio

$$U_{\eta\lambda} := \frac{\lambda_\eta}{\lambda_1} = \frac{\sum_j \lambda_j}{\lambda_1} = \frac{\pi^{5/2}}{4\,[\Gamma(3/4)]^2 c_1} \approx 2.39\,, \tag{4.95}$$

where the analytical result for the sum of the relaxation times of the Zimm model has been used [32].

The time scale λ_η is closely related to the intrinsic viscosity,

$$[\eta]_0 := \lim_{\rho_p \to 0} \frac{\eta_p}{\rho_p\,\eta_s}\,, \tag{4.96}$$

where ρ_p is the polymer mass density. The index 0 at the square brackets is a reminder that this quantity has to be evaluated in the limit of vanishing shear rate. The relationships $\eta_p \sim \lambda_\eta n_p \sim \lambda_1 n_p$ and $\rho_p \sim N\,n_p$ imply $[\eta]_0 \sim N^{1/2}$ for the Zimm model, whereas $[\eta]_0 \sim N$ for the Rouse model (the exponent of N in such a relation is known as the *Mark-Houwink exponent*). The Zimm prediction for the exponent is found to be in better agreement with experimental results (see §3.6 of [10]). The prefactor in the power law for the intrinsic viscosity can be characterized in terms of another universal ratio,

$$U_{\eta\mathrm{R}} := \lim_{n_p \to 0} \frac{\eta_p}{n_p\,\eta_s\,(4\pi \overline{R}_g^3/3)} = \frac{9}{2}\,\frac{U_{\eta\lambda}}{U_{\mathrm{RD}}}\,\frac{D_h\,\lambda_1}{\overline{R}_g^2} = \frac{9\pi}{8\sqrt{2}\,[\Gamma(3/4)]^2} \approx 1.66425\,. \tag{4.97}$$

If the polymer solution is interpreted as a suspension of spheres with volume fraction $n_p\,4\pi\overline{R}_g^3/3$, then the famous Einstein formula for the viscosity of suspensions gives $U_{\eta\mathrm{R}} = 2.5$. A carefully measured experimental value for polystyrene in cyclohexane at the Θ temperature is $U_{\eta\mathrm{R}} = 1.49(6)$ [42]. From Monte Carlo simulations, one can obtain lower and upper bounds for $U_{\eta\mathrm{R}}$ for configuration-dependent hydrodynamic interactions (no equilibrium averaging) [43]; a lower bound is 1.36(5), and the less precise upper bound is consistent with the experimental value quoted above.

The general definition of other universal ratios involving viscometric functions and their numerically determined values for the Zimm model are [32]

$$U_{\Psi\eta} := \frac{n_p\,k_B T\,\Psi_1}{\eta_p^2} \approx 0.413865 \tag{4.98}$$

$$U_{\Psi\Psi} := \frac{\Psi_2}{\Psi_1} = 0\,. \tag{4.99}$$

Again it is understood that these ratios are defined in the limits $n_p \to 0$ and $\dot\gamma \to 0$. The most reliable experimental results for $U_{\Psi\eta}$ can be obtained from flow birefringence when assuming the validity of the stress-optical law. For a narrowly distributed polystyrene solution in a Θ solvent, $U_{\Psi\eta} = 0.46$ has been measured [44]. However, the correct value of $U_{\Psi\eta}$ may be smaller than 0.46 because even very small polydispersity can lead to a considerable increase in $U_{\Psi\eta}$. This strong sensitivity to polydispersity has been explained and verified in [45] where, for a good solvent, the value $U_{\Psi\eta} = 0.51$ has been extracted from experimental data.

The determination of the universal ratios $U_{\rm RD}$, $U_{\eta\lambda}$, $U_{\eta\rm R}$, $U_{\Psi\eta}$, and $U_{\Psi\Psi}$ is a major challenge for Brownian dynamics simulations. Fixman's simulations suggest that the ratio $U_{\rm RD}$ for configuration-dependent hydrodynamic interactions should be about 3%–5% smaller than the prediction (4.92) of the Zimm model [46]. For the ratio $U_{\eta\rm R}$, no universal behavior could be extracted from Fixman's simulations. If the viscometric functions depend on shear rate, then one can furthermore investigate the universal functions of the reduced shear rate $\lambda_\eta\dot{\gamma}$ associated with the various material properties.

The different molecular weight dependences of the intrinsic viscosity imply that, for long chains, the polymer contribution to the viscosity predicted by the Zimm model is smaller than for the Rouse model. Exercise 4.25 shows that the situation for short chains is the other way round. Short chains with hydrodynamic interaction are hence very untypical. For long chains it should be observed that h^* occurs only in the combination ζ/h^* in all material properties (in the diffusion coefficient, in the relaxation times, and hence in the constitutive equation and all rheological properties). Therefore, the parameter h^* has no observable effect on the material properties of long chains—it can be absorbed in the basic time constant and it does not occur in dimensionless ratios. Like the Rouse model, the Zimm model is also essentially parameter-free (cf. the discussion at the end of Subsection 4.1.2). It should hence be considered as an artifact if a given set of experimental data can be fitted best with a certain combination of parameters N and h^*. Since beads are not well-defined chemical entities, it is purely accidental if a fit with $N = 50$ works better than with $N = 49$ or 51.

Exercise 4.26 Discuss the universal long chain properties of the Rouse model.

4.2.4 Gaussian Approximation

The molecular weight dependence predicted by the Zimm model for the diffusion coefficient, the relaxation times, and the viscometric functions constitute a remarkable improvement over the corresponding Rouse predictions. On the other hand, the Zimm model fails to explain the experimentally observed shear rate dependence of the viscometric functions and the occurrence of a nonzero second normal stress difference. All these failures result from the equilibrium averaging of the hydrodynamic-interaction tensor. A more rigorous treatment shows that hydrodynamic interaction leads to shear rate dependent viscometric functions and to a negative second normal stress difference.

The simplest idea for improving the Zimm model consists in averaging the hydrodynamic interaction not at equilibrium but rather in the flow situation of interest [32, 47–49]. While the equilibrium averaging of the Zimm model can be carried out before the diffusion equation is solved in a flow field ("preaveraging"), the improved description of hydrodynamic interaction requires the self-consistent determination of the distribution function (which depends on the averaged hydrodynamic-interaction tensors) and of the hydrodynamic-interaction tensors (which are averaged with the distribution function to be determined). This idea is hence known as the *consistent averaging method*. The consistently averaged hydrodynamic-interaction tensors make the diffusion equation nonlinear in the probability density so that a process with mean field interactions results (see Sect. 3.3.4).

The replacement of the hydrodynamic-interaction tensors with averages leads to linear stochastic differential equations of motion, so that the solution is a Gaussian process. The fact that averages occur in the equations of motion leads to a nonlinear consistency condition for the moments (which is actually a system of order N^2 coupled nonlinear equations). This nonlinearity makes solution of the model considerably more complicated, however, it also leads to more realistic material functions. A very successful scheme for an approximate decoupling of the nonlinear consistency equations by means of the eigenmodes of the Rouse or Zimm model has been suggested [50, 51]. With increasing shear rate, the polymer contribution to the viscosity and the first normal stress coefficient are predicted to decrease at low shear rates ("shear thinning"); for sufficiently long chains (around $N = 5$), these material functions go through a minimum and increase at higher shear rates. This "shear thickening" effect arises because hydrodynamic interactions are partially switched off, and more and more beads become exposed to the flow field when the polymers become more stretched at high shear rates (it was pointed out at the end of the preceding subsection that the viscosity of long free-draining chains is much larger than for non-free-draining chains). A power-law dependence on shear rate found for very long chains indicates that this switching off is only partial and persists to very high shear rates [52].

The consistent averaging method leads to a non-vanishing second normal stress difference. However, its sign at small shear rates is positive, whereas Brownian dynamics simulations indicate that without any approximations a negative sign results (see Fig. 4.8). The comparison with simulations furthermore shows that even where the qualitative behavior of the predictions of the consistent averaging method is realistic, the quantitative agreement is not completely satisfactory. Typical deviations are of the order of 10% to 30% for the viscosity and of the order of 20% to 40% for the first normal stress coefficient.

By replacing the hydrodynamic-interaction tensors with their self-consistent averages, the consistent averaging method neglects fluctuation effects of the hydrodynamic interactions. Much better quantitative agreement is obtained for the so-called *Gaussian approximation* [53–55] (see Fig. 4.8). Since both equilibrium averaging and self-consistent averaging lead to Gaussian processes one can try to avoid such averaging assumptions and directly postulate a Gaussian distribution which is chosen such that the time-evolution equations for the lowest moments are of exactly the same form as for the full model. We here describe the idea of the Gaussian approximation in a more general context.

The Gaussian approximation for the general stochastic differential equation considered in Sect. 3.3,

$$\mathrm{d}\boldsymbol{X}_t = \boldsymbol{A}(t, \boldsymbol{X}_t)\,\mathrm{d}t + \mathbf{B}(t, \boldsymbol{X}_t) \cdot \mathrm{d}\boldsymbol{W}_t\,, \tag{4.100}$$

is constructed as follows. By averaging (4.100) and the differential equation for $\boldsymbol{X}_t\boldsymbol{X}_t$ obtained by the Itô formula, we obtain the time-evolution equations

$$\frac{\mathrm{d}}{\mathrm{d}t}\langle \boldsymbol{X}_t \rangle = \langle \boldsymbol{A}(t, \boldsymbol{X}_t) \rangle\,, \tag{4.101}$$

$$\frac{\mathrm{d}}{\mathrm{d}t}\langle \boldsymbol{X}_t\boldsymbol{X}_t \rangle = \langle \boldsymbol{A}(t, \boldsymbol{X}_t)\boldsymbol{X}_t \rangle + \langle \boldsymbol{X}_t\boldsymbol{A}(t, \boldsymbol{X}_t) \rangle + \langle \mathbf{D}(t, \boldsymbol{X}_t) \rangle\,, \tag{4.102}$$

where, as before, $\mathbf{D}(t, \boldsymbol{x}) = \mathbf{B}(t, \boldsymbol{x}) \cdot \mathbf{B}^T(t, \boldsymbol{x})$. In general, the right-hand sides of (4.101) and (4.102) involve complicated averages so that the system of equations for the first and second moments is not closed. This situation clearly occurs when $\boldsymbol{A}(t, \boldsymbol{x})$ is a nonlinear function of $\boldsymbol{x}$. However, if we assume that the distribution of $\boldsymbol{X}_t$ is Gaussian, all averages can be expressed in terms of the first and second moments which are the only free parameters in a Gaussian distribution. Under this assumption, (4.101) and (4.102) yield a closed system of equations for the parameters of a time-dependent Gaussian. If the initial moments $\langle \boldsymbol{X}_0 \rangle$ and $\langle \boldsymbol{X}_0\boldsymbol{X}_0 \rangle$ are known, we can then solve (4.101) and (4.102) in order to obtain $\boldsymbol{\alpha}_t := \langle \boldsymbol{X}_t \rangle$ and $\boldsymbol{\Theta}_t := \langle \boldsymbol{X}_t\boldsymbol{X}_t \rangle - \boldsymbol{\alpha}_t\boldsymbol{\alpha}_t$. Of course, application of the Gaussian approximation is meaningful only if $\boldsymbol{\Theta}_t$ turns out to be positive-semidefinite.

The above procedure yields the first and second moments and hence the Gaussian distribution at any time t, but it does not define a stochastic process yet. This is good enough for many applications, for example, for determining the time evolution of the stress tensor. In order to define a Gaussian process we assume that the drift term in (4.100) can naturally be written in the form $\boldsymbol{A}(t, \boldsymbol{X}_t) = \boldsymbol{a}(t) + \mathbf{M}(t, \boldsymbol{X}_t) \cdot \boldsymbol{X}_t$, where $\mathbf{M}$ is a $d \times d$-matrix. A Gaussian approximation to the stochastic differential equation (4.100) can then be introduced as

$$\mathrm{d}\boldsymbol{X}_t = \left[\boldsymbol{a}^{\mathrm{GA}}(t) + \mathbf{M}^{\mathrm{GA}}(t) \cdot \boldsymbol{X}_t\right]\,\mathrm{d}t + \mathbf{B}^{\mathrm{GA}}(t) \cdot \mathrm{d}\boldsymbol{W}_t\,, \tag{4.103}$$

where $\boldsymbol{a}^{\mathrm{GA}}$, $\mathbf{M}^{\mathrm{GA}}$, and $\mathbf{B}^{\mathrm{GA}}$ are given by

$$\boldsymbol{a}^{\mathrm{GA}}(t) := \boldsymbol{a}(t) - \boldsymbol{\alpha}_t \cdot \langle [\boldsymbol{\nabla}\mathbf{M}(t, \boldsymbol{X})] \cdot \boldsymbol{X} \rangle_{\boldsymbol{\alpha}_t\boldsymbol{\Theta}_t} + \boldsymbol{\Theta}_t : \left\langle \boldsymbol{\nabla}\mathbf{M}^T(t, \boldsymbol{X}) \right\rangle_{\boldsymbol{\alpha}_t\boldsymbol{\Theta}_t}\,,$$

$$\mathbf{M}^{\mathrm{GA}}(t) := \langle \mathbf{M}(t, \boldsymbol{X}) \rangle_{\boldsymbol{\alpha}_t\boldsymbol{\Theta}_t} + \langle [\boldsymbol{\nabla}\mathbf{M}(t, \boldsymbol{X})] \cdot \boldsymbol{X} \rangle^T_{\boldsymbol{\alpha}_t\boldsymbol{\Theta}_t}\,,$$

and

$$\mathbf{B}^{\mathrm{GA}}(t) \cdot [\mathbf{B}^{\mathrm{GA}}(t)]^T := \langle \mathbf{D}(t, \boldsymbol{X}) \rangle_{\boldsymbol{\alpha}_t\boldsymbol{\Theta}_t}\,.$$

The symbol $\langle g(\boldsymbol{X}) \rangle_{\boldsymbol{\alpha}_t\boldsymbol{\Theta}_t}$ represents an expectation in which the (dummy) random variable $\boldsymbol{X}$ has a Gaussian distribution with parameters $\boldsymbol{\alpha}_t$, $\boldsymbol{\Theta}_t$, so that

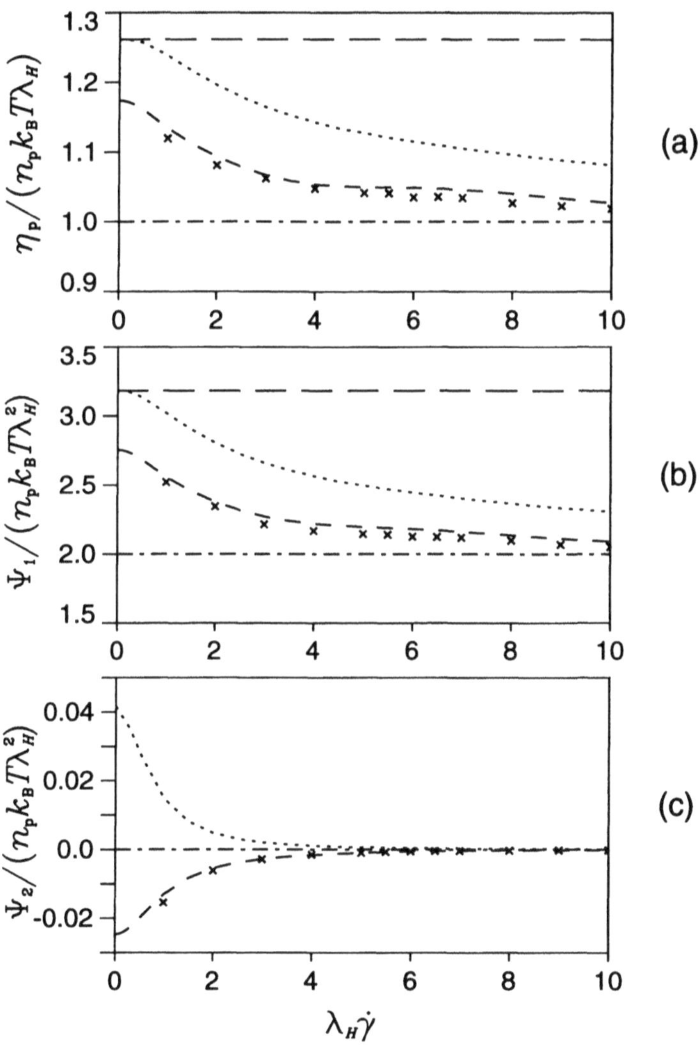

Fig. 4.8. Comparison of various predictions for (a) the polymer contribution to the viscosity, (b) the first normal stress coefficient, and (c) the second normal stress coefficient as functions of shear rate for Hookean dumbbells with hydrodynamic interaction in steady shear flow. The regularized Oseen-Burgers hydrodynamic-interaction tensor (4.67) with $h^* = 0.15$ is assumed. The *long-dashed horizontal* and the *dotted lines* represent the results predicted by the preaveraging and the consistent averaging approximations, respectively. The *short-dashed lines* are calculated from the Gaussian approximation. The results of Brownian dynamics simulations are represented by the *crosses*, which are larger than the statistical error bars. The *chain-dotted horizontal line* represents the result for Hookean dumbbells without hydrodynamic interaction. [Results of W. Zylka (1988), diploma thesis, University of Freiburg.]

all these expectations can be considered as given functions of t. By writing down the time-evolution equations for the first and second moments obtained from (4.103) and comparing to (4.101), (4.102) one can verify that the Gaussian random variable $\boldsymbol{X}_t$ obtained by solving (4.103) is characterized by $\boldsymbol{\alpha}_t$, $\boldsymbol{\Theta}_t$ (this can be seen with the help of (2.61)). One may hence replace the dummy random variable $\boldsymbol{X}$ in the definitions of the coefficients occurring in (4.103) by $\boldsymbol{X}_t$. Therefore, (4.103) is an example for a stochastic differential equation with mean field interactions (see Sect. 3.3.4). If the diffusion matrices in (4.100) and (4.103) are decomposed in the same manner then (4.103) may be regarded as a strong (trajectory-wise) Gaussian approximation to (4.100).

If we assume that $\boldsymbol{a}(t) = \langle \boldsymbol{X}_0 \rangle = \boldsymbol{0}$ and that $\mathbf{M}(t, \boldsymbol{x}) = \mathbf{M}(t, -\boldsymbol{x})$, then $\boldsymbol{\alpha}_t = \boldsymbol{0}$, and the Gaussian approximation has the simpler form

$$\mathrm{d}\boldsymbol{X}_t = \mathbf{M}^{\mathrm{GA}}(t) \cdot \boldsymbol{X}_t \, \mathrm{d}t + \mathbf{B}^{\mathrm{GA}}(t) \cdot \mathrm{d}\boldsymbol{W}_t \,, \tag{4.104}$$

with

$$M_{jk}^{\mathrm{GA}}(t) = \langle M_{jk}(t, \boldsymbol{X}) \rangle_{\alpha_t \Theta_t} + \sum_{l,n=1}^{d} (\Theta_t)_{ln} \, \langle \nabla_n \nabla_k M_{jl}(t, \boldsymbol{X}) \rangle_{\alpha_t \Theta_t} \,. \tag{4.105}$$

In the consistent averaging approximation, $\mathbf{M}(t, \boldsymbol{X}_t)$ should be replaced by its average, so that the gradient terms in (4.105) will be absent. These terms can hence be identified as the *contribution stemming from fluctuations*.

The Gaussian approximation has previously been applied in treating hydrodynamic interaction [53–55]. However, so far only the time-dependent second moments have been considered; the full definition of a stochastic process through the stochastic differential equation (4.104) is new. This attempt to develop a more complete formulation shows also some problems associated with the Gaussian approximation. The equilibrium dynamics for chains with Hookean springs coincides for the Zimm model, the consistent averaging method, and the Gaussian approximation. In particular, we recover the Boltzmann distribution at equilibrium for all these approximations. However, the Green-Kubo formula[10] for the viscosity is violated within the Gaussian approximation. This must be so because equilibrium stress correlations remain unaffected whereas the linear viscoelastic behavior, and in particular the zero-shear-rate viscosity, changes [41, 53]. The Gaussian approximation for two-time joint distributions based on the stochastic differential equation (4.103) for the stochastic process is less successful than for the time-evolution of the one-time probability density obtained from the moment equations (4.101), (4.102). This can also be seen if one attempts to formulate a Gaussian approximation for transition probabilities between any two times because then the Chapman-Kolmogorov equation is violated. One has to introduce a distinct initial time, which is here assumed to be $t = 0$. This is not a problem for steady flows or for start-up of steady flows at

[10] Expressions for the diffusion coefficient, the viscosity, and other kinetic coefficients in terms of suitable two-time equilibrium correlation functions are often referred to as Green-Kubo formulas.

a given initial time; however, this indicates that the formulation of a Gaussian approximation to a stochastic differential equation is not fully satisfactory.

Application of the Gaussian approximation to Hookean chains with hydrodynamic interaction leads to a considerable improvement over the Zimm model [53–55]. The predictions of the Gaussian approximation for the zero-shear-rate properties of dumbbells and short chains are in excellent quantitative agreement with Brownian dynamics simulations [29, 56] (see also Fig. 4.8). In particular, the second normal stress difference at small shear rates becomes negative. Rather than presenting more detailed numerical results we just mention some values of the universal ratios introduced in the preceding subsection as obtained by extrapolation of numerical results for short chains (cf. (4.97), (4.98)),

$$U_{\eta\mathrm{R}} = 1.213(3)\,, \quad U_{\Psi\eta} = 0.560(3)\,, \tag{4.106}$$

where the numbers in parentheses indicate the uncertainty in the last figure. All the material properties entering such universal ratios should be determined from the more reliable moment equations for nonequilibrium systems rather than via Green-Kubo formulas from the equilibrium two-time correlations which can be determined from the approximate stochastic differential equation.

4.2.5 Simulation of Dumbbells

Brownian dynamics simulations are the only tool available for treating chains with hydrodynamic interaction rigorously. Even though excellent approximation schemes for models with hydrodynamic interaction have been discussed in the foregoing subsection, the quality of these approximations can be judged only by comparisons with simulation results such as those given in Fig. 4.8. Once a successful approximation scheme has been established, like the Gaussian approximation for hydrodynamic interaction, a phenomenon may be considered as well understood, and Brownian dynamics simulations become less important from a fundamental point of view (they may still be very important for practical purposes such as flow calculations).

The purpose of this subsection is to explore various of the different numerical integration methods introduced in Sect. 3.4 and to discuss their respective advantages and disadvantages, including the difficulty of implementation. A systematic investigation of the influence of hydrodynamic interaction on the rheological properties is deliberately not intended. Bead-spring models with hydrodynamic interaction constitute an ideal playground for developing sophisticated Brownian dynamics simulations, and we restrict ourselves mainly to Hookean dumbbells for reasons of clarity. However, it is pointed out when an approach cannot be generalized in a straightforward manner to the simulation of chain models. In the presence of hydrodynamic interaction, the decomposition (4.65) of the diffusion matrix for long chains is very expensive, so that one is usually limited to $N \leq 20$ [57]. Fixman tried various ideas for performing the decomposition with high efficiency [39, 58], where the most efficient one is based

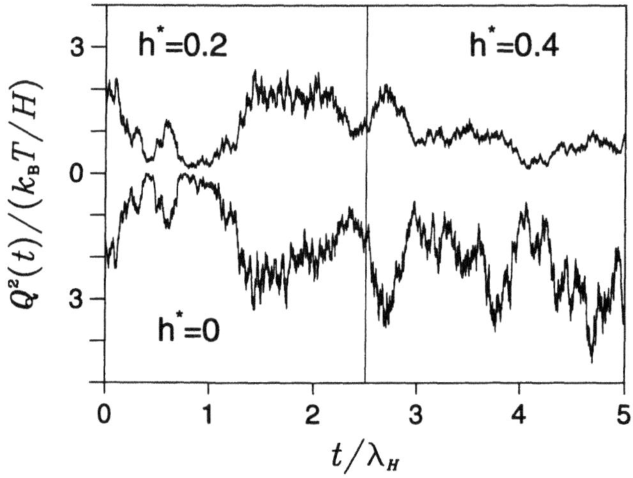

Fig. 4.9. Time-dependent dumbbell size at equilibrium constructed from a single trajectory of the Wiener process with and without hydrodynamic interaction. To facilitate comparison, the trajectory found in the absence of hydrodynamic interaction is shown in a mirror imaged form. At $t = 2.5\,\lambda_H$, the hydrodynamic-interaction parameter is changed from 0.2 to 0.4 in order to bring out the effect of hydrodynamic interaction more clearly.

on an approximation of the square root function in Chebyshev polynomials [58]. We first give an illustrative example of the usefulness of strong approximation schemes, and we then discuss various weak approximation schemes for determining rheological properties in steady shear flow.

Example 4.27 Trajectories at Equilibrium

The comparison of trajectories may be very helpful in understanding the consequences of some physical effect or approximation. We illustrate this point by considering trajectories of $\boldsymbol{Q}^2(t)$ for Hookean dumbbells at equilibrium. In Fig. 4.9, the results obtained for $\boldsymbol{Q}^2(t)$ with and without hydrodynamic interaction from a single trajectory of the Wiener process are shown in a mirror imaged form. Rather large values of the hydrodynamic-interaction parameter h^* have been chosen so that one immediately realizes two important features of hydrodynamic interaction: its attractive nature and its tendency to reduce fluctuations. Important insights into the consequences of a physical effect can thus be gained in a very simple manner. In the same way it is possible to obtain very direct information about the consequences of making an approximation.

For the pointwise comparison of trajectories, strong solutions of stochastic differential equations are required. In Fig. 4.10, results for the time-dependent size of a dumbbell with hydrodynamic interaction ($h^* = 0.2$) at equilibrium ob-

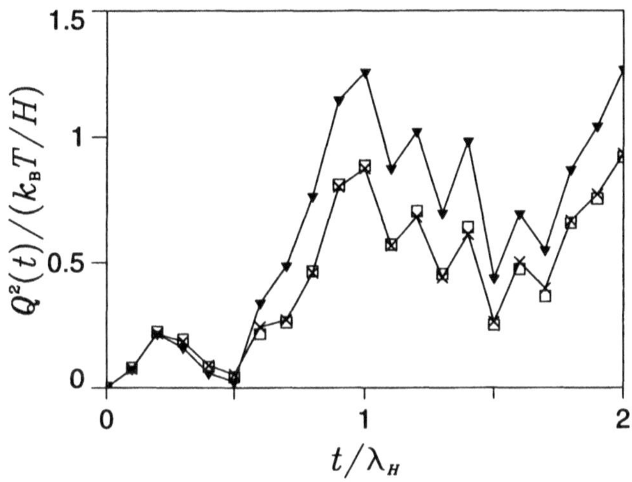

Fig. 4.10. Time-dependent size of a dumbbell with hydrodynamic interaction ($h^* = 0.2$) at equilibrium constructed from a single trajectory of the Wiener process. The *triangles* and *crosses* indicate the results constructed by means of the Euler and Mil'shtein schemes, respectively, where $\Delta t = 0.05\,\lambda_H$ has been used (not all data are shown in the figure). The *squares* indicate the "exact" results constructed by means of the Euler scheme with a much smaller time step ($\Delta t = 0.001\,\lambda_H$).

tained by the Euler and Mil'shtein schemes are shown. The stochastic integrals required in the Mil'shtein scheme have been obtained by introducing $n = 50$ subintervals (see the remarks in the end of Sect. 3.4.2). The derivative-free version of the Mil'shtein scheme constructed by means of (3.126) has been used. In this example, the different orders of strong convergence, $\upsilon = 1/2$ for the Euler scheme (triangles) and $\upsilon = 1$ for the Mil'shtein scheme (crosses), have the obvious consequence that the crosses stay much closer to the exact results which are indicated by the squares. □

If we want to calculate material functions for steady shear flow, it is possible to do this by using the ideas of Sect. 4.1.4. Starting from suitable initial conditions, for example from equilibrium initial conditions, we could construct an ensemble of trajectories over several relaxation times until a steady-state situation is reached. From this ensemble of steady-state configurations we could evaluate any property determined by the polymer configurations. The problem with this approach is that it is necessary to decide over how many relaxation times the simulation has to be performed, and this decision certainly depends on the accuracy required for the results.

For stationary problems there is a more natural alternative approach. Simulation of a single, very long trajectory is sufficient because time-independent steady-state expectations can be evaluated as long-time averages from a single trajectory. This approach, which is based on the assumption of *ergodicity*, can be justified in a heuristic manner as follows. If the total simulation time after reaching the steady state (at $t_0 := 0$) is $t_{\max}$, we can evaluate partial time averages of some quantity of interest over the n successive time intervals $[(j-1)t_{\text{block}}, jt_{\text{block}}]$ for $j = 1, 2, \ldots, n$, where $t_{\text{block}} := t_{\max}/n$. In other words, we split the total number of simulation data generated for some quantity into n subblocks. For a very long trajectory, both t_{block} and n can be chosen to be large. If t_{block} is large compared to the characteristic relaxation time of the system then these partial time averages may be regarded as independent realizations of a random variable which does not depend on the particular trajectory.[11] The strong law of large numbers (see Example 2.67) then implies that the arithmetic mean of the partial time averages, which is equal to the total time average, converges almost surely. Since the limit does not depend on the particular trajectory it must coincide with the desired expectation. Rigorous results such as a sufficient condition for ergodicity and statements concerning the order of weak convergence can be found in Theorem 4.8.8 and Sect. 17.2 of [18]. The order of weak convergence for all integration schemes introduced in Sect. 3.4 is expected to remain unchanged when ergodicity is employed to evaluate steady-state averages.

Even the statistical error bars can be estimated reliably from a simulation of only a single trajectory (see Sect. 8.3 of [59] or [60]). In order to do so, we first assume that t_{block} is large enough to obtain independent partial time averages (which are also referred to as group mean values in [59] or subblock means in [60]). As before, the variance of these independent data and hence the statistical error of the total time average can be estimated according to Exercise 2.68. In practice, however, it may not be known whether t_{block} is large enough to guarantee independence. If t_{block} is chosen too small, then correlations in the partial time averages typically reduce the fluctuations and lead to an underestimation of the statistical errors. If, for a given trajectory, several values of t_{block} are considered, the error estimate in general increases with t_{block} until a constant value is reached when t_{block} becomes large enough to guarantee the independence of the partial time averages (which is required for a proper and consistent error estimation). It is hence possible to check the consistency of the error estimate without knowing t_{block} in advance by plotting the error estimate versus t_{block}. For all our simulations of dumbbells, the trajectories are so long that t_{block} can easily be chosen to be beyond good and evil.

If the evaluation of the quantity of interest requires a considerable amount of computer time compared to carrying out a time step then it is advantageous not to calculate this quantity after every time step. Due to the correlation in the data not much accuracy is lost if only data after a certain number of time steps

[11] In order to obtain the independence of the partial time averages it is clearly necessary that any two configurations be connected by a continuous trajectory in configuration space. If a region of low probability in configuration space needs to be traversed for reaching one configuration from another one then it is important that, in a simulation, t_{block} is larger than the transition time.

are used for calculating time averages (remember that the correlation time is of the order of the characteristic relaxation time, and a time step should typically be much smaller).

For Hookean dumbbells, the results of Exercise 4.18 imply the following stochastic differential equation of motion for the connector vector,

$$\mathrm{d}\boldsymbol{Q}(t) = \left\{\boldsymbol{\kappa}(t)\cdot\boldsymbol{Q}(t) - \frac{2H}{\zeta}[\boldsymbol{\delta} - \zeta\boldsymbol{\Omega}(\boldsymbol{Q}(t))]\cdot\boldsymbol{Q}(t)\right\}\mathrm{d}t + \sqrt{\frac{4k_\mathrm{B}T}{\zeta}}\,\mathbf{B}(\boldsymbol{Q}(t))\cdot\mathrm{d}\boldsymbol{W}_t\,, \tag{4.107}$$

where we assume that the hydrodynamic-interaction tensor is of the regularized type (4.67), and that $\mathbf{B}(\boldsymbol{Q})\cdot\mathbf{B}^T(\boldsymbol{Q}) = \boldsymbol{\delta} - \zeta\boldsymbol{\Omega}(\boldsymbol{Q})$. Since the tensor $\boldsymbol{\delta} - \zeta\boldsymbol{\Omega}(\boldsymbol{Q})$ is of the form $g(\boldsymbol{Q})\,\boldsymbol{\delta} + \tilde{g}(\boldsymbol{Q})\,\boldsymbol{Q}\boldsymbol{Q}/Q^2$ with suitable real-valued functions $g(\boldsymbol{Q})$ and $\tilde{g}(\boldsymbol{Q})$, the tensor $\mathbf{B}(\boldsymbol{Q})$ can most conveniently be chosen as the symmetric square root of the diffusion tensor,

$$\mathbf{B}(\boldsymbol{Q}) = \sqrt{g(\boldsymbol{Q})}\,\boldsymbol{\delta} + \left(\sqrt{g(\boldsymbol{Q}) + \tilde{g}(\boldsymbol{Q})} - \sqrt{g(\boldsymbol{Q})}\right)\frac{\boldsymbol{Q}\boldsymbol{Q}}{Q^2}\,. \tag{4.108}$$

Equation (4.107) is the starting point for applying numerical integration schemes in this subsection. The only new feature compared to the dumbbell simulation of Sect. 4.1.4 is the straightforward evaluation of the diffusion tensor and its square root.

Exercise 4.28 Develop a routine for Brownian dynamics simulations of Hookean dumbbells with hydrodynamic interaction in steady shear flow, based on the Euler scheme for (4.107). Carry out steady-state simulations in order to verify that the second normal stress difference at $\lambda_H\dot{\gamma} = 1$ is negative.

According to the general theory of numerical integration schemes, the order of weak convergence for the simulation algorithm of Exercise 4.28 is $\upsilon = 1$. Simulation results suggest that, as in the absence of hydrodynamic interactions, the corrections for the steady-state viscosity obtained from the Euler scheme are actually of order $(\Delta t)^2$. With the idea of a full second-order scheme in mind, Iniesta and García de la Torre [61] suggested a predictor-corrector scheme which, for the general stochastic differential equation (4.100), can be formulated in the notation of Sect. 3.4 as

$$\overline{\boldsymbol{Y}}_{j+1} = \widetilde{\boldsymbol{Y}}_j + \boldsymbol{A}(t_j, \widetilde{\boldsymbol{Y}}_j)\,\Delta t_j + \mathbf{B}(t_j, \widetilde{\boldsymbol{Y}}_j)\cdot\Delta\boldsymbol{W}_j\,, \tag{4.109}$$

$$\widetilde{\boldsymbol{Y}}_{j+1} = \widetilde{\boldsymbol{Y}}_j + \frac{1}{2}\left[\boldsymbol{A}(t_{j+1}, \overline{\boldsymbol{Y}}_{j+1}) + \boldsymbol{A}(t_j, \widetilde{\boldsymbol{Y}}_j)\right]\Delta t_j + \overline{\mathbf{B}}\cdot\Delta\boldsymbol{W}_j\,, \tag{4.110}$$

$$\overline{\mathbf{B}}\cdot\overline{\mathbf{B}}^T = \frac{1}{2}\left[\mathbf{D}(t_{j+1}, \overline{\boldsymbol{Y}}_{j+1}) + \mathbf{D}(t_j, \widetilde{\boldsymbol{Y}}_j)\right], \tag{4.111}$$

where $\mathbf{B}(t_j, \widetilde{\boldsymbol{Y}}_j)$ and $\overline{\mathbf{B}}$ are lower triangular matrices constructed in the same manner from $\mathbf{D}(t_j, \widetilde{\boldsymbol{Y}}_j)$ and (4.111), respectively, so that a smooth dependence of $\mathbf{B}$ on $\mathbf{D}$ is ensured. For additive noise, this scheme is very similar to the predictor-corrector scheme discussed in Exercise 3.47, so that the order of weak convergence is indeed $\upsilon = 2$. If, for multiplicative noise, the same $\Delta \boldsymbol{W}_j$ is used in (4.109) and (4.110) (as implied by the notation) then the Itô-Stratonovich problem implies that *this scheme does not lead to the correct solution* of the underlying stochastic differential equation. In order to obtain the correct solution in the limit of vanishing time-step width one could use a modified drift term. Alternatively, independent sets of random numbers $\Delta \boldsymbol{W}_j$ can be used in (4.109) and (4.110) (even though not stated explicitly, this was actually done in [61]). Then, one obtains a weak first-order scheme, and actually quite a few second-order terms are treated correctly, namely those in the first two lines of (3.121).

In order to obtain a general second-order scheme we should employ (3.121). Note that the action of the infinitesimal generator on (4.108) implies second-order derivatives of the square root of rational functions, and many other "ugly" terms occur. Since the development of a second-order scheme is straightforward and simple in principle but extremely tedious in practice this is an ideal problem for symbolic mathematical computation [62]. Ideally, a tool for symbolic computation should create the required computer code for a second-order algorithm, for example in FORTRAN, once the functions $\boldsymbol{A}$ and $\mathbf{B}$ are specified.

Exercise 4.29 Develop a second-order simulation algorithm for Hookean dumbbells with hydrodynamic interaction in steady shear flow by means of suitable software for symbolic mathematical computation (e.g., by means of Maple).

Even if a second-order algorithm can be implemented rather conveniently by exploiting software for symbolic mathematical computation, the evaluation of all the required derivatives is expensive, and the resulting computer program is hence rather inefficient. Furthermore, this procedure cannot be generalized to the case of longer bead-spring chains because in that case the analytical form of $\mathbf{B}$ is unknown. We hence use a combination of the ideas of predictor-corrector and of Runge-Kutta schemes to avoid derivatives. More precisely, we consider the following scheme. For general stochastic differential equations with time-independent diffusion terms, we first construct a predictor step and several supporting values,

$$\overline{\boldsymbol{Y}}_{j+1} = \widetilde{\boldsymbol{Y}}_j + \boldsymbol{A}(t_j, \widetilde{\boldsymbol{Y}}_j)\, \Delta t_j + \mathbf{B}(\widetilde{\boldsymbol{Y}}_j) \cdot \Delta \boldsymbol{W}_j \,, \tag{4.112}$$

$$\boldsymbol{\Upsilon}_j^n = \widetilde{\boldsymbol{Y}}_j + \frac{1}{d'}\, \boldsymbol{A}(t_j, \widetilde{\boldsymbol{Y}}_j)\, \Delta t_j + \boldsymbol{b}_n(\widetilde{\boldsymbol{Y}}_j) \sqrt{\Delta t_j} \,, \tag{4.113}$$

$$\overline{\boldsymbol{\Upsilon}}_j^n = \widetilde{\boldsymbol{Y}}_j + \frac{1}{d'}\, \boldsymbol{A}(t_j, \widetilde{\boldsymbol{Y}}_j)\, \Delta t_j - \boldsymbol{b}_n(\widetilde{\boldsymbol{Y}}_j) \sqrt{\Delta t_j} \,, \tag{4.114}$$

where $\boldsymbol{b}_n$, $n = 1, \ldots, d'$, is the nth column vector of the matrix $\mathbf{B}$, and we then set

$$\widetilde{\boldsymbol{Y}}_{j+1} \;=\; \widetilde{\boldsymbol{Y}}_j + \frac{1}{2}\left[\boldsymbol{A}(t_{j+1}, \overline{\boldsymbol{Y}}_{j+1}) + \boldsymbol{A}(t_j, \widetilde{\boldsymbol{Y}}_j)\right] \Delta t_j + \frac{2-d'}{2}\, \mathbf{B}(\widetilde{\boldsymbol{Y}}_j) \cdot \Delta \boldsymbol{W}_j \,,$$

$$+\frac{1}{4}\sum_{n=1}^{d'}\left[\mathbf{B}(\boldsymbol{\Upsilon}_j^n)+\mathbf{B}(\overline{\boldsymbol{\Upsilon}}_j^n)\right]\cdot\Delta\boldsymbol{W}_j+\Delta\widetilde{\boldsymbol{Y}}_j\,, \tag{4.115}$$

where the kth component of the d-dimensional column vector $\Delta\widetilde{\boldsymbol{Y}}_j$ is given by

$$\frac{1}{4\sqrt{\Delta t_j}}\sum_{m,n=1}^{d'}(\Delta\boldsymbol{W}_j\Delta\boldsymbol{W}_j-\Delta t_j\,\boldsymbol{\delta}+\Delta\mathbf{V}_j)_{mn}\left[B_{km}(\boldsymbol{\Upsilon}_j^n)-B_{km}(\overline{\boldsymbol{\Upsilon}}_j^n)\right].$$

For Hookean dumbbells with hydrodynamic interaction, this algorithm has been implemented for $d = d' = 3$. The implementation of this derivative-free second order scheme is not much more complicated than for the Euler scheme. A detailed comparison of the results for the first normal stress difference obtained by various simulation algorithms discussed in this subsection can be found in Fig. 4.11. At least for this material property, the improved first-order scheme (4.109)–(4.111) performs much better than the Euler scheme; in fact, it can hardly be distinguished from a second-order scheme. The explicit second-order scheme generated by means of the symbolic computation software Maple is very expensive. Therefore, the derivative-free second-order scheme (4.112)–(4.115) seems to be best suited for the simulation of dumbbells with hydrodynamic interaction. All the extrapolated results shown for $\Delta t = 0$ are consistent. The first-order nature of the predictor-corrector scheme (4.109)–(4.111) becomes obvious when the mean-square dumbbell size is considered. The mean-square sizes obtained from the improved first-order scheme and from the explicit second-order scheme are compared in Fig. 4.12.

In performing simulations with the explicit second-order scheme for large Δt, a remarkable problem is observed: the error bars obtained from the standard procedure described in Sect. 4.1.4 do not decrease like $1/\sqrt{n}$, where n is the number of independent configurations. In Fig. 4.13, the distribution of results for the mean-square size, obtained by averaging over periods of $10\,000\lambda_H$ is shown for the explicit second-order simulation algorithm for $\lambda_H\dot{\gamma} = 1$ and $\Delta t = 0.4\lambda_H$. The probability density is constructed from 100 000 averages over periods of $10\,000\lambda_H$. When averaging over such a large number of relaxation times, according to the central limit theorem, one would expect the distribution to be Gaussian. However, the distribution in Fig. 4.13 is clearly skewed, and it has a slowly decreasing tail for large values of the mean-square size. For a power-law decrease of the tail with exponent $-(1+\alpha_s)$ with $1 < \alpha_s < 2$, the mean is finite but the variance is infinite; the central limit theorem (see Example 2.71) is hence not applicable. Indeed, possible limit distributions with slowly decreasing tails, so-called α-stable distributions, exist for all values of α_s with $0 < \alpha_s < 2$ (see, e.g., Sect. 2.4 of [63] and Chap. 1 of [64]). For such α-stable distributions, the width of the distribution decreases only as $1/n^{(\alpha_s-1)/\alpha_s}$, so that the usual behavior is recovered for $\alpha_s = 2$. Moreover, the width cannot be described by the square root of the variance. The error may be characterized by an alternative description of the width, for example, by choosing $\overline{\sigma}$ such that one has $p([\alpha-\overline{\sigma},\alpha+\overline{\sigma}]) = 0.6827$, where α is the mean. With this definition,

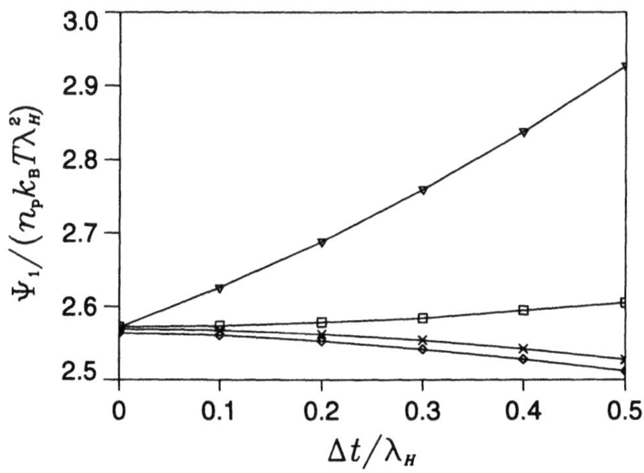

Fig. 4.11. A detailed comparison of the results for the first normal stress difference obtained by various simulation algorithms for $\lambda_H \dot{\gamma} = 1$ for different time steps. The different symbols, which are larger than the error bars, represent the results of the Euler scheme (*triangles*), the improved first-order scheme (4.109)–(4.111) (*diamonds*), the explicit second-order scheme implemented by means of Maple (*squares*), and the derivative-free second-order scheme (4.112)–(4.115) (*crosses*).

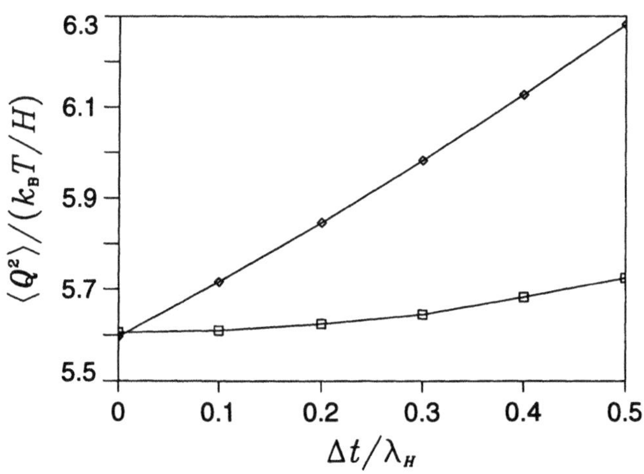

Fig. 4.12. A detailed comparison of the results for the mean-square polymer size obtained by two simulation algorithms for $\lambda_H \dot{\gamma} = 1$ for different time steps. The different symbols, which are larger than the statistical error bars, represent the results of the improved first-order scheme (4.109)–(4.111) (*diamonds*) and the explicit second-order scheme implemented by means of Maple (*squares*).

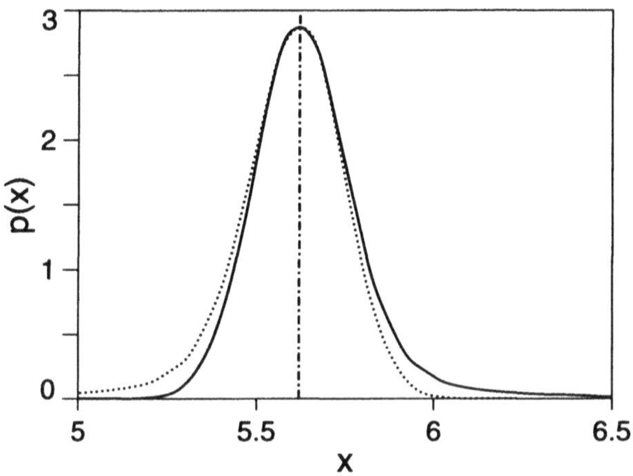

Fig. 4.13. Probability density for the mean-square size obtained from the explicit second-order simulation algorithm for $\lambda_H \dot{\gamma} = 1$ and $\Delta t = 0.4\lambda_H$ when averaging over $10\,000\lambda_H$ (*continuous line*). The mirror image (*dotted line*) of the probability density has been included to make the skewness of the distribution more obvious.

$\overline{\sigma}$ coincides with the width σ for Gaussian distributions. For a simulation based on the derivative-free second-order scheme, the distribution obtained under the same conditions as in Fig. 4.13 is perfectly symmetric, indistinguishable from a Gaussian, and somewhat narrower than the one for the explicit scheme in Fig. 4.13.

In order to find out whether the distribution of the simulation results from the explicit second-order algorithm is approximately of the α-stable type we compare curves obtained from several simulations with $\Delta t = 0.4\lambda_H$, where we average over time periods of $100\lambda_H$, $1\,000\lambda_H$, and $10\,000\lambda_H$. For the the α-stable type, the resulting probability densities should all have the same shape as the one in Fig. 4.13, differing only in width. More precisely, if the probability density is scaled around the mean by a scale factor of $n^{(\alpha_s-1)/\alpha_s}$ with $n = 10$ ($n = 100$) for averages over $1\,000\lambda_H$ ($100\lambda_H$), we should obtain the same curve as in Fig. 4.13. Figure 4.14 shows the comparison for $\alpha_s = 1.85$. Each probability density is constructed from 100 000 averages over the quoted time periods. Indeed, all the curves almost coincide; only when averaging over $100\lambda_H$, the distribution is somewhat more skewed. There seem to be only small deviations from α-stable behavior, and there is no indication of a crossover to standard Gaussian behavior. The unusual distribution of the simulation results obtained from the explicit second-order algorithm is presumably caused by the weakly singular behavior of the hydrodynamic-interaction tensor at small distances. Although the regularized hydrodynamic-interaction tensor is proportional to the length Q for small $\boldsymbol{Q}$, its second derivatives diverge as $1/Q$.

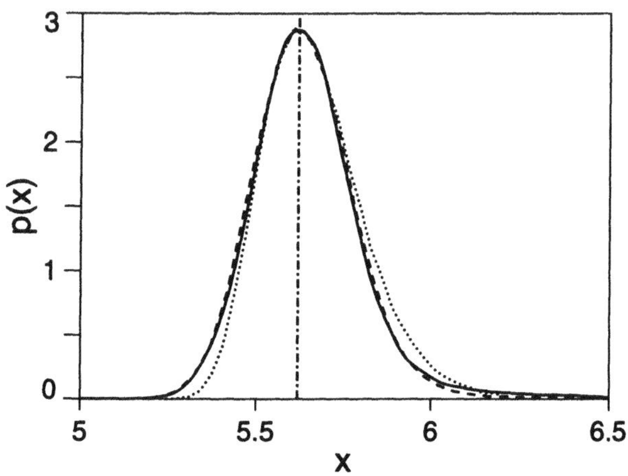

Fig. 4.14. Suitably scaled probability densities for the mean-square size obtained from the explicit second-order simulation algorithm for $\lambda_H \dot{\gamma} = 1$ and $\Delta t = 0.4\lambda_H$. When averaging over $100\lambda_H$, $1\,000\lambda_H$, and $10\,000\lambda_H$, we obtain the *dotted*, *dashed*, and *continuous lines*, respectively.

We have considered here only *nonequilibrium* Brownian dynamics simulations for Hookean dumbbells with hydrodynamic interaction. The linear viscoelastic behavior of the model can also be estimated from *equilibrium* simulations since, according to a Green-Kubo formula, the relaxation modulus is proportional to the time correlation function for the shear stress. This approach has frequently been used in Brownian dynamics simulations of polymer molecules, since it was proposed in connection with a simulation of charged bead-spring chains [65]. By applying this idea, the linear viscoelastic behavior can be obtained without dividing small simulation results by a small strain rate. However, accurate simulations for the small correlations at large time differences are very time consuming. One usually fits the sum of a few exponentials to the simulation data, and one proceeds by using the result of this fitting procedure in the time integrals involved in the calculation of various material properties.

4.3 Nonlinear Forces

Various nonlinear effects play an important role in the understanding of the polymer dynamics in dilute solutions: hydrodynamic interaction, excluded volume, finite extensibility, and internal viscosity [37, 66]. Since none of these effects can be treated rigorously, Brownian dynamics simulations constitute the only tool for studying how all these nonlinear effects influence the properties of bead-spring chains. While hydrodynamic interaction has been treated in the foregoing section, we discuss here two nonlinear interaction force laws, namely excluded-volume forces and finitely extensible nonlinear elastic (FENE) spring forces. We will not discuss models with internal viscosity here, even though there has recently been considerable interest in studying this effect by Brownian dynamics simulations [67, 68]. Noninteracting bead-spring chains with configuration-dependent anisotropic friction, which have been suggested as models for concentrated solutions and melts, can also be treated by simulations [69, 70] (see also models with anisotropic friction depending on averages of the configuration [71–75]).

The repulsive excluded-volume interactions between different segments of a polymer molecule are of fundamental interest because they are known to have an important effect on the universal exponents and ratios characterizing polymer properties in the long chain limit. The interest in FENE chains is of a more practical origin because—even though negligible in the long chain limit—the finite extensibility strongly affects the behavior of real polymer molecules in most experiments and processing situations. While excluded-volume forces occur for any pair of beads, the FENE forces constitute a replacement for the Hookean forces joining successive beads to form a chain.

4.3.1 Excluded Volume

Like hydrodynamic interaction, the excluded-volume effect has an important influence on the properties of dilute polymer solutions, even in the limit of extremely long chains. Unlike for hydrodynamic interaction, however, there are no well-established and carefully tested approximation schemes for calculating the influence of excluded volume on rheological properties of dilute solutions with relatively small efforts, but still good accuracy.

The universal character of excluded-volume effects is best known for static properties. Most famous is the chain length dependence of the root-mean-square radius of gyration $\overline{R}_{\mathrm{g}} \sim N^{\nu_{\mathrm{ev}}}$, with the exponent $\nu_{\mathrm{ev}} \approx 0.588$. One similarly expects scaling laws for dynamic properties such as $D_{\mathrm{h}} \sim N^{-\nu_{\mathrm{ev}}}$, $\lambda_1 \sim N^{3\nu_{\mathrm{ev}}}$, $[\eta]_0 \sim N^{3\nu_{\mathrm{ev}}-1}$, and $\Psi_1 \sim \Psi_2 \sim N^{6\nu_{\mathrm{ev}}}$ (for fixed n_{p}), so that in the presence of excluded volume the ratios introduced in Sect. 4.2.3 are also universal numbers. Like the exponent ν_{ev}, these universal ratios depend on whether or not excluded volume is considered (good solvents or Θ solvents), but not on other chemical details of the monomers or solvent. In the presence of shear flow, the shear rate dependence of various material properties is found to be determined by universal master curves when plotted versus the reduced shear rate $\beta := \lambda_\eta \dot{\gamma}$.

One of the most important tasks of kinetic theory in treating excluded-volume interactions is the calculation of universal ratios and master curves. The repertoire of methods available for this purpose is very limited. Kinetic theory models with excluded volume are very complicated, in particular, since hydrodynamic interaction should also always be taken into account in bead-spring models of dilute solutions. Excluded-volume effects can be built into the bead-spring models with hydrodynamic interaction described by the equation of motion (4.64) in a straightforward manner by adding all pairwise interaction forces between bead μ and other beads to the potential force $\boldsymbol{F}_\mu$ exerted by the Hookean springs pulling on bead μ.

A relatively simple and successful approach to the dynamic behavior of models with excluded volume and hydrodynamic interaction was developed by Fixman [76]. He introduced a quadratic repulsive excluded-volume potential, where the coefficients depend on the average chain configuration. The assumption of a quadratic potential leading to a Gaussian distribution is crucial for the tractability of the model. When it is assumed that the essence of excluded volume can be captured by means of a uniform expansion factor then one can formulate a self-consistency condition for the expansion factor. Within this approximation one finds $\nu_{\mathrm{ev}} = 0.6$. Hydrodynamic interactions are treated by Fixman on a level which lies between preaveraging and consistent averaging (see Sect. 8.3.2.1 of [37]).

Another approach to hydrodynamic interaction and excluded volume is based on renormalization group theory, which is used as a method for analyzing and refining the results of a perturbative treatment of these effects [34, 52, 77, 78]. Various universal ratios and the master curves describing the shear rate dependence of the viscometric functions have been calculated. Renormalization group calculations show that excluded volume contributes to shear thinning because it increases the viscosity mostly at low shear rates. However, at large shear rates this tendency is overpowered by the strong shear-thickening effect due to hydrodynamic interaction. Analogous statements hold for the first normal stress coefficient.

In renormalization group calculations, the potential characterizing the excluded-volume interaction is usually assumed to be of the δ-function type. When constructing a perturbation theory, singularities resulting from products of δ-functions with the same argument can be avoided if a the potential is smeared out to a narrow Gaussian, that is,

$$V_{\mathrm{ev}}(\boldsymbol{r}) = \frac{v_{\mathrm{ev}}\, k_{\mathrm{B}}T}{(2\pi\, a_{\mathrm{ev}}^2)^{3/2}} \exp\left\{-\frac{1}{2}\frac{\boldsymbol{r}^2}{a_{\mathrm{ev}}^2}\right\}, \tag{4.116}$$

for each pair of beads separated by a distance $\boldsymbol{r}$, where v_{ev} and a_{ev} determine the strength and the range of the potential, and the δ-potential is recovered for $a_{\mathrm{ev}} \to 0$. Actually, the perturbation theory can be constructed very nicely for the potential (4.116) without letting a_{ev} go to zero, because one has to evaluate expectations with Gaussian distributions anyway. Compact expressions for the first-order effect of excluded volume on the viscometric functions can be derived; these might be helpful in testing simulation algorithms. It is interesting to note that the indirect contribution (increased tension in springs due to coil expansion) and the direct contribution (momentum exchange due to excluded-volume forces) to the second normal stress coefficient cancel exactly. This observation underlines the importance of the direct excluded-volume contribution to the stress tensor, which is usually neglected in renormalization group calculations because it is isotropic for a δ-type potential.

In the theoretical investigation of static properties of excluded-volume chains, Monte Carlo simulations constitute an indispensable tool. Brownian dynamics simulations play a similar role when it comes to dynamic properties. It is therefore very important to develop efficient simulation algorithms so that long chains can be handled. The unavoidable influence of hydrodynamic interaction on the dynamic behavior of macromolecules in dilute solution implies that one has to deal with multiplicative noise, so that the full power of the theory for the numerical integration of stochastic differential equations is required.

In Brownian dynamics simulations, excluded-volume interactions are usually incorporated through hard-sphere or Lennard-Jones type potentials [79–83]. While the use of Hookean springs for establishing the chain connectivity is crucial for all other methods discussed in this subsection on excluded volume, this is not true for Brownian dynamics simulations. The universal long chain behavior might be detectable for shorter chains if a connector force

$$\boldsymbol{F}^{\mathrm{c}} = H \, |\boldsymbol{Q}|^{\nu_{\mathrm{ev}}/(1-\nu_{\mathrm{ev}})} \, \frac{\boldsymbol{Q}}{|\boldsymbol{Q}|} \tag{4.117}$$

is used, which incorporates excluded-volume effects in addition to entropic effects on the sub-bead scale. The selection of the excluded-volume potential should also be guided by the idea of formulating the basic bead-spring model as close as possible to the self-similar structure which leads to universal long chain behavior.

Most of the current simulations of models with excluded volume and hydrodynamic interaction are for rather short chains. Conceptually, there is not much new when excluded volume is added, and we hence omit examples of simple simulations. The importance of the long chain limit is a strong motivation for developing more efficient integration techniques for stochastic differential equations. Higher-order schemes and variance reduction methods are of special importance.

Exercise 4.30 Calculate the mean-square extension of a dumbbell with connector force (4.117) at equilibrium.

4.3.2 Finite Polymer Extensibility

Whereas the chains with Hookean springs in the models considered so far are infinitely extensible, real polymers can certainly be extended to their fully stretched length at most (if they do not break). For large extensions of a polymer, the linear spring-force law is hence a poor approximation which can be improved upon by introducing a finitely extensible nonlinear spring force. Peterlin [84, 85] obtained a tractable linearization of a dumbbell model with a nonlinear spring-force law by introducing a linear force with a flow-field-dependent, effective spring constant which he evaluated in a self-consistent manner. He replaced the configuration dependent "spring constant" by a spring constant that depends on the self-consistently averaged configuration. This procedure is very similar to the consistent-averaging method for the hydrodynamic interaction described in Sect. 4.2.4. While Peterlin applied his idea to the inverse Langevin spring-force law [86], the same idea was used by Bird et al. [1, 87] to study Warner's FENE spring-force law [88],

$$\boldsymbol{F}^{\mathrm{c}} = \frac{H\boldsymbol{Q}}{1 - \boldsymbol{Q}^2/(b\,k_{\mathrm{B}}T/H)} \tag{4.118}$$

where $b\,k_{\mathrm{B}}T/H$ is the square of the maximum possible spring extension and b is a dimensionless parameter describing the finite extensibility of the FENE springs. In the limit $b \to \infty$, we obtain a Hookean spring-force law with spring constant H. The $\boldsymbol{Q}^2$-dependent prefactor of $\boldsymbol{Q}$ in (4.118) is sometimes referred to as an effective spring constant.

When $\boldsymbol{Q}^2$ in the denominator of (4.118) is averaged, the resulting model is known as the FENE-P model ("P" stands for Peterlin, who first suggested this kind of simplification). Because even this expression does not permit simple analytical solutions (except for dumbbells), still another simplification has been proposed, called FENE-PM ("modified Peterlin") [89]. Here, the averaged denominator expression for the spring force in the jth link is replaced by the average of the expressions for all the links in the chain. The constitutive equation for this model is particularly simple: a slight modification of a generalized Maxwell model, which allows for some mode-coupling.

All models with consistently averaged nonlinear spring-force laws predict, for instance, shear-rate-dependent viscometric functions, and, for extensional flow, finite material functions for all strain rates. While steady-state properties obtained from the FENE-P model are in rather good agreement with those of the FENE model, the time-dependent behavior predicted by these models exhibits remarkable differences [90, 91]. Brownian dynamics simulations are the only useful tool for detecting such deviations. A model that combines the consistent-averaging method for the hydrodynamic interaction and for a nonlinear spring-force law for chains of arbitrary length has been developed and studied in great detail [92, 93]. Fixman's approach to models with hydrodynamic interaction and excluded volume [76] has been generalized by including FENE-P effects [94].

Note that the consistent-averaging method for the nonlinear springs and for the hydrodynamic interaction are not on exactly the same footing. One replaces $\boldsymbol{Q}^2$ in the denominator of (4.118) by its average rather than averaging the effective spring constant, which would be the precise analog of the the consistent-averaging method for the hydrodynamic interaction. However, for spring-force laws with singularities at finite extensions, like the FENE spring-force law (4.118), the average of the effective spring constant does not exist. For springs with no singularities, like a spring-force law that is a sum of a linear and a cubic term [95, 96], both averages exist and the two averaging methods lead, in general, to different results. The arguments of this paragraph imply also that the idea of the Gaussian approximation cannot be applied to FENE springs.

In general, approximations in which a term is replaced by a self-consistent average are potentially problematic. The physics of the underlying problem may change fundamentally because the Fokker-Planck equation changes from an equation that is linear in the probability density to a nonlinear equation in p (due to the occurrence of p through averages in the drift or/and diffusion terms). In particular, the uniqueness of solutions of the approximate model becomes questionable unless the coefficient functions satisfy Lipschitz conditions (see Sect. 3.3.4). In situations where there are no problems with the existence or uniqueness of solutions, these solutions of self-consistently averaged models can be obtained from simulations of large ensembles of trajectories, where the dynamics of each trajectory depends on averages calculated over the whole ensemble. For large ensembles, the interaction between any two trajectories is very weak. For the FENE-P model, the occurrence of averages results from an approximation which is better avoided in a simulation. The situation is similar for the consistent averaging method and the Gaussian approximation for hydrodynamic interactions. For different reasons, averages come in when a many-chain system is reduced to a single-chain theory with effective interactions (see, for example, Sect. 5.2.2 on liquid crystal polymers), or when a full-chain theory is reduced to a one-segment theory (see, for example, Sect. 6.3.2 on a special reptation model). The mathematical background for stochastic differential equations involving averages in the coefficient functions has been sketched in Sect. 3.3.4.

The value of b in the FENE spring-force law (4.118) cannot be chosen arbitrarily. For a chain with a pure carbon backbone one knows the bond angles and one can derive the estimate [92]

$$b \approx \frac{N_{\rm C}}{\sigma_{\rm sf}^2 \, N}, \tag{4.119}$$

where $N_{\rm C}$ is the number of carbon atoms in the backbone of the polymer molecule, $\sigma_{\rm sf}$ is an empirical steric factor [97], and b has been assumed to be a large number (this assumption is consistent with the b values found below). Typical values of the steric factor are [97]:

For polyethylene at $T = 100°{\rm C}$: $\sigma_{\rm sf} = 1.77$,
For polystyrene at $T = 70°{\rm C}$: $\sigma_{\rm sf} = 2.35$.

For polystyrene molecules with some 10 000 monomer units per polymer, if we take $N = 25$, then we obtain $b = 150$ as a typical value. An important conclusion is that b is not a free parameter but can be estimated from an equation like (4.119), which shows that b is roughly equal to the number of monomer units represented by a bead, and therefore b should be a large number. Note that in the derivation of (4.119) the parameter b has not been estimated from small deformations of a chain but rather from the length of the fully stretched chain. Therefore, this estimation of the parameter b might not be completely adequate for predicting material functions for weak flows.

Exercise 4.31 Derive (4.119).

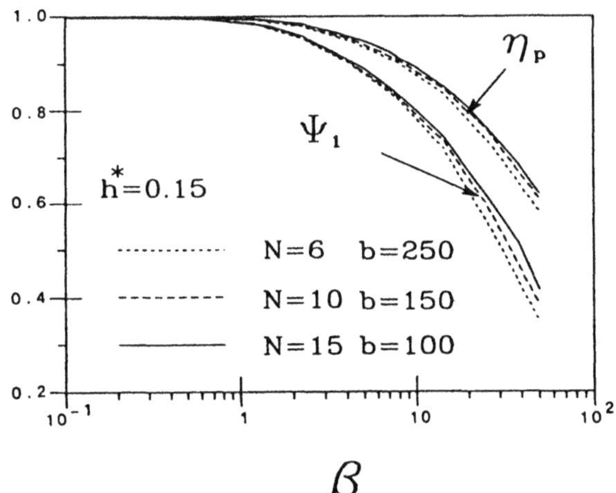

Fig. 4.15. The normalized polymer contribution to the viscosity and first normal stress coefficient for FENE parameter and chain length values which satisfy the relationship $b \sim 1/N$ with a constant value for the hydrodynamic-interaction parameter of $h^* = 0.15$ (from [93], reproduced with permission from Elsevier Science). The dimensionless quantity $\beta := \lambda_\eta \dot{\gamma}$ is the reduced shear rate, where the time scale λ_η has been introduced in (4.94).

From a fundamental point of view, the use of simple mechanical models is much harder to justify in discussing FENE effects than for understanding universal hydrodynamic-interaction or excluded-volume effects. Equation (4.119) suggests that the FENE effects on global polymer properties disappear in the long chain limit. On the other hand, (4.119) implies that the parameters in Fig. 4.15 correspond to macromolecules of considerable molecular weight, and drastic FENE effects are visible anyway. Finite extensibility leads to a strong shear thinning tendency for real polymer solutions. Figure 4.15 can furthermore be used to justify simple mechanical models: if a given polymer system is modeled by chains of various numbers of beads N, then (4.119) suggests $b \sim 1/N$, and the curves in Fig. 4.15 for different pairs of b and N nearly overlap. The model predictions depend only weakly on the local modeling of polymer molecules. In the spirit of renormalization group theory, FENE effects constitute a correction to scaling of practical importance. They do not contribute to the universal long chain behavior, but they have rather general and unambiguous consequences for polymer molecules in the experimentally accessible range of molecular weights.

After describing the idea behind the FENE model and establishing the relevance of FENE effects, simulation of this model is an important task. The singularity in the FENE spring-force law requires some special precautions and, strictly speaking, the theorems establishing the uniqueness of solutions and the convergence of approximation schemes are not applicable.

If the Euler scheme is applied to FENE dumbbells, a new problem occurs. For any finite time step, there is a certain probability that the maximum allowed spring extension is exceeded. A simple possibility of avoiding this problem is to reject moves in which the maximum extension is exceeded. However, with a finite time step, it is also dangerous to have an extension very close to the maximum allowed value because a very large force and hence a very large displacement in the next step would result. We therefore suggest to reject all moves which lead to a value of $\boldsymbol{Q}^2$ larger than

$$\left(1-\sqrt{\frac{\Delta t}{\lambda_H}}\right) b\,\frac{k_B T}{H}\,. \tag{4.120}$$

In spite of the explicit occurrence of $\sqrt{\Delta t}$, we do not expect corrections of order $\sqrt{\Delta t}$ for the material functions because the rejection of configurations occurs very rarely. A first-order weak algorithm should result. In formulating (4.120) flow effects have not been taken into account. The time steps in (4.120) should be chosen so small that no large changes in configuration occur during a single time step even in the presence of flow.

Exercise 4.32 In order to justify the criterion (4.120), determine $\bar{b}(\Delta t) < b$ such that for $\boldsymbol{Q}^2 = \bar{b}(\Delta t)$ the displacement due to the spring force in a step of size Δt is equal to $\sqrt{b} - \sqrt{\bar{b}(\Delta t)}$ (consider units of time and length such that $\lambda_H = k_B T/H = 1$).

Another possibility of avoiding configurations in the unphysical range is to use an implicit algorithm. Moreover, larger time steps may be used for such an algorithm. We here consider a semi-implicit algorithm in which only the spring-force law is treated implicitly,

$$\overline{\boldsymbol{Q}}(t_{j+1}) = \boldsymbol{Q}(t_j) + \left[\boldsymbol{\varkappa}(t_j)\cdot\boldsymbol{Q}(t_j) - \frac{1}{2}\frac{\boldsymbol{Q}(t_j)}{1-\boldsymbol{Q}^2(t_j)/b}\right]\Delta t_j + \Delta\boldsymbol{W}_j\,, \tag{4.121}$$

and

$$\begin{aligned}\left[1 + \frac{1}{4}\frac{\Delta t_j}{1-\boldsymbol{Q}^2(t_{j+1})/b}\right]\boldsymbol{Q}(t_{j+1}) &= \boldsymbol{Q}(t_j)\\ &+ \frac{1}{2}\left[\boldsymbol{\varkappa}(t_{j+1})\cdot\overline{\boldsymbol{Q}}(t_{j+1}) + \boldsymbol{\varkappa}(t_j)\cdot\boldsymbol{Q}(t_j) - \frac{1}{2}\frac{\boldsymbol{Q}(t_j)}{1-\boldsymbol{Q}^2(t_j)/b}\right]\Delta t_j + \Delta\boldsymbol{W}_j\,.\end{aligned} \tag{4.122}$$

The direction of $\boldsymbol{Q}(t_{j+1})$ is given by the direction of the right-hand side of (4.122). The length of $\boldsymbol{Q}(t_{j+1})$ can be determined from a cubic equation which, for arbitrary length of the vector on the right-hand side of (4.122), has a unique solution between 0 and $\sqrt{b}$. The same random numbers $\Delta\boldsymbol{W}_j$ occur in (4.121) and (4.122).

Simulation results obtained from the Euler scheme with rejection of unphysical moves and from the semi-implicit scheme (4.121), (4.122) for $b = 50$ and

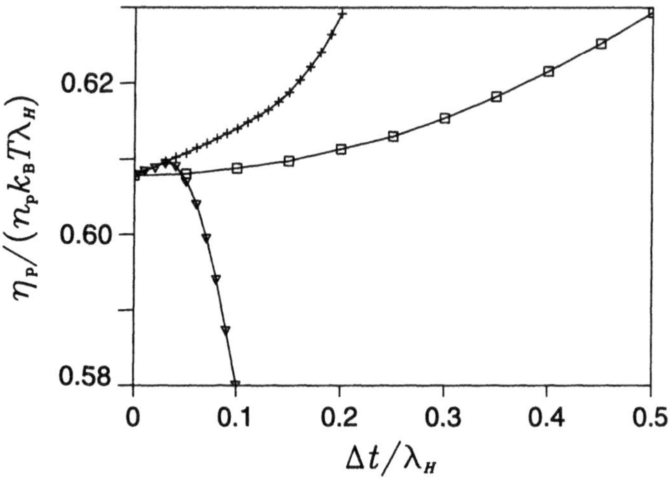

Fig. 4.16. Polymer contribution to the viscosity for $b = 50$, $\lambda_H \dot{\gamma} = 5$, and different time steps. The different symbols, which are larger than the statistical error bars, represent the results of the Euler scheme with rejection of unphysical moves according to the criterion (4.120) (*triangles*) and of the semi-implicit scheme (4.121), (4.122) (*squares*). The *crosses* represent the results for a modified rejection rule (see text).

$\lambda_H \dot{\gamma} = 5$ are compared in Fig. 4.16. For this rather high shear rate the time step should be small compared to $0.2\,\lambda_H$. Actually, the simulation based on the rejection criterion (4.120) failed for $\Delta t \geq 0.1\,\lambda_H$. This criterion should be modified because the rather large shear flow effects for $\lambda_H \dot{\gamma} = 5$ have not been taken into account in its derivation. For example, by changing the maximum value for $\boldsymbol{Q}^2$ allowed in the simulations from (4.120) to $[1 - \sqrt{\Delta t/(50\lambda_H)}]\, b\, k_B T/H$, somewhat larger time steps can be used for the simulations (see Fig. 4.16). Time steps considerably larger than $0.2\,\lambda_H$ can be used in the semi-implicit algorithm. The first and second order of weak convergence expected for the rejection and semi-implicit algorithms, respectively, is confirmed by Fig. 4.16. Whereas the second-order scheme requires almost 70% more computer time per time step, significantly larger time steps can be used for the second-order scheme in order to achieve a given level of accuracy. In obtaining the extrapolated results at zero time-step width from our simulations, the second-order algorithm has been found to be four to fifteen times more efficient.

In view of the singular behavior of the drift term, the FENE model constitutes a very natural problem for developing algorithms with stochastic time-step width control. A simple realization of this idea has been used in a Brownian dynamics simulation of FENE chains modeling adsorbed polymer molecules in shear flow [98] and polymer molecules in extensional flow [99]. The same idea has also been used by Fixman to remove problems in an approximate treatment of rapid vibrational motions by an implicit algorithm [100].

In the mathematical literature, a method of controlling the time step for all trajectories of an ensemble in a deterministic manner has been developed for weak integration schemes [101]. Actually, also the order of the algorithm is controlled, where higher orders are achieved by extrapolating the simulation results for several different time-step widths. Based on the information about discretization errors for a given quantity of interest, as obtained from various extrapolations in the current time step, one selects the optimum order and time-step width for the next time step [101].

Exercise 4.33 Show that in $[0, \sqrt{b}]$ there exists a unique solution of the cubic equation for the length of $\boldsymbol{Q}(t_{j+1})$ following from (4.122).

Exercise 4.34 Develop a routine for Brownian dynamics simulations of FENE dumbbells in steady shear flow, based on the semi-implicit scheme (4.121), (4.122). Carry out steady-state simulations for $b = 50$, $\lambda_H \dot{\gamma} = 5$, and compare your results to those in Fig. 4.16.

References

1. Bird RB, Curtiss CF, Armstrong RC, Hassager O (1987) *Dynamics of Polymeric Liquids, Vol 2, Kinetic Theory*, 2nd Edn. Wiley-Interscience, New York Chichester Brisbane Toronto Singapore
2. Zwanzig R (1969) Adv. Chem. Phys. 15: 325
3. Fixman M (1978) J. Chem. Phys. 69: 1527, 1538
4. Ermak DL, McCammon JA (1978) J. Chem. Phys. 69: 1352
5. Rouse PE Jr (1953) J. Chem. Phys. 21: 1272
6. Hermans JJ (1943) Physica 10: 777
7. Kuhn W, Kuhn H (1943) Helv. Chim. Acta 26: 1394
8. Doi M, Edwards SF (1986) *The Theory of Polymer Dynamics*. Clarendon, Oxford (The International Series of Monographs on Physics)
9. Schieber JD, Öttinger HC (1988) J. Chem. Phys. 89: 6972
10. Bird RB, Armstrong RC, Hassager O (1987) *Dynamics of Polymeric Liquids, Vol 1, Fluid Mechanics*, 2nd Edn. Wiley-Interscience, New York Chichester Brisbane Toronto Singapore
11. Lodge AS, Wu Y (1971) Rheol. Acta 10: 539
12. Öttinger HC (1986) J. Non-Newtonian Fluid Mech. 19: 357
13. Knuth DE (1981) *The Art of Computer Programming, Vol 2, Seminumerical Algorithms*, 2nd Edn. Addison-Wesley, Reading Menlo Park London Amsterdam Don Mills Sydney
14. Press WH, Teukolsky SA, Vetterling WT, Flannery BP (1992) *Numerical Recipes in FORTRAN. The Art of Scientific Computing*, 2nd Edn. Cambridge University Press, Cambridge
15. Vattulainen I, Ala-Nissila T, Kankaala K (1994) Phys. Rev. Lett. 73: 2513
16. Petersen WP (1994) Intl. J. High Speed Comp. 3: 387

17. Bouleau N, Lépingle D (1994) *Numerical Methods for Stochastic Processes*. Wiley, New York Chichester Brisbane Toronto Singapore (Wiley Series in Probability and Mathematical Statistics. Applied Probability and Statistics)
18. Kloeden PE, Platen E (1992) *Numerical Solution of Stochastic Differential Equations*. Springer, Berlin Heidelberg New York London Paris Tokyo Hong Kong Barcelona Budapest (Applications of Mathematics, Vol 23)
19. Öttinger HC (1994) Macromolecules 27: 3415
20. Melchior M, Öttinger HC (1994) in: Gallegos C (ed) *Progress and Trends in Rheology IV*: Proceedings of the Fourth European Rheology Conference. Steinkopff, Darmstadt, p 111
21. Kallianpur G (1980) *Stochastic Filtering Theory*. Springer, New York Heidelberg Berlin (Applications of Mathematics, Vol 13)
22. Melchior M, Öttinger HC (1995) *Variance Reduced Simulations of Stochastic Differential Equations*. To appear in J. Chem. Phys.
23. Melchior M, Öttinger HC (1995) *Variance Reduced Simulations of Polymer Dynamics*. Preprint
24. Kirkwood JG, Riseman J (1948) J. Chem. Phys. 16: 565
25. Zimm BH (1956) J. Chem. Phys. 24: 269; a corrected version of that paper can be found in J. J. Hermans (1978) *Polymer Solutions Properties, Part II: Hydrodynamics and Light Scattering*, Dowden, Hutchinson & Ross, Stroudsburg, pp. 73-84
26. Öttinger HC, Rabin Y (1989) J. Rheol. 33: 725
27. Rotne J, Prager S (1969) J. Chem. Phys. 50: 4831
28. Yamakawa H (1970) J. Chem. Phys. 53: 436
29. Zylka W, Öttinger HC (1989) J. Chem. Phys. 90: 474
30. Felderhof BU (1977) Physica A 89: 373
31. Mazur P, van Saarloos W (1982) Physica A 115: 21
32. Öttinger HC (1987) J. Chem. Phys. 86: 3731; a list of errors can be found in J. Chem. Phys. 87: 1460
33. Öttinger HC, Rabin Y (1989) J. Non-Newtonian Fluid Mech. 33: 53
34. Öttinger HC (1989) Phys. Rev. A 40: 2664
35. Liu TW, Öttinger HC (1987) J. Chem. Phys. 87: 3131
36. Schümmer P, Otten B (1984) in: Mena B, García-Rejón A, Rangel-Nafaile C (eds) *Advances in Rheology, Vol 1, Theory*: Proc. IX Intl. Congress on Rheology. Universidad Nacional Autónoma de México, México, p 399
37. Larson RG (1988) *Constitutive Equations for Polymer Melts and Solutions*. Butterworths, Boston London Singapore Sydney Toronto Wellington (Butterworths Series in Chemical Engineering)
38. Öttinger HC (1987) J. Chem. Phys. 87: 6185
39. Fixman M (1981) Macromolecules 14: 1710
40. Zimm BH, Roe GM, Epstein LF (1956) J. Chem. Phys. 24: 279
41. Öttinger HC, Zylka W (1992) J. Rheol. 36: 885
42. Miyaki Y, Einaga Y, Fujita H, Fukuda M (1980) Macromolecules 13: 588
43. Rubio AM, Freire JJ, Clarke JHR, Yong CW, Bishop M (1995) J. Chem. Phys. 102: 2277
44. Janeschitz-Kriegl H, (1969) Adv. Polym. Sci. 6: 170
45. Bossart J, Öttinger HC (1995) Macromolecules 28: 5852
46. Fixman M (1983) J. Chem. Phys. 78: 1594

47. Öttinger HC (1985) J. Chem. Phys. 83: 6535
48. Öttinger HC (1986) J. Chem. Phys. 85: 1669
49. Öttinger HC (1987) Coll. & Polym. Sci. 265: 101
50. Magda JJ, Larson RG, Mackay ME (1988) J. Chem. Phys. 89: 2504
51. Kishbaugh AJ, McHugh AJ (1990) J. Non-Newtonian Fluid Mech. 34: 181
52. Öttinger HC (1990) Phys. Rev. A 41: 4413
53. Öttinger HC (1989) J. Chem. Phys. 90: 463
54. Öttinger HC (1989) Coll. & Polym. Sci. 267: 1
55. Wedgewood LE (1989) J. Non-Newtonian Fluid Mech. 31: 127
56. Zylka W (1991) J. Chem. Phys. 94: 4628
57. López Cascales JJ, Navarro S, García de la Torre J (1992) Macromolecules 25: 3574
58. Fixman M (1986) Macromolecules 19: 1204
59. Honerkamp J (1993) *Stochastic Dynamical Systems*. VCH Publishers, New York Weinheim Cambridge
60. Bishop M, Frinks S (1987) J. Chem. Phys. 87: 3675
61. Iniesta A, García de la Torre J (1990) J. Chem. Phys. 92: 2015
62. Kloeden PE, Scott WD (1993) Maple Technical Newsletter Issue 10: 60
63. Janicki A, Weron A (1994) *Simulation and Chaotic Behavior of α-Stable Stochastic Processes*. Marcel Dekker, New York Basel Hong Kong (Monographs and Textbooks in Pure and Applied Mathematics, Vol 178)
64. Samorodnitsky G, Taqqu MS (1994) *Stable Non-Gaussian Random Processes*. Chapman & Hall, New York London
65. Fujimori S, Nakajima H, Wada Y, Doi M (1975) J. Polym. Sci. Polym. Phys. Ed. 13: 2135
66. Bird RB, Öttinger HC (1992) Annu. Rev. Phys. Chem. 43: 371
67. Wedgewood LE (1993) Rheol. Acta 32: 405
68. Hua CC, Schieber JD (1995) J. Non-Newtonian Fluid Mech. 56: 307
69. Biller P, Petruccione F (1988) J. Chem. Phys. 89: 2412
70. Biller P (1989) Continuum Mech. Thermodyn. 1: 53
71. Giesekus H (1966) Rheol. Acta 5: 29
72. Giesekus H (1982) J. Non-Newtonian Fluid Mech. 11: 69
73. Giesekus H (1982) Rheol. Acta 21: 366
74. Giesekus H (1983) J. Non-Newtonian Fluid Mech. 12: 367
75. Honerkamp J, Öttinger HC (1986) J. Chem. Phys. 84: 7028
76. Fixman M (1966) J. Chem. Phys. 45: 785, 793
77. Baldwin P, Helfand E (1990) Phys. Rev. A 41: 6772
78. Zylka W, Öttinger HC (1991) Macromolecules 24: 484
79. Díaz FG, García de la Torre J, Freire JJ (1989) Polymer 30: 259
80. Rey A, Freire JJ, García de la Torre J (1989) J. Chem. Phys. 90: 2035
81. López Cascales JJ, García de la Torre J (1991) Polymer 32: 3359
82. Rudisill JW, Cummings PT (1992) J. Non-Newtonian Fluid Mech. 41: 275
83. Rudisill JW, Fetsko SW, Cummings PT (1993) Comp. Polym. Sci. 3: 23
84. Peterlin A (1961) Makromol. Chem. 44-46: 338
85. Peterlin A (1966) J. Polym. Sci. Polym. Lett. B4: 287
86. Kuhn W, Grün F (1942) Kolloid-Z. 101: 248
87. Bird RB, Dotson PJ, Johnson NL (1980) J. Non-Newtonian Fluid Mech. 7: 213; corrigenda were published in J. Non-Newtonian Fluid Mech. 8: 193 (1981) and

15: 255 (1984); in addition, because of an error in (58), pointed out by H. H. Saab, pp. 224–234 are invalidated
88. Warner HR Jr (1972) Ind. Eng. Chem. Fundam. 11: 379
89. Wedgewood LE, Ostrov DN, Bird RB (1991) J. Non-Newtonian Fluid Mech. 40: 119
90. Laso M, Öttinger HC (1993) J. Non-Newtonian Fluid Mech. 47: 1
91. van den Brule BHAA (1993) J. Non-Newtonian Fluid Mech. 47: 357
92. Öttinger HC (1987) J. Non-Newtonian Fluid Mech. 26: 207
93. Wedgewood LE, Öttinger HC (1988) J. Non-Newtonian Fluid Mech. 27: 245
94. Ahn KH, Lee SJ (1992) J. Non-Newtonian Fluid Mech. 43: 143
95. Armstrong RC, Ishikawa S (1980) J. Rheol. 24: 143
96. Honerkamp J, Öttinger HC (1986) J. Non-Newtonian Fluid Mech. 21: 157
97. Rodriguez F (1989) *Principles of Polymer Systems*, 3rd Edn. Hemisphere, New York Washington Philadelphia London, Sect. 7-2
98. Atkinson J, Goh CJ, Phan-Thien N (1984) J. Chem. Phys. 80: 6305
99. Fetsko SW, Cummings PT (1995) J. Rheol. 39: 285
100. Fixman M (1986) Macromolecules 19: 1195
101. Hofmann N (1994) *Beiträge zur schwachen Approximation stochastischer Differentialgleichungen*. Ph.D. Thesis, Humboldt-Universität zu Berlin

5. Models with Constraints

Mechanical models of polymers involving constraints have a long tradition in polymer kinetic theory (see Chap. 11 of [1]). The *Kirkwood-Riseman chain* proposed in 1948 consists of N beads and $N-1$ rigid rods of length L with a fixed angle θ between successive links; links can rotate freely, that is, no rotatory potentials are included. This model is also called the freely rotating bead-rod chain. The *Kramers chain* proposed in 1944 is a freely jointed bead-rod chain with N beads and $N-1$ rigid rods of length L.

Although models with constraints often impose fixed bond lengths or bond angles, they usually do not pretend to portray details of the chemical structure. The idea is rather to model molecules that exhibit considerable rigidity on much larger scales, such as some biological macromolecules or liquid crystal polymers. The fact that large-scale persistence phenomena rather than local chemical structures are modeled by constraints is particularly obvious in the Kratky-Porod, or wormlike, chain. The *Kratky-Porod chain* is the continuous chain obtained by taking the limits $N \to \infty$, $L \to 0$, $\theta \to \pi$ in the Kirkwood-Riseman chain under the restrictions that the contour length $(N-1)L$ and the persistence length $L/(1+\cos\theta)$ remain constant.

There are *many approaches to models with constraints* on various levels (Fokker-Planck equations, stochastic differential equations, numerical integration schemes). In view of the subtleties and pitfalls of stochastic calculus—which we keep emphasizing throughout this book—it should be expected that some of the approaches are incorrect; this is indeed so. It is not even obvious whether approaches formulated on different levels coincide in general or, at least, in special cases. For example, Ref. 1 contains a Fokker-Planck equation for the configurational distribution function in the space of generalized coordinates, and in Ref. 2 a simulation algorithm in Cartesian space is proposed; are these approaches equivalent? (This question is answered on p. 244.)

In this chapter, we first describe the various levels on which models with constraints can be formulated, and how the formulations on different levels can be translated and compared. The dynamics for the general bead-rod-spring model for dilute solutions with hydrodynamic interaction, introduced through a Fokker-Planck equation in the space of generalized coordinates in [1], is first rewritten in terms of stochastic differential equations in Cartesian space, and a numerical integration scheme is then developed. The equations of motion and simulation algorithms of this chapter include the bead-spring models discussed in Chap. 4 as special cases. The general formulation of models with constraints is finally applied to obtain simulation algorithms for dilute solutions of rod-like molecules and for liquid crystal polymers.

5.1 General Bead-Rod-Spring Models

In the preceding chapter on bead-spring models for dilute solutions, we started with the Rouse model and then introduced various nonlinear effects to obtain more realistic and hence also more complicated models. In contrast, this chapter is opened with the formulation and reformulation of a very general bead-rod-spring model, and the discussion of simpler special cases is postponed to the subsequent section. In the general presentation given in this section, we first discuss some fundamental issues concerning models with constraints. The kinetic theory formulation of models with constraints is adopted as a safe starting point, and various formulations of the stochastic differential equations of motion are then derived. This section concludes with remarks on simulation algorithms and the appropriate stress tensor expressions for models with constraints. A condensed version of this section has been published in [3].

5.1.1 Philosophy of Constraints

Models with constraints can be formulated on the levels of Fokker-Planck equations, stochastic differential equations, or numerical integration schemes. For each of these levels one has the additional choice of describing the polymer configurations either by generalized coordinates or by Cartesian coordinates supplemented with constraint conditions. This leads to the six different approaches which are summarized in Fig. 5.1; we refer to these approaches as A1 to A6 (see Fig. 5.1).

All the approaches A1 to A6 can be used equivalently. Here the term "equivalent" is used in a somewhat loose sense. We know from Chap. 3 that a proper approximation scheme leads to a weak or strong solution of a stochastic differential equation but, of course, there are many possible approximation schemes for a given stochastic differential equation. Fokker-Planck equations characterize only the distribution of a stochastic process so that they can determine only weak solutions of stochastic differential equations; if we are interested in strong solutions then Fokker-Planck equations do not provide the required information. With these provisions, we here discuss how one can go back and forth between all the different approaches A1 to A6.

In Chap. 3 we learned how to move vertically through the diagram in Fig. 5.1. When working in Cartesian space the constraining forces must be chosen such that the solution of the resulting stochastic differential equation satisfies the constraints at all times. The Fokker-Planck equation then admits solutions that stay in a subspace for all times (containing δ-functions as factors

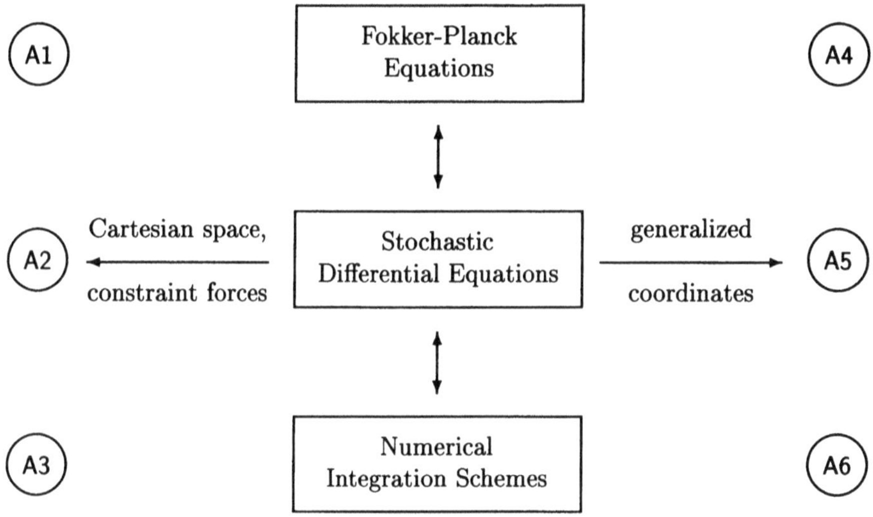

Fig. 5.1. Classification of different approaches to models with constraints.

which require an appropriate interpretation when inserted into a partial differential equation), and numerical integration schemes should be designed such that the constraints are rigorously satisfied.

When working with generalized coordinates, the main problem is the introduction of these coordinates, and the formulation and handling of the stochastic differential equations for the generalized coordinates.

Moving horizontally through the diagram is more problematic and requires additional considerations. On the level of stochastic differential equations, there is a transformation between the Cartesian coordinates for all position vectors and the generalized coordinates; if the transformation is sufficiently smooth then the Itô formula can be applied to jump back and forth between the respective stochastic differential equations (A2 $\leftrightarrow$ A5). If the numerical approximation scheme of A3 strictly satisfies the constraints then going back and forth between A3 and A6 is a matter of deterministic coordinate transformations for a given trajectory constructed by the one or other approach.

The horizontal step between A1 and A4 is complicated by the above-mentioned fact that the solution of the Fokker-Planck equation in Cartesian space is restricted to some subspace so that δ-functions arise as factors in the transition probability densities. All such subtleties are avoided by taking the path A1 $\leftrightarrow$ A2 $\leftrightarrow$ A5 $\leftrightarrow$ A4.

The classification scheme 5.1 allows us to identify those steps that are dangerous due to the peculiarities of stochastic calculus. Some readers might have the hope that the introduction of constraints would be unproblematic when done for Stratonovich stochastic differential equations. As a motivation for reading the following subsections, which actually follow the steps A4 → A5 → A2 → A3, this hope must better be destroyed immediately by considering the problems caused by spurious drift terms (see Sect. 3.3.6). The following exercise serves that purpose.

Exercise 5.1 Consider the Stratonovich stochastic differential equation $\mathrm{d}\boldsymbol{X}_t = \boldsymbol{A}\,\mathrm{d}t + \mathbf{B}(\boldsymbol{X}_t) \odot \mathrm{d}\boldsymbol{W}_t$ for a two-dimensional process $\boldsymbol{X}$, where

$$\boldsymbol{A} = -\frac{1}{2}\begin{pmatrix} 0 \\ 1 \end{pmatrix}, \qquad \mathbf{B}(\boldsymbol{x}) = \begin{pmatrix} \cos x_1 & -\sin x_1 \\ \sin x_1 & \cos x_1 \end{pmatrix} \quad \text{for} \quad \boldsymbol{x} = \begin{pmatrix} x_1 \\ x_2 \end{pmatrix},$$

and $\boldsymbol{W}$ is a two-dimensional Wiener process. Show that the Fokker-Planck equation associated with this process coincides with the one for $\mathrm{d}\boldsymbol{X}_t = \mathrm{d}\boldsymbol{W}_t$ (cf. Exercise 3.44). Both stochastic differential equations hence describe a two-dimensional Wiener process, so that $\boldsymbol{A}$ is a spurious drift term. Now, in order to obtain diffusion on a circle, the constraint of constant length of $\boldsymbol{X}_t$ is introduced by means of the projection operator $\mathbf{P}(\boldsymbol{x}) := \boldsymbol{\delta} - (\boldsymbol{x}\boldsymbol{x}/x^2)$. Show that the Fokker-Planck equations associated with $\mathrm{d}\boldsymbol{X}_t = \mathbf{P}(\boldsymbol{X}_t) \cdot \boldsymbol{A}\,\mathrm{d}t + [\mathbf{P}(\boldsymbol{X}_t) \cdot \mathbf{B}(\boldsymbol{X}_t)] \odot \mathrm{d}\boldsymbol{W}_t$ and $\mathrm{d}\boldsymbol{X}_t = \mathbf{P}(\boldsymbol{X}_t) \odot \mathrm{d}\boldsymbol{W}_t$ do not coincide.

There are other problems in addition to the pitfalls of stochastic calculus which make the approach to models with constraints even more difficult. For a very general and enlightening discussion of problems associated with constraints, even in classical mechanics, see [4]. The most important and well-known difficulty results from the observation that in statistical mechanics *rigid bonds are essentially different from very stiff elastic bonds*, even in the limit of infinite stiffness. The deeper reason for this discrepancy has been worked out very clearly by van Kampen [5]. There have been many attempts to establish a relationship between rigid and infinitely stiff systems, where the key idea is to introduce a corrective potential force for going back and forth between the two types of systems. We here mention only three important contributions by Fixman [6], Rallison [7], and Hinch [8].

Hinch considers the stochastic differential equations of motion for a constrained system in the absence of hydrodynamic interaction and flow within the approach A2 [8]. Since he extensively uses the familiar rules of deterministic calculus, he implicitly assumes a Stratonovich interpretation of his differential equations with multiplicative noise. The delicate step is then the construction of the appropriate Fokker-Planck equation, from which he can unambiguously read off the corrective force. The resulting corrective force depends on whether or not bead inertia is taken into account in deriving expressions for the constraint forces. This ambiguity in the treatment of bead inertia when relating rigid and very stiff systems has previously been emphasized by Fixman [6] who, in contrast to most other authors, introduces the constraint forces only *after* neglecting bead inertia. In the presence of hydrodynamic interaction and flow, Fixman formulates a stochastic differential equation on the level A5, where he verifies that his tentatively assumed noise term leads to the desired diffusion equation when going to the level A4. Moreover, Fixman suggests a simulation algorithm for models without hydrodynamic interaction, formulated on the level A3 [6].

Rallison [7] worked on the level of probability densities and Fokker-Planck equations (A4). At equilibrium, he compares the Boltzmann distributions for infinitely stiff and rigid systems, and the discrepancy between the two results may be used to readily identify the potential of the corrective force required for going back and forth between rigid and very stiff systems. In addition, Rallison considers a quantum mechanical approach to very stiff systems, and he finds still another, third result for the equilibrium probability density in which the corresponding corrective force is affected by the configuration-dependent zero point energy of the frozen degrees of freedom that are quantized. Moreover, an argument is offered why, for nonequilibrium systems, the correction forces may simply be added to the interaction forces occurring in the Fokker-Planck equation (or, equivalently, in the stochastic differential equations of motion). The formulation of mobility tensors in this argument is so general that hydrodynamic interactions are admitted.

It may be surprising that the relationship between rigid and very stiff systems has been used in both directions. Various authors have replaced rigid bonds by very stiff bonds, usually even without including proper corrective forces, because reliable simulation algorithms for elastic models are much better known, and because algorithms for rigid models are anticipated to be problematic [9] or time consuming [10]. This direction is not recommended because the rapid oscillations occurring for very stiff bonds require extremely small time steps; most of the computer time is wasted on resolving these rapid oscillations which are usually of no physical relevance. It seems to be much better to avoid such rapid motions in a real stiff system by going to a rigid system. Of course, this recommendation presumes the availability of suitable simulation algorithms for constrained systems. Such simulation algorithms are introduced below in this chapter.

Once the relationship between rigid and very stiff systems is established, one may ask "What is the correct approach?" One can try to rule out rigid constraints with the remark that nothing is rigid in real life. However, there are no beads and springs in real life either, and such mechanical contrivances including rods or other constraints may nevertheless model the behavior of real polymer systems very successfully, maybe even better than mechanical models without rigid constraints. One can try to add sophistication to the discussion by bringing quantum mechanics into the game. Rallison [7] showed that one cannot obtain a rigid classical system as a limiting case of a quantized system, whereas very stiff classical systems can be obtained in a certain limit. Since he generally finds the influence of quantum effects on rheological properties to be small, Rallison strongly suggests using infinitely stiff springs rather than rigid rods. However, quantum mechanics ought to be applied on the level of nuclei and electrons (or other well-defined entities), and it is hence questionable whether quantum mechanics on the level of beads may be used for answering the question "rigid or very stiff?" In other words, it is not clear whether the steps of coarse-graining and quantization commute. Another fundamental problem with infinitely stiff systems has been pointed out by van Kampen [5]: the limit of infinite stiffness depends on the choice of the confining potential so that a unique definition of the "correct" model requires the determination of the "correct" potential for the stiff springs and other constraining interactions (see also Appendix A of [7]). In the spirit of this book, we do not want to discuss the question about the "correct" approach any further. We are interested in understanding and simulating *given* models, and in the relationship between different approaches. Understanding these relationships is certainly an important first step in the direction of formulating the "correct" models. As the safest starting point we here rely on the kinetic theory formulation of [1]. The step from rigid to infinitely stiff can be done in a very simple manner once simulation algorithms for rigid systems have been developed.

We conclude this subsection with some remarks on the relationship between models with constraints and *stochastic differential equations on manifolds*. Since the type of constraints introduced in the next subsection implies that the polymer configuration space can be regarded as a manifold naturally embedded into Cartesian space, consideration of stochastic differential equations on manifolds seems to be very natural. Since there exists an extensive literature on the latter subject [11, 12] establishing a relationship appears to be even more attractive (a very nice "overture to stochastic differential geometry" can be found in Sect. V.5 of [13]). However, such a geometric formulation is not really helpful for several reasons. For example, the calculation of stresses is related to configurations and interactions in Cartesian space which cannot be expressed naturally in terms of intrinsic properties of the manifold of constrained configurations. Even worse is the effect of hydrodynamic interactions which clearly propagate in Cartesian space and have nothing to do with the manifold of constrained configurations. Since hydrodynamic interactions affect the diffusive properties they even lead to a modification of the metric matrix which is then no longer of truly geometric origin. Formally, a geometric formulation is still possible, but it is not very illuminating because rather obscure corrective forces need to be introduced.

5.1.2 Formulation of Stochastic Differential Equations

A diffusion equation for general bead-rod-spring models of dilute solutions including hydrodynamic interaction has been derived by C. F. Curtiss in [1] in the framework of polymer kinetic theory. We take that diffusion equation, which is formulated on the level of generalized coordinates, as the starting point for the discussion of models with constraints in this chapter (approach A4 of Fig. 5.1).

Before we can write down the fundamental diffusion equation, we first need to introduce some notation (we essentially follow the notation and terminology of [1]). Since the constraints are assumed to restrict the internal configurations only it is useful to introduce the position vector of bead μ with mass M_μ ($\mu = 1, 2, \ldots, N$) with respect to the center of mass by

$$\boldsymbol{R}_\mu := \boldsymbol{r}_\mu - \boldsymbol{r}_c = \sum_\nu \left(\delta_{\mu\nu} - \frac{M_\nu}{M_p}\right) \boldsymbol{r}_\nu\,, \qquad \boldsymbol{r}_c := \frac{1}{M_p} \sum_\mu M_\mu \boldsymbol{r}_\mu\,, \tag{5.1}$$

where $\boldsymbol{r}_\mu$ is the bead position vector with respect to an arbitrary point fixed in space, $\boldsymbol{r}_c$ is the center of mass position, and $M_p := \sum_\mu M_\mu$ is the total mass of a chain. One then has

$$\sum_\mu M_\mu \boldsymbol{R}_\mu = \boldsymbol{0}\,. \tag{5.2}$$

If the internal motions are restricted by d' time-independent constraints then we need $d = 3N - 3 - d'$ generalized coordinates $Q_1, Q_2, \ldots, Q_d$ in order to characterize the internal configurations. We assume that there exist smooth functions $\boldsymbol{R}_\mu(Q_1, Q_2, \ldots, Q_d)$ which specify the constrained internal configurations in terms of the generalized coordinates.

We allow for beads having not only different masses M_μ but also different frictional properties described by symmetric friction tensors $\boldsymbol{\zeta}_\mu$. As in Sect. 4.2.1, hydrodynamic interactions are described in terms of the symmetric tensors $\boldsymbol{\Omega}_{\mu\nu} := \boldsymbol{\Omega}(\boldsymbol{r}_\mu - \boldsymbol{r}_\nu) = \boldsymbol{\Omega}(\boldsymbol{R}_\mu - \boldsymbol{R}_\nu)$, and we introduce the *effective friction tensors* $\boldsymbol{\zeta}_{\mu\nu}$ by

$$\sum_\nu \boldsymbol{\zeta}_{\mu\nu} \cdot (\delta_{\nu\mu'} \boldsymbol{\zeta}_\nu^{-1} + \boldsymbol{\Omega}_{\nu\mu'}) = \delta_{\mu\mu'}\, \boldsymbol{\delta}\,. \tag{5.3}$$

Modified effective friction tensors are introduced as

$$\tilde{\boldsymbol{\zeta}}_{\mu\nu} := \boldsymbol{\zeta}_{\mu\nu} - \boldsymbol{\lambda}_\mu^T \cdot \mathbf{Z} \cdot \boldsymbol{\lambda}_\nu\,, \tag{5.4}$$

where the *total effective friction tensor* $\mathbf{Z}$ and the *weight tensors* $\boldsymbol{\lambda}_\mu$ are defined by

$$\mathbf{Z} := \sum_{\mu\nu} \boldsymbol{\zeta}_{\mu\nu}\,. \tag{5.5}$$

and

$$\boldsymbol{\lambda}_\nu := \mathbf{Z}^{-1} \cdot \sum_\mu \boldsymbol{\zeta}_{\mu\nu}\,, \tag{5.6}$$

A further useful quantity in describing the dynamics of the generalized coordinates for models with hydrodynamic interaction is the *modified metric matrix*, which is a $d \times d$-matrix with elements

$$\tilde{g}_{jk} := \sum_{\mu\nu} \frac{\partial \boldsymbol{R}_\mu}{\partial Q_j} \cdot \tilde{\boldsymbol{\zeta}}_{\mu\nu} \cdot \frac{\partial \boldsymbol{R}_\nu}{\partial Q_k}\,, \tag{5.7}$$

which is defined only in the manifold characterized by the constraints (that is, in the space of constrained configurations). The modified metric matrix can be considered as a function of the generalized coordinates because $\tilde{\boldsymbol{\zeta}}_{\mu\nu}$ depends only on the relative position vectors $\boldsymbol{R}_\mu(Q_1, Q_2, \ldots, Q_d)$.

The elements of the inverse matrix, which is also defined in the space of constrained configurations only, are denoted by $\tilde{G}_{jk}$,

$$\sum_{l=1}^{d} \tilde{g}_{jl}\, \tilde{G}_{lk} = \delta_{jk}\,. \tag{5.8}$$

Furthermore, we need the determinant of a *metric matrix* which is not affected by hydrodynamic interactions; this matrix has the elements

$$\mathbf{g}_{jk} := \sum_{\mu} M_{\mu} \frac{\partial \boldsymbol{R}_{\mu}}{\partial Q_j} \cdot \frac{\partial \boldsymbol{R}_{\mu}}{\partial Q_k}, \tag{5.9}$$

and we denote its determinant by $\mathbf{g}$.

Exercise 5.2 Prove the following identities:

$$\boldsymbol{\zeta}_{\mu\nu}^T = \boldsymbol{\zeta}_{\nu\mu}, \tag{5.10}$$

$$\tilde{\boldsymbol{\zeta}}_{\mu\nu}^T = \tilde{\boldsymbol{\zeta}}_{\nu\mu}, \tag{5.11}$$

$$\mathbf{Z}^T = \mathbf{Z}, \tag{5.12}$$

$$\sum_{\nu} \boldsymbol{\lambda}_{\nu} = \boldsymbol{\delta}, \tag{5.13}$$

$$\sum_{\mu} \tilde{\boldsymbol{\zeta}}_{\mu\nu} = \sum_{\nu} \tilde{\boldsymbol{\zeta}}_{\mu\nu} = \mathbf{0}, \tag{5.14}$$

$$\sum_{\mu} \boldsymbol{\lambda}_{\mu} \cdot (\delta_{\mu\nu} \boldsymbol{\zeta}_{\mu}^{-1} + \boldsymbol{\Omega}_{\mu\nu}) = \mathbf{Z}^{-1} \tag{5.15}$$

$$\sum_{\nu} \tilde{\boldsymbol{\zeta}}_{\mu\nu} \cdot (\delta_{\nu\mu'} \boldsymbol{\zeta}_{\nu}^{-1} + \boldsymbol{\Omega}_{\nu\mu'}) = \delta_{\mu\mu'} \boldsymbol{\delta} - \boldsymbol{\lambda}_{\mu}^T, \tag{5.16}$$

$$\tilde{\mathbf{g}}_{jk} = \tilde{\mathbf{g}}_{kj}, \qquad \tilde{\mathsf{G}}_{jk} = \tilde{\mathsf{G}}_{kj}. \tag{5.17}$$

Exercise 5.3 Evaluate the following derivatives, where x stands for any component of a position vector $\boldsymbol{r}_{\mu}$ or $\boldsymbol{R}_{\mu}$, or for any generalized coordinate Q_j (if $\boldsymbol{R}_{\mu}$ is regarded as a function of the generalized coordinates):[1]

$$\sum_{\nu} \frac{\partial}{\partial \boldsymbol{r}_{\nu}} \boldsymbol{\Omega}_{\mu\mu'} = \mathbf{0}, \tag{5.18}$$

$$\frac{\partial}{\partial x} \mathbf{Z} = -\sum_{\mu\nu} \mathbf{Z} \cdot \boldsymbol{\lambda}_{\mu} \cdot \left(\frac{\partial}{\partial x} \boldsymbol{\Omega}_{\mu\nu} \right) \cdot \boldsymbol{\lambda}_{\nu}^T \cdot \mathbf{Z}, \tag{5.19}$$

$$\frac{\partial}{\partial x} \mathbf{Z}^{-1} = \sum_{\mu\nu} \boldsymbol{\lambda}_{\mu} \cdot \left(\frac{\partial}{\partial x} \boldsymbol{\Omega}_{\mu\nu} \right) \cdot \boldsymbol{\lambda}_{\nu}^T, \tag{5.20}$$

$$\frac{\partial}{\partial x} \boldsymbol{\lambda}_{\nu} = -\sum_{\mu\mu'} \boldsymbol{\lambda}_{\mu} \cdot \left(\frac{\partial}{\partial x} \boldsymbol{\Omega}_{\mu\mu'} \right) \cdot \tilde{\boldsymbol{\zeta}}_{\mu'\nu}, \tag{5.21}$$

$$\frac{\partial}{\partial x} \tilde{\boldsymbol{\zeta}}_{\mu\nu} = -\sum_{\mu'\nu'} \tilde{\boldsymbol{\zeta}}_{\mu\mu'} \cdot \left(\frac{\partial}{\partial x} \boldsymbol{\Omega}_{\mu'\nu'} \right) \cdot \tilde{\boldsymbol{\zeta}}_{\nu'\nu}. \tag{5.22}$$

The diffusion equation (16.2-6) of [1] for the probability density $p = p(t, Q_1, Q_2, \ldots, Q_d)$ for only one species of polymer molecules in a solvent that is regarded as a continuum can be written in the form

[1]Throughout this chapter we assume that the friction tensors $\boldsymbol{\zeta}_{\mu}$ are independent of the polymer configuration; otherwise, these tensors can easily be reintroduced where derivatives of hydrodynamic-interaction tensors occur.

$$\frac{\partial p}{\partial t} = -\sum_{j=1}^{d} \frac{\partial}{\partial Q_j} \left\{ \sum_{k=1}^{d} \tilde{\mathsf{G}}_{jk} (\mathcal{F}_k + \mathcal{F}_k^{(\mathrm{e})} + \mathbf{M}_k : \boldsymbol{\varkappa}) \, p \right. \tag{5.23}$$
$$\left. + \, k_{\mathrm{B}}T \frac{1}{\sqrt{\mathsf{g}}} \sum_{k=1}^{d} \left[\frac{\partial}{\partial Q_k} \left(\tilde{\mathsf{G}}_{jk} \sqrt{\mathsf{g}} \right) \right] p \right\} + k_{\mathrm{B}}T \sum_{j,k=1}^{d} \frac{\partial^2}{\partial Q_j \partial Q_k} (\tilde{\mathsf{G}}_{jk} \, p) \, .$$

In that equation, the generalized intramolecular and external forces, $\mathcal{F}_k$ and $\mathcal{F}_k^{(\mathrm{e})}$, are given by

$$\mathcal{F}_k := \sum_{\mu} \frac{\partial \boldsymbol{R}_\mu}{\partial Q_k} \cdot \boldsymbol{F}_\mu \, , \tag{5.24}$$

$$\mathcal{F}_k^{(\mathrm{e})} := \sum_{\mu} \frac{\partial \boldsymbol{R}_\mu}{\partial Q_k} \cdot \boldsymbol{F}_\mu^{(\mathrm{e})} \, , \tag{5.25}$$

where it has been assumed in [1] that $\mathcal{F}_k^{(\mathrm{e})}$ can be expressed in terms of the generalized coordinates, that is, the external forces $\boldsymbol{F}_\mu^{(\mathrm{e})}$ do not depend on the center of mass position. The effect of the flow field is described by the tensor

$$\mathbf{M}_k := \sum_{\mu\nu} \boldsymbol{R}_\nu \frac{\partial \boldsymbol{R}_\mu}{\partial Q_k} \cdot \tilde{\zeta}_{\mu\nu} \, . \tag{5.26}$$

As in Chap. 16 of [1], we assume that the velocity gradient tensor $\boldsymbol{\varkappa}$ does not depend on position.[2]

The terms on the right side of (5.23) can be interpreted as follows. The terms in the first line represent the deterministic effects of the intramolecular forces, the external forces and the flow field transformed from Cartesian space to generalized coordinates by means of the transformation $\boldsymbol{R}_\mu(Q_1, Q_2, \ldots, Q_d)$. The terms in the second line describe the effects of Brownian motion in the manifold defined by the constraints. The occurrence of the metric matrix $\tilde{\mathsf{G}}_{jk}$ in the second-order derivative term and of the additional drift term is a well-known characteristic of diffusions on manifolds [14, 15]. Unusual is the modification of the metric matrix due to the presence of hydrodynamic interactions [14] (note that the determinant g is not affected by hydrodynamic interactions).

Exercise 5.4 The internal configuration of a rigid dumbbell can be described by the polar angles θ and ϕ. Consider the generalized coordinates $Q_1 = z = \cos\theta$ and $Q_2 = \phi$ with $-1 \le Q_1 \le 1$, $0 \le Q_2 < 2\pi$. In the absence of hydrodynamic interaction, calculate $\tilde{\mathsf{G}}_{jk}$ and g for rigid dumbbells with identical beads with mass M and scalar friction coefficient ζ which are separated by the fixed distance L. Write down the diffusion equation in shear flow for a given external potential $k_{\mathrm{B}}T\, V^{(\mathrm{e})}(z, \phi)$ (i.e., $V^{(\mathrm{e})}$ is a dimensionless function), and verify that the corresponding stochastic differential equations can be written as[3]

[2] Alternatively, we may make the weaker assumption that $\boldsymbol{\varkappa}$ is a slowly varying function of position so that it can be regarded as constant on the polymer length scale.

[3] Throughout this section on general bead-rod-spring models we suppress the time argument of stochastic processes. Since, as in Chap. 4, we use the same symbols for dummy

$$\mathrm{d}z = -\left\{\frac{1}{2}\dot{\gamma}\, z(1-z^2)\,\sin 2\phi + \frac{1}{6\lambda}\left[(1-z^2)\frac{\partial V^{(\mathrm{e})}}{\partial z} + 2z\right]\right\}\mathrm{d}t + \sqrt{\frac{1-z^2}{3\lambda}}\,\mathrm{d}W^z\,, \tag{5.27}$$

and

$$\mathrm{d}\phi = -\left[\dot{\gamma}\sin^2\phi + \frac{1}{6\lambda(1-z^2)}\frac{\partial V^{(\mathrm{e})}}{\partial \phi}\right]\mathrm{d}t + \frac{1}{\sqrt{3\lambda(1-z^2)}}\,\mathrm{d}W^\phi\,, \tag{5.28}$$

where W^z and W^ϕ are independent Wiener processes, and $\lambda := \zeta L^2/(12 k_{\mathrm{B}} T)$ is a time constant.

We next move on from level A4 to A5 in Fig. 5.1. The Fokker-Planck equation (5.23) is in a form that allows immediate reformulation as an Itô stochastic differential equation. The diffusion matrix can be decomposed in such a manner that a d-dimensional Wiener process is sufficient for formulating the stochastic differential equation (see Exercise 5.4). However, since the evaluation of all the coefficients in (5.23) as functions of the generalized coordinates is in general a very complicated task we here establish a connection to the bead positions in real space.

We can easily jump back and forth between bead positions and generalized coordinates if we define in the space of constrained configurations

$$\frac{\partial Q_j}{\partial \boldsymbol{R}_\mu} := \sum_{k=1}^{d} \tilde{\mathsf{G}}_{jk} \sum_\nu \tilde{\boldsymbol{\zeta}}_{\mu\nu} \cdot \frac{\partial \boldsymbol{R}_\nu}{\partial Q_k}\,. \tag{5.29}$$

This definition is designed such that the (formal) chain rule

$$\sum_\mu \frac{\partial Q_j}{\partial \boldsymbol{R}_\mu} \cdot \frac{\partial \boldsymbol{R}_\mu}{\partial Q_k} = \sum_{l=1}^{d} \tilde{\mathsf{G}}_{jl}\, \tilde{\mathsf{g}}_{lk} = \delta_{jk}\,, \tag{5.30}$$

holds. Furthermore, from (5.14) follows

$$\sum_\mu \frac{\partial Q_j}{\partial \boldsymbol{R}_\mu} = \boldsymbol{0}\,. \tag{5.31}$$

We can introduce a Wiener process with $3N > d$ components to obtain the following stochastic differential equation corresponding to the Fokker-Planck equation (5.23),

$$\begin{aligned} \mathrm{d}Q_j \;=\; & \sum_{k=1}^{d}\sum_\mu \tilde{\mathsf{G}}_{jk}\frac{\partial \boldsymbol{R}_\mu}{\partial Q_k}\cdot\Big\{\Big[\sum_\nu(\delta_{\mu\nu}\boldsymbol{\delta} - \boldsymbol{\lambda}_\mu^T)\cdot(\boldsymbol{F}_\nu + \boldsymbol{F}_\nu^{(\mathrm{e})}) + \sum_\nu \tilde{\boldsymbol{\zeta}}_{\mu\nu}\cdot\boldsymbol{\kappa}\cdot\boldsymbol{R}_\nu\Big]\,dt \\ & + \sqrt{2k_{\mathrm{B}}T}\sum_{\nu\nu'}\tilde{\boldsymbol{\zeta}}_{\mu\nu}\cdot\mathbf{B}_{\nu\nu'}\cdot\mathrm{d}\boldsymbol{W}_{\nu'}\Big\} + k_{\mathrm{B}}T\,\frac{1}{\sqrt{\mathsf{g}}}\sum_{k=1}^{d}\frac{\partial}{\partial Q_k}\left(\tilde{\mathsf{G}}_{jk}\sqrt{\mathsf{g}}\right)\,\mathrm{d}t\,, \end{aligned} \tag{5.32}$$

---

variables in probability densities and for the corresponding stochastic processes, the distinction between those two different kinds of objects becomes clear only through the context. In view of the involved notation in this chapter this inconvenience appears to be acceptable. It may be comforting that in Sect. 5.2 and in Chap. 6 we shall be able to afford the luxury of a clear distinction between dummy variables and stochastic processes.

where, for all $\mu, \mu' = 1, 2, \ldots, N$,

$$\delta_{\mu\mu'}\boldsymbol{\zeta}_\mu^{-1} + \boldsymbol{\Omega}_{\mu\mu'} = \sum_\nu \mathbf{B}_{\mu\nu} \cdot \mathbf{B}_{\mu'\nu}^T . \tag{5.33}$$

Notice that we have introduced the weight tensors $\boldsymbol{\lambda}_\mu^T$ in front of the force terms. For the intramolecular forces this has no effect since the additional terms vanish when the summation over ν is carried out. The same argument applies when $\boldsymbol{F}_\nu^{(e)}$ depends only on the internal configuration of the chain. If, however, the external forces depend on the center of mass position then (18.2-23) of [1] implies that the weight tensors need to be introduced in the above manner. The appropriateness of the decomposition of the diffusion matrix for the noise term in (5.32), which can be written as $\sqrt{2k_BT}\sum_{\mu\nu}(\partial Q_j/\partial \boldsymbol{R}_\mu) \cdot \mathbf{B}_{\mu\nu} \cdot \mathrm{d}\boldsymbol{W}_\nu$, is established in the following exercise.

Exercise 5.5 Verify the identity

$$\tilde{\mathsf{G}}_{jk} = \sum_{\mu\mu'\nu} \frac{\partial Q_j}{\partial \boldsymbol{R}_\mu} \cdot \mathbf{B}_{\mu\nu} \cdot \mathbf{B}_{\mu'\nu}^T \cdot \frac{\partial Q_k}{\partial \boldsymbol{R}_{\mu'}} . \tag{5.34}$$

Exercise 5.6 Prove the following rule for differentiating the determinant M of an arbitrary regular $d \times d$-matrix with elements M_{jk} that depend on some parameter x,

$$\frac{2}{\sqrt{\mathsf{M}}} \frac{\partial}{\partial x} \sqrt{\mathsf{M}} = \frac{\partial}{\partial x} \ln \mathsf{M} = \sum_{j,k=1}^{d} \mathsf{M}_{kj}^{-1} \frac{\partial \mathsf{M}_{jk}}{\partial x} . \tag{5.35}$$

When applied to the metric matrix (5.9), this rule is very useful for an explicit evaluation of the last term in (5.32). In this context, we write $\ln \mathsf{M}$ even if M is not dimensionless; when a derivative is taken then the units of M are irrelevant.

In the next step, we move on from level A5 to A2 in Fig. 5.1. By applying the Itô formula to the nonlinear transformation from the generalized coordinates Q_j to the bead position vectors $\boldsymbol{R}_\mu$ we obtain the following stochastic differential equation

$$\begin{aligned} \mathrm{d}\boldsymbol{R}_\mu &= \sum_{\mu'} \mathbf{H}_{\mu\mu'} \cdot \Big\{ \Big[\sum_\nu (\delta_{\mu'\nu}\boldsymbol{\delta} - \boldsymbol{\lambda}_{\mu'}^T) \cdot (\boldsymbol{F}_\nu + \boldsymbol{F}_\nu^{(e)}) + \sum_\nu \tilde{\boldsymbol{\zeta}}_{\mu'\nu} \cdot \boldsymbol{\kappa} \cdot \boldsymbol{R}_\nu \Big] \mathrm{d}t \\ &\qquad + \sqrt{2k_BT} \sum_{\nu\nu'} \tilde{\boldsymbol{\zeta}}_{\mu'\nu} \cdot \mathbf{B}_{\nu\nu'} \cdot \mathrm{d}\boldsymbol{W}_{\nu'} \Big\} \\ &\quad + k_BT \sum_{j,k=1}^{d} \frac{1}{\sqrt{\mathsf{g}}} \frac{\partial}{\partial Q_j} \left(\sqrt{\mathsf{g}}\, \tilde{\mathsf{G}}_{jk} \frac{\partial \boldsymbol{R}_\mu}{\partial Q_k} \right) \mathrm{d}t , \end{aligned} \tag{5.36}$$

where

$$\mathbf{H}_{\mu\nu} := \sum_{j,k=1}^{d} \tilde{\mathsf{G}}_{jk} \frac{\partial \boldsymbol{R}_\mu}{\partial Q_j} \frac{\partial \boldsymbol{R}_\nu}{\partial Q_k} . \tag{5.37}$$

Exercise 5.7 Prove the following identities for $\mathbf{H}_{\mu\nu}$:

$$\sum_{\mu'\nu'} \mathbf{H}_{\mu\mu'} \cdot \tilde{\boldsymbol{\zeta}}_{\mu'\nu'} \cdot \frac{\partial \boldsymbol{R}_{\nu'}}{\partial Q_j} = \frac{\partial \boldsymbol{R}_\mu}{\partial Q_j}, \tag{5.38}$$

$$\sum_{\mu'\nu'} \mathbf{H}_{\mu\mu'} \cdot \tilde{\boldsymbol{\zeta}}_{\mu'\nu'} \cdot \mathbf{H}_{\nu'\nu} = \mathbf{H}_{\mu\nu}. \tag{5.39}$$

The above stochastic differential equations for the bead position vectors $\boldsymbol{R}_\mu$ with respect to the center of mass have been derived rigorously from the diffusion equation (5.23).[4] We next propose a system of stochastic differential equations for the position vectors $\boldsymbol{r}_\mu$ which lead to (5.36) for $\boldsymbol{R}_\mu$,

$$\begin{aligned} d\boldsymbol{r}_\mu = {} & \sum_\nu \mathbf{P}_{\mu\nu} \cdot \Big\{ \Big[\boldsymbol{v}_0 + \boldsymbol{\kappa} \cdot \boldsymbol{r}_\nu + \sum_{\nu'} (\delta_{\nu\nu'} \boldsymbol{\zeta}_\nu^{-1} + \boldsymbol{\Omega}_{\nu\nu'}) \cdot (\boldsymbol{F}_{\nu'} + \boldsymbol{F}_{\nu'}^{(e)}) \Big] dt \\ & \qquad + \sqrt{2 k_B T} \sum_{\nu'} \mathbf{B}_{\nu\nu'} \cdot d\boldsymbol{W}_{\nu'} \Big\} \\ & + k_B T \sum_\nu (\delta_{\mu\nu} \boldsymbol{\delta} - \boldsymbol{\lambda}_\nu) \cdot \sum_{j,k=1}^{d} \frac{1}{\sqrt{g}} \frac{\partial}{\partial Q_j} \left(\sqrt{g}\, \tilde{G}_{jk} \frac{\partial \boldsymbol{R}_\nu}{\partial Q_k} \right) dt \\ & + k_B T \sum_{\mu'\nu\nu'} \mathbf{P}_{\mu'\nu}^T : \left(\frac{\partial}{\partial \boldsymbol{r}_{\mu'}} \boldsymbol{\Omega}_{\nu\nu'} \right) \cdot \boldsymbol{\lambda}_{\nu'}^T \, dt, \end{aligned} \tag{5.40}$$

where

$$\mathbf{P}_{\mu\nu} := \boldsymbol{\lambda}_\nu + \sum_{\mu'} (\delta_{\mu\mu'} \boldsymbol{\delta} - \boldsymbol{\lambda}_{\mu'}) \cdot \sum_{j=1}^{d} \frac{\partial \boldsymbol{R}_{\mu'}}{\partial Q_j} \frac{\partial Q_j}{\partial \boldsymbol{R}_\nu}, \tag{5.41}$$

is a projection operator in the following sense,

$$\sum_\nu \mathbf{P}_{\mu\nu} \cdot \mathbf{P}_{\nu\mu'} = \mathbf{P}_{\mu\mu'}. \tag{5.42}$$

The latter identity follows from (5.13), (5.30) and (5.31).

The derivation of (5.36) from (5.40) can be based on the identity

$$\sum_{\mu'} \mathbf{H}_{\mu\mu'} \cdot \tilde{\boldsymbol{\zeta}}_{\mu'\nu} = \sum_{j=1}^{d} \frac{\partial \boldsymbol{R}_\mu}{\partial Q_j} \frac{\partial Q_j}{\partial \boldsymbol{R}_\nu}, \tag{5.43}$$

which follows directly from the definitions (5.29) and (5.37). This identity implies

[4]Strictly speaking, these equations involve coefficient functions depending on the generalized coordinates. It is assumed that these coefficient functions can be expressed in a closed form in terms of the vectors $\boldsymbol{R}_\mu$, so that we obtain a stochastic differential equation in the Cartesian space into which the space of constrained configurations is embedded. An explicit construction of such a closed-form expression in terms of constraint conditions is given in the following subsection. We are interested in solutions which remain in the space of constrained configurations.

$$\sum_{\mu'} \left(\delta_{\mu\mu'} - \frac{M_{\mu'}}{M_{\rm p}} \right) \mathbf{P}_{\mu'\nu} = \sum_{\mu'} \mathbf{H}_{\mu\mu'} \cdot \tilde{\boldsymbol{\zeta}}_{\mu'\nu} \, . \tag{5.44}$$

In summary, we have seen that the stochastic differential equations (5.32) and (5.40) lead to the same equation of motion for the internal configurations $\boldsymbol{R}_\mu$ (these equations are equivalent in the strong sense, that is, they lead to the same trajectories). The term in the last line of (5.40) and the term containing $\boldsymbol{\lambda}_\nu$ in the next-to-last line are independent of μ and hence do not contribute to the internal motions. We have introduced these terms in (5.40) such that the center of mass motion is also consistent with the kinetic theory results of [1]. If only internal motions are of interest then these terms may be neglected.

Exercise 5.8 Prove the following identities for $\mathbf{P}_{\mu\nu}$:

$$\begin{aligned}
\sum_\nu \mathbf{P}_{\mu\nu} &= \boldsymbol{\delta} \, , \\
\sum_\mu \boldsymbol{\lambda}_\mu \cdot \mathbf{P}_{\mu\nu} &= \boldsymbol{\lambda}_\nu \, , \\
\sum_{\mu'} \tilde{\boldsymbol{\zeta}}_{\mu\mu'} \cdot \mathbf{P}_{\mu'\nu} &= \sum_{\mu'\nu'} \tilde{\boldsymbol{\zeta}}_{\mu\mu'} \cdot \mathbf{H}_{\mu'\nu'} \cdot \tilde{\boldsymbol{\zeta}}_{\nu'\nu} \, , \\
\sum_{\mu'} \tilde{\boldsymbol{\zeta}}_{\mu\mu'} \cdot \mathbf{P}_{\mu'\nu} &= \sum_{\mu'} \mathbf{P}^T_{\mu'\mu} \cdot \tilde{\boldsymbol{\zeta}}_{\mu'\nu} \, , \\
\sum_\mu \frac{M_\mu}{M_{\rm p}} \mathbf{P}_{\mu\nu} &= \boldsymbol{\lambda}_\nu - \sum_{\mu\mu'} \boldsymbol{\lambda}_\mu \cdot \mathbf{H}_{\mu\mu'} \cdot \tilde{\boldsymbol{\zeta}}_{\mu'\nu} \, , \\
\sum_{\mu\mu'} \frac{M_\mu}{M_{\rm p}} \mathbf{P}_{\mu\mu'} \cdot (\delta_{\mu'\nu} \zeta_{\mu'}^{-1} + \boldsymbol{\Omega}_{\mu'\nu}) &= \mathbf{Z}^{-1} + \sum_{\mu\mu'} \boldsymbol{\lambda}_\mu \cdot \mathbf{H}_{\mu\mu'} \cdot \boldsymbol{\lambda}^T_{\mu'} - \sum_\mu \boldsymbol{\lambda}_\mu \cdot \mathbf{H}_{\mu\nu} \, , \\
\sum_{\mu'} \mathbf{P}_{\mu\mu'} \cdot (\delta_{\mu'\nu} \zeta_{\mu'}^{-1} + \boldsymbol{\Omega}_{\mu'\nu}) &= \mathbf{H}_{\mu\nu} - \sum_{\mu'} \boldsymbol{\lambda}_{\mu'} \cdot \mathbf{H}_{\mu'\nu} - \sum_{\nu'} \mathbf{H}_{\mu\nu'} \cdot \boldsymbol{\lambda}^T_{\nu'} \\
&\quad + \sum_{\mu'\nu'} \boldsymbol{\lambda}_{\mu'} \cdot \mathbf{H}_{\mu'\nu'} \cdot \boldsymbol{\lambda}^T_{\nu'} + \mathbf{Z}^{-1} \, , \\
\sum_{\mu'} \mathbf{P}_{\mu\mu'} \cdot (\delta_{\mu'\nu} \zeta_{\mu'}^{-1} + \boldsymbol{\Omega}_{\mu'\nu}) &= \sum_{\mu'} (\delta_{\mu\mu'} \zeta_\mu^{-1} + \boldsymbol{\Omega}_{\mu\mu'}) \cdot \mathbf{P}^T_{\nu\mu'} \\
&= \sum_{\mu'\nu'} \mathbf{P}_{\mu\mu'} \cdot (\delta_{\mu'\nu'} \zeta_{\mu'}^{-1} + \boldsymbol{\Omega}_{\mu'\nu'}) \cdot \mathbf{P}^T_{\nu\nu'} \, .
\end{aligned}$$

Exercise 5.9 Consider the rigid dumbbells with identical beads of Exercise 5.4. Determine $\mathbf{H}_{\mu\nu}$ and $\mathbf{P}_{\mu\nu}$. Use the equations of motion (5.40) in order to derive the following stochastic differential equation for $\boldsymbol{U} = (\boldsymbol{r}_2 - \boldsymbol{r}_1)/L$,

$$\boxed{\mathrm{d}\boldsymbol{U} = (\boldsymbol{\delta} - \boldsymbol{U}\boldsymbol{U}) \cdot \left\{ \left[\boldsymbol{\kappa} \cdot \boldsymbol{U} + \frac{1}{\zeta L} \left(\boldsymbol{F}_2^{(\rm e)} - \boldsymbol{F}_1^{(\rm e)} \right) \right] \mathrm{d}t + \frac{1}{\sqrt{3\lambda}} \mathrm{d}\boldsymbol{W} \right\} - \frac{1}{3\lambda} \boldsymbol{U} \, \mathrm{d}t \, .}$$

Combine the Itô formula with the results of Exercise 5.4 to obtain an alternative derivation of this equation.

Exercise 5.10 Prove that the stochastic differential equations (5.40) lead to the same equation of motion for the center of mass as is obtained from kinetic theory [1]. (Warning: Solution of this exercise requires rather lengthy and formal developments.)

5.1.3 Generalized Coordinates Versus Constraint Conditions

If (5.40) is used as a starting point for a simulation, then the evaluation of the coefficients requires an explicit parametrization of the vectors $\boldsymbol{R}_\mu$ in terms of generalized coordinates. In a sense, we have not reached the level A2 of Fig. 5.1 yet because we still need generalized coordinates. In many applications it is simpler to work directly with time-independent constraints of the form

$$g_j(\boldsymbol{r}_1, \boldsymbol{r}_2, \ldots, \boldsymbol{r}_N) = 0 \qquad \text{for } j = 1, 2, \ldots, d', \tag{5.45}$$

where, for example, for constrained bond lengths and angles, only two or three bead positions are involved in each constraint.

For developing simulation algorithms for models with constraints it is useful to express all coefficients in the stochastic differential equations of motion in terms of these constraints; this has previously been emphasized by Fixman [6], who gives a useful equation for eliminating generalized coordinates in favor of constraints (see (3.35) of his paper), and by Hinch [8]. In doing so, one obtains a continuation around the space of constrained configurations (which, of course, is not unique). We assume that only the internal configurations are constrained, that is,

$$\sum_\mu \frac{\partial g_j}{\partial \boldsymbol{r}_\mu} = \boldsymbol{0}. \tag{5.46}$$

We define a further *modified metric matrix* with $d' \times d'$ elements given by

$$\overline{\mathsf{G}}_{jk} := \sum_{\mu\nu} \frac{\partial g_j}{\partial \boldsymbol{r}_\mu} \cdot (\delta_{\mu\nu} \zeta_\mu^{-1} + \boldsymbol{\Omega}_{\mu\nu}) \cdot \frac{\partial g_k}{\partial \boldsymbol{r}_\nu}, \tag{5.47}$$

and its inverse with elements $\overline{\mathsf{g}}_{jk}$,

$$\sum_{l=1}^{d'} \overline{\mathsf{g}}_{jl} \, \overline{\mathsf{G}}_{lk} = \delta_{jk}. \tag{5.48}$$

We also need a *metric matrix* formed from the constraint conditions which is not modified by hydrodynamic interactions,

$$\hat{\mathsf{G}}_{jk} := \sum_\mu \frac{1}{M_\mu} \frac{\partial g_j}{\partial \boldsymbol{r}_\mu} \cdot \frac{\partial g_k}{\partial \boldsymbol{r}_\mu}. \tag{5.49}$$

We denote the determinant of this matrix by $\hat{\mathsf{G}}$, and the elements of the inverse matrix by $\hat{\mathsf{g}}_{jk}$. Finally, we define the quantities

$$\frac{\partial \boldsymbol{R}_\mu}{\partial g_j} := \sum_{k=1}^{d'} \overline{\mathsf{g}}_{jk} \sum_\nu (\delta_{\mu\nu} \zeta_\mu^{-1} + \boldsymbol{\Omega}_{\mu\nu}) \cdot \frac{\partial g_k}{\partial \boldsymbol{r}_\nu}. \tag{5.50}$$

Exercise 5.11 Prove the following identities:

$$\sum_\mu \frac{\partial g_j}{\partial \boldsymbol{r}_\mu} \cdot \frac{\partial \boldsymbol{R}_\mu}{\partial g_k} = \delta_{jk}\,, \tag{5.51}$$

$$\sum_\mu \frac{\partial g_j}{\partial \boldsymbol{r}_\mu} \cdot \frac{\partial \boldsymbol{R}_\mu}{\partial Q_k} = \sum_\mu \frac{\partial Q_j}{\partial \boldsymbol{R}_\mu} \cdot \frac{\partial \boldsymbol{R}_\mu}{\partial g_k} = 0\,, \tag{5.52}$$

$$\sum_\mu \boldsymbol{\lambda}_\mu \cdot \frac{\partial \boldsymbol{R}_\mu}{\partial g_k} = \boldsymbol{0}\,. \tag{5.53}$$

As a first step in the course of eliminating all generalized coordinates, the projection operator $\mathbf{P}_{\mu\nu}$ can now be expressed in the form

$$\mathbf{P}_{\mu\nu} = \delta_{\mu\nu}\,\boldsymbol{\delta} - \sum_{j=1}^{d'} \frac{\partial \boldsymbol{R}_\mu}{\partial g_j}\,\frac{\partial g_j}{\partial \boldsymbol{r}_\nu}\,. \tag{5.54}$$

This expression for $\mathbf{P}_{\mu\nu}$ involves only the constraint conditions and hydrodynamic-interaction tensors; consideration of generalized coordinates is not necessary. Actually, all the definitions of this subsection are meaningful even outside the space of constrained configurations (so far, only the result (5.52) is restricted to the space of constrained configurations). The equivalence of the expressions (5.41) and (5.54) in the space of constrained configurations can be shown by comparing

$$\sum_\nu \mathbf{P}_{\mu\nu} \cdot \frac{\partial \boldsymbol{R}_\nu}{\partial Q_j}\,, \qquad \sum_\nu \mathbf{P}_{\mu\nu} \cdot \frac{\partial \boldsymbol{R}_\nu}{\partial g_j}\,, \qquad \sum_\nu \mathbf{P}_{\mu\nu} \cdot \boldsymbol{\delta}\,, \tag{5.55}$$

corresponding to checking the equivalent action on $d+d'+3 = 3N$ independent base vectors. In the absence of hydrodynamic interaction, the linear independence of the base vectors follows from the assumption that the d' constraints are independent, that is, that $3N-3-d'$ generalized coordinates are sufficient for characterizing the internal configurations. The linear independence must hold also for weak hydrodynamic interaction and hence constitutes the generic case. The equivalence then follows from (5.30), (5.31), (5.46) and the results of Exercise 5.11.

What remains to be done is to eliminate the generalized coordinates from the term in the next-to-last line in (5.40). As one should expect for a sensible formulation of the dynamics of models with constraints, this term is independent of the choice of generalized coordinates. Inspired by a very common and helpful procedure in the general theory of relativity, namely the introduction of locally inertial coordinates [16], we choose a set of generalized coordinates in which this term can be handled most conveniently, so that a relationship with the constraint conditions can readily be established. More precisely, we introduce generalized coordinates such that at a given point of the space of constrained configurations the following conditions hold,

$$\mathrm{g}_{jk} = \delta_{jk}\,, \tag{5.56}$$

$$\sum_\mu M_\mu \frac{\partial^2 \boldsymbol{R}_\mu}{\partial Q_j \partial Q_k} \cdot \frac{\partial \boldsymbol{R}_\mu}{\partial Q_l} = 0 . \tag{5.57}$$

Exercise 5.12 Prove that a set of generalized coordinates can be introduced such that at a given point of the space of constrained configurations the conditions (5.56) and (5.57) are satisfied. (Hint: The proof can be based on a Cholesky decomposition of the metric matrix for given generalized coordinates; such a decomposition can be used to locally transform to a special set of coordinates with the desired properties.)

The elimination of second-order derivatives with respect to generalized coordinates in favor of second-order derivatives of the constraint conditions with respect to bead positions is based on the identity

$$\sum_\mu \frac{\partial g_l}{\partial \boldsymbol{r}_\mu} \cdot \frac{\partial^2 \boldsymbol{R}_\mu}{\partial Q_j \partial Q_k} = -\sum_{\mu\nu} \frac{\partial \boldsymbol{R}_\mu}{\partial Q_j} \cdot \frac{\partial^2 g_l}{\partial \boldsymbol{r}_\mu \partial \boldsymbol{r}_\nu} \cdot \frac{\partial \boldsymbol{R}_\nu}{\partial Q_k} , \tag{5.58}$$

which is obtained by taking the derivative of the first part of (5.52) with respect to Q_j and applying the chain rule to one of the resulting terms. If the generalized coordinates satisfy the conditions (5.56) and (5.57) then an explicit representation for $\partial^2 \boldsymbol{R}_\mu/(\partial Q_j \partial Q_k)$ can be derived. This is done in the following exercise.

Exercise 5.13 Derive the representation

$$\frac{\partial^2 \boldsymbol{R}_\mu}{\partial Q_j \partial Q_k} = -\sum_{l,n=1}^{d'} \tilde{\mathsf{g}}_{ln} \left(\sum_{\mu'\nu'} \frac{\partial \boldsymbol{R}_{\mu'}}{\partial Q_j} \cdot \frac{\partial^2 g_n}{\partial \boldsymbol{r}_{\mu'} \partial \boldsymbol{r}_{\nu'}} \cdot \frac{\partial \boldsymbol{R}_{\nu'}}{\partial Q_k} \right) \frac{1}{M_\mu} \frac{\partial g_l}{\partial \boldsymbol{r}_\mu} , \tag{5.59}$$

at a point where the generalized coordinates satisfy (5.56) and (5.57).

With the result of the preceding exercise we obtain after carrying out the summation over k by means of (5.41) and after using (5.54) and the result of Exercise 5.6

$$\sum_\mu \sum_{k=1}^{d} \frac{\partial^2 \boldsymbol{R}_\mu}{\partial Q_j \partial Q_k} \cdot \frac{\partial Q_k}{\partial \boldsymbol{R}_\mu} = \sum_{k=1}^{d'} \sum_{\mu'\nu'} \frac{\partial \boldsymbol{R}_{\mu'}}{\partial Q_j} \cdot \frac{\partial^2 g_k}{\partial \boldsymbol{r}_{\mu'} \partial \boldsymbol{r}_{\nu'}} \cdot \frac{\partial \boldsymbol{R}_{\nu'}}{\partial g_k} - \frac{1}{2} \sum_\mu \frac{\partial \boldsymbol{R}_\mu}{\partial Q_j} \cdot \frac{\partial}{\partial \boldsymbol{r}_\mu} \ln \hat{\mathsf{G}} . \tag{5.60}$$

We can now evaluate the term in the next-to-last line of (5.40). A direct calculation in a coordinate system satisfying the conditions (5.56) and (5.57) at a given point, which imply that the derivatives of g vanish there, gives for this term

$$\begin{aligned}
&- k_B T \sum_{\nu\nu'} \sum_{j,k,l=1}^{d} (\delta_{\mu\nu} \boldsymbol{\delta} - \boldsymbol{\lambda}_\nu) \cdot \tilde{\mathsf{G}}_{lj} \frac{\partial \boldsymbol{R}_\nu}{\partial Q_l} \frac{\partial^2 \boldsymbol{R}_{\nu'}}{\partial Q_j \partial Q_k} \cdot \frac{\partial Q_k}{\partial \boldsymbol{R}_{\nu'}} \, dt \\
&+ k_B T \sum_{j,k=1}^{d} \tilde{\mathsf{G}}_{jk} \sum_\nu \sum_{l=1}^{d'} \frac{\partial \boldsymbol{R}_\mu}{\partial g_l} \frac{\partial g_l}{\partial \boldsymbol{r}_\nu} \cdot \frac{\partial^2 \boldsymbol{R}_\nu}{\partial Q_j \partial Q_k} \, dt \\
&+ k_B T \sum_{\mu'\nu\nu'} \mathbf{P}^T_{\mu'\nu} : \left(\frac{\partial}{\partial \boldsymbol{r}_{\mu'}} \boldsymbol{\Omega}_{\nu\nu'} \right) \cdot (\mathbf{P}^T_{\mu\nu'} - \boldsymbol{\lambda}^T_{\nu'}) \, dt .
\end{aligned}$$

In deriving the above equation, the derivative of $\tilde{\mathsf{G}}_{jk}$ has been expressed in terms of the derivative of its inverse matrix $\tilde{\mathsf{g}}_{jk}$. When the expression (5.7) for $\tilde{\mathsf{g}}_{jk}$ is differentiated then three terms result according to the product rule; the derivative of $\tilde{\boldsymbol{\zeta}}_{\mu\nu}$ has been evaluated by means of (5.22).

After eliminating the remaining derivatives with respect to generalized coordinates from this last expression by means of (5.58) and (5.60) and using it in (5.40) we obtain the final result

$$
\boxed{
\begin{aligned}
\mathrm{d}\boldsymbol{r}_\mu = {} & \sum_\nu \mathbf{P}_{\mu\nu} \cdot \Big\{ [\boldsymbol{v}_0 + \boldsymbol{\varkappa} \cdot \boldsymbol{r}_\nu] \mathrm{d}t \\
& \qquad + \sum_{\nu'} (\delta_{\nu\nu'} \zeta_\nu^{-1} + \boldsymbol{\Omega}_{\nu\nu'}) \cdot (\boldsymbol{F}_{\nu'} + \boldsymbol{F}_{\nu'}^{(\mathrm{e})} + \boldsymbol{F}_{\nu'}^{(\mathrm{m})}) \, \mathrm{d}t \\
& \qquad + \sqrt{2k_\mathrm{B}T} \sum_{\nu'} \mathbf{B}_{\nu\nu'} \cdot \mathrm{d}\boldsymbol{W}_{\nu'} \Big\} \\
& - k_\mathrm{B}T \sum_{j=1}^{d'} \sum_{\mu'\nu\nu'} \left[\mathbf{P}_{\nu\mu'} \cdot (\delta_{\mu'\nu'} \zeta_{\mu'}^{-1} + \boldsymbol{\Omega}_{\mu'\nu'}) \right] : \frac{\partial^2 g_j}{\partial \boldsymbol{r}_{\nu'} \partial \boldsymbol{r}_\nu} \frac{\partial \boldsymbol{R}_\mu}{\partial g_j} \, \mathrm{d}t \\
& - k_\mathrm{B}T \sum_{j=1}^{d'} \sum_{\mu'\nu\nu'} \mathbf{P}_{\mu\mu'} \cdot (\delta_{\mu'\nu'} \zeta_{\mu'}^{-1} + \boldsymbol{\Omega}_{\mu'\nu'}) \cdot \frac{\partial^2 g_j}{\partial \boldsymbol{r}_{\nu'} \partial \boldsymbol{r}_\nu} \cdot \frac{\partial \boldsymbol{R}_\nu}{\partial g_j} \, \mathrm{d}t \\
& + k_\mathrm{B}T \sum_{\mu'\nu\nu'} \mathbf{P}_{\mu'\nu}^T : \left(\frac{\partial}{\partial \boldsymbol{r}_{\mu'}} \boldsymbol{\Omega}_{\nu\nu'} \right) \cdot \mathbf{P}_{\mu\nu'}^T \, \mathrm{d}t \, .
\end{aligned}
}
\tag{5.61}
$$

In the second line of this equation we have introduced an extra "metric" force due to the constraints which depends on the determinant of the metric matrix in the space of constrained configurations,

$$
\boldsymbol{F}_\mu^{(\mathrm{m})} := \frac{1}{2} k_\mathrm{B}T \frac{\partial}{\partial \boldsymbol{r}_\mu} \ln \hat{\mathsf{G}} \, . \tag{5.62}
$$

Equation (5.61), together with the representation (5.54) for $\mathbf{P}_{\mu\nu}$, constitutes the desired reformulation of (5.40) in which generalized coordinates are eliminated in favor of the constraint conditions. Equation (5.61) constitutes a closed-form expression for the equations of motion in the Cartesian space into which the space of constrained configurations is naturally embedded (level A2 of Fig. 5.1). In the manifold of constrained configurations, the form of this equation is independent of the detailed way in which the constraints are formulated. In the absence of a flow field, the Fokker-Planck equation associated with (5.61) coincides with the one given at the bottom of p. 229 of [8] for inertial dynamics.[5]

[5]Hinch does not account for hydrodynamic interactions because he uses a diagonal friction matrix. If his ζ_{mn}^{-1} is formally replaced by $(\delta_{\mu\nu} \zeta_\mu^{-1} + \boldsymbol{\Omega}_{\mu\nu})$, then, surprisingly, his diffusion equation is of exactly the same form as the one associated with (5.61).

Exercise 5.14 Show that the last three lines of (5.61) can be written in the more compact form

$$k_{\mathrm{B}}T \sum_{\mu'\nu\nu'} \frac{\partial}{\partial \boldsymbol{r}_\nu} \cdot \left[\mathbf{P}_{\nu\nu'} \cdot (\delta_{\nu'\mu'} \zeta_{\nu'}^{-1} + \boldsymbol{\Omega}_{\nu'\mu'}) \cdot \mathbf{P}^T_{\mu\mu'} \right] \mathrm{d}t \,.$$

By using this expression, the Fokker-Planck equation associated with (5.61) can be written in a very elegant form.

Exercise 5.15 Determine the effect of the metric force on the relative motion of two identical beads by calculating $F^{(\mathrm{m})}_{\mathrm{rel}} := \left| \sum_{\mu\nu} (-1)^\mu \mathbf{P}_{\mu\nu} \cdot \boldsymbol{F}^{(\mathrm{m})}_\nu \right|$. Assume that there is only one constraint $g(\boldsymbol{r}_2 - \boldsymbol{r}_1) = 0$ and neglect hydrodynamic interaction. Evaluate $F^{(\mathrm{m})}_{\mathrm{rel}}$ for $g(\boldsymbol{r}_2 - \boldsymbol{r}_1) = (\boldsymbol{r}_2 - \boldsymbol{r}_1)^2 - L^2$.

5.1.4 Numerical Integration Schemes

There are various possibilities for simulating models with constraints. A first possibility is based on the numerical integration of the stochastic differential equations for the generalized coordinates Q_j. However, this possibility is feasible only if the number of generalized coordinates involved is small, that is, for almost rigid molecules. If there are many internal degrees of freedom and hence many generalized coordinates then, in general, handling them is very complicated. A closely related idea of formulating and simulating stochastic differential equations of motion for the unconstrained variables of a system of linked rigid bodies at equilibrium has been developed by Pear and Weiner [17, 18].

A second possibility for simulating models with constraints is based on the stochastic differential equations (5.61) in Cartesian space. The numerical integration schemes of Sect. 3.4 can be applied in order to obtain simulation algorithms. The Itô formula implies that $g_j(\boldsymbol{r}_1, \boldsymbol{r}_2, \ldots, \boldsymbol{r}_N)$ is an Itô process with $\mathrm{d}g_j = 0$. Of course, the approximate solution constructed from a numerical integration scheme will not satisfy the constraints rigorously. The order of strong convergence determines the order of the violation of the constraint conditions. Since all coefficients in (5.61) are defined also outside the space of constrained configurations it is not necessary to restore the constraints rigorously after each time step. For simulations over a large time interval, however, the accumulation of deviations from the space of constrained configurations should be avoided by intermediate steps in which the constraints are enforced.

We now focus our attention on a third possibility for simulating models with constraints which is related to the SHAKE-HI algorithm introduced in [2]. If the bead positions at some initial time are given by $\boldsymbol{r}_\mu$, then the positions after a time step of width Δt are constructed in two steps. First an unconstrained move is taken,

$$\begin{aligned} \bar{\boldsymbol{r}}_\mu &= \boldsymbol{r}_\mu + \left[\boldsymbol{v}_0 + \boldsymbol{\varkappa} \cdot \boldsymbol{r}_\mu + \sum_\nu (\delta_{\mu\nu} \zeta_\mu^{-1} + \boldsymbol{\Omega}_{\mu\nu}) \cdot (\boldsymbol{F}_\nu + \boldsymbol{F}^{(\mathrm{e})}_\nu + \boldsymbol{F}^{(\mathrm{m})}_\nu) \right] \Delta t \\ &\quad + \sqrt{2k_{\mathrm{B}}T} \sum_\nu \mathbf{B}_{\mu\nu} \cdot \Delta \boldsymbol{W}_\nu \,, \end{aligned} \qquad (5.63)$$

where $\Delta \boldsymbol{W}_\nu$ is the increment of the Wiener process $\boldsymbol{W}_\nu$ for the time step under consideration, and $\mathbf{B}_{\mu\nu}$ was introduced in (5.33). All the coefficients in (5.63) are evaluated with the configuration at the beginning of the time step. Then, the final bead positions are obtained as

$$\tilde{\boldsymbol{r}}_\mu = \bar{\boldsymbol{r}}_\mu - \sum_{j=1}^{d'} \gamma_j \left[\sum_\nu (\delta_{\mu\nu} \boldsymbol{\zeta}_\mu^{-1} + \boldsymbol{\Omega}_{\mu\nu}) \cdot \frac{\partial g_j}{\partial \boldsymbol{r}_\nu} \right]_c , \tag{5.64}$$

where $[\ldots]_c$ indicates that the corresponding term is evaluated at the positions $(1-c)\boldsymbol{r}_\mu + c\bar{\boldsymbol{r}}_\mu$ with $c \in [0,1]$. The set of Lagrange multipliers γ_j is to be determined such that all the constraints are satisfied rigorously, that is $g_j(\tilde{\boldsymbol{r}}_1, \tilde{\boldsymbol{r}}_2, \ldots, \tilde{\boldsymbol{r}}_N) = 0$.

In order to derive the stochastic differential equation corresponding to the iteration scheme (5.63) and (5.64) we do not need the exact solution for γ_j but only the contributions to γ_j which are of order $(\Delta t)^{1/2}$ and Δt. The precise form of these contributions can be obtained by expanding $g_j(\tilde{\boldsymbol{r}}_1, \tilde{\boldsymbol{r}}_2, \ldots, \tilde{\boldsymbol{r}}_N)$. If the results are combined with (5.63) and (5.64), then it is found that the above construction provides a numerical integration scheme for the stochastic differential equation

$$\begin{aligned} \mathrm{d}\boldsymbol{r}_\mu \; = \; & \sum_\nu \mathbf{P}_{\mu\nu} \cdot \Big\{ \Big[\boldsymbol{v}_0 + \boldsymbol{\varkappa} \cdot \boldsymbol{r}_\nu + \sum_{\nu'} (\delta_{\nu\nu'} \boldsymbol{\zeta}_\nu^{-1} + \boldsymbol{\Omega}_{\nu\nu'}) \cdot (\boldsymbol{F}_{\nu'} + \boldsymbol{F}_{\nu'}^{(\mathrm{e})} + \boldsymbol{F}_{\nu'}^{(\mathrm{m})}) \Big] \, \mathrm{d}t \\ & \qquad + \sqrt{2 k_\mathrm{B} T} \sum_{\nu'} \mathbf{B}_{\nu\nu'} \cdot \mathrm{d}\boldsymbol{W}_{\nu'} \Big\} \\ & - k_\mathrm{B} T \sum_{j=1}^{d'} \sum_{\mu'\nu\nu'} \Big[\mathbf{P}_{\nu\mu'} \cdot (\delta_{\mu'\nu'} \boldsymbol{\zeta}_{\mu'}^{-1} + \boldsymbol{\Omega}_{\mu'\nu'}) \Big] : \frac{\partial^2 g_j}{\partial \boldsymbol{r}_{\nu'} \partial \boldsymbol{r}_\nu} \frac{\partial \boldsymbol{R}_\mu}{\partial g_j} \, \mathrm{d}t \\ & - 2c \, k_\mathrm{B} T \sum_{\mu'\nu\nu'} \mathbf{P}_{\mu\nu} \cdot \sum_{j=1}^{d'} \left[\frac{\partial g_j}{\partial \boldsymbol{r}_{\mu'}} \cdot (\delta_{\mu'\nu'} \boldsymbol{\zeta}_{\mu'}^{-1} + \boldsymbol{\Omega}_{\mu'\nu'}) \cdot \frac{\partial}{\partial \boldsymbol{r}_{\nu'}} \right] \frac{\partial \boldsymbol{R}_\nu}{\partial g_j} \, \mathrm{d}t \, . \end{aligned} \tag{5.65}$$

Equation (5.65) is very similar to (5.61). More precisely, if the derivative of $\partial \boldsymbol{R}_\nu / \partial g_j$ in the last line of this equation is evaluated one recovers the stochastic differential equation (5.61) provided that $c = \frac{1}{2}$. The choice $c = \frac{1}{2}$ is crucial for obtaining terms involving derivatives of hydrodynamic-interaction tensors. The order of weak convergence of the algorithm (5.63) and (5.64) is $\upsilon = 1$. In the scheme of Fig. 5.1, we have now reached the lower left corner A3, starting from the upper right corner A4.

In the absence of hydrodynamic interactions and for identical beads with isotropic friction tensors one has

$$k_\mathrm{B} T \sum_{j=1}^{d'} \sum_\nu \frac{\partial^2 g_j}{\partial \boldsymbol{r}_{\nu'} \partial \boldsymbol{r}_\nu} \cdot \frac{\partial \boldsymbol{R}_\nu}{\partial g_j} = \boldsymbol{F}_{\nu'}^{(\mathrm{m})} , \tag{5.66}$$

so that the next-to-last line in (5.61) cancels the term involving $\boldsymbol{F}_{\nu'}^{(\mathrm{m})}$. In this situation, a valid simulation algorithm is obtained when the metric force $\boldsymbol{F}_{\nu'}^{(\mathrm{m})}$

in (5.63) is neglected and (5.64) is used with $c = 0$. Such an algorithm was employed by Liu for Kramers chains [19]. The SHAKE-HI algorithm suggested in [2] corresponds to setting $c = 0$ and neglecting $\boldsymbol{F}_{\nu'}^{(\mathrm{m})}$ even in the presence of hydrodynamic interactions. For this choice, (5.61) and (5.65) do not generally coincide, so that the SHAKE-HI algorithm does not reproduce the corresponding kinetic theory models.

The Lagrange multipliers γ_j in (5.64) are to be determined from a set of nonlinear equations. These equations need to be solved by an iterative procedure. For example, one can construct an iteration scheme by writing the constraints in the form

$$\gamma_j = \gamma_j + \sum_{k=1}^{d'} \left[\bar{\mathsf{g}}_{jk}\right]_{c'} g_k(\tilde{\boldsymbol{r}}_1, \tilde{\boldsymbol{r}}_2, \ldots, \tilde{\boldsymbol{r}}_N) \,, \tag{5.67}$$

where $[\ldots]_{c'}$ indicates that the corresponding term is evaluated at the positions $(1-c')\boldsymbol{r}_\mu + c'\tilde{\boldsymbol{r}}_\mu$ with $c' \in [0,1]$. Starting with $\gamma_j = 0$, successive approximations for γ_j can be generated by evaluating the right-hand side of (5.67) with the current approximate Lagrange multipliers γ_j until all constraints are satisfied within a specified tolerance (the vectors $\tilde{\boldsymbol{r}}_\mu$ depend on γ_j according to the definition (5.64)). Note that $[\bar{\mathsf{g}}_{jk}]_{c'}$ needs to be calculated only once in this iterative procedure. Liu [19] observed rapid convergence when he employed a special case of this iterative scheme in a simulation of Kramers chains (with $c = c' = 0$ and $\boldsymbol{F}_{\nu}^{(\mathrm{m})} = \boldsymbol{0}$). Fixman [6] had previously used the same idea for solving the quadratic constraint equations for models with fixed bond lengths and bond angles. The fact that $\overline{\mathsf{G}}_{jk}$ typically has a narrow band structure can be exploited to design very efficient algorithms for calculating metric forces and Lagrange multipliers [6].

A well-known alternative for solving the set of nonlinear equations for the Lagrange multipliers is used in the SHAKE-HI algorithm [2]. In that approach, only one Lagrange multiplier γ_j is determined at a time, such that the corresponding constraint $g_j = 0$ is enforced. Since the enforcement of a particular constraint partially destroys constraints that were enforced previously, it is necessary to repeat the cycle of enforcing all constraints until all constraint equations are satisfied within the specified tolerance [2].

We conclude this subsection with some heuristic or mnemonic remarks. The unconstrained move (5.63) is of the Itô type because all coefficients are evaluated at the beginning of the time step. We have seen in Sect. 4.2.1 that the Itô approach is most natural for models with hydrodynamic interaction. Therefore, only the occurrence of the metric forces $\boldsymbol{F}_{\nu}^{(\mathrm{m})}$ needs to be justified. To this end, it is important to realize that the metric force (5.62) is exactly the negative of the corrective force which needs to be applied in order to make a rigid system behave like a very stiff system when bead inertia is taken into account in evaluating the constraint forces [8].[6] In other words, if the forces

[6] Also in the kinetic theory derivation of the diffusion equation [1], the effects of constraints on the dynamics of the system are formulated before the momentum variables, that is the effects of bead inertia, are eliminated. This procedure is at variance with Fixman's approach [6], in which the constraints are introduced after eliminating bead inertia.

$\boldsymbol{F}_\nu^{(\mathrm{m})}$ in (5.63) are omitted then we obtain a simulation of an infinitely stiff system in the presence of external forces and hydrodynamic interaction;[7] the truly rigid system necessitates the occurrence of the metric forces $\boldsymbol{F}_\nu^{(\mathrm{m})}$ in the unconstrained move (5.63). In this sense, it is slightly easier to simulate infinitely stiff rather than rigid systems.

Use of $c = \frac{1}{2}$ in (5.64) corresponds to a Stratonovich type restoration of constraints; one may thus think of the mechanism for restoring constraints as being governed by the rules of deterministic calculus. It is most remarkable that these simple heuristic arguments on the level of a simulation algorithm reproduce the full complexity of the kinetic theory equations.

Exercise 5.16 Consider the rigid dumbbells with identical beads of Exercise 5.4. Determine the stochastic differential equation for $\boldsymbol{U} = (\boldsymbol{r}_2 - \boldsymbol{r}_1)/L$, which was previously obtained in Exercise 5.9, from (5.65). Show that $\boldsymbol{F}_{\nu'}^{(\mathrm{m})}$ and c drop out in the stochastic differential equation for $\boldsymbol{U}$, so that the metric forces $\boldsymbol{F}_{\nu'}^{(\mathrm{m})}$ can be ignored and c can be chosen arbitrarily. For $c = 1$, this observation implies that an unconstrained move followed by a rescaling step provides a valid simulation algorithm for dumbbells with equal beads and no hydrodynamic interaction.

Exercise 5.17 Consider the two-bead system of Exercise 5.15 with the constraint $g(\boldsymbol{Q}) = \tilde{g}(\boldsymbol{Q})\,[(Q_1^2/a^2) + (Q_2^2/b^2) - 1] = 0$ for $\boldsymbol{Q} := \boldsymbol{r}_2 - \boldsymbol{r}_1$. Show that $F_{\mathrm{rel}}^{(\mathrm{m})}$ depends on the choice of $\tilde{g}(\boldsymbol{Q})$. Calculate $F_{\mathrm{rel}}^{(\mathrm{m})}$ for $\tilde{g}(\boldsymbol{Q}) = 1$ and for $\tilde{g}(\boldsymbol{Q}) = [(Q_1^2/a^4) + (Q_2^2/b^4)]^{-1/2}$.

The results of this exercise show that, if the metric force in (5.61) is omitted in order to obtain an infinitely stiff system, the resulting model in general depends on the particular formulation of the constraints.

5.1.5 Stress Tensor

As for obtaining the diffusion equation, we rely on kinetic theory for obtaining the appropriate stress tensor for general bead-rod-spring models. If we specialize the *modified Kramers expression* for the stress tensor, (16.3-11) of [1], to a solution containing only one polymer species, the polymer contribution to the stress tensor $\boldsymbol{\tau}^p = \boldsymbol{\tau}^p(t)$ may be written in the form

$$\begin{aligned}\boldsymbol{\tau}^p &= n_\mathrm{p} k_\mathrm{B} T \sum_{j=1}^{d} \sum_{\mu} \left\langle \frac{1}{\sqrt{\mathrm{g}}} \frac{\partial}{\partial Q_j} \left(\sqrt{\mathrm{g}}\, \boldsymbol{R}_\mu \frac{\partial Q_j}{\partial \boldsymbol{R}_\mu} \right) \right\rangle \\ &\quad + n_\mathrm{p} \sum_{\mu\nu} \left\langle \boldsymbol{R}_\mu \left(\boldsymbol{F}_\nu + \boldsymbol{F}_\nu^{(\mathrm{e})} \right) \cdot \mathbf{P}_{\nu\mu} \right\rangle \end{aligned} \tag{5.68}$$

[7] By construction, the solutions of the stochastic differential equation (5.61) in the space of constrained configurations are independent of the formulation of the constraints, that is, independent of the choice of the functions g_j. When the metric forces are neglected, the solutions may depend on the particular formulation of the constraints. This unpleasant ambiguity is related to the previously mentioned fact that the limit of infinitely stiff systems depends on the details of the assumed confining potential.

$$-n_{\mathrm{p}} \sum_{\mu\mu'\nu} \left\langle \boldsymbol{R}_\mu \boldsymbol{R}_\nu \cdot \boldsymbol{\varkappa}^T \cdot \tilde{\boldsymbol{\zeta}}_{\nu\mu'} \cdot (\delta_{\mu'\mu} \boldsymbol{\delta} - \mathbf{P}_{\mu'\mu}) \right\rangle .$$

In order to obtain an expression for the stress tensor which is more useful in computer simulations we need to eliminate the generalized coordinates from the first line of (5.68). We can be avoid repeating the lengthy reformulations of Sect. 5.1.3 if we express the stress tensor in terms of the drift velocity for $\boldsymbol{R}_\mu$ which is defined as the deterministic contribution to the stochastic differential equation (5.36),

$$\begin{aligned} \boldsymbol{A}_\mu &:= \sum_{\mu'\nu} \mathbf{H}_{\mu\mu'} \cdot \left[(\delta_{\mu'\nu} \boldsymbol{\delta} - \boldsymbol{\lambda}_{\mu'}^T) \cdot (\boldsymbol{F}_\nu + \boldsymbol{F}_\nu^{(\mathrm{e})}) + \tilde{\boldsymbol{\zeta}}_{\mu'\nu} \cdot \boldsymbol{\varkappa} \cdot \boldsymbol{R}_\nu \right] \\ &\quad + k_{\mathrm{B}} T \sum_{j,k=1}^{d} \frac{1}{\sqrt{g}} \frac{\partial}{\partial Q_j} \left(\sqrt{g}\, \tilde{G}_{jk} \frac{\partial \boldsymbol{R}_\mu}{\partial Q_k} \right) . \end{aligned} \tag{5.69}$$

By combining (5.68) and (5.69) we can eliminate the terms involving generalized coordinates to obtain

$$\boxed{\begin{aligned} \boldsymbol{\tau}^{\mathrm{p}} &= n_{\mathrm{p}} \sum_{\mu\nu} \left\langle \boldsymbol{R}_\mu \tilde{\boldsymbol{\zeta}}_{\mu\nu} \cdot (\boldsymbol{A}_\nu - \boldsymbol{\varkappa} \cdot \boldsymbol{R}_\nu) \right\rangle \\ &+ n_{\mathrm{p}} k_{\mathrm{B}} T \left[\left(\sum_{\mu} \langle \mathbf{P}_{\mu\mu} \rangle \right) - \boldsymbol{\delta} - \sum_{\mu\mu'\nu\nu'} \left\langle \boldsymbol{R}_\mu \mathbf{P}_{\mu'\nu}^T : \left(\frac{\partial}{\partial \boldsymbol{r}_{\mu'}} \boldsymbol{\Omega}_{\nu\nu'} \right) \cdot \tilde{\boldsymbol{\zeta}}_{\nu'\mu} \right\rangle \right] . \end{aligned}}$$

(5.70)

We refer to (5.70) as the *modified Giesekus expression* for the stress tensor. Next, some comments on the limitations and usefulness of the modified Giesekus expression seem to be appropriate.

While the incorporation of *external forces* into the equations of motion for the polymer molecules is very natural, their presence is more problematic for the stress tensor. The deeper reason for this problem is that, in general, external forces lead to migration phenomena and inhomogeneous concentrations. In the terminology of [1], concentration gradients lead to Brownian forces, and a complicated interplay between the total Brownian and external forces on the polymer molecules results. Since these phenomena are not considered in the usual formulation of kinetic theory [1], (5.70) can safely be used only if the total external force on each molecule vanishes (indeed, this has been assumed in going from (5.68) to (5.70)). The fact that external forces do not explicitly occur in (5.70) might indicate that this expression could also be useful in the presence of arbitrary external forces. The modified Giesekus expression is closely related to (18.4-4) of [1] in which, however, the total Brownian and external forces are kept.

The last term in (5.70) involves derivatives of hydrodynamic-interaction tensors and hence vanishes if *hydrodynamic-interaction effects* are absent (or if they are treated by some averaging approximation). The remaining terms in the last line of (5.70) constitute a generalization of the term $(N-1)n_{\mathrm{p}}\,k_{\mathrm{B}}T\,\boldsymbol{\delta}$ which results in the absence of constraints; the *constraints* can lead to anisotropic momentum exchange due to internal bead motions.

The combination $\boldsymbol{A}_\nu - \boldsymbol{\kappa}\cdot\boldsymbol{R}_\nu$ occurring in the first term of (5.70) may be interpreted as a *convected derivative* of $\boldsymbol{R}_\nu$ because $\boldsymbol{A}_\nu$ was introduced as the drift velocity for $\boldsymbol{R}_\nu$. While in the usual Giesekus expression a convected derivative of an average tensor occurs (see Example 5.18 below), the convected derivative in the above modification acts only on a particular building block of the tensor to be averaged.

There are two important reasons why the modified Giesekus expression for the stress tensor is very useful. First, the results of Sect. 5.1.3 immediately allow us to express the drift velocity $\boldsymbol{A}_\mu$, and hence the stress tensor (5.70), in terms of the constraint conditions; there is no need to introduce generalized coordinates. Second, in a numerical integration scheme the drift velocity $\boldsymbol{A}_\nu$ in (5.70) may be replaced by the discrete approximation $[\boldsymbol{R}_\nu(t+\Delta t)-\boldsymbol{R}_\nu(t)]/\Delta t$, provided that all other terms in the average involving the drift velocity are evaluated at time t. This statement follows from the fact that, except for terms that vanish as $\Delta t \to 0$, the difference between these expressions is proportional to increments of the Wiener process which are independent of the polymer configurations at time t.

Example 5.18 Giesekus Expression for the Stress Tensor

For models with equilibrium averaged hydrodynamic interaction or no hydrodynamic interaction, the tensors $\tilde{\boldsymbol{\zeta}}_{\mu\nu} = \tilde{\zeta}_{\mu\nu}\,\boldsymbol{\delta}$ are independent of configuration and time. The Itô formula implies the following time-evolution equation for the *averaged structure tensor* $\mathbf{K} := \sum_{\mu\nu}\tilde{\zeta}_{\mu\nu}\,\langle\boldsymbol{R}_\mu\boldsymbol{R}_\nu\rangle$,

$$\frac{\mathrm{d}}{\mathrm{d}t}\mathbf{K} = \sum_{\mu\nu}\tilde{\zeta}_{\mu\nu}\left(\langle\boldsymbol{A}_\mu\boldsymbol{R}_\nu\rangle + \langle\boldsymbol{R}_\mu\boldsymbol{A}_\nu\rangle + 2k_{\mathrm{B}}T\,\langle\mathbf{H}_{\mu\nu}\rangle\right)\,. \tag{5.71}$$

Equation (5.71) establishes a relationship between the time derivative of an average tensor and averages involving the drift velocities $\boldsymbol{A}_\mu$. Since in the absence of external forces the stress tensor is symmetric [1], we can use (5.71) to rewrite the modified Giesekus expression of the stress tensor in the form

$$\boldsymbol{\tau}^{p} = \frac{1}{2}n_{\mathrm{p}}\left(\frac{\mathrm{d}}{\mathrm{d}t}\mathbf{K} - \boldsymbol{\kappa}\cdot\mathbf{K} - \mathbf{K}\cdot\boldsymbol{\kappa}^{T}\right)\,, \tag{5.72}$$

where the relationship between $\mathbf{H}_{\mu\nu}$ and $\mathbf{P}_{\mu\nu}$ has been used. Equation (5.72) is the familiar Giesekus expression for the stress tensor [20]. This expression is particularly useful for evaluating stresses in steady state flows. □

The theory of models with constraints in this chapter has been developed only for homogeneous flows. For slowly varying velocity gradients and concentrations, the formalism presented in Chap. 18 of [1] corresponds to introducing a position-dependence of the polymer contribution to the stress tensor in (5.68) and (5.70) by (i) replacing n_p by $n_p(\boldsymbol{r}, t)$, (ii) evaluating the velocity gradients $\boldsymbol{\varkappa}$ at the center of mass position $\boldsymbol{r}_c$, and (iii) replacing all averages $\langle \cdots \rangle$ by conditional expectations $E(\cdots \mid \boldsymbol{r}_c = \boldsymbol{r})$.

Exercise 5.19 Obtain a constitutive equation for the polymer contribution to the stress tensor for Hookean dumbbells without hydrodynamic interaction by expressing the averaged structure tensor in (5.72) in terms of the stress tensor. Compare the result to (4.20).

5.2 Rigid Rod Models

The simplest models with constraints are those which represent completely rigid objects. We here consider only linear rigid objects, or rodlike molecules, such as isotactic polypropylene, proteins in helical forms, DNA in its helix configuration, or the tobacco mosaic virus. Here, the rigid dumbbell model is of central interest because its generalization to multibead rods consisting of identical, equally spaced beads is possible through a simple change of time constants. Even the effect of hydrodynamic interactions on internal motions can be introduced through a change of time constants where, in this case, different time constants appear in the diffusion equation and in one of the contributions to the stress tensor (see Chap. 14 of [1]). All these time constants related to the rotational behavior of multibead rods are proportional to N^3, with logarithmic corrections in the presence of hydrodynamic interaction (see Sect. 8.2 of [21]).

Although only two generalized coordinates are required for characterizing the internal configuration of rigid objects, that is their global orientation, the rigid dumbbell model is so complex that only few of its properties can be determined analytically. In general, one has to resort to numerical methods. Since only two degrees of freedom are involved, numerical solution of the underlying diffusion equation is feasible; this has been done for both types of models considered in the following, namely for dilute solutions of rigid dumbbells [22] and for liquid crystal polymers [23]. Brownian dynamics simulations provide a much simpler alternative for treating these rigid rod models and, for less constrained models with more internal degrees of freedom, they constitute the only available, rigorous approach. In the analytical approach, the risk of introducing approximations to obtain tractable models has been experienced in modeling liquid crystal polymers, where decoupling approximations for certain moments can change the predicted behavior drastically. Due to the importance of models with constraints for biopolymers and for high-performance polymeric materials, the crucial role of Brownian dynamics simulations is self-evident.

5.2.1 Dilute Solutions of Rod-like Molecules

The most important exact results for dilute solutions of rigid dumbbells are expansions for small strain rates, which exist even for time-dependent flows (memory-integral expansion) [1]. In particular, the linear viscoelastic behavior is known to be characterized by a relaxation modulus which is a single exponential. In simple extensional flow, there exists a closed-form expression for the extensional viscosity [1]. In steady shear flow, asymptotic power laws for the viscometric functions have been established [22].

The diffusion equation for dilute solutions of rigid dumbbells without hydrodynamic interaction in an external force field in terms of generalized coordinates has been found in Exercise 5.4, where the stochastic differential equations for the generalized coordinates have also been obtained from the diffusion equation. In order to obtain a more compact formulation of the diffusion equation in generalized coordinates one can introduce a shorthand vector notation (see Chap. 14 of [1]). The stochastic differential equation for the orientation vector of rigid dumbbells in Cartesian space has been derived in Exercise 5.9. A numerical integration scheme of first order of weak convergence is obtained when an unconstrained move is combined with a subsequent rescaling step (see Exercise 5.16).

The stochastic differential equation for rigid dumbbells is a special case of the following system of Itô equations in d-dimensional space,

$$\begin{aligned} \mathrm{d}\boldsymbol{X}_t &= \left[\left(\boldsymbol{\delta} - \frac{\boldsymbol{X}_t\boldsymbol{X}_t}{X_t^2}\right)\cdot\boldsymbol{A}(t,\boldsymbol{X}_t) - \frac{d-1}{2}\,B^2\,\frac{\boldsymbol{X}_t}{X_t^2}\right]\,\mathrm{d}t \\ &+ B\left(\boldsymbol{\delta} - \frac{\boldsymbol{X}_t\boldsymbol{X}_t}{X_t^2}\right)\cdot\mathrm{d}\boldsymbol{W}_t, \end{aligned} \tag{5.73}$$

for which $\mathrm{d}\boldsymbol{X}_t^2 = 0$ so that the constraint of constant length L is satisfied. More precisely, the rigid dumbbell model introduced in Exercises 5.4 and 5.9 is obtained for $d = 3$,

$$\begin{aligned} \boldsymbol{A}(t,\boldsymbol{x}) &= \boldsymbol{\kappa}(t)\cdot\boldsymbol{x} + \frac{1}{\zeta}\left(\boldsymbol{F}_2^{(\mathrm{e})} - \boldsymbol{F}_1^{(\mathrm{e})}\right) = \boldsymbol{\kappa}(t)\cdot\boldsymbol{x} - \frac{2\,k_\mathrm{B}T}{\zeta}\frac{\partial V^{(\mathrm{e})}(\boldsymbol{x})}{\partial\boldsymbol{x}}, \\ B^2 &= \frac{L^2}{3\lambda}. \end{aligned}$$

The polymer contribution to the stress tensor for this three-dimensional model in terms of the orientation vector $\boldsymbol{U}_t := \boldsymbol{X}_t/L$ is

$$\begin{aligned} \boldsymbol{\tau}^\mathrm{p}(t) &= n_\mathrm{p}k_\mathrm{B}T\left[\boldsymbol{\delta} - 3\left\langle\boldsymbol{U}_t\boldsymbol{U}_t\right\rangle - 6\lambda\boldsymbol{\kappa}(t) : \left\langle\boldsymbol{U}_t\boldsymbol{U}_t\boldsymbol{U}_t\boldsymbol{U}_t\right\rangle\right] \\ &+ \frac{1}{2}n_\mathrm{p}L\left\langle\boldsymbol{U}_t\,(\boldsymbol{\delta} - \boldsymbol{U}_t\boldsymbol{U}_t)\cdot\left(\boldsymbol{F}_2^{(\mathrm{e})} - \boldsymbol{F}_1^{(\mathrm{e})}\right)\right\rangle. \end{aligned} \tag{5.74}$$

Exercise 5.20 Derive the expression (5.74) for the polymer contribution of the stress tensor in dilute rigid dumbbell solutions.

Exercise 5.21 In the absence of external forces, derive the Giesekus expression,

$$\boldsymbol{\tau}^{p}(t) = 3n_{\mathrm{p}}k_{\mathrm{B}}T\,\lambda\left[\frac{\mathrm{d}}{\mathrm{d}t}\langle \boldsymbol{U}_t\boldsymbol{U}_t\rangle - \boldsymbol{\kappa}(t)\cdot\langle \boldsymbol{U}_t\boldsymbol{U}_t\rangle - \langle \boldsymbol{U}_t\boldsymbol{U}_t\rangle\cdot\boldsymbol{\kappa}^T(t)\right],$$

for the stress tensor in dilute rigid dumbbell solutions.

A first-order weak integration scheme for the general problem (5.73) is obtained by applying the Euler scheme to the corresponding unconstrained equation

$$\mathrm{d}\boldsymbol{X}_t = \boldsymbol{A}(t,\boldsymbol{X}_t)\,\mathrm{d}t + B\,\mathrm{d}\boldsymbol{W}_t,$$

and rescaling the result after every step. We next address the problem of developing a second-order scheme. Of course, we can simply apply the general second-order scheme (3.121) to the system (5.73) with multiplicative noise. However, the result obtained after a discrete time step does not strictly satisfy the constraint of constant length. It can be shown rigorously that, if the result of the second-order scheme (3.121) applied to (5.73) is rescaled, one obtains a second-order scheme which strictly satisfies the constraints. It seems to be possible to construct a simpler second-order scheme for (5.73) consisting of the following steps: (i) application of a second-order predictor-corrector scheme to the unconstrained problem with additive noise; (ii) addition of suitable terms of order $\Delta\boldsymbol{W}_j\,\Delta t_j$ and Δt_j^2; (iii) rescaling.

In the absence of external forces, the second-order scheme (3.121) to be followed by rescaling for rigid dumbbells in a time-independent flow is,

$$\begin{aligned}
\widetilde{\boldsymbol{U}}_{j+1} &= \widetilde{\boldsymbol{U}}_j + \frac{1}{\sqrt{3\lambda}}(\boldsymbol{\delta} - \widetilde{\boldsymbol{U}}_j\widetilde{\boldsymbol{U}}_j)\cdot\Delta\boldsymbol{W}_j + (\boldsymbol{\delta} - \widetilde{\boldsymbol{U}}_j\widetilde{\boldsymbol{U}}_j)\cdot\boldsymbol{\kappa}\cdot\widetilde{\boldsymbol{U}}_j\,\Delta t_j \\
&- \frac{1}{6\lambda}\Big[(\Delta\boldsymbol{W}_j)^2\widetilde{\boldsymbol{U}}_j + (\widetilde{\boldsymbol{U}}_j\cdot\Delta\boldsymbol{W}_j)\,\Delta\boldsymbol{W}_j \\
&\qquad\qquad - 2\,(\widetilde{\boldsymbol{U}}_j\cdot\Delta\boldsymbol{W}_j)^2\,\widetilde{\boldsymbol{U}}_j - \Delta\mathbf{V}_j\cdot\widetilde{\boldsymbol{U}}_j\Big] \qquad (5.75)\\
&+ \frac{1}{2\sqrt{3\lambda}}\Big[-\frac{2}{3\lambda}\boldsymbol{\delta} + \frac{4}{3\lambda}\widetilde{\boldsymbol{U}}_j\widetilde{\boldsymbol{U}}_j + \boldsymbol{\kappa} - \widetilde{\boldsymbol{U}}_j\widetilde{\boldsymbol{U}}_j\cdot\boldsymbol{\kappa} - 2\boldsymbol{\kappa}\cdot\widetilde{\boldsymbol{U}}_j\widetilde{\boldsymbol{U}}_j \\
&- 2\widetilde{\boldsymbol{U}}_j\widetilde{\boldsymbol{U}}_j\cdot\boldsymbol{\kappa}^T - (\widetilde{\boldsymbol{U}}_j\cdot\boldsymbol{\kappa}\cdot\widetilde{\boldsymbol{U}}_j)\,\boldsymbol{\delta} + 5\,(\widetilde{\boldsymbol{U}}_j\cdot\boldsymbol{\kappa}\cdot\widetilde{\boldsymbol{U}}_j)\,\widetilde{\boldsymbol{U}}_j\widetilde{\boldsymbol{U}}_j\Big]\cdot\Delta\boldsymbol{W}_j\,\Delta t_j \\
&+ \frac{1}{2}\Big\{\frac{1}{9\lambda^2}\widetilde{\boldsymbol{U}}_j - \frac{1}{3\lambda}(3\boldsymbol{\kappa} + \boldsymbol{\kappa}^T)\cdot\widetilde{\boldsymbol{U}}_j \\
&\qquad + \frac{7}{3\lambda}(\widetilde{\boldsymbol{U}}_j\cdot\boldsymbol{\kappa}\cdot\widetilde{\boldsymbol{U}}_j)\,\widetilde{\boldsymbol{U}}_j + \boldsymbol{\kappa}^2\cdot\widetilde{\boldsymbol{U}}_j - 2\,(\widetilde{\boldsymbol{U}}_j\cdot\boldsymbol{\kappa}\cdot\widetilde{\boldsymbol{U}}_j)\,\boldsymbol{\kappa}\cdot\widetilde{\boldsymbol{U}}_j \\
&\qquad + 3\,(\widetilde{\boldsymbol{U}}_j\cdot\boldsymbol{\kappa}\cdot\widetilde{\boldsymbol{U}}_j)^2\,\widetilde{\boldsymbol{U}}_j - [\widetilde{\boldsymbol{U}}_j\cdot(\boldsymbol{\kappa} + \boldsymbol{\kappa}^T)\cdot\boldsymbol{\kappa}\cdot\widetilde{\boldsymbol{U}}_j]\,\widetilde{\boldsymbol{U}}_j\Big\}\,\Delta t_j^2,
\end{aligned}$$

where $\widetilde{\boldsymbol{U}}_j := \boldsymbol{U}_{t_j}$. Although (5.75) looks rather lengthy, its numerical simulation can be handled quite efficiently.

Exercise 5.22 Develop a routine for a second-order Brownian dynamics simulation of rigid dumbbells in steady shear flow based on (5.75) and rescaling. For $\lambda\dot{\gamma} = 1$,

Table 5.1. Estimates for the numerical coefficients in (5.76)–(5.78) obtained from Brownian dynamics simulations by assuming that these asymptotic results for the viscometric functions hold at the given shear rates (the numbers in parentheses represent the statistical error in the last figure displayed)

$\lambda_h\dot{\gamma}$	c_0	c_1	c_2
100	0.6771(1)	1.1930(2)	0.908(1)
200	0.6706(1)	1.1986(2)	0.972(1)
500	0.6649(2)	1.2006(5)	1.042(1)
1000	0.6616(4)	1.2029(5)	1.079(2)
2000	0.6596(5)	1.2022(6)	1.109(2)
5000	0.6584(3)	1.2037(5)	1.138(2)
10000	0.6584(3)	1.2031(5)	1.157(2)

compare the results for the viscometric functions obtained from the expression (5.74) and from the Giesekus expression for the stress tensor (see Exercise 5.21).

Brownian dynamics simulations are a simple and valuable tool for calculating the viscometric functions for rigid dumbbells. Compared to the numerical solution of the diffusion equation by Galerkin's method [22], the implementation is very simple and, moreover, simulations give much better results at high shear rates. We illustrate the latter point by considering the asymptotic results for the viscometric functions at high shear rates [22],

$$\frac{\eta_{\mathrm{p}}}{n_{\mathrm{p}} k_{\mathrm{B}} T \lambda_h} \approx c_0 \frac{1-h}{1-2h} \frac{1}{(\lambda_h \dot{\gamma})^{1/3}} \tag{5.76}$$

$$\frac{\Psi_1}{n_{\mathrm{p}} k_{\mathrm{B}} T \lambda_h^2} \approx c_1 \frac{1-h}{1-2h} \frac{1}{(\lambda_h \dot{\gamma})^{4/3}} \tag{5.77}$$

$$\frac{\Psi_2}{n_{\mathrm{p}} k_{\mathrm{B}} T \lambda_h^2} \approx c_2 \frac{h}{1-2h} \frac{1}{(\lambda_h \dot{\gamma})^{7/3}} \tag{5.78}$$

where $h := \zeta/(8\pi\eta_{\mathrm{s}}L)$ is the hydrodynamic-interaction parameter and $\lambda_h := \lambda/(1-h)$ is a modified time constant. For the numerical coefficients c_0, c_1, and c_2, Stewart and Sørensen [22] give the extrapolated values

$$c_0 = 0.678\,, \qquad c_1 = 1.20\,, \qquad c_2 = 0.93\,. \tag{5.79}$$

We here consider hydrodynamic interaction because, for rigid dumbbells, all the effects of hydrodynamic interaction can be incorporated into the definition of time constants so that no additional complications arise. In particular, the numerical coefficients c_0, c_1, and c_2 in (5.76)–(5.78) can be determined by simulating rigid dumbbells without hydrodynamic interaction [1].

The numerical calculations in [22] were limited to $\lambda_h\dot{\gamma} \leq 100$, where the highest precision was obtained for $\lambda_h\dot{\gamma} \leq 30$. Via Brownian dynamics simulations, it is no problem to obtain accurate results for $\lambda_h\dot{\gamma} \geq 100$. For the determination of the numerical coefficients c_0, c_1, and c_2 we employed the second-order algorithm (5.75) with rescaling, time-step extrapolation, and the expression (5.74) for the stress tensor. In Table 5.1 the simulation results for the coefficients in (5.76)–(5.78) evaluated by assuming that the asymptotic results hold at a given shear rate are listed for $100 \leq \lambda_h\dot{\gamma} \leq 10000$. For comparison, Stewart and Sørensen give the values 0.681, 1.2, and 0.9 for $\lambda_h\dot{\gamma} = 100$, where the last digit of each of these values had to be obtained by extrapolation of results for smaller shear rates. Obviously, Brownian dynamics allows us a much more precise determination of the coefficients in (5.76)–(5.78) because very precise results can be obtained at very high shear rates. By extrapolating the results of Table 5.1 we obtain

$$c_0 = 0.6584(2)\,, \qquad c_1 = 1.2034(4)\,, \qquad c_2 = 1.223(2)\,. \tag{5.80}$$

These extrapolations have been performed with the subroutine TEXTRA (see solution to Exercise 4.11). We replaced Δt in the subroutine TEXTRA by $(\lambda_h\dot{\gamma})^{-1}$ or by $(\lambda_h\dot{\gamma})^{-1/3}$. For c_0 and c_1, we obtained the same results for both choices; for c_2, the extrapolation worked only for the latter choice. The evaluation of the coefficient $c_1 = \pi^{1/2}/[2^{2/3}\Gamma(7/6)] \approx 1.20357$ [24], which is the only coefficient known analytically, seems to be least subtle.

An alternative construction of higher-order algorithms which rigorously satisfy the constraints can be based on the stochastic differential equations for the generalized coordinates. For the rigid dumbbell model, such equations have been formulated in Exercise 5.4. The singular behavior of (5.28) for $z = 1$ may lead to problems in developing higher-order algorithms. A simple way around such problems is to use two different spherical coordinate systems, where the poles of one are in the equatorial plane of the other one.

5.2.2 Liquid Crystal Polymers

Liquid crystal polymers are of great interest in material science because extremely strong fibers and films can be produced from such materials. This interest is underlined by the numerous and extensive sessions on liquid crystal polymers in general rheology meetings (for recent proceedings see, e.g., J. Rheol. 38 (1994) 1471–1638). In general, liquid crystals are very complex materials intermediate between liquids, which have no long-range order, and crystalline solids, which possess long-range three-dimensional positional order. The simplest liquid crystals have long-range orientational order, but no long-range positional order [25]. The average molecular orientation is usually characterized by a *director*, where the average is taken spatially over a region large enough to contain many molecules.

Most of the recent work on liquid crystal polymers pertains to solutions and melts of stiff, rod-like molecules. Doi formulated a theory of the dynamics of liquid crystal polymers in flow by incorporating a mean field, excluded-volume interaction potential between different rods into the equation of motion for noninteracting rigid rods discussed in the preceding subsection [26]. Since the interpretation of diffusion equations on the unit sphere is delicate and an elegant shorthand notation is often used in a somewhat uncritical manner, we here prefer to give the stochastic differential equation of the Doi model for a configuration vector $\boldsymbol{X}_t$ in three-dimensional Cartesian space,

$$\begin{aligned} \mathrm{d}\boldsymbol{X}_t &= \left(\boldsymbol{\delta} - \frac{\boldsymbol{X}_t\boldsymbol{X}_t}{X_t^2}\right) \cdot \left\{ \left[\boldsymbol{\kappa}(t) \cdot \boldsymbol{X}_t - \frac{1}{2} B^2(\boldsymbol{X}_t) \frac{\partial V_{\mathrm{ev}}(\boldsymbol{X}_t)}{\partial \boldsymbol{x}} + \frac{1}{2} \frac{\partial B^2(\boldsymbol{X}_t)}{\partial \boldsymbol{x}} \right] dt \right. \\ &\quad \left. + B(\boldsymbol{X}_t)\,\mathrm{d}\boldsymbol{W}_t \right\} - B^2(\boldsymbol{X}_t) \frac{\boldsymbol{X}_t}{X_t^2}\,\mathrm{d}t \,. \qquad (5.81) \end{aligned}$$

This Itô stochastic differential equation is formulated such that $\mathrm{d}\boldsymbol{X}_t^2 = 0$ and that, for $\boldsymbol{\kappa}(t) = \boldsymbol{0}$, $V_{\mathrm{ev}}(\boldsymbol{x}) = 0$ and arbitrary $B(\boldsymbol{x})$, any probability distribution depending only on the length of the configuration vector solves the time-independent diffusion equation.

The physical content of (5.81) is most easily understood by comparing it to the equation of motion for rigid dumbbells of length L in dilute solution. The parameter $B = L/\sqrt{3\lambda}$ is replaced by the configuration-dependent quantity $B(\boldsymbol{X}_t)$, and the role of the dimensionless external potential $V^{(\mathrm{e})}$ is taken by V_{ev}. The potential V_{ev} describes the effective interaction of a rigid rod with other rods, which is certainly important in liquid crystal polymers. The configuration dependence of B accounts for the fact that the rotational diffusivity of an individual rod can depend on its orientation with respect to the other rods in the system. In the Doi theory, the form of these coefficient functions can most conveniently be expressed in terms of the dimensionless auxiliary function

$$g(\boldsymbol{x}) := \frac{\langle |\boldsymbol{x} \times \boldsymbol{X}_t| \rangle}{L^2} \,. \qquad (5.82)$$

One then has (see Sects. 9.2.4 and 10.2.2 of [21] or 10.2.1 and 10.3.1 of [25])

$$V_{\mathrm{ev}}(\boldsymbol{x}) := U_{\mathrm{ev}}\, g(\boldsymbol{x}) \,, \qquad (5.83)$$

$$B(\boldsymbol{x}) := \frac{\pi L \sqrt{2 D_{\mathrm{r}}}}{4 g(\boldsymbol{x})} \,, \qquad (5.84)$$

where the dimensionless strength of the potential U_{ev}, which is proportional to the rod concentration, and the rotational diffusivity D_{r} occur as new parameters. Notice that the coefficient functions $V_{\mathrm{ev}}(\boldsymbol{x})$ and $B(\boldsymbol{x})$ occurring in (5.81) involve averages, thus underlining the mean field type incorporation of the interactions with other rods, and of the hindrance of rotational motion due to other rods. The so-called *Onsager potential* (5.83) is sometimes replaced by the *Maier-Saupe potential* (see Sect. 10.3.3 of [25])

$$V_{\mathrm{ev}}(\boldsymbol{x}) := -\frac{3}{2L^4}\, U_{\mathrm{ev}}\, \langle \boldsymbol{X}_t \boldsymbol{X}_t \rangle : \boldsymbol{x}\boldsymbol{x} \,, \qquad (5.85)$$

and the configuration dependence of $B(\boldsymbol{x})$ is usually avoided by replacing $g(\boldsymbol{x})$ in (5.84) by the average $\langle g(\boldsymbol{X}_t)\rangle$, or even by a given time-independent constant. For the Maier-Saupe potential, a self-consistent field argument shows that the stress tensor of the Doi theory can be written in the form

$$\boldsymbol{\tau}(t) = n_{\mathrm{p}} k_{\mathrm{B}} T \left\{ \boldsymbol{\delta} - 3\left\langle \boldsymbol{U}_t \boldsymbol{U}_t \right\rangle - \left\langle \boldsymbol{U}_t \left(\boldsymbol{\delta} - \boldsymbol{U}_t \boldsymbol{U}_t\right) \cdot \frac{\partial V_{\mathrm{ev}}(\boldsymbol{U}_t)}{\partial \boldsymbol{u}} \right\rangle \right\} , \qquad (5.86)$$

where $\boldsymbol{U}_t := \boldsymbol{X}_t / L$ characterizes the rod orientation. The expression (5.86) can be obtained from (5.74) by neglecting the solvent contribution (that is, removing the superscript p) and the viscous contribution (that is, the quadrad term involving $\boldsymbol{\kappa}(t)$), and by letting the mean field potential again play the role of an external potential. Although this seems to be a natural assumption for semidilute and more concentrated rod systems (see Chap. 10 of [25]), experiments show that the viscous contribution is essential, in particular at high shear rates [27, 28]. A systematic kinetic theory derivation of the equation of motion and the expression for the stress tensor shows that the viscous contribution can occur in (5.86) with an arbitrary prefactor between 0 and 1 [27] (this prefactor is called σ in [27]).

The full model, consisting of the equation of motion (5.81) and the stress tensor expression (5.86), is very complicated. Because of the occurrence of averages in the excluded-volume potential and in the diffusion term, the corresponding Fokker-Planck equation is even nonlinear in the probability density and hence leads to the complications previously discussed in Sects. 3.3.4 and 4.3.2. Of course, such complications are a reasonable price to pay for simplifying a many-rod theory in a mean field type manner. Originally, Doi derived closed approximate equations for the second moments and the stress tensor by assuming a constant diffusion term, using the Maier-Saupe potential, and making a decoupling approximation for the fourth moments. The decoupling approximation qualitatively changes some model predictions. Only when avoiding this approximation has it been found that, in steady shear flow, with increasing shear rate the director first performs time-periodic rotations ("tumbling"), then oscillates between two limiting angles ("wagging"), and only at very high shear rates reaches a time-independent steady state orientation ("flow aligning") [23, 29]. The occurrence of tumbling and wagging is crucial for understanding the experimentally observed fact that the first normal stress difference is positive at low shear rates, negative at intermediate shear rates, and positive again at high shear rates. When compared to the first normal stress difference, the second normal stress difference has been found to be usually of the opposite sign [23].

In view of the drastic consequences of decoupling approximations, numerical integration of (5.81) provides an important alternative approach to investigating the dynamic behavior of liquid crystal polymers. Actually, we constructed a rough phase diagram for the model with Maier-Saupe potential and constant rotational diffusion coefficient in steady shear flow by means of simulations in early 1990. The simulation of weakly interacting processes (see Sect. 3.3.4)) was performed in generalized coordinates and, as an initial condition, the director was always assumed to be in the plane of shear. When the numerical solution of the diffusion equation for a model with Onsager potential and self-consistently averaged diffusion coefficient appeared [23], the construction of a detailed phase diagram by means of simulations was not continued. A much earlier simulation of liquid crystal polymers in shear flow, based on an iteration scheme for treating mean field interactions rather than on the simulation of weakly interacting processes, was unfortunately restricted to the flow aligning regime [30]. In a simulation, it may be difficult to distinguish precisely between tumbling and wagging (an oscillating director may flip over from time to time and hence resemble a tumbling one) and between wagging and flow aligning (random director fluctuations around a fixed direction need to be distinguished from oscillations).

The phase diagram of the full Doi model in shear flow is even richer than discussed above. The existence of stable "log rolling" and "kayaking" states, for which the director is not in the plane of shear, becomes obvious when initial configurations are used for which the director is not limited to the plane of shear. The existence of such states, even when the self-consistent averaging of the diffusion coefficient is avoided, has been established for the model with Onsager potential by stochastic simulations [31]. These simulations are much more time consuming than simulations for the model with Maier-Saupe potential and configuration-independent diffusion coefficient because, in the latter situation, only a single ensemble average is required for constructing all trajectories of the ensemble. For the Onsager potential or configuration-dependent diffusion, an ensemble average needs to be evaluated for each trajectory, and the simulation of only a few 10 000 trajectories becomes extremely expensive.

Simulation techniques can easily handle more complicated models which cannot be treated by other methods. For example, a tensorial description of the configuration-dependent rotational diffusivity is unproblematic. Also models of flexible liquid crystal polymers can be treated by simulation methods after introducing some internal degrees of freedom into models with constraints.

In a simulation, the finite size of the ensemble may have a systematic effect on the results. For example, at start-up of steady shear flow, otherwise undamped oscillations may be damped due to a loss of synchronization. Such a damping is also observed experimentally, and it is believed to be associated with the defect texture in liquid crystal polymers. The defects have been neglected in the model presented here although it is known that the interplay between flow behavior and texture is not only complicated but also very important. In other words, simulations suggest that finite ensemble size (or, fluctuations in the director orientation) might be related to texture effects. A similar idea has been developed by Marrucci and Maffettone in formulating a polydomain theory by introducing a Fokker-Planck equation for the fluctuating director [32]. The ensemble size or the amplitude of the director fluctuations is related to the elastic energy associated with spatial variations in the director orientation (*Frank elasticity*). A mesoscopic theory for the rheological properties of textured liquid crystal polymers at low flow rates was developed by Larson and Doi [33].

References

1. Bird RB, Curtiss CF, Armstrong RC, Hassager O (1987) *Dynamics of Polymeric Liquids, Vol 2, Kinetic Theory*, 2nd Edn. Wiley-Interscience, New York Chichester Brisbane Toronto Singapore
2. Allison SA, McCammon JA (1984) Biopolymers 23: 167
3. Öttinger HC (1994) Phys. Rev. E 50: 2696
4. van Kampen NG, Lodder JJ (1984) Am. J. Phys. 52: 419
5. van Kampen NG (1981) Appl. Sci. Res. 37: 67
6. Fixman M (1978) J. Chem. Phys. 69: 1527
7. Rallison JM (1979) J. Fluid Mech. 93: 251
8. Hinch EJ (1994) J. Fluid Mech. 271: 219
9. Díaz FG, García de la Torre J (1988) J. Chem. Phys. 88: 7698
10. Allison SA (1986) Macromolecules 19: 118
11. Elworthy KD (1982) *Stochastic Differential Equations on Manifolds.* Cambridge University Press, Cambridge London New York New Rochelle Melbourne Sydney (London Mathematical Society Lecture Note Series, Vol 70)
12. Elworthy D (1988) in: Hennequin PL (ed) *École d'Été de Probabilités de Saint-Flour XV-XVII, 1985-87.* Springer, Berlin Heidelberg New York, p 277 (Lecture Notes in Mathematics, Vol 1362)
13. Rogers LCG, Williams D (1987) *Diffusions, Markov Processes, and Martingales, Vol 2, Itô Calculus.* Wiley, Chichester New York Brisbane Toronto Singapore (Wiley Series in Probability and Mathematical Statistics)
14. Graham R A 109, 209-212 (1985).
15. Öttinger HC (1991) J. Rheol. 35: 1275
16. Weinberg S (1972) *Gravitation and Cosmology, Principles and Applications of the General Theory of Relativity.* Wiley, New York London Sydney Toronto, Sect. 3.2
17. Pear MR, Weiner JH (1979) J. Chem. Phys. 71: 212
18. Pear MR, Weiner JH (1980) J. Chem. Phys. 72: 3939
19. Liu TW (1989) J. Chem. Phys. 90: 5826
20. Giesekus H (1962) Rheol. Acta 2: 50
21. Doi M, Edwards SF (1986) *The Theory of Polymer Dynamics.* Clarendon Press, Oxford (International Series of Monographs on Physics, Vol 73)
22. Stewart WE, Sørensen JP (1972) Trans. Soc. Rheol. 16: 1
23. Larson RG (1990) Macromolecules 23: 3983
24. Öttinger HC (1988) J. Rheol. 32: 135
25. Larson RG (1988) *Constitutive Equations for Polymer Melts and Solutions.* Butterworths, Boston London Singapore Sydney Toronto Wellington (Butterworths Series in Chemical Engineering)
26. Doi M (1981) J. Polym. Sci. Polym. Phys. Ed. 19: 229
27. Bhave AV, Menon RK, Armstrong RC, Brown RA (1993) J. Rheol. 37: 413
28. Smyth SF, Mackay ME (1994) J. Rheol. 38: 1549
29. Marrucci G, Maffettone PL (1989) Macromolecules 22: 4076
30. Honerkamp J, Seitz R (1987) J. Chem. Phys. 87: 3120
31. Larson RG, Öttinger HC (1991) Macromolecules 24: 6270
32. Marrucci G, Maffettone PL (1990) J. Rheol. 34: 1231
33. Larson RG, Doi M (1991) J. Rheol. 35: 539

6. Reptation Models for Concentrated Solutions and Melts

Concentrated polymer solutions and polymer melts are extremely complex many-particle systems. For that reason, it is unquestionably important to make both ingenious and far-reaching assumptions in describing the dynamics of polymers in such undiluted systems, and to make use of computer simulations. A widely and successfully applied class of molecular models for the polymer dynamics in concentrated solutions and melts relies on the notion of *reptational motion* [1]. The first reptation theory for the rheology of undiluted polymers was developed in a series of papers by Doi and Edwards published in 1978 and 1979 [2–5]. The *Doi-Edwards model* is based on the assumption that each polymer in a highly entangled system moves ("reptates") in a tube formed by other polymers. Several further assumptions need to be made, and several different representations of the reptating polymer are used in order to derive the final diffusion equation describing the dynamics of the polymers. For the stress tensor required in deriving rheological properties, the Doi-Edwards model employs a formula from rubber elasticity.

Since 1981 another series of papers on a kinetic theory for polymer melts has been published by Curtiss, Bird, and coworkers [6–11]. In their systematic kinetic theory derivation of a diffusion equation for the polymer dynamics, Curtiss and Bird use anisotropic friction tensors to describe the hindrance of sideway motions of the polymers in concentrated systems. These anisotropic friction tensors express essentially the same physical idea as the constraining tube in the Doi-Edwards model. After making several other well-defined assumptions, the diffusion equation obtained by Curtiss and Bird has exactly the same form as in the Doi-Edwards model (however, the two models differ in the interpretation of the time constant appearing in the diffusion equation). In the *Curtiss-Bird model*, the stress tensor is also derived in a systematic manner by resorting to the same kind of approximations previously made in the derivation of the diffusion equation. One term in the resulting expression has the same form as the stress tensor assumed in the Doi-Edwards model.

The Doi-Edwards and Curtiss-Bird models are usually described in terms of the respective basic ideas and mathematical approximations underlying these models (for a detailed description and discussion of these models see the textbooks by Doi and Edwards [12] and by Bird, Curtiss, Armstrong, and Hassager [13]). In contrast, the interest of this chapter is focused on the final model equations—the diffusion equation for the polymer dynamics and the expression for the stress tensor—which are used for concrete model predictions. By writing down the equivalent stochastic differential equation, which has a very simple structure, the *content of the final equations* characterizing the Doi-Edwards and Curtiss-Bird models is worked out in an unambiguous manner without any reference to the derivation of these equations. The benefit of such an approach is twofold. First, this approach helps to clarify the physics behind the final model equations which result from a long and certainly necessary process of simplifying an extremely complicated problem by means of a series of more or less far-reaching and interfering assumptions. Second, by elaborating the stochastic content of the fundamental diffusion equation of the Doi-Edwards and Curtiss-Bird models it is possible to design an elegant computer simulation algorithm for these models, which can eventually be applied also to more complicated reptation theories.

An older class of models for concentrated polymer solutions and melts is based on the idea of modifying ("liquefying") the network theory of rubber elasticity by allowing for temporary network junctions; the junctions represent strong polymer-polymer interactions at isolated points along chains. These models are discussed in Chap. 20 of [13], with a special emphasis on the Lodge network model and its modifications. Even though there is no molecular description of the central processes of junction formation and destruction, the classical network models are still used quite often as a theoretical framework for discussing experimental results.

Stochastic simulation techniques have been useful for understanding network models with configuration-dependent rates of formation and destruction of network junctions and with non-Gaussian network strands [14–18]. However, the formation and destruction of junctions lead to a time-evolution equation for the configurational distribution function that is not of the Fokker-Planck type (see, e.g., Sect. VII.(E) of [19] for a rigorous discussion of destruction terms). We therefore restrict ourselves in this chapter on concentrated solutions and melts to the discussion of reptation models.

In the first section of this chapter, it is our main intention to demonstrate the general ideas behind simulation algorithms for reptation models and to check their efficiency rather than reproducing detailed results for the Doi-Edwards and Curtiss-Bird models which have previously been determined by other numerical methods [8–11]. As an illustrative example for applying the simulation method to more complicated reptation models and for modifying reptation theories on the level of stochastic processes, the reptating-rope model developed by Jongschaap and Geurts [20–23] is discussed in the second section of this chapter on reptation models. In a final section, we discuss a new reptation model and the stochastic formulation of one of the most important modifications of the original Doi-Edwards reptation model, namely a model avoiding the independent alignment assumption.

6.1 Doi-Edwards and Curtiss-Bird Models

In this section, we discuss the basic equations of the Doi-Edwards and Curtiss-Bird models without examining their respective derivations and their justification. Since these famous models have so often been described in the conventional framework of diffusion equations we will jump directly into the stochastic formulation of reptation models. First, we construct the stochastic process associated with the diffusion equation which we here refer to as the *reptation process*. Second, the stress tensor is expressed as an ensemble average over the trajectories of this reptation process. The reformulation of the basic equations is then used to develop increasingly more sophisticated simulation algorithms for the Doi-Edwards and Curtiss-Bird models, where the boundary conditions require some special considerations in developing algorithms of higher order in the time-step width.

6.1.1 Polymer Dynamics

The original versions of the Doi-Edwards and Curtiss-Bird models are based on the following diffusion equation for a probability density $p(\boldsymbol{u}, s, t)$ [4, 6],

$$\boxed{\frac{\partial p(\boldsymbol{u}, s, t)}{\partial t} = -\frac{\partial}{\partial \boldsymbol{u}} \cdot \left\{ \left[\boldsymbol{\kappa}(t) \cdot \boldsymbol{u} - \boldsymbol{\kappa}(t) : \boldsymbol{u}\boldsymbol{u}\boldsymbol{u} \right] p(\boldsymbol{u}, s, t) \right\} + \frac{1}{\lambda} \frac{\partial^2 p(\boldsymbol{u}, s, t)}{\partial s^2}}\,, \tag{6.1}$$

where $\boldsymbol{u}$ is a three-dimensional vector, $s \in [0, 1]$, the quantity λ is a characteristic time constant, namely the so-called *disengagement or reptation time*, and the transposed velocity-gradient tensor $\boldsymbol{\kappa}(t)$ characterizes the given homogeneous flow field. One is interested in a solution of (6.1) which is concentrated on the unit sphere defined by $|\boldsymbol{u}| = 1$; it will become clear below why (6.1) admits such solutions. The unit vector $\boldsymbol{u}$ describes the direction of the polymer chain at the position s within the chain, where the label s varies from 0 to 1 in going from one chain end to the other. The boundary conditions for $s = 0$ and $s = 1$ supplementing the diffusion equation (6.1) are,

$$p(\boldsymbol{u}, 0, t) = p(\boldsymbol{u}, 1, t) = \frac{1}{4\pi}\, \delta(|\boldsymbol{u}| - 1)\,. \tag{6.2}$$

According to Sect. 3.3.3, the diffusion equation or Fokker-Planck equation (6.1) is equivalent to a stochastic differential equation. Since, apart from time, there occur a unit vector $\boldsymbol{u}$ and a real number s in the list of arguments of the probability density p, the equivalent stochastic process is of the form $(\boldsymbol{U}_t, S_t)_{t \in \mathbb{T}}$ with a vector $\boldsymbol{U}_t$ and a real number S_t from the interval $[0, 1]$, where $\mathbb{T}$ is the time range of interest. In other words, the interpretation of (6.1) as a Fokker-Planck equation suggests that $p(\boldsymbol{u}, s, t)$ is not to be regarded as a probability density in $\boldsymbol{u}$ which parametrically depends on the chain-position label s, but rather as the joint probability density of two dynamic variables $\boldsymbol{u}$ and s. From the general equivalence between Fokker-Planck equations and stochastic differential equations derived in Sect. 3.3.3 one then finds the following time evolution of the processes $\boldsymbol{U}$ and S:

- Since there occurs only a first-order derivative with respect to $\boldsymbol{u}$ in (6.1), the process $\boldsymbol{U}$ satisfies the following deterministic differential equation,

$$\mathrm{d}\boldsymbol{U}_t = [\boldsymbol{\kappa}(t) \cdot \boldsymbol{U}_t - \boldsymbol{\kappa}(t) : \boldsymbol{U}_t\boldsymbol{U}_t\boldsymbol{U}_t]\,\mathrm{d}t = (\boldsymbol{\delta} - \boldsymbol{U}_t\boldsymbol{U}_t) \cdot \boldsymbol{\kappa}(t) \cdot \boldsymbol{U}_t\,\mathrm{d}t\,. \quad (6.3)$$

  The term $\boldsymbol{\kappa}(t) \cdot \boldsymbol{U}_t$ in (6.3) expresses the fact that the vector $\boldsymbol{U}_t$ follows the flow field. The transverse projection operator $(\boldsymbol{\delta} - \boldsymbol{U}_t\boldsymbol{U}_t)$ makes sure that the dynamics preserves the property of being a unit vector, so that $\boldsymbol{U}_t$ can only rotate with the flow field.

- The occurrence of only a second-order derivative with respect to s in (6.1) indicates that S is a pure diffusion process which is completely characterized by the diffusion constant $1/\lambda$. In other words, the process S is the solution of the following stochastic differential equation

$$\mathrm{d}S_t = \sqrt{\frac{2}{\lambda}}\,\mathrm{d}W_t\,, \quad (6.4)$$

  where W is the Wiener process.

- The only coupling between the two processes $\boldsymbol{U}$ and S arises through the boundary conditions (6.2) which have been neglected in writing down the differential equations (6.3) and (6.4) for $\boldsymbol{U}_t$ and S_t. When the process S reaches one of the boundaries, $\boldsymbol{U}$ no longer follows the flow field but rather is to be chosen as a randomly oriented unit vector. The boundary conditions (6.2) can be shown to imply that 0 and 1 constitute reflecting boundaries for the process S.

It is mainly the nontrivial coupling between the processes $\boldsymbol{U}$ and S through the boundary conditions which leads to mathematical complications in solving the Fokker-Planck equation for the Doi-Edwards and Curtiss-Bird models. The fact that the boundaries 0 and 1 of the range of S are of the reflecting type can be shown as follows. By integrating the well-known formal solution of (6.1) and (6.2) [4, 6] one obtains the marginal probability density $\hat{p}(s,t) = \int \mathrm{d}^3u\, p(\boldsymbol{u},s,t) = 1$ and hence the condition $\partial\hat{p}(s,t)/\partial s = 0$ for all s and, in particular, at the boundaries. Since there is no drift term in (6.4), and since the diffusion coefficient is constant, this condition is characteristic of reflecting boundaries (see Sect. 3.3.5). Notice that the boundary conditions (6.2) are formulated for the joint distribution of $\boldsymbol{U}_t$ and S_t rather than for the conditional probability density of $\boldsymbol{U}_t$ at the boundaries and hence they contain information going beyond the isotropy of $\boldsymbol{U}_t$ at the boundaries, namely about the reflecting boundary conditions for S.

Just as in the preceding chapters, the translation of the diffusion equation (6.1) into the reptation process $(\boldsymbol{U}, S)$ is the key idea for developing a computer-simulation algorithm for the Doi-Edwards and Curtiss-Bird models. The initial conditions for $\boldsymbol{U}_0$ and S_0 depend on the given flow history. For example, equilibrium initial conditions correspond to a random unit vector $\boldsymbol{U}_0$ and a uniform random number $S_0 \in [0,1]$.

Although it is not at all necessary for designing simulation algorithms, we cannot resist giving a brief interpretation of the reptation process $(\boldsymbol{U}, S)$. In concentrated solutions and melts, the motion of each polymer is restricted by the surrounding polymers, and this situation may be taken into account by introducing the notion of a confining tube (or, alternatively, by introducing anisotropic frictional properties). The unit vector $\boldsymbol{U}_t$ may be interpreted as the direction of a particular tube segment (or as the direction of lowest friction), and the number S_t describes which segment of a particular polymer chain is currently in the tube segment with the orientation $\boldsymbol{U}_t$. The tube and hence also the polymer in the tube are deformed by the flow field (cf. (6.3) for $\boldsymbol{U}_t$). As a consequence of the reptational motion of the polymer in the tube, the orientation of the tube segment is forced on a permanently changing segment within the particular reptating chain (cf. (6.4) for S_t). These dynamic processes are illustrated in Fig. 6.1. When one of the chain ends leaves the particular tube segment, a new orientation is created at random (cf. the boundary condition). The fact that the polymer dynamics in a melt as described in the Doi-Edwards and Curtiss-Bird models can be represented by the stochastic time evolution of a single unit vector $\boldsymbol{U}_t$ and a number $S_t \in [0,1]$ makes clear that these models may be regarded as "one-segment theories." These models contain information only about the orientation of a single segment, $\boldsymbol{U}_t$, and its position, S_t, in a single chain of the many-chain system and no information about other segments or chains. All the effects of chain connectivity and interactions between different chains are incorporated only in an effective, mean-field-type manner. This should be kept in mind in discussing the reliability of various model predictions.

6.1.2 Stress Tensor

In order to calculate the rheological properties of polymeric liquids one needs not only an equation for the polymer dynamics but also an expression for the stress tensor. In the Doi-Edwards and Curtiss-Bird models, the stress tensor is usually evaluated as an integral performed with the solution $p(\boldsymbol{u}, s, t)$ of the Fokker-Planck equation (6.1). According to the equivalence between Fokker-Planck equations and stochastic differential equations, the stress tensor of the Curtiss-Bird model can be expressed as the following expectation of the reptation process $(\boldsymbol{U}, S)$ [7],

$$\boxed{\boldsymbol{\tau}(t) = N n_{\mathrm{p}} k_{\mathrm{B}} T \left\{ \frac{1}{3}\boldsymbol{\delta} - \left\langle \boldsymbol{U}_t \boldsymbol{U}_t \right\rangle - \epsilon \lambda\, \boldsymbol{\kappa}(t) : \left\langle S_t(1-S_t)\, \boldsymbol{U}_t \boldsymbol{U}_t \boldsymbol{U}_t \boldsymbol{U}_t \right\rangle \right\},} \quad (6.5)$$

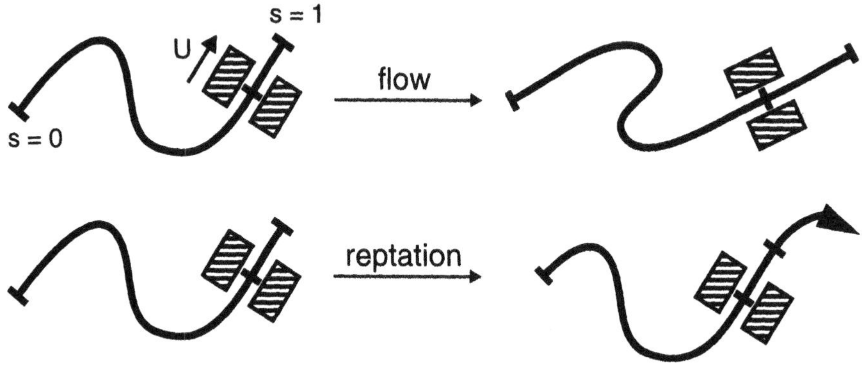

Fig. 6.1. Illustration of reptation dynamics. The *lines* represent a probe chain, and the *bars* indicate the constraining effect of the matrix material via a tube segment or anisotropic frictional properties. The upper part of the figure shows how the flow deforms the matrix material and hence changes the orientation of the probe chain (rotation of the tangent vector $\boldsymbol{U}$). The lower part of the figure illustrates how the probe chain reptates through the matrix material (displacement of the position label S).

where ϵ is the link tension coefficient ($0 \leq \epsilon \leq 1$). The link tension coefficient characterizes the anisotropic frictional properties of the polymer chains in a melt. As before, n_p is the number density of polymers, and N is the number of beads of the Kramers chains which are used as a starting point for the kinetic theory derivation of the Curtiss-Bird model. In the special case $\epsilon = 0$, one recovers the stress tensor of the Doi-Edwards model.[1] In the original publication [7], (6.5) has been derived in the framework of kinetic theory. The first two terms are similar to the expression for the stress tensor (4.18) for the Rouse model. The double occurrence of the unit vector $\boldsymbol{U}_t$ is related to the inextensibility of the rods in the underlying Kramers chains and to a uniform tension in the chains (cf. also (5.74) for rigid dumbbells). A phenomenological understanding of the last term in (6.5), including an estimate of the magnitude of ϵ, can be gained within the reptating-rope model (see Sect. 6.2).

[1]In the formulation of the Doi-Edwards model, a Rouse chain consisting of N beads with friction coefficient ζ is employed. These beads are supposed to be much smaller objects than the beads of the Kramers chains underlying the Curtiss-Bird models and, accordingly, N is a much larger number for Doi and Edwards than for Curtiss and Bird. The N of Curtiss and Bird is actually closely related to the number of steps in the "primitive chain", Z, which Doi and Edwards define as the contour length of equilibrium chains divided by the tube width. Furthermore, Doi and Edwards use the symbols c and τ_d for the number density of beads and for the disengagement time, respectively. The equations of Doi and Edwards [12] can be translated into the language of Curtiss and Bird, which is adopted in this book, if the following dictionary is used: $c/N \to n_\mathrm{p}$, $Z \to N/3$, $\tau_\mathrm{d} \to \lambda/\pi^2$.

In the following, we show how the properties of the reptation process can be used to reformulate the expression for the stress tensor (6.5). The alternative formulation in terms of memory integrals is particularly useful for obtaining analytical results such as the linear viscoelastic behavior (see Example 6.8 for the discussion of linear viscoelasticity for a more general model). The fact that, away from the boundaries 0 and 1 of the range of S, the processes $\boldsymbol{U}$ and S evolve independently can be used to rewrite (6.5). It is helpful to use the results concerning the last reflection time for a Wiener process in $[0,1]$ which were derived in Example 3.37.

Furthermore, it is very convenient to introduce the quantity $\widehat{\boldsymbol{u}} = \widehat{\boldsymbol{u}}(\boldsymbol{u}, t, t')$ as the result of the deterministic time evolution of the unit vector $\boldsymbol{u}$ in the time interval from t' to t, that is, as the solution of (6.3) with the initial condition $\widehat{\boldsymbol{u}} = \boldsymbol{u}$ at $t = t'$. The quantity $\widehat{\boldsymbol{u}}$ can be written more explicitly as $\widehat{\boldsymbol{u}}(\boldsymbol{u}, t, t') = \mathbf{E}(t,t') \cdot \boldsymbol{u} / |\mathbf{E}(t,t') \cdot \boldsymbol{u}|$, where the tensor $\mathbf{E}(t,t')$ has been introduced in (4.14). The average of an arbitrary function $g(\boldsymbol{u})$ can be expressed as the time-integral of the probability that the last reflection before t happened at time t', that is $\overline{\mu}(t-t')$, times the average of $g(\boldsymbol{u})$ under the condition that the last reflection occurred at t'. Since under this condition $\boldsymbol{U}_t$ results from the deterministic time evolution of a random unit vector in the time interval $[t', t]$, one obtains

$$\langle g(\boldsymbol{U}_t) \rangle = \int_{-\infty}^{t} \mathrm{d}t' \, \overline{\mu}(t-t') \int \mathrm{d}^3u \, \frac{\delta(|\boldsymbol{u}|-1)}{4\pi} \, g(\widehat{\boldsymbol{u}}(\boldsymbol{u}, t, t')) \,. \tag{6.6}$$

Formally, (6.6) can be derived from (2.60) when $\boldsymbol{U}_t$ and T_t^{LR} play the roles of X and Y and the transformation formula (2.29) is used. The identity (6.6) can be used to rewrite the average $\langle \boldsymbol{U}_t \boldsymbol{U}_t \rangle$ in the expression (6.5) for the stress tensor as a memory integral. In order to rewrite the last term in (6.5) in the same fashion, one needs to generalize the identity (6.6) to the case of functions $g(\boldsymbol{u}, s)$. To obtain this generalization, we consider the probability distribution of s under the condition that the last reflection before t happened at t', which was also derived in Example 3.37. By again expressing the desired average in terms of conditional averages (under the condition that the last reflection happened at the time t') one obtains the following result,

$$\langle g(\boldsymbol{U}_t, S_t) \rangle = \int_{-\infty}^{t} \mathrm{d}t' \, \overline{\mu}(t-t') \int_0^1 \mathrm{d}s \int \mathrm{d}^3u \, \frac{\delta(|\boldsymbol{u}|-1)}{4\pi} \, \overline{\pi}(s, t-t') \, g(\widehat{\boldsymbol{u}}(\boldsymbol{u}, t, t'), s) \,. \tag{6.7}$$

Notice that the identities (6.6) and (6.7) have been obtained directly from the properties of the reptation process $(\boldsymbol{U}, S)$ without solving the diffusion equation (6.1). Only the deterministic time evolution of $\boldsymbol{U}$, the properties of the Wiener process S in the interval $[0,1]$, and the coupling of these two processes through the boundary conditions have been used.

Exercise 6.1 Determine $\widehat{\boldsymbol{u}}(\boldsymbol{u}, t, t')$ for simple shear flow.

Exercise 6.2 Derive (6.7) in a formal manner by using (2.60).

Exercise 6.3 Use (6.6) and (6.7) in order to derive an explicit expression for the stress tensor (6.5) of the Curtiss-Bird model in terms of memory integrals. Verify that the result agrees with (19.6-9) of [13].

We here refer to the diffusion equation (6.1) with boundary conditions (6.2), or the equivalent reptation process $(\boldsymbol{U}, S)$, together with the stress tensor expression (6.5) as the *original Curtiss-Bird and Doi-Edwards reptation models*, where in the latter case $\epsilon = 0$ is implied. These equations yield an unambiguous description of the rheological behavior of melts. Often the term "reptation model" is used much more loosely and vaguely. In particular, the equations, ideas, assumptions and pictures occurring in various stages of the derivation of the final rheological equations for the Doi-Edwards model are employed to predict many other properties of melts. Most frequently used are the conclusions about chain motions on various length and time scales which are based on a combination of the dynamic behavior of the Rouse model on short and long time scales with the idea of one-dimensional motion within tubes on the intermediate length scales, that is, on length scales larger than the tube diameter and shorter than the polymer size. Another type of argument, based on the notion of reptational motion in random-walk shaped tubes, leads to the prediction of the molecular-weight dependence of various measurable quantities. For example, the longest relaxation time and the zero-shear-rate viscosity are predicted to be proportional to N^3, and the diffusion coefficient is predicted to decrease as N^{-2}. An evaluation of the success of reptation models, in this loose sense, by careful examination of existing experimental data can be found in a review article by Lodge, Rotstein, and Prager [24]. A particular challenge for theoreticians is the explanation of the $N^{3.4}$-dependence of the viscosity observed in many experiments on melts of high-molecular-weight polymers.

In their derivation of the diffusion equation (6.1), Curtiss and Bird give a diffusion equation for the complete chain which may be used as a starting point when knowledge of joint probabilities of several segments is required. We here do not consider any such intermediate results of the derivation of the Doi-Edwards and Curtiss-Bird models.

All stochastic simulations based on the reptation process are used as a tool to determine the properties predicted by the original reptation models—they have nothing to say about the existence of reptational motion in real systems.

6.1.3 Simulations in Steady Shear Flow

The reformulation of the diffusion equation (6.1) in terms of stochastic differential equations for the reptation process $(\boldsymbol{U}, S)$ leads to a simulation algorithm for the Doi-Edwards and Curtiss-Bird models [25]. We here discuss steady shear flow (see Sect. 1.2.1). Note that at a given value of $\dot{\gamma}$, the viscometric functions of the Curtiss-Bird model for any value of the link tension coefficient ϵ can according to (6.5) be expressed as [9]

$$\eta(\dot{\gamma}) = \eta_\Omega(\dot{\gamma}) + \epsilon\, \eta_\Upsilon(\dot{\gamma}), \tag{6.8}$$
$$\Psi_1(\dot{\gamma}) = \Psi_{1,\Omega}(\dot{\gamma}) + \epsilon\, \Psi_{1,\Upsilon}(\dot{\gamma}), \tag{6.9}$$
$$\Psi_2(\dot{\gamma}) = \Psi_{2,\Omega}(\dot{\gamma}) + \epsilon\, \Psi_{2,\Upsilon}(\dot{\gamma}), \tag{6.10}$$

where the functions with subscript Ω are the respective viscometric functions of the Doi-Edwards model. In [9], the identity $\Psi_{2,\Omega}(\dot{\gamma}) = -\Psi_{2,\Upsilon}(\dot{\gamma})$ has been derived, thereby reducing the number of unknown functions of one variable in (6.8)–(6.10) from six to five; the zero-shear-rate limits of these functions are

$$\eta_\Omega(0) = \frac{3}{2}\, \eta_\Upsilon(0) = \frac{1}{60}\, N n_\mathrm{p} k_\mathrm{B} T \lambda\,, \tag{6.11}$$

$$\Psi_{1,\Omega}(0) = -\frac{7}{2}\, \Psi_{2,\Omega}(0) = \frac{1}{300}\, N n_\mathrm{p} k_\mathrm{B} T \lambda^2\,, \tag{6.12}$$

and $\Psi_{1,\Upsilon}(0)$ vanishes. These equations imply that the parameter N must have a rather well-defined physical significance because its order of magnitude can be measured,

$$\left(1 + \frac{2}{3}\epsilon\right)^2 N = \frac{6\,[\eta(0)]^2}{\Psi_1(0)\, n_\mathrm{p} k_\mathrm{B} T}\,. \tag{6.13}$$

The situation is very different for bead-spring models of dilute solutions, where the number of beads N has no physical meaning because the beads are purely fictitious objects. Various equivalent ways of expressing the right side of (6.13) in terms of linear viscoelastic functions can be found in Table 5.3-1 of [26].

The occurrence of the parameter ϵ in estimating N can be avoided by considering the relaxation modulus in the limit of small times, $G(0+)$, which allows us an independent determination of N,

$$N = \frac{5\, G(0+)}{n_\mathrm{p} k_\mathrm{B} T}\,. \tag{6.14}$$

The value $G(0+)$ can equivalently be obtained as the constant plateau value of the storage modulus at high frequencies, where the frequencies should not be so high that other processes than reptation become relevant. If the validity of the Curtiss-Bird model is assumed, the combination of (6.13) and (6.14) can be used for a direct measurement of ϵ.

The physical significance of N is related to the notion of entanglements. Actually, $N/5$ is usually defined as the number of entanglements in a chain, and this corresponds to saying that the molecular weight between entanglements is

$$\boxed{M_\mathrm{e} := \frac{\rho_\mathrm{p} R T}{G(0+)}}\,, \tag{6.15}$$

where ρ_p is the polymer mass density, and R is the gas constant. For highly flexible polymers, the number of backbone atoms between entanglements usually is 120 to 350 (see Sect. 13.B of [27]).

In the remainder of this subsection, it is shown how the five independent functions in (6.8)–(6.10) can be estimated by means of computer simulations of the reptation process. In a first step, we discuss the simulation algorithm obtained by naive time-discretization of the reptation process, and in two further steps we develop more efficient simulation techniques.

Step 1: If the numerical integration of the stochastic differential equations for the reptation process proceeds in time steps of a fixed width Δt, then the deterministic differential equation (6.3) for the process $\boldsymbol{U}$ may be replaced with the following discretized equation for $\widetilde{\boldsymbol{U}}_j = \boldsymbol{U}_{j\Delta t}$, $j = 0, 1, 2\ldots$,

$$\widetilde{\boldsymbol{U}}_{j+1} = \frac{\widetilde{\boldsymbol{U}}_j + \boldsymbol{\kappa} \cdot \widetilde{\boldsymbol{U}}_j \, \Delta t}{|\widetilde{\boldsymbol{U}}_j + \boldsymbol{\kappa} \cdot \widetilde{\boldsymbol{U}}_j \, \Delta t|}, \tag{6.16}$$

where for a time-dependent flow field the tensor $\boldsymbol{\kappa}$ is evaluated at the time $t = j\Delta t$. To first order in Δt, one recovers the differential equation (6.3) after expanding the denominator in (6.16). The advantage of the discretization (6.16) compared to other possible discretizations of the differential equation (6.3) lies in the fact that, in the course of the simulation, the length of the vector $\widetilde{\boldsymbol{U}}_j$ is always exactly equal to unity. In steady shear flow, the integration scheme (6.16) happens to be exact for arbitrarily large time steps (see Exercise 6.1).

By applying the weak Euler scheme (3.119) to (6.4), one obtains the following difference equation for $\widetilde{S}_j = S_{j\Delta t}$, $j = 0, 1, 2, \ldots$,

$$\widetilde{S}_{j+1} = \widetilde{S}_j + \sqrt{(2/\lambda)} \, \Delta W_j. \tag{6.17}$$

The quantities ΔW_j are independent Gaussian random variables with mean 0 and variance Δt. Notice that the Euler scheme for the stochastic differential equation (6.4) is exact.

The simplest simulation algorithm for the reptation process proceeds as follows. If we assume equilibrium initial conditions at $t_0 = 0$, then $\widetilde{S}_0$ is to be chosen as a random number from the interval $[0, 1]$, and also the unit vector $\widetilde{\boldsymbol{U}}_0$ is to be chosen at random. In each time step, there is a Gaussian random number $(2/\lambda)^{1/2} \, \Delta W_j$ added to $\widetilde{S}_j$ in order to obtain $\widetilde{S}_{j+1}$. If the resulting value for $\widetilde{S}_{j+1}$ lies outside the interval $[0, 1]$, it is replaced by the value obtained via reflection at the boundary passed in leaving the interval $[0, 1]$, that is, $\widetilde{S}_{j+1} \to -\widetilde{S}_{j+1}$ for $\widetilde{S}_{j+1} < 0$, and $\widetilde{S}_{j+1} \to 2 - \widetilde{S}_{j+1}$ for $\widetilde{S}_{j+1} > 1$. If no reflection occurs, then (6.16) is used to construct $\widetilde{\boldsymbol{U}}_{j+1}$ from $\widetilde{\boldsymbol{U}}_j$; otherwise, the boundary condition (6.2) requires that the unit vector $\widetilde{\boldsymbol{U}}_{j+1}$ be chosen at random. In the following, we refer to this procedure as the "naive algorithm."

In order to demonstrate that the naive simulation algorithm described in the preceding paragraph can be employed to obtain very precise estimates of the exact results for the Doi-Edwards and Curtiss-Bird models, we have performed simulations for steady shear flow at dimensionless shear rate $\lambda\dot{\gamma} = 1$ on a CRAY-2 computer. Intermediate shear rates are particularly interesting because for low and high shear rates exact series expansions are available [9, 28]. For various values of the time-step width Δt, a large number of steps has been carried out in the simulations: a total number of 10×10^9 time steps for $\Delta t = 0.002\,\lambda$, 2×10^9 time steps for $\Delta t = 0.01\,\lambda$, 10^9 time steps for $\Delta t = 0.02\,\lambda$ and $0.03\,\lambda$, and 0.5×10^9 time steps for $\Delta t/\lambda = 0.04, 0.05, \ldots, 0.09, 0.10$. For a steady flow, ergodicity implies that simulation of a single trajectory over sufficiently long time yields the correct averages (see Sect. 4.2.5). However, by simulating 1000 trajectories of the discretized reptation process in the innermost loops of the simulation program and by vectorizing these loops, a factor of almost 5 (3.5) in computer time has been saved for $\Delta t = 0.01\,\lambda$ ($\Delta t = 0.05\,\lambda$). For smaller time-step width, the vectorization is more efficient because on average it takes more time steps before a nonvectorizable reflection at $S = 0$ or $S = 1$ occurs.

All the 1000 trajectories constructed in the innermost loops are started with a random unit vector $\boldsymbol{U}$ and $S = 0$. Then each trajectory is evolved over a total time of 2λ in order to reach a typical steady-state configuration before the time-independent moments occurring in the various components of the stress tensor (6.5) are estimated from the simulations. For each time-step width, the number of time steps over which the "measurement" of the stress tensor is performed for a single trajectory is one thousandth of the total number of time steps listed above.

The simulation results obtained for the contribution $\eta_\Omega(\dot\gamma)$ to the viscosity in (6.8) for $\lambda\dot\gamma = 1$ and various time steps are indicated by the triangles in Fig. 6.2. In this figure, the symbols are more than ten times larger than the statistical errors in the simulation results. Figure 6.2 contains a very important warning: one can see that without careful time step extrapolation one obtains very misleading results because the convergence of the corrections due to finite Δt is very slow. Even at $\Delta t = 0.002\,\lambda$, the deviation from the exact result is more than 20%. This slow convergence in Δt is a consequence of the important role played by the boundary conditions. It is known that the error introduced by time discretization for an equivalent first-passage-time problem is of the order $\sqrt{\Delta t}$ [29]. For that reason, we have fitted polynomials of various degrees in $\sqrt{\Delta t}$ to the simulation data in Fig. 6.2 by a least squares method. For a given degree of the polynomial, the results for the large time steps have been discarded such that the remaining data points are consistent with a polynomial curve of the given degree in $\sqrt{\Delta t}$, where the consistency has been checked by a χ^2 test (this is the same procedure as described in the solution to Exercise 4.11, where $\sqrt{\Delta t}$ now takes the role previously played by Δt). It was possible to obtain the zero-time-step result reliably with a second-order fit (continuous line through the triangles in Fig. 6.2); the upper dashed line in Fig. 6.2 indicates that a first-order fit in $\sqrt{\Delta t}$ is insufficient for a trustworthy extrapolation. In the first line of Table 6.1, the result of the second-order extrapolation is compared to the "exact" result obtained by numerical evaluation of the Curtiss-Bird memory integral expression for the stress tensor in [9]. Within the statistical error bars, perfect agreement between naive simulation and exact results is found (in Table 6.1, the numbers in parentheses represent the statistical error in the last figure displayed).

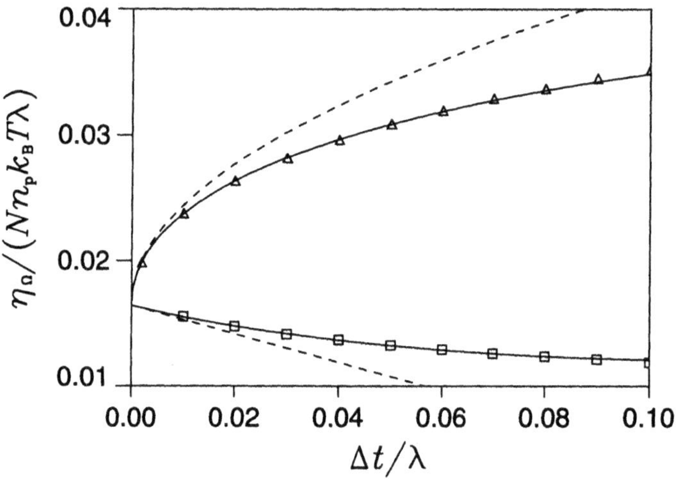

Fig. 6.2. Simulation results for the function $\eta_\Omega(\dot{\gamma})$ to be used in (6.8) for calculating the viscosity at dimensionless shear rate $\lambda\dot{\gamma} = 1$ as a function of the time-step width Δt. The *triangles* and *squares* mark the results of the naive and improved simulation algorithms, respectively. The *continuous lines* represent the results of a second-order fit in $\sqrt{\Delta t}$ for the naive algorithm and of a fit in Δt with $\sqrt{\Delta t}^3$ corrections for the improved algorithm. For these fits obtained by higher-order regression, the respective contributions linear in $\sqrt{\Delta t}$ and Δt are indicated by the *dashed lines*.

Table 6.1. Functions to be used in (6.8)–(6.10) for calculating the viscometric functions at dimensionless shear rate $\lambda\dot{\gamma} = 1$

	exact	naive algorithm	improved algorithm
$10 \times \dfrac{\eta_\Omega}{Nn_pk_BT\lambda}$	0.1643	0.1645(6)	0.1643(4)
$10 \times \dfrac{\eta_\Upsilon}{Nn_pk_BT\lambda}$	0.1121	0.11200(2)	0.11198(1)
$10^2 \times \dfrac{\Psi_{1,\Omega}}{Nn_pk_BT\lambda^2}$	0.3242	0.330(9)	0.319(6)
$10^4 \times \dfrac{\Psi_{1,\Upsilon}}{Nn_pk_BT\lambda^2}$	0.4455	0.49(4)	0.42(2)
$10^3 \times \dfrac{\Psi_{2,\Upsilon}}{Nn_pk_BT\lambda^2}$	0.9195	0.920(5)	0.921(3)

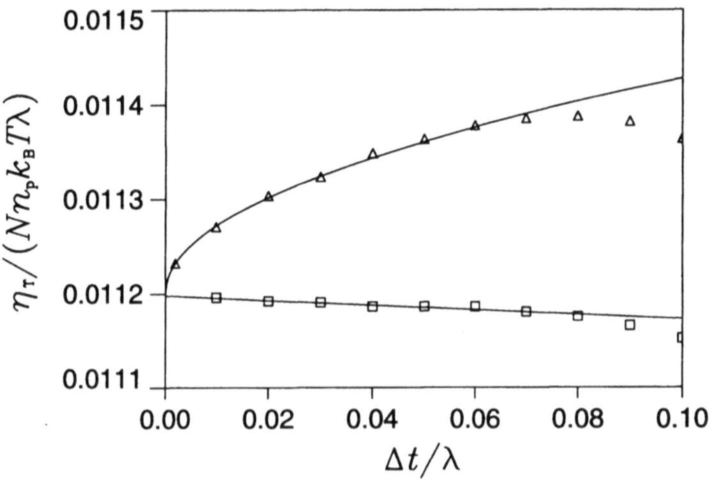

Fig. 6.3. Simulation results for the function $\eta_\Upsilon(\dot\gamma)$ to be used in (6.8) for calculating the viscosity at dimensionless shear rate $\lambda\dot\gamma = 1$ as a function of the time-step width Δt. The *triangles* and *squares* mark the results of the naive and improved simulation algorithms, respectively. The *continuous lines* represent the results of first-order fits in $\sqrt{\Delta t}$ and Δt for the naive and improved algorithms, respectively.

The simulation results obtained for the contribution $\eta_\Upsilon(\dot\gamma)$ to the viscosity in (6.8) for $\lambda\dot\gamma = 1$ and various time steps are indicated by the triangles in Fig. 6.3. In this figure, the symbols are between three and five times larger than the statistical errors in the simulation results. One again finds a $\sqrt{\Delta t}$ behavior for small time steps. However, for $\eta_\Upsilon(\dot\gamma)$ the first-order fit in $\sqrt{\Delta t}$ is sufficient for a reliable extrapolation to zero time-step width. Note that the variation of the data in Fig. 6.3 for different time steps is far less than in Fig. 6.2. For $\Delta t = 0.002\,\lambda$, the deviation from the exact result is only about 0.3%. As a remarkable consequence, five significant figures for η_Υ at $\lambda\dot\gamma = 1$ can be obtained from our simulation of the reptation process. The results for the other functions introduced in (6.8)–(6.10) to characterize the viscometric functions can be obtained in a similar manner. All these results obtained from the naive simulation algorithm are summarized in Table 6.1; they are found to be in good agreement with the exact numbers displayed in [9]. The largest relative error is obtained for $\Psi_{1,\Upsilon}$; note, however, that for $\lambda\dot\gamma = 1$ the contribution $\Psi_{1,\Upsilon}$ causes only a minor modification of the first normal stress coefficient $\Psi_{1,\Omega}$ predicted by the Doi-Edwards model.

Exercise 6.4 Develop a routine for simulating the polymer dynamics of the Doi-Edwards and Curtiss-Bird models in steady shear flow based on the naive simulation algorithm.

Step 2: A severe drawback of the naive simulation algorithm developed in the preceding step is its slow convergence for small time steps. Therefore, expensive simulations for very small time steps are necessary for reliably extrapolating for $\Delta t \to 0$. For example, the simulation for the data points at $\Delta t = 0.002$ in Figs. 6.2, 6.3 require almost the same amount of computer time as all the other data points together. In order to avoid corrections of the order $\sqrt{\Delta t}$ for small time steps, we employ a method recently suggested in connection with first-passage-time problems [29]. To this end, one needs to understand the origin of the $\sqrt{\Delta t}$ corrections which may be unexpected for the weak Euler algorithm.

In the naive algorithm, the simulation of the process S yields the exact properties of this process at the discrete times $j\Delta t$. For steady shear flow, the deterministic time-evolution (6.16) of the process $\boldsymbol{U}$ is also exact (see Exercise 6.1; in general, the error in the deterministic time-evolution will be of order Δt). Corrections of order $\sqrt{\Delta t}$ can hence originate only from the treatment of the reflecting boundary conditions. Even if both $\tilde{S}_j$ and $\tilde{S}_{j+1} = \tilde{S}_j + (2/\lambda)^{1/2}\,\Delta W_j$ lie in the interval $[0,1]$, there is a certain nonzero probability that the process S left the interval $[0,1]$ between $j\Delta t$ and $(j+1)\Delta t$, especially when the values of S are close to the boundaries at 0 or 1. Hence, the situation can occur that reflections that should trigger the process $\boldsymbol{U}$ to start with a new random unit vector go unnoticed in a simulation with discrete time step. It has been recognized in [29] that such unobserved reflections at the boundaries (or passages of the boundaries) happen with a probability of order $\sqrt{\Delta t}$.

The conditional probability for an unobserved reflection at $b = 0$ or $b = 1$ for given values of $\tilde{S}_j \in [0,1]$ and $\tilde{S}_{j+1} = \tilde{S}_j + (2/\lambda)^{1/2}\,\Delta W_j \in [0,1]$ is

$$P_{\mathrm{u}} = \exp\left\{-\frac{\lambda}{\Delta t}\left(b - \tilde{S}_j\right)\left(b - \tilde{S}_{j+1}\right)\right\}. \tag{6.18}$$

Obviously, this probability is exponentially small and hence completely negligible, except when the distance between the boundary b and $\tilde{S}_j$ is at most of order $\sqrt{\Delta t}$ (then the distance between b and $\tilde{S}_{j+1}$ is of the same order, because the difference between $\tilde{S}_j$ and $\tilde{S}_{j+1}$ is also of order $\sqrt{\Delta t}$); for that reason, one needs to consider only one boundary in each small time step.

Now, the obvious thing to do is to improve the naive algorithm by choosing a new random unit vector $\tilde{\boldsymbol{U}}_{j+1}$ with probability P_{u} even when no reflection at a boundary has been observed in the time interval $[j\Delta t, (j+1)\Delta t]$. The boundary b used to calculate P_{u} is taken as 0 if $\tilde{S}_j + \tilde{S}_{j+1} \leq 1$ and 1 if $\tilde{S}_j + \tilde{S}_{j+1} > 1$. With this modification of the naive algorithm one would expect the leading-order corrections to be of the order Δt, because the only error is due to the fact that the precise time at which the last observed or unobserved reflection in the interval $[j\Delta t, (j+1)\Delta t]$ occurred is unknown.

The simulation results obtained with the improved algorithm for the contributions to the viscosity in (6.8) for $\lambda\dot{\gamma} = 1$ and various time-step widths are indicated by the squares in Figs. 6.2 and 6.3. These figures show that $\sqrt{\Delta t}$ corrections are indeed absent and that the systematic errors caused by finite time steps are considerably reduced. In Fig. 6.2, it is necessary to also consider

$\sqrt{\Delta t^3}$ corrections to the leading-order Δt behavior for reliably extrapolating the results for η_Ω to zero time-step width, whereas for the extrapolation of the results for η_Υ a first-order fit in Δt is sufficient (see Fig. 6.3).

In spite of the fact that the calculation of the probability P_u in (6.18) and the selection of a random number necessary to decide whether an unobserved reflection should occur make the simulation programs run longer by a factor of about 1.9, the improved algorithm is more efficient because the simulations may be performed at larger time steps. Even without performing the very expensive simulations at $\Delta t = 0.002\,\lambda$ the results from the improved algorithm compiled in Table 6.1 are somewhat more accurate than those from the naive algorithm. Again, all the simulation results are consistent with the exact results of [9].

Exercise 6.5 Derive the probability (6.18) for unobserved reflections. (Hint: The derivation can be based on the reflection principle for the Wiener process. This principle states that, if the trajectories of a Wiener process are reflected at some time, even at a trajectory-dependent first-passage time, then the resulting process is also a Wiener process.)

Exercise 6.6 Develop a routine for simulating the polymer dynamics of the Doi-Edwards and Curtiss-Bird models in steady shear flow based on the improved simulation algorithm.

Step 3: In this step, we develop a further improved simulation algorithm for the reptation process in which the error due to time-discretization can be neglected. For the algorithm suggested below, it is unnecessary to perform simulations for various time steps Δt and to extrapolate the results to zero time step [30].

The leading-order corrections for the algorithm suggested in Step 2 are of the order Δt, because the only remaining error is due to the fact that the precise time at which the last reflection in the interval $[j\Delta t, (j+1)\Delta t]$ occurred, that is, at which the last replacement of $\boldsymbol{U}$ by a random unit vector was triggered, is unknown. The algorithm proposed in the following is based on the precise determination of the random time at which the last reflection at a boundary (or equivalently, passage of a boundary) occurred. This random time may be found to arbitrarily high precision by repeated bisection of each time interval in which an observed or unobserved reflection occurs.

Suppose that $S = \tilde{S}_j \in [0,1]$ and $S' = \tilde{S}_j + (2/\lambda)^{1/2}\,\Delta W_j$. We consider the possibility of a reflection at the boundary $b = 0$ ($b = 1$) for $S + S' \leq 1$ ($S + S' > 1$). For $S' \notin [0,1]$, a reflection was observed. If S' lies in the interval $[0,1]$, but an unobserved reflection at the boundary b occurred in the interval $[j\Delta t, (j+1)\Delta t]$ according to the criterion given in Step 2, S' is replaced by $2b - S'$ in order to make the reflection obvious. For observed or unobserved reflections, we determine the precise time of the last reflection at b (the last passage of b in going from S to S') in the following manner. We first calculate the random position S'' at the intermediate time $(j+\frac{1}{2})\Delta t$ which, according to Example 2.58, is $S'' = \frac{1}{2}(S+S') + \frac{1}{2}(2/\lambda)^{1/2}\,\Delta W$, where ΔW is another Gaussian random number with mean 0 and variance Δt that is independent of all the other random numbers used in the simulation. If $S'' \in [0,1]$, the boundary must be passed in the second half of the interval $[j\Delta t, (j+1)\Delta t]$; in other words, the last reflection for the Wiener process in the interval $[0,1]$ with reflecting boundaries must have occurred in the interval $[(j+\frac{1}{2})\Delta t, (j+1)\Delta t]$, and we use S'' as the new initial value at time $t = (j+\frac{1}{2})\Delta t$. If $S'' \notin [0,1]$, it is not clear whether the final value $S' \notin [0,1]$ is reached from the intermediate value S'' with or without further passage of the boundary by the underlying continuous-time process. The probability for an unobserved passage in the interval $[(j+\frac{1}{2})\Delta t, (j+1)\Delta t]$ is given by (6.18), with $\tilde{S}_j$ and $\tilde{S}_{j+1}$ replaced by S'' and S', and Δt replaced by $\Delta t/2$ for the smaller time step. If there is no further passage of the boundary the last reflection must have occurred in the first half of the interval $[j\Delta t, (j+1)\Delta t]$, otherwise the reflection occurred in the second half of that interval. The same procedure of bisection is then repeated for the resulting, shorter time intervals with the corresponding initial and final values S and S'' (or, S'' and S') until the time of the last reflection is determined with the desired precision. For this algorithm, the only error due to the finite time step Δt is caused by reflections at both boundaries in a single time step. The probability for such double reflections is exponentially small and hence completely negligible for sufficiently small time steps.

The simulations for all values of $\lambda\dot{\gamma}$ listed in Table 6.2 were performed with time step $\Delta t = 0.01\,\lambda$. For this time step, the probability for the difference $|\tilde{S}_{j+1} - \tilde{S}_j|$ calculated according to (6.17) to be larger than one is about 1.5×10^{-12}. Therefore, reflections at both boundaries of the interval $[0,1]$ will practically not occur in a simulation of at most 2×10^9 time steps, and a simulation with $\Delta t = 0.01\,\lambda$ produces no error due to time discretization. For low and intermediate shear rates, the simulation for each value of the shear rate ($\lambda\dot{\gamma} = 0.2, 0.5, 1, 2, 5, 10, 20, 50, 100$) consisted of a total number of 2×10^9 time steps, and the times at which reflections occurred were calculated with an error of at most $1.9 \times 10^{-8}\,\lambda$. For high shear rates, the simulation for each value of the shear rate ($\lambda\dot{\gamma} = 200, 500, 1000, 2000, 5000$) consisted of 1.5×10^9 time steps, and the times at which reflections occurred were calculated with an error of at most $1.9 \times 10^{-11}\,\lambda$. By simulating an ensemble of 10000 trajectories in the innermost loops the most time consuming parts of the simulation program could be vectorized. Each run for a single value of the shear rate required between 3.5 and 4 hours of CPU time on a CRAY-2 computer.

In Table 6.2, we give a detailed comparison between our simulation results for η_Ω and η_Γ and the "exact" results obtained by numerical integration of the Curtiss-Bird expression for the stress tensor (in Table 6.2, the numbers in parentheses represent the statistical error in the last figure displayed as estimated from the fluctuations in the results for the 10000 trajectories simulated). The values referred to as "exact" in Table 6.2 were taken from [9] where four significant figures have been displayed. For $\lambda\dot{\gamma} \le 1$ and $\lambda\dot{\gamma} \ge 100$, more precise results obtained from asymptotic series expansions at low and high shear rates [28] have been included for η_Ω.

Table 6.2. The two contributions to the dimensionless viscosity predicted by the Curtiss-Bird model

$\lambda\dot{\gamma}$	$\frac{\eta_\Omega}{N n_p k_B T \lambda}$ exact	$\frac{\eta_\Omega}{N n_p k_B T \lambda}$ simulation	$\frac{\eta_\Upsilon}{N n_p k_B T \lambda}$ exact	$\frac{\eta_\Upsilon}{N n_p k_B T \lambda}$ simulation
0.2	1.665705×10^{-2}	$1.66(1) \times 10^{-2}$	1.112×10^{-2}	$1.1113(2) \times 10^{-2}$
0.5	1.660717×10^{-2}	$1.660(5) \times 10^{-2}$	1.114×10^{-2}	$1.1136(2) \times 10^{-2}$
1	1.643674×10^{-2}	$1.643(2) \times 10^{-2}$	1.121×10^{-2}	$1.1202(2) \times 10^{-2}$
2	1.585×10^{-2}	$1.583(1) \times 10^{-2}$	1.140×10^{-2}	$1.1414(2) \times 10^{-2}$
5	1.337×10^{-2}	$1.337(1) \times 10^{-2}$	1.202×10^{-2}	$1.2024(2) \times 10^{-2}$
10	9.867×10^{-3}	$9.861(2) \times 10^{-3}$	1.219×10^{-2}	$1.2182(2) \times 10^{-2}$
20	6.017×10^{-3}	$6.0188(7) \times 10^{-3}$	1.112×10^{-2}	$1.1108(1) \times 10^{-2}$
50	2.417×10^{-3}	$2.4169(2) \times 10^{-3}$	7.917×10^{-3}	$7.9016(9) \times 10^{-3}$
100	1.053694×10^{-3}	$1.0537(1) \times 10^{-3}$	5.280×10^{-3}	$5.2742(6) \times 10^{-3}$
200	4.263571×10^{-4}	$4.2636(3) \times 10^{-4}$	3.212×10^{-3}	$3.2140(4) \times 10^{-3}$
500	1.203950×10^{-4}	$1.2038(1) \times 10^{-4}$	1.521×10^{-3}	$1.5218(2) \times 10^{-3}$
1000	4.483609×10^{-5}	$4.4840(4) \times 10^{-5}$	8.274×10^{-4}	$0.8267(1) \times 10^{-3}$
2000	1.642285×10^{-5}	$1.6423(2) \times 10^{-5}$	4.364×10^{-4}	$4.3751(8) \times 10^{-4}$
5000	4.283247×10^{-6}	$4.2834(5) \times 10^{-6}$	1.841×10^{-4}	$1.8394(4) \times 10^{-4}$

The agreement between the simulation and exact results for η_Ω and η_Υ in Table 6.2 is very satisfactory. The simulations yield remarkably precise results. For practical purposes, one could certainly sacrifice the last significant figure. When ten times larger error bars are acceptable, the required computer time is reduced by a factor of 100 and is hence of the order of two or three minutes for each shear rate. The computer simulation is problematic at low shear rates where a small off-diagonal component of the stress tensor has to be estimated and divided by the small shear rate in order to obtain η_Ω. On the other hand, the results obtained by numerical integration are very reliable at low shear rates, whereas the evaluation of the required integrals at high shear rates is rather difficult.

Similarly, the agreement between the simulation and exact results for the functions characterizing the normal stress coefficients in the definitions (6.8)–(6.10) is roughly as good as for the viscosity. Furthermore, the relationship $\Psi_{2,\Omega}(\dot{\gamma}) = -\Psi_{2,\Upsilon}(\dot{\gamma})$ is corroborated by the simulation results.

6.1.4 Efficiency of Simulations

We conclude this section on the Doi-Edwards and Curtiss-Bird models with a few general remarks on the efficiency of simulations. In order to describe an improved algorithm for small and large strain rates we consider shear flow as an illustrative example. In the low and high shear-rate regimes, two very different time scales (namely, λ and $1/\dot{\gamma}$) are involved in the problem. This fact should be expected to cause problems, because the time-step width Δt should be small compared to the smaller of the above physical time scales. On the other hand, statistically independent configurations are obtained only on the larger time scale. It is important to have as many as possible independent configurations in order to reduce the statistical error bars. Fortunately, the time scale λ occurs only in the stochastic differential equation (6.4) for the process S, and the time scale $1/\dot{\gamma}$ appears only in the deterministic differential equation (6.3) for the process $\boldsymbol{U}$. This observation suggests that it might be advantageous to introduce different time steps for the simulation of these two processes when $\lambda\dot{\gamma}$ is a very small or large number. In the case of steady shear flow, the analytical solution of (6.3) for $\boldsymbol{U}_t$ is available. In general, $\boldsymbol{U}_t$ can be expressed as $\boldsymbol{U}_t = \mathbf{E}(t,t')\cdot\boldsymbol{U}_{t'}/|\mathbf{E}(t,t')\cdot\boldsymbol{U}_{t'}|$, so that only the tensor $\mathbf{E}(t,t')$ defined in (4.14) must be evaluated by numerical integration. For the Doi-Edwards model, where the process S does not occur in the stress tensor but only triggers the selection of new randomly oriented unit vectors $\boldsymbol{U}$, the reflection times can be obtained by selecting them in accordance with their known distribution [31].

Although the main motivation for developing a simulation algorithm for reptation models was not the investigation of the original Doi-Edwards and Curtiss-Bird models, for which the material functions in many flow situations have been determined by other, often more efficient numerical methods [8–11], the transparency of the simulation algorithm, which makes the implementation on a computer very easy, might prompt computer simulations as a convenient tool for data comparisons with these models in time-dependent flow situations. In order to illustrate the power and flexibility of computer simulations for complex time-dependent problems, investigations of the constrained recovery after shear flow, where the shear rate is controlled in such a way that the shear stress vanishes [32], have been carried out for reptation models [33, 34]. Also recoil after simple extensional flow has been studied by means of simulations [34]. In these problems, simple programming is not the only advantage of stochastic simulations: for high shear rates, simulations are also computationally more efficient than traditional methods. The reason for this is that, when small time steps Δt are required, computer time increases as $(\Delta t)^{-1}$ for simulations and as $(\Delta t)^{-2}$ when memory integrals need to be evaluated in each time step. Similar advantages of simulations at high strain rates are also observed in steady flow situations.

Another situation in which simulations become very advantageous is when a wide range of time scales is involved in a problem, for example for polydisperse melts. In order to study polydispersity effects on the Curtiss-Bird model [13, 35–37], one needs to simulate independent copies of the reptation process for each chain length occurring in the polydisperse system, where each of these copies has its own characteristic time constant λ. The connection between the distribution of chain lengths and the distribution of time constants is given by (19.3-25) of [13]. The recoil for polydisperse melts after steady shear flow has been simulated in [38].

We show in the following sections that, when generalizations of the original Doi-Edwards and Curtiss-Bird models are considered, the advantages of computer simulations compared to other numerical methods become self-evident. Furthermore, simulations also provide the framework and show the directions in which reptation models should be generalized and improved.

6.2 Reptating-Rope Model

A very interesting generalized reptation theory is the *reptating-rope model* developed by Jongschaap and Geurts [20–23]. In this model, an attempt is made to introduce effects of correlations between the chain orientations at different positions within a single chain. It is therefore not surprising that computer simulation of the reptating-rope model requires the simulation of two copies of the reptation process [39].

The goal of this section is to describe a simple and efficient algorithm for simulating the reptating-rope model, and it serves us to illustrate that simulations are a valuable tool for studying more complicated reptation models. The reptating-rope model was developed by Jongschaap and Geurts to describe the rheological behavior of concentrated polymer solutions and melts in homogeneous flows. This reptation model formally contains the well-known Doi-Edwards and Curtiss-Bird models as special cases. The stochastic reformulation of the model required for developing a simulation algorithm not only yields high-precision results for the material functions in steady shear flow but also provides a stochastic reformulation of the reptating-rope model in which the physical content and the simplicity of the model is not obscured by lengthy mathematical derivations or complicated equations. We discuss some results for the viscometric functions of the reptating-rope model.

6.2.1 Basic Model Equations

Rather than deriving the simulation algorithm from the final equations characterizing the reptating-rope model we here give a rederivation of these equations in a form which directly leads to a simulation algorithm. In doing so, we not only obtain a self-contained stochastic description of the reptating-rope model but also a more transparent formulation of the underlying equations.

While the description of the polymer dynamics is still based on the reptation process $(\boldsymbol{U}, S)$, the most important new feature of the reptating-rope model is the derivation of the stress tensor. In this model, the polymers are represented as elastic ropes moving in confining tubes, and the stress tensor is expressed in the form

$$\boldsymbol{\tau}(t) = -N n_{\mathrm{p}} k_{\mathrm{B}} T \left\langle \boldsymbol{U}_t \boldsymbol{U}_t \right\rangle + \Delta\boldsymbol{\tau}(t) , \tag{6.19}$$

where all symbols have the same meaning as in the preceding section. The contribution $\Delta\boldsymbol{\tau}(t)$ is given by

$$\Delta\tau(t) = -n_{\rm p} L_{\rm C} \left\langle \sigma_{\rm F}(S_t, t)\, \boldsymbol{U}_t \boldsymbol{U}_t \right\rangle, \tag{6.20}$$

where L_C is the contour length of the rope and $\sigma_F(s,t)$ is the rope tension caused by the flow field at position $s \in [0,1]$ and time t. The rope tension has the dimension of a force. In (6.19), isotropic terms have been omitted because they do not contribute to the rheological behavior of incompressible liquids.

The first contribution to the stress tensor in (6.19) has the same form as the Doi-Edwards expression for the stress tensor. This contribution may be regarded to be of the form (6.20) with a constant rope tension $\sigma_B = N k_B T / L_C$ caused by the Brownian forces [20]. The tension σ_B corresponds to the force required to keep a one-dimensional elastic chain at a fixed average contour length L_C [3].

In writing down the second contribution to the stress tensor (6.20) we followed [20] (however, in [20], the variable s is assumed to vary in the interval $[0, L_C]$ rather than $[0,1]$). In later papers on the reptating-rope model [21–23] Jongschaap and Geurts introduced an additional factor of N in their expression for the stress tensor (see, for example, (4) of [20] in contrast to (1) of [21]). Since for a constant rope tension σ_F the stress tensor should be proportional to chain length rather than to the square of the chain length we use the original expression (6.20).

Notice that (6.20) requires more than local information about the orientation of the chain, because the rope tension $\sigma_F(s,t)$ in general depends on the state of the entire rope, even at previous times. This is the reason why it turns out that a simulation of a single copy of the reptation process is not sufficient for the reptating-rope model.

In order to obtain an approximate expression for the rope tension $\sigma_F(s,t)$ to be used in (6.20) one has to consider the properties of the rope, the confining tube, and the interactions between rope and tube in more detail. We assume that the elasticity of the rope is characterized by the parameter κ which is the elasticity modulus times the cross section of the rope and hence has the dimension of force. The parameter κ is the hypothetical force required in order to double the length of the rope (for this interpretation it has been assumed that the linear relationship between the applied force and the resulting elongation of the rope could be extended to large deformations). By equating the frictional and the elastic forces on the rope, one obtains the following partial differential equation,

$$\frac{1}{\kappa}\frac{\partial \sigma_F(s,t)}{\partial t} - g(s,t) = \frac{L_C}{\lambda k_B T}\frac{\partial^2 \sigma_F(s,t)}{\partial s^2}, \tag{6.21}$$

where $\sigma_F(s,t)/\kappa$ is the relative elongation of the rope and $g(s,t)$ is the time derivative of the relative elongation of the tube resulting from the applied flow. If the direction of the tube at position s and time t is given by the unit vector $\boldsymbol{u}$, then $\boldsymbol{\varkappa}(t) \cdot \boldsymbol{u}$ describes the direction and the rate for the relative elongation of the tube, and $g(s,t) = \boldsymbol{u} \cdot \boldsymbol{\varkappa}(t) \cdot \boldsymbol{u}$ its component along the tube in which the rope is confined. The time constant λ occurring in (6.21) is according to (6.4) related to the diffusive, reptational motion of the rope in the tube and hence, via the Nernst-Einstein equation, proportional to the friction coefficient between rope and tube. Equation (6.21) thus equates the frictional force due to the difference in the rate of change of the relative elongation for rope and tube and the corresponding variation of the elastic force in the rope. By using a constant value L_C in (6.21) it is tacitly assumed that the contour lengths of the polymer molecules are constant. This assumption is not only questionable for high velocity gradients but also neglects the dynamics of changes in the contour length of the rope and hence possibly eliminates time scales much shorter than the disengagement time for which there is direct experimental evidence, for example, at start-up of steady shear flow [40]. We later take the limit of inextensible ropes, $\kappa \to \infty$, in which the assumption of constant L_C is natural.

From the above explanation it is clear that (6.21) is a force balance for the rope modeling a typical polymer in a concentrated solution or melt. On the other hand, the kinetic theory derivation of the reptation dynamics is also based on a force balance for the polymers [6]. This observation immediately raises the question about the consistency of the assumed reptation dynamics and the time-evolution equation for the rope tension in the reptating-rope model. In spite of possible physical inconsistencies, which should certainly be avoided by using a common starting point for the derivation of the equations for the polymer dynamics and for the stress tensor, the final equations for the reptating-rope model are mathematically consistent. Therefore, the construction of a sound simulation algorithm is possible.

Exercise 6.7 Derive the time-evolution equation (6.21) for the rope tension from a force balance.

In the following, we restrict ourselves to the inextensible rope. However, a simulation algorithm could also be developed for the extensible rope model; computer simulations are presumably the only tool suitable for treating the very complicated extensible rope model. In the limit of an inextensible rope, $\kappa \to \infty$, the rope tension instantaneously responds to the external deformation. By using the static Green's function for the one-dimensional Laplace operator $\partial^2/\partial s^2$ occurring in (6.21), one obtains the following solution to (6.21),

$$\sigma_F(s,t) = \frac{\lambda k_B T}{2L_C} \int_0^1 [s + s' - |s - s'| - 2ss']\, g(s',t)\, ds' , \tag{6.22}$$

which obviously satisfies the boundary condition $\sigma_F(0,t) = \sigma_F(1,t) = 0$.

In order to evaluate the rope tension at position s and time t one hence needs to know the time derivative of the relative elongation of the tube, $g(s',t)$, at the same time t and a different position s' where $g(s',t)$ depends on the chain orientation at the position s'. The occurrence of an integral over s' in (6.22) suggests that we introduce another copy of the reptation process, $(\boldsymbol{U}', S')$, such that $\int_0^1 ds'$ can be regarded as an average over S'_t and the time derivative of the relative tube elongation at $s' = S'_t$ is given by $g(S'_t, t) = \boldsymbol{U}'_t \cdot \boldsymbol{\kappa}(t) \cdot \boldsymbol{U}'_t$. Then, (6.20) can be rewritten as

$$\Delta\tau(t) = -\frac{1}{2} n_{\mathrm{p}} k_{\mathrm{B}} T \lambda\, \boldsymbol{\kappa}(t) : \left\langle \left[S_t + S'_t - |S_t - S'_t| - 2S_t S'_t\right] \boldsymbol{U}'_t \boldsymbol{U}'_t \boldsymbol{U}_t \boldsymbol{U}_t \right\rangle . \quad (6.23)$$

As long as the difference $|S_t - S'_t|$ is large, the directions $\boldsymbol{U}_t$ and $\boldsymbol{U}'_t$ of the polymer at the positions S_t and S'_t, respectively, may be assumed to be independent. However, if $|S_t - S'_t|$ becomes small one needs to be worried about correlations in the chain orientation because $\boldsymbol{U}_t$ and $\boldsymbol{U}'_t$ then describe the directions of neighboring portions of a chain. In the reptating-rope model, a correlation between the two processes $(\boldsymbol{U}, S)$ and $(\boldsymbol{U}', S')$ is introduced in the following way. One first considers two stochastically independent copies of the reptation process, say $(\boldsymbol{U}, S)$ and $(\widehat{\boldsymbol{U}}, \widehat{S})$. The process $(\boldsymbol{U}', S')$ is then defined by $S'_t = \widehat{S}_t$ and

$$\boldsymbol{U}'_t = \begin{cases} \widehat{\boldsymbol{U}}_t & \text{for } |S_t - S'_t| > \Delta \\ \boldsymbol{U}_t & \text{for } |S_t - S'_t| \leq \Delta \end{cases} . \quad (6.24)$$

In other words, the processes $\boldsymbol{U}$ and $\boldsymbol{U}'$ describing the chain orientations at the positions S and S' are independent if the labels S and S' are separated at least by a "correlation length parameter" Δ $(0 \leq \Delta \leq 1)$, and the orientations $\boldsymbol{U}$ and $\boldsymbol{U}'$ coincide otherwise. This assumption is closely related to the "independent alignment assumption" of the Doi-Edwards model [12]. It is reasonable to assume $\Delta \approx 1/N$ for the correlation length parameter, where N is the number of segments in a polymer chain. Since the correlation length might be larger than the length of a single segment we write

$$\Delta = \frac{1}{2\epsilon N} , \quad (6.25)$$

where the parameter ϵ is unity over twice the correlation length in units of segment lengths and should hence satisfy the inequalities $0 \leq \epsilon \leq 1/2$.

Notice that according to the definition (6.24) the processes $(\boldsymbol{U}, S)$ and $(\boldsymbol{U}', S')$ are not on a completely equal footing. However, in the average occurring in (6.23), $(\boldsymbol{U}, S)$ and $(\boldsymbol{U}', S')$ may be interchanged without affecting the result for $\Delta\tau(t)$. Furthermore, the introduction of correlations between $(\boldsymbol{U}, S)$ and $(\boldsymbol{U}', S')$ according to (6.24) is somewhat ambiguous. The idea that a chain is composed of N segments would suggest that we should choose $\boldsymbol{U}'_t = \boldsymbol{U}_t$ when S_t and S'_t belong to the same segment, that is, when S_t and S'_t lie in the same interval $[(i-1)/N, i/N]$ for some $i \in \{1, 2, \ldots, N\}$.

It is obvious how the correlated reptation processes $(\boldsymbol{U}, S)$ and $(\boldsymbol{U}', S')$ can be simulated on a computer, and how the simulated trajectories can be used to estimate the moments in (6.23) for the contribution $\Delta\tau(t)$ to the stress tensor. The correlation between the processes $(\boldsymbol{U}, S)$ and $(\boldsymbol{U}', S')$ is specified in (6.24). A memory-integral expression for the stress tensor, which reveals the equivalence between the stochastic and conventional approaches to the reptating-rope model, can be derived from (6.19) and (6.23) in a similar manner as for the Doi-Edwards and Curtiss-Bird models in Sect. 6.1.2 [39]. The stochastic formulation is far more compact and transparent than the lengthy and complicated expressions in terms of single and two-time memory integrals given in the original papers on the reptating-rope model [22, 23].

6.2.2 Results for Steady Shear Flow

We here consider only steady shear flow with constant shear rate $\dot{\gamma}$. The viscometric functions of the reptating-rope model can according to (6.19) and (6.23) be expressed as

$$\eta(\dot{\gamma}) = \eta_{\Omega}(\dot{\gamma}) + \Delta\,\epsilon\,\eta_{\Upsilon}(\dot{\gamma})\,, \tag{6.26}$$

$$\Psi_1(\dot{\gamma}) = \Psi_{1,\Omega}(\dot{\gamma}) + \Delta\,\epsilon\,\Psi_{1,\Upsilon}(\dot{\gamma})\,, \tag{6.27}$$

$$\Psi_2(\dot{\gamma}) = \Psi_{2,\Omega}(\dot{\gamma}) + \Delta\,\epsilon\,\Psi_{2,\Upsilon}(\dot{\gamma})\,, \tag{6.28}$$

where the functions with subscript Ω are the respective viscometric functions of the Doi-Edwards model, and the factor 1/2 in (6.23) has, according to (6.25), been replaced with $\Delta\,\epsilon\,N$ in order to have a factor of N in the definition of the material functions with subscript Υ. Notice that the functions $\eta_{\Upsilon}(\dot{\gamma})$, $\Psi_{1,\Upsilon}(\dot{\gamma})$, and $\Psi_{2,\Upsilon}(\dot{\gamma})$ depend also on the correlation length parameter Δ.

The general expressions for the viscometric functions, (6.26)–(6.28), contain two important special cases:

- By formally setting $\Delta\,\epsilon = 0$ and hence neglecting all correlations between different links one recovers the Doi-Edwards model. According to (6.25), the Doi-Edwards model appears as the natural limit for $N \to \infty$. The only free parameters are the time constant λ and an overall factor N multiplying the viscometric functions.

- The occurrence of two different tangent vectors $\boldsymbol{U}_t$ and $\boldsymbol{U}'_t$ in (6.23) can, according to (6.24), be avoided by choosing $\Delta = 1$. One can then perform the average over S'_t in (6.23) and, after doing so, one recovers the Curtiss-Bird expressions for the functions $\eta_{\Upsilon}(\dot{\gamma})$, $\Psi_{1,\Upsilon}(\dot{\gamma})$, and $\Psi_{2,\Upsilon}(\dot{\gamma})$. In the Curtiss-Bird model, the parameter ϵ is treated as a free parameter rather than as fixed by the condition $\epsilon = 1/(2N)$ following from (6.25). By assuming that the orientational correlation extends over the entire polymer ($\Delta = 1$) the chains are forced to behave essentially like rigid rods (this is closely related to the "mild-curvature approximation" in the Curtiss-Bird theory). It is consistent with this observation that N has to be of order unity if the corrections to the Doi-Edwards results for the "natural" choice $\epsilon = 1/(2N)$ should produce a noticeable effect.

The zero-shear-rate results for the viscometric functions (6.26)–(6.28) which can be used to check the simulation program and to normalize the simulation results were derived in [23],

$$\eta(0) = \frac{1}{60}\left\{1 + \frac{2}{3}\Delta^2\,\epsilon\,(4 - 6\Delta + 4\Delta^2 - \Delta^3)\right\} N n_{\mathrm{p}} k_{\mathrm{B}} T \lambda\,,$$

$$\Psi_1(0) = \frac{1}{300} N n_{\mathrm{p}} k_{\mathrm{B}} T \lambda^2\,,$$

$$\Psi_2(0) = -\frac{1}{1050}\left\{1 - \Delta^2\,\epsilon\,(4 - 5\Delta + 5\Delta^3 - 4\Delta^4 + \Delta^5)\right\} N n_{\mathrm{p}} k_{\mathrm{B}} T \lambda^2\,.$$

The high-shear-rate limit of the viscosity for the reptating-rope model is [39]

$$\eta(\dot{\gamma}) \approx N n_{\mathrm{p}} k_{\mathrm{B}} T \lambda \frac{\epsilon \Delta^2 (2-\Delta)}{\lambda \dot{\gamma}} \qquad \text{for } \lambda\dot{\gamma} \to \infty . \tag{6.29}$$

Equation (6.29), which is in good agreement with the simulation results, shows that the limiting slope of the viscosity curves for $\Delta > 0$ is -1. For $\Delta = 0$ (Doi-Edwards model), the limiting slope is $-3/2$. Notice, however, that the viscosity curves for small Δ may at intermediate shear rates have slopes more negative than -1 before they reach their limiting slopes for $\lambda\dot{\gamma} \to \infty$.

The viscometric functions at intermediate shear rates can be obtained from stochastic simulation techniques. Adapting the algorithms previously developed for the Doi-Edwards and Curtiss-Bird models to the simulation of the reptating-rope model is a very simple task [39]. The total number of independent, simulated trajectories of the reptation process can be divided into two halves, one of which represents the trajectories of the process $(\boldsymbol{U}, S)$ and the other half represents the trajectories of the independent process $(\widehat{\boldsymbol{U}}, \widehat{S})$. Then, the definition (6.24) can be used to construct new pairs of trajectories for the correlated processes $(\boldsymbol{U}, S)$ and $(\boldsymbol{U}', S')$ from pairs of trajectories for the independent processes $(\boldsymbol{U}, S)$ and $(\widehat{\boldsymbol{U}}, \widehat{S})$ thus taking orientational correlations into account. With these trajectories for the two correlated processes $(\boldsymbol{U}, S)$ and $(\boldsymbol{U}', S')$ at hand, the ensemble averages in the equations (6.19) and (6.23) for the stress tensor can be estimated, and the functions introduced in (6.26)–(6.28) to characterize the viscometric functions for the reptating-rope model can be evaluated. Since the correlation length parameter Δ does not affect the stochastic time-evolution of the independent reptation processes $(\boldsymbol{U}, S)$ and $(\widehat{\boldsymbol{U}}, \widehat{S})$, the same ensemble of trajectories may be used to calculate the viscometric functions for various values of the parameter Δ. Detailed simulations of the reptating-rope model have been carried out in exactly the same way as described in Step 3 of Sect. 6.1.3, that is, with exponentially small errors due to time discretization.

Within the reptating-rope model, the viscosity and the first normal stress coefficient decrease monotonically with increasing shear rate [39]. Figure 6.4 shows the shear-rate dependence of the total viscosity for $\epsilon = 1/2$ and several values of Δ. The uppermost curve corresponds to the Curtiss-Bird model (with $\epsilon = 1/2$), whereas the lowest curve corresponds to the Doi-Edwards model. The variation of the viscometric functions for $0 \leq \Delta \leq 1$ is very similar to the variation of the corresponding Curtiss-Bird results for $0 \leq \epsilon \leq 1/2$ [9].

Simulation of the reptating-rope model requires only minor modifications of the computer programs developed for the Doi-Edwards and Curtiss-Bird models. The simulation results obtained for the viscometric functions over a wide range of shear rates are subject only to statistical errors which can easily be estimated. Except for very small shear rates, the simulations yield high-precision results for the viscometric functions. In particular, the simulations provide very precise results in the high-shear-rate regime where the numerical methods usually applied in investigating reptation theories become inefficient and may lead to incorrect results unless the required numerical integrations are

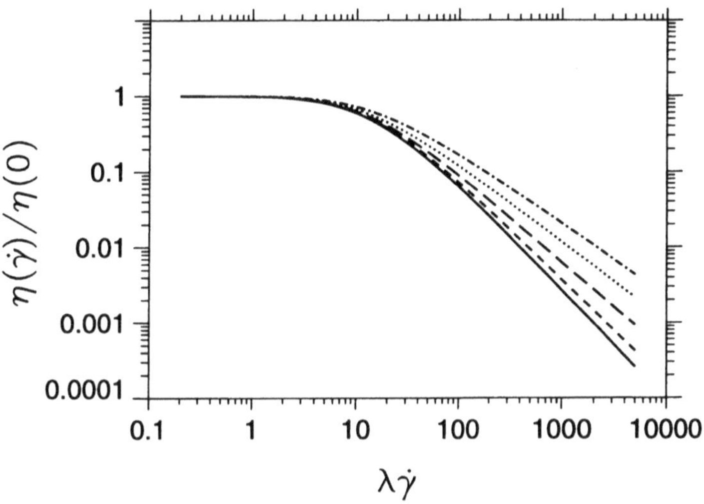

Fig. 6.4. Simulation results for the normalized viscosity as a function of the dimensionless shear rate $\lambda\dot{\gamma}$ for $\epsilon = 1/2$ and for five different values of the correlation length parameter Δ (from bottom to top: Δ = 0, 0.1, 0.25, 0.5, 1). The curve for $\Delta = 0$ corresponds to the Doi-Edwards model, and the curve for $\Delta = 1$ represents the Curtiss-Bird result for the link tension coefficient $\epsilon = 1/2$.

done extremely carefully [41]. Even more complicated reptation theories, for example the reptating-rope model for extensible ropes, can also be handled by computer simulations.

6.3 Modified Reptation Models

The simulation algorithms described in the preceding subsections can easily be modified to treat various generalizations of the Doi-Edwards and Curtiss-Bird models. In particular, the reptation process $(\boldsymbol{U}, S)$ can be modified by introducing a stochastic term into the differential equation for $\boldsymbol{U}$ or a deterministic term into the differential equation for S. Both approaches are described in this section.

For example, if the "Brownian motion" is assumed to be a superposition of the expressions obtained from the equilibration-in-momentum-space and reptation approximations (see §19.2 of [13] for details), a term involving second-order derivatives with respect to $\boldsymbol{u}$ appears in the generalization of the diffusion equation (6.1). Therefore, an additional, stochastic force on $\boldsymbol{U}_t$ needs to be introduced into the simulation algorithm. The resulting model is closely related to the modified reptation model which is discussed in Sect. 6.3.1.

On the other hand, the introduction of a deterministic term into the stochastic differential equation for S_t is necessary in order to simulate a "rigorous" constitutive equation for the reptation model that contains a first-order derivative with respect to s. Such an equation, which is considered in more detail in Sect. 6.3.2, was derived by Doi without using the "independent alignment approximation." However, in this case there is also a term proportional to the probability density p in the diffusion equation (with no derivatives) which can be simulated only by creating and annihilating segment vectors $\boldsymbol{U}$ at positions S in the same manner as in network models [14]. A drift term in the equation for S, and possibly a S-dependent diffusion term, has also been introduced by des Cloizeaux in order to obtain a broader, more realistic spectrum than from the Doi-Edwards model [42]. In particular, des Cloizeaux recommends a "basic diffusion model" which is most conveniently formulated by a Stratonovich stochastic differential equation of the form $\mathrm{d}S_t = B(S_t) \circ \mathrm{d}W_t$.

Another important modification of the original Doi-Edwards reptation model will not be discussed here in detail: the incorporation of tube-length fluctuations [43]. In the modified model, the dynamics of the chain ends, which is modeled by a non-Markovian Gaussian process, plays an important role. A variational calculation [43] and simulations of the chain end dynamics [44, 45] have shown that tube-length fluctuations lead to a molecular-weight dependence of the viscosity which is very similar to the experimentally observed 3.4 power law in the range of experimentally accessible molecular weights. From the simulation point of view, efficient methods for the simulation of memory effects and for the determination of minima and maxima over trajectories are particularly interesting [44]. While the simulation of tube-length fluctuations can be based on the theory of stochastic integrals, we here restrict ourselves to the discussion of Markovian solutions of stochastic differential equations.

In spite of the development of many modifications of the original Doi-Edwards and Curtiss-Bird models, there still is no reptation model, nor any other kinetic theory model, that can describe the linear and nonlinear rheological behavior of a real polymer melt in a fully satisfactory manner. For a more realistic description of melts, for example in flow calculations, phenomenological models, such as the integral models of the Rivlin-Sawyers or K-BKZ type [26, 46], are still much more useful than reptation models.

6.3.1 A Model Related to "Double Reptation"

In this subsection, we introduce a modified reptation model [47] which accounts for the effect of constraint release (Doi-Edwards picture) or partially anisotropic Brownian motion (Curtiss-Bird picture). The formulation of the model is based exclusively on simplicity and various consistency criteria, such as the fluctuation-dissipation theorem and thermodynamic consistency (see [48] for a discussion of such criteria in connection with models for dilute polymer solutions). While the proposed dynamics has been considered before, the expression for the stress tensor deviates from that previously suggested. Actually, the model presented here was developed as a simple example for illustrating one of the main themes of this book: the idea of modifying existing models in the framework of stochastic processes in order to make them more realistic without sacrificing their tractability.

After introducing the model and discussing the basic assumptions we derive a memory-integral expression for the stress tensor from which we obtain the linear viscoelastic behavior and the zero-shear-rate viscosity. The expression for the relaxation modulus is similar to the one obtained from the idea of double reptation [49] so that the new model may be regarded as an extension of double reptation into the regime of nonlinear viscoelasticity. As an example of a nonlinear viscoelastic property, we discuss high-precision simulation results for the shear-rate-dependent viscosity and its asymptotic behavior at high shear rates. Furthermore, it is described how polydisperse melts can be modeled.

In the notation of the foregoing sections of this chapter, the simplest modification of the traditional reptation models is obtained by adding a stochastic term to the usually deterministic differential equation for $\boldsymbol{U}_t$. Then, the process $\boldsymbol{U}$ satisfies the Itô stochastic differential equation

$$d\boldsymbol{U}_t = (\boldsymbol{\delta} - \boldsymbol{U}_t\boldsymbol{U}_t) \cdot \left[\boldsymbol{\kappa}(t) \cdot \boldsymbol{U}_t \, dt + \frac{\pi\bar{\epsilon}}{(3\lambda)^{1/2}} \, d\boldsymbol{W}_t\right] - \frac{\pi^2\bar{\epsilon}^2}{3\lambda} \boldsymbol{U}_t \, dt \,, \qquad (6.30)$$

where $\bar{\epsilon}$ is a real number related to the reptation coefficient ϵ' of [13]. In this equation, a randomizing effect, which is expressed in terms of the three-dimensional Wiener process $\boldsymbol{W}_t$, is superimposed on the deterministic rotation of $\boldsymbol{U}_t$ with the flow field. Equation (6.30) is the time-evolution equation for a rigid dumbbell with time constant $\lambda^{\rm rd} := \lambda/(\pi^2\bar{\epsilon}^2)$ previously formulated in Exercise 5.9. The usual deterministic differential equation for $\boldsymbol{U}$ is recovered for $\bar{\epsilon} = 0$. The remaining dynamical equation (6.4) for S and the boundary conditions (6.2) are chosen exactly as in the original Doi-Edwards and Curtiss-Bird models (where the Wiener processes in (6.4) and (6.30) are independent).

On the level of diffusion equations, the idea of introducing noise into the time-evolution equation for $\boldsymbol{U}$ has been suggested previously. Indeed, after a suitable redefinition of parameters, the stochastic differential equations formulated here are equivalent to the diffusion equation (19.3-26) of [13]. In the spirit of the Curtiss-Bird model, the new term involving $\bar{\epsilon}$ results from a superposition of the equilibration-in-momentum-space and reptation assumptions for the Brownian forces on a chain in a concentrated system. In the spirit of the Doi-Edwards model, this term may be regarded as a constraint release mechanism which allows configurational relaxation to take place not only at the chain ends (for other descriptions of constraint release, see [49] and the references given on p. 238 and p. 282 of [12]). Since the release of constraints occurs through the reptational motion of surrounding chains forming a tube, and hence occurs on the same time scale as the reptation of a probe chain, the parameter $\bar{\epsilon}$ should be a number of order unity. In particular, we assume $\bar{\epsilon}$ to be independent of molecular weight. For polymers which are long enough to be massively entangled we then expect $\bar{\epsilon}$ to be a universal number.

In the absence of flow, $\boldsymbol{U}_t$ and S_t are uniform random variables on the unit sphere and on the interval $[0,1]$, respectively. In other words, $\boldsymbol{U}_t$ and S_t are distributed according to the Boltzmann distribution; this observation may be regarded as a verification of the fluctuation-dissipation theorem.

While the reptation dynamics formulated above has previously been introduced in an equivalent form in [13], the stress tensor expression considered here,

$$\begin{aligned} \boldsymbol{\tau}(t) &= N n_{\mathrm{p}} k_{\mathrm{B}} T \Big\{ \frac{1}{3}\boldsymbol{\delta} - \big\langle \boldsymbol{U}_t \boldsymbol{U}_t \big\rangle - \epsilon\lambda\, \boldsymbol{\kappa}(t) : \big\langle S_t(1-S_t)\, \boldsymbol{U}_t\boldsymbol{U}_t\boldsymbol{U}_t\boldsymbol{U}_t \big\rangle \\ &\quad - \bar{\bar{\epsilon}}\lambda\, \boldsymbol{\kappa}(t) : \big\langle \boldsymbol{U}_t\boldsymbol{U}_t\boldsymbol{U}_t\boldsymbol{U}_t \big\rangle \Big\}, \end{aligned} \tag{6.31}$$

deviates significantly from the one of [13]. In (6.31), ϵ is the link tension coefficient of the Curtiss-Bird model, and $\bar{\bar{\epsilon}}$ is another dimensionless parameter.

In [13], the anisotropic term $-\frac{1}{2} N n_{\mathrm{p}} k_{\mathrm{B}} T\, \pi^2 \bar{\epsilon}^2 \langle S_t(1-S_t)\, \boldsymbol{U}_t\boldsymbol{U}_t \rangle$ appears in the stress tensor instead of the term above involving $\bar{\bar{\epsilon}}$. Such a contribution, however, is inconsistent with a very general, systematic formulation of the time-evolution equations for nonequilibrium systems developed by Jongschaap [50, 51] (this problem does not occur for the original Curtiss-Bird model for which $\bar{\epsilon} = 0$). On the other hand, additional terms proportional to $\boldsymbol{\kappa}(t)$ are not ruled out by Jongschaap's formalism. Although the validity of Jongschaap's formalism may be questioned we here adopt it as a thermodynamic consistency criterion. We hence keep the term proportional to the link tension coefficient ϵ of the Curtiss-Bird model, and we add a simple new term proportional to $\bar{\bar{\epsilon}}$ such that for $\bar{\bar{\epsilon}} = 2/(\pi^2\bar{\epsilon}^2)$ the stress tensor contribution of rigid dumbbells with dynamics given by (6.30) is recovered. Note that stress tensor contributions proportional to $\boldsymbol{\kappa}(t)$ lead to a violation of the stress-optical law, and that the coefficients ϵ and $\bar{\bar{\epsilon}}$ should hence be small. For the link tension coefficient ϵ, dynamic viscosity measurements indeed suggest that ϵ should be small [13], and it has been argued above that ϵ should be of the order of $1/N$ (see Sect. 6.2.2). While experimental data for the shear-rate dependence of the viscosity suggest rather large values of ϵ (in the range from 0.15 to 1), data for the extensional viscosity in simple extension can be fitted best with a small value of ϵ ($\epsilon = 0.05$) [37]. Experimental or theoretical estimates of the small parameter $\bar{\bar{\epsilon}}$ still need to be determined. Since the presence of the noise term in (6.30) is expected to reduce the decay of the viscosity with shear rate and to avoid the large recoil effects of the Doi-Edwards model, small values of ϵ and $\bar{\bar{\epsilon}}$ should be sufficient for predicting realistic material behavior.

In the same way as in Sect. 6.1.2, one can formulate the stress tensor in the form of a memory-integral expression,

$$\begin{aligned} \boldsymbol{\tau}(t) &= N n_{\mathrm{p}} k_{\mathrm{B}} T \Big\{ \frac{1}{3}\boldsymbol{\delta} - \int_{-\infty}^{t} dt' \Big[\bar{\mu}(t-t') \big\langle \boldsymbol{U}_t\boldsymbol{U}_t \big\rangle_{t'}^{\mathrm{rd}} \\ &\quad + \big(\epsilon\, \bar{\nu}(t-t') + \bar{\bar{\epsilon}}\, \bar{\mu}(t-t') \big) \lambda\, \boldsymbol{\kappa}(t) : \big\langle \boldsymbol{U}_t\boldsymbol{U}_t\boldsymbol{U}_t\boldsymbol{U}_t \big\rangle_{t'}^{\mathrm{rd}} \Big] \Big\}, \end{aligned} \tag{6.32}$$

where $\overline{\mu}(t)$ was introduced in (3.91),

$$\overline{\nu}(t) := \frac{16}{\pi^2\lambda} \sum_{\substack{n=1 \\ n \text{ odd}}}^{\infty} \frac{1}{n^2} \, \mathrm{e}^{-n^2\pi^2 t/\lambda} , \tag{6.33}$$

previously occurred in the solution to Exercise 6.3, and $\langle \ldots \rangle_{t'}^{\mathrm{rd}}$ denotes an expectation for rigid dumbbells which were at equilibrium at the time t' and have been exposed to the flow after t' (the time constant for these dumbbells is $\lambda^{\mathrm{rd}} = \lambda/(\pi^2\overline{\epsilon}^2)$). Note that the expression (6.32) is a rather formal result because the evaluation of the rigid dumbbell averages $\langle \ldots \rangle_{t'}^{\mathrm{rd}}$ is nontrivial. The importance of rigid dumbbell results for a modified reptation model with the same dynamics as considered here has previously been emphasized in [13].

Example 6.8 Linear Viscoelastic Behavior
Equation (6.32) is a convenient starting point for evaluating the relaxation modulus $G(t)$. Due to the presence of the prefactor $\mathsf{x}(t)$, the fourth moments of $\boldsymbol{U}_t$ can be replaced by their equilibrium values, and the linear viscoelastic limit of the second moments of rigid dumbbells can be found in [13] (from the results of §14.5 of [13], even the second-order memory-integral expansion can be constructed in a rigorous manner). One obtains for the relaxation modulus introduced in (1.15)

$$G(t) = N n_{\mathrm{p}} k_{\mathrm{B}} T \lambda \left[\frac{1}{45} (\epsilon + 6\overline{\overline{\epsilon}}) \delta(t) + \frac{1}{10} \overline{\nu}(t) \mathrm{e}^{-\overline{\epsilon}^2 \pi^2 t/\lambda} \right] . \tag{6.34}$$

The relaxation times following from (6.34) are of the form

$$\lambda_n = \frac{1+\overline{\epsilon}^2}{n^2+\overline{\epsilon}^2} \lambda_1, \quad \lambda_1 = \frac{\lambda}{(1+\overline{\epsilon}^2)\pi^2}, \quad n \text{ odd} . \tag{6.35}$$

For increasing $\overline{\epsilon}$, the shorter relaxation times are closer to the first one. In that sense, the longest relaxation time becomes less dominant, and the spectrum hence becomes more realistic. However, the total weight of all the higher relaxation times relative to the weight of the first relaxation time, $(\pi^2/8) - 1 \approx 0.23$, is not influenced by the presence of the new term (in contrast, double reptation gives $(\pi^4/64) - 1 \approx 0.52$). Note that the expression (6.34) for the relaxation modulus is closely related to the idea of double reptation which accounts for the fact that a constraint imposed by a surrounding chain on a probe chain can be removed by the reptational motion of either of the two chains [49]. According to double reptation, the non-instantaneous contribution to the modulus $G(t)$ of the original Doi-Edwards model obtained for $\epsilon = \overline{\overline{\epsilon}} = 0$ is multiplied by $G(t)/G(0)$. A similar additional factor is here given by the time dependence of the relaxation modulus for rigid dumbbells with a comparable relaxation time, which is $\exp\{-\overline{\epsilon}^2\pi^2 t/\lambda\}$. The success of double reptation suggests that a more realistic behavior should also result for the present modification of the original reptation models. While the idea of double reptation is limited to linear viscoelasticity, the model introduced in this section allows the prediction of the rheological behavior in the nonlinear regime as well. □

The success of the double reptation idea is truly spectacular when it is applied to *polydisperse melts* [49]. In that case, the time constants λ in (6.30) and (6.4) should be sampled independently according to suitable probability distributions. The time constant λ of the surrounding chains occurring in (6.30) must be sampled according to the mass fraction of chains with the corresponding molecular weight (see (19.3-25) of [13] for this correspondence between molecular weight and time constant). The time constant λ of the reptating probe chain in (6.4) must be selected randomly according to the number fraction of chains with the corresponding molecular weight (notice that the selected molecular weight of the probe chain enters through the factor N in the stress tensor expression (6.31), too). This simple procedure results in a rather involved mixing rule for nonlinear rheological properties due to the coupling of the time constants of molecules with different molecular weights in the stochastic differential equations of motion. The mixing rule given here differs crucially from the one implied by (19.3-26) of [13], which corresponds to having equal time constants λ in (6.4) and (6.30). The development of successful mixing rules is an extremely important task because they provide the key to material characterization via polymer melt rheology.

The nonlinear rheological behavior of the new model suggested here satisfies the *principle of material objectivity*. This principle—also associated with the key words *frame invariance* or *rheological invariance*—states the following: if two deformation histories of a material differ only by time-dependent overall rotations and translations, then the history-dependent stress tensors at any given time differ only by the rotational transformation corresponding to that time (see § 9.1 of [26] or Sect. 3.3 of [46]). The material objectivity of the modified reptation model can be shown by analyzing the transformation behavior of the equation of motion (6.30) and of the stress tensor expression (6.31) under time-dependent rotations (cf. the procedure for dilute solution models in [48]).

From the relaxation modulus (6.34) we obtain the zero-shear-rate viscosity

$$\eta(0) = \frac{1}{60} N n_{\mathrm{p}} k_{\mathrm{B}} T \lambda \left\{ \frac{12}{\pi^2 \bar{\epsilon}^2} \left[1 - \frac{\tanh(\pi\bar{\epsilon}/2)}{\pi\bar{\epsilon}/2} \right] + \frac{2}{3}\epsilon + 4\bar{\bar{\epsilon}} \right\} . \tag{6.36}$$

For $\bar{\epsilon} = \bar{\bar{\epsilon}} = 0$, we recover (6.11).

Exercise 6.9 Derive (6.36).

In order to find the shear-rate-dependent viscosity we have simulated the reptation process $(\boldsymbol{U}, S)$ for the modified model. Following the ideas of Step 2 of Sect. 6.1.3, a first-order algorithm has been developed for integrating (6.30). In Fig. 6.5, the viscosity of the modified reptation model for $\bar{\epsilon} = 1$, $\epsilon = \bar{\bar{\epsilon}} = 0$ obtained by high-precision simulations as a function of dimensionless shear rate $\lambda\dot{\gamma}$ is compared to the corresponding result for the Doi-Edwards model ($\epsilon = \bar{\epsilon} = \bar{\bar{\epsilon}} = 0$). The relative statistical errors are of the order of 10^{-3}. The fact that increasing $\bar{\epsilon}$ leads to a smaller relaxation time λ_1 (see (6.35)) implies that, for the modified model, deviations from the zero-shear-rate viscosity appear only at

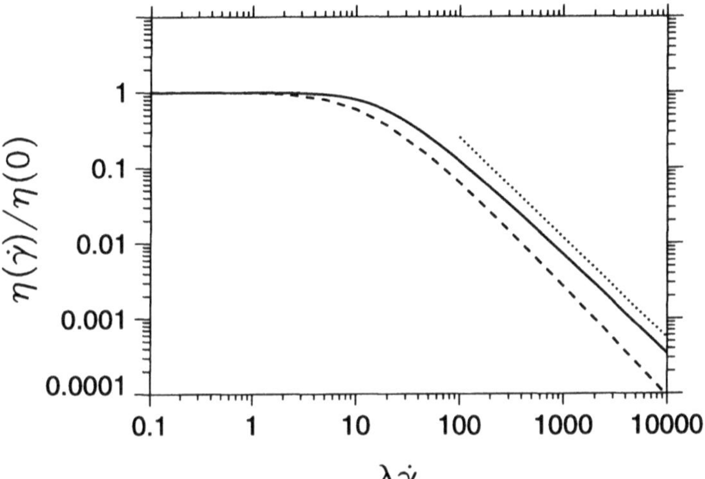

Fig. 6.5. The normalized viscosity as a function of the dimensionless shear rate $\lambda\dot\gamma$ for the Doi-Edwards model (*dashed line*) and the new reptation model for $\bar\epsilon = 1$, $\epsilon = \bar{\bar\epsilon} = 0$ (*continuous line*). At high shear rates, the asymptotic viscosity curve for the new reptation model should possess the slope $-4/3$ in the double logarithmic presentation; this slope is indicated by the *dotted line*.

higher dimensionless shear rates $\lambda\dot\gamma$. In addition, the decay of the viscosity with shear rate is slower for the modified model. While for the Doi-Edwards model the asymptotic behavior of the viscosity is given by $\eta(\dot\gamma) \sim \dot\gamma^{-3/2}$, rigid-rod results imply [52, 53] $\eta(\dot\gamma) \sim \dot\gamma^{-4/3}$ for the modified model. For the new model, smaller values of ϵ (or $\bar{\bar\epsilon}$) than usually assumed in the Curtiss-Bird model should hence prevent the viscosity from decreasing more rapidly than $\dot\gamma^{-1}$ or, in other words, prevent the shear stress from being a nonmonotonic function of shear rate.

In summary, the modified reptation model considered in this subsection is more realistic than the original reptation models. The linear viscoelastic behavior of the new model is less dominated by a single relaxation time. In the nonlinear viscoelastic regime, which can easily be studied by computer simulations, the decay of the viscosity with shear rate is weakened and thus more realistic. In view of the additional mechanism for orientational relaxation, which acts not only at the chain ends, the new model may be expected to perform better than the original reptation models when flows with strain reversal or recoil experiments are considered. The mixing rules for polydisperse melts are another promising application of the new model.

6.3.2 Doi-Edwards Model Without Independent Alignment

A generalization of the Doi-Edwards model, which has been derived without making use of the independent alignment approximation, is sometimes referred to as "rigorous constitutive equation for the reptation model" (see Sect. 7.9 of [12]). Within this more rigorous approach, the diffusion equation (6.1) for the probability density $p(\boldsymbol{u}, s, t)$ is modified as follows,

$$\begin{aligned}\frac{\partial p(\boldsymbol{u}, s, t)}{\partial t} &= -\frac{\partial}{\partial \boldsymbol{u}} \cdot \left\{ [\boldsymbol{\varkappa}(t) \cdot \boldsymbol{u} - \boldsymbol{\varkappa}(t) : \boldsymbol{u}\boldsymbol{u}\boldsymbol{u}] \, p(\boldsymbol{u}, s, t) \right\} + \frac{1}{\lambda} \frac{\partial^2 p(\boldsymbol{u}, s, t)}{\partial s^2} \\ &\quad - \frac{\partial}{\partial s} [\boldsymbol{\varkappa}(t) : \hat{\boldsymbol{\tau}}(s, t) \, p(\boldsymbol{u}, s, t)] + \boldsymbol{\varkappa}(t) : \boldsymbol{u}\boldsymbol{u} \, p(\boldsymbol{u}, s, t) \,, \end{aligned} \tag{6.37}$$

where, for $s \leq 1/2$,

$$\hat{\boldsymbol{\tau}}(s, t) := - \int_s^{1/2} \mathrm{d}s' \int \mathrm{d}^3 u \, p(\boldsymbol{u}, s', t) \, \boldsymbol{u}\boldsymbol{u} \,, \tag{6.38}$$

and $\hat{\boldsymbol{\tau}}(s, t) := -\hat{\boldsymbol{\tau}}(1 - s, t)$ for $s > 1/2$. The boundary conditions (6.2) remain unchanged. The symmetry around $s = 1/2$ can actually be exploited to restrict the process S to the interval $[0, 1/2]$, where $1/2$ is a reflecting boundary for S which has no effect on $\boldsymbol{U}$. If this symmetry is exploited, the doubled probability for finding certain S values should be corrected by introducing a factor of $1/2$ in (6.38).

The first-order derivative term in the second line of (6.37) corresponds to a systematic drift. The physical significance of this term is as follows. If a probe chain is trying to follow the flow field then its contour length tends to be changed. Such changes in contour length relax very rapidly compared to the reptation process. During relaxation the chain deforms within the tube (the chain typically retracts into the deformed tube), and the tube orientation is felt by other segments; as a result, the process S exhibits a drift depending on the flow rate and the local orientation in the same way as the rate of change of the relative elongation occurring in (6.21). The resulting change of probability at the chain ends is actually compensated by the derivative-free last term in (6.37). Such a term can be interpreted as a formation or destruction of configurations [14]; depending on the sign of this term, configurations have to be introduced or removed with a proper probability. The more rigorous version of the Doi-Edwards model is hence mathematically similar to network models, in which the formation and destruction of network strands plays an important role, and it can be simulated accordingly [14].

Notice that (6.37) is nonlinear in p because the drift term involves an average. If the symmetry of the problem is used to restrict S to the interval $[0, 1/2]$, then one can express this average as

$$\hat{\boldsymbol{\tau}}(s, t) := -\frac{1}{2} \left\langle \boldsymbol{U}_t \boldsymbol{U}_t \, \chi_{S_t^{-1}([s, 1/2])} \right\rangle \,. \tag{6.39}$$

This situation is very similar to what we found for approximate descriptions of hydrodynamic interaction, for the FENE-P model, and for rod models of liquid crystal polymers. Equation (6.39) also shows how the stress tensor expression for the Doi-Edwards model, which is not modified in the more rigorous model, can be expressed in terms of $\hat{\boldsymbol{\tau}}(s,t)$,

$$\boldsymbol{\tau}(t) = N n_{\rm p} k_{\rm B} T \left[\frac{1}{3}\,\boldsymbol{\delta} + 2\,\hat{\boldsymbol{\tau}}(0,t) \right] . \tag{6.40}$$

Simulation of the model discussed here is rather involved. It is best to introduce an approximation to the above model before simulating it. We divide the interval $[0, 1/2]$ into n bins of equal length, and we approximate $\hat{\boldsymbol{\tau}}(s,t)$ by a piecewise linear function $\hat{\boldsymbol{\tau}}^n(s,t)$ for which the values at the ends of each bin are determined from (6.38). Then, only a finite number of averages is involved in the approximate model. Furthermore, in each bin, the derivative-free formation and destruction term is averaged over the bin. By this construction we achieve the following: (i) the full problem can be approximated with any desired degree of accuracy by choosing n sufficiently large, and (ii) the probability density for S_t, that is $\int \mathrm{d}^3u\, p(\boldsymbol{u}, s, t)$, is independent of s and t (for proper initial conditions).

While the simulation of the drift and diffusion terms is straightforward, for example by numerical integration of

$$\mathrm{d}S_t = \boldsymbol{\varkappa}(t) : \hat{\boldsymbol{\tau}}^n(S_t, t)\, \mathrm{d}t + \sqrt{\frac{2}{\lambda}}\, \mathrm{d}W_t \,, \tag{6.41}$$

where $\hat{\boldsymbol{\tau}}^n(S_t, t)$ is estimated from an ensemble of configurations, the formation and destruction term is considerably more delicate. When restoring the proper fixed number of configurations in each bin, the formation or destruction of unit vectors is determined in a complicated manner by all the orientations of the existing configurations in that bin. Most delicate is the bin $[0, 1/(2n)]$ because, in an interval of size proportional to $\sqrt{\Delta t}$, also randomly oriented configurations have to be introduced which correspond to the reflected configurations in the original Doi-Edwards model.

The fact that even the more rigorous Doi-Edwards model can be simulated shows how powerful stochastic simulation techniques are. The fact that the simulation becomes rather involved indicates that there may be other, more natural ways of modifying the original model. Since the drift due to contour length relaxation depends only on averaged configurations it would, from a simulation point of view, be more natural to replace the last term in (6.37) by the following formation and destruction term,

$$\boldsymbol{\varkappa}(t) : \left[\int \mathrm{d}^3u\, p(\boldsymbol{u}, s, t)\, \boldsymbol{u}\boldsymbol{u} \right]\, p(\boldsymbol{u}, s, t)$$

(cf. footnote on p. 277 of [12]). This replacement would simplify the simulations considerably. Even more natural would be the simulation of chains rather than segments. The contour length relaxation could then depend on the actual chain configuration, and the occurrence of averages in the drift and formation/destruction terms, which are difficult to simulate, could be avoided. At the same time, more aspects of the full polymer dynamics could be captured in chain models. The goal of developing models which are easier to simulate automatically leads in the direction of more realistic models.

References

1. de Gennes PG (1971) J. Chem. Phys. 55: 572
2. Doi M, Edwards SF (1978) J. Chem. Soc. Faraday Trans. II 74: 1789
3. Doi M, Edwards SF (1978) J. Chem. Soc. Faraday Trans. II 74: 1802
4. Doi M, Edwards SF (1978) J. Chem. Soc. Faraday Trans. II 74: 1818
5. Doi M, Edwards SF (1979) J. Chem. Soc. Faraday Trans. II 75: 38
6. Curtiss CF, Bird RB (1981) J. Chem. Phys. 74: 2016
7. Curtiss CF, Bird RB (1981) J. Chem. Phys. 74: 2026
8. Bird RB, Saab HH, Curtiss CF (1982) J. Phys. Chem. 86: 1102
9. Bird RB, Saab HH, Curtiss CF (1982) J. Chem. Phys. 77: 4747
10. Saab HH, Bird RB, Curtiss CF (1982) J. Chem. Phys. 77: 4758
11. Fan XJ, Bird RB (1984) J. Non-Newtonian Fluid Mech. 15: 341
12. Doi M, Edwards SF (1986) *The Theory of Polymer Dynamics*. Clarendon Press, Oxford (International Series of Monographs on Physics, Vol 73)
13. Bird RB, Curtiss CF, Armstrong RC, Hassager O (1987) *Dynamics of Polymeric Liquids, Vol 2, Kinetic Theory*, 2nd Edn. Wiley-Interscience, New York Chichester Brisbane Toronto Singapore
14. Petruccione F, Biller P (1988) J. Chem. Phys. 89: 577
15. Petruccione F, Biller P (1988) Rheol. Acta 27: 557
16. Petruccione F (1989) Continuum Mech. Thermodyn. 1: 97
17. Biller P, Petruccione F (1990) J. Chem. Phys. 92: 6322
18. Hermann W, Petruccione F (1992) J. Rheol. 36: 1461
19. Øksendal B (1985) *Stochastic Differential Equations*. Springer, Berlin Heidelberg New York Tokyo
20. Jongschaap RJJ (1988) in: Giesekus H, Hibberd MF (eds) *Progress and Trends in Rheology II*: Proceedings of the Second Conference of European Rheologists. Steinkopff, Darmstadt, p 99 (Supplement to Rheologica Acta)
21. Geurts BJ, Jongschaap RJJ (1988) J. Rheol. 32: 353
22. Geurts BJ (1988) J. Non-Newtonian Fluid Mech. 28: 319
23. Geurts BJ (1989) J. Non-Newtonian Fluid Mech. 31: 27
24. Lodge TP, Rotstein NA, Prager S (1990) Adv. Chem. Phys. 79: 1
25. Öttinger HC (1989) J. Chem. Phys. 91: 6455
26. Bird RB, Armstrong RC, Hassager O (1987) *Dynamics of Polymeric Liquids, Vol 1, Fluid Mechanics*, 2nd Edn. Wiley-Interscience, New York Chichester Brisbane Toronto Singapore
27. Ferry JD (1980) *Viscoelastic Properties of Polymers*, 3rd Edn. Wiley, New York Chichester Brisbane Toronto Singapore
28. Öttinger HC (1990) J. Non-Newtonian Fluid Mech. 37: 265
29. Strittmatter W (1988) *Numerische Behandlung von Stochastischen Dynamischen Systemen*. Ph.D. Thesis, University of Freiburg

30. Öttinger HC (1991) in: Roe RJ (ed) *Computer Simulation of Polymers*. Prentice-Hall, Englewood Cliffs, p 188 (Polymer Science and Engineering Series)
31. Petruccione F, Biller P (1990) J. Chem. Phys. 92: 6327
32. Meissner J (1975) Rheol. Acta 14: 201
33. Öttinger HC (1990) in: Oliver DR (ed) *Proceedings of the Third European Rheology Conference and Golden Jubilee Meeting of the British Society of Rheology*. Elsevier, London New York, p 381
34. Borgbjerg U, de Pablo JJ, Öttinger HC (1994) J. Chem. Phys. 101: 7144
35. Schieber JD, Curtiss CF, Bird RB (1986) Ind. Eng. Chem. Fundam. 25: 471
36. Schieber JD (1987) J. Chem. Phys. 87: 4917
37. Schieber JD (1987) J. Chem. Phys. 87: 4928
38. Borgbjerg U, de Pablo JJ (1995) Macromolecules 28: 4540
39. Öttinger HC (1990) J. Chem. Phys. 92: 4540
40. Pearson D, Herbolzheimer E, Grizzuti N, Marrucci G (1991) J. Polym. Sci. Polym. Phys. Ed. 29: 1589
41. Geurts BJ (1989) J. Non-Newtonian Fluid Mech. 33: 349
42. des Cloizeaux J (1990) Macromolecules 23: 3992
43. Doi M (1983) J. Polym. Sci. Polym. Phys. Ed. 21: 667
44. Ketzmerick R, Öttinger HC (1989) Continuum Mech. Thermodyn. 1: 113
45. O'Connor NPT, Ball RC (1992) Macromolecules 25: 5677
46. Larson RG (1988) *Constitutive Equations for Polymer Melts and Solutions*. Butterworths, Boston London Singapore Sydney Toronto Wellington (Butterworths Series in Chemical Engineering)
47. Öttinger HC (1994) Phys. Rev. E 50: 4891
48. Schieber JD, Öttinger HC (1994) J. Rheol. 38: 1909
49. des Cloizeaux J (1988) Europhys. Lett. 5: 437
50. Jongschaap RJJ (1991) in: Casas-Vázquez J, Jou D (eds) *Rheological Modelling: Thermodynamical and Statistical Approaches*. Springer, Berlin Heidelberg New York London Paris Tokyo Hong Kong Barcelona Budapest, p 215 (Lecture Notes in Physics, Vol 381)
51. Jongschaap RJJ (1991) Appl. Sci. Res. 48: 117
52. Stewart WE, Sørensen JP (1972) Trans. Soc. Rheol. 16: 1
53. Öttinger HC (1988) J. Rheol. 32: 135

Landmark Papers and Books

In the following, some landmark contributions to the theory of Brownian motion, to polymer kinetic theory, to stochastic calculus, and to simulations based on stochastic differential equations are compiled. When writing this manuscript, the author became increasingly fascinated by the historical development of all these field that form the background for this book. Particularly fascinating are the deep physical and mathematical insights found in many of the early papers, and both the fruitful interrelations and the missing links between the different fields. Of course, such a selection of landmark papers and books must be rather subjective, and it may express mainly the author's personal taste and background. Nevertheless, the author wanted to offer this personally biased survey to those readers who are interested in the history and development of science.

1905 The cosmopolitan physicist Albert Einstein (1879–1955) offers a kinetic theory interpretation of Brownian motion and translational diffusivity, and he shows how this interpretation can be used to determine Avogadro's number (Ann. d. Physik 17, 549–560 [in German]). In the course of 1905, Einstein supplements his work by a more general approach based on the theory of fluctuations at equilibrium which can be applied also to rotational diffusion, and he discusses the limitations of his results for short times; his supplementary results were published in 1906 (Ann. d. Physik 19, 371–381 [in German]).

1906 The Polish physicist Marian von Smoluchowski (1872–1917), born and educated in Austria, publishes his kinetic theory approach to Brownian motion (Ann. d. Physik 21, 756–780 [in German]). He considers the frequent small changes in the direction of motion of a Brownian particle due to collisions with small particles, and he compares his results to all the experimental observations he could find in the literature. Although he knew Einstein's papers, von Smoluchowski decided to publish his independently developed approach, "da mir meine Methode direkter, einfacher und darum vielleicht auch überzeugender zu sein scheint als jene Einsteins."

1908 The French physicist Paul Langevin (1872–1946) gives "une démonstration infiniment plus simple" of Einstein's theory by formulating a stochastic differential equation for a Brownian particle (Comptes rendus 146, 530-533 [in French]). By considering the stochastic differential equation for the square of the position he can directly use the principle of equipartition of energy, and hence does not need to specify the second moments of the stochastic force.

1915 Marian von Smoluchowski formulates the diffusion equation for the probability density of a Brownian particle in position space in the presence of external forces (Ann. d. Physik 48, 1103–1112 [in German]). He discusses the solutions for various special cases, including absorbing and reflecting boundary conditions.

1917 The German physicist Max Planck (1858–1947) derives the multivariate Fokker-Planck equation in an abstract configuration space (Sitzber. Königl. Preuß. Akad. Wiss., 323–341 [in German]). Furthermore, he discusses the concept of a probability current, the condition of vanishing probability current which leads to stationary solutions, the principle of detailed balance, and generalizations to jump processes which are required in describing quantum phenomena. Planck's work was motivated by a paper on electric dipoles rotating in a radiation field published by the Dutch physicist Adriaan Daniël Fokker (1887–1972) in 1914 (Ann. d. Physik 43, 810–820 [in German]) in which the condition of vanishing probability current is given without proof (as a general tool for solving stochastic problems which exhibit "eine Ähnlichkeit mit dem Problem der B r o w n schen Bewegung").

1919 Starting from the Langevin equation for the position and velocity of a Brownian particle, the Dutch physicist Leonard Salomon Ornstein (1880–1941) derives separate Fokker-Planck equations for the probability densities in velocity space and position space (Proceedings Royal Acad. Amsterdam 21, 96–108). In his discussion of an objection to the Langevin treatment of Brownian motion raised by J. D. van der Waals Jr. and A. Snethlage, he notices the necessity of the Stratonovich interpretation of the noise in the stochastic differential equation for the square of the velocity (in the notation of Sect. 3.1.1, he uses the assumption $\langle V_t F_t^{\mathrm{B}} \rangle = k_{\mathrm{B}} T \zeta / M$).

1921 The Swedish physicist Oskar Klein (1894–1977) derives the Fokker-Planck equation for a system of interacting Brownian particles in the presence of external forces (Arkiv för matematik, astronomi, o. fysik 16, N:o 5, 1–51 [in German]). He treats both position and momentum variables, and he describes two different ways of eliminating the momentum variables; hydrodynamic interactions are neglected.

1923 In connection with Brownian motion, the US-American mathematician Norbert Wiener (1894–1964) introduces the concept of "differential-space" (J. Math. and Phys. 2, 131–174). In modern terms, his "differential-space" corresponds to the probability space $(\mathbb{R}^{\mathbb{T}}, \mathcal{B}^{\mathbb{T}}, P^W)$, where $\mathbb{T} = [0, t_{\mathrm{max}}]$ and P^W is the measure induced by the Wiener process. The main purpose of Wiener's paper is to find classes of functions from $\mathbb{R}^{\mathbb{T}}$ to $\mathbb{R}$ (that is, functionals) for which expectations with respect to P^W can be defined. In his measure theoretical approach to Brownian motion, Wiener shows that "it is infinitely improbable, under our distribution of functions $f(t)$, that $f(t)$ be a continuous function of limited total variation, and in particular that it have a bounded derivative." He also investigates the continuity of the functions $f(t)$.

1933 The polymath Russian scientist Andrei Nikolaevich Kolmogorov (1903–1987) introduces the rigorous axiomatic formulation of mathematical stochastics. His elegant and amazingly modern booklet *Grundbegriffe der Wahrscheinlichkeitsrechnung* (Springer, Berlin [in German]) immediately elevates probability theory to the level of modern mathematics.

1934 The Swiss chemist and physicist Werner Kuhn (1899-1963) develops the random coil model for polymers in dilute solution (Kolloid-Z. 68, 2–15 [in German]). He derives the probability distribution for the end-to-end vector and the average size and shape of random coils. He expresses the size of statistically independent segments in terms of bond lengths and bond angles ("Kuhn step", "Kuhn length"). Realizing the importance of excluded volume effects, Kuhn introduces the idea of an exponent ν for the molecular weight dependence of the polymer size, and with some intuition he estimates $\nu \approx 0.61$. The close relationship between the assumption of a power law and self-similarity is expressed in beautiful clarity. From his results for the polymer size he infers the Mark-Houwink exponent $3\nu - 1 \approx 0.5 \ldots 0.84$ for the molecular weight dependence of the intrinsic viscosity (at the time, he unfortunately was supposed to explain the exponent 1 suggested by Staudinger's measurements).

1943 The US-American physicist Subrahmanyan Chandrasekhar (1910–1995), who is of Indian origin, reviews applications of stochastic methods in physics and astronomy (Rev. Mod. Phys. 15, 1–89). Even today, this classical review article is often cited in connection with Brownian dynamics simulations, although it was written before the advent of stochastic calculus. The discussion of Langevin and diffusion equations is, of course, limited to the case of additive noise, and solving the Langevin equation "has to be understood rather in the sense of specifying a probability distribution $W(\boldsymbol{u}, t; \boldsymbol{u}_0)$ which governs the probability of occurrence of the velocity $\boldsymbol{u}$ at time t given that $\boldsymbol{u} = \boldsymbol{u}_0$ at $t = 0$."

1944 The Dutch physicist Hendrik Anthony Kramers (1894–1952) develops a kinetic theory for dilute polymer solutions undergoing potential flows (Physica 11, 1–19 [in Dutch]; two years later, a translation into English was published in J. Chem. Phys. 14, 415–424). He shows how the viscosity at zero shear rate can be obtained by studying a planar potential flow. Kramers considers very general polymer models involving constraints, including freely jointed bead-rod chains, rings and branched molecules. The terms "Kramers chain" or "Kramers model," "Kramers matrix," and "Kramers expression for the stress tensor" (including various modifications) are all based on this key contribution to polymer kinetic theory.

1944 The Japanese mathematician Kiyosi Itô (1915–) introduces stochastic integrals with respect to the Wiener process and derives their basic properties (Proc. Imp. Acad. Tokyo 20, 519-524). He characterizes important classes of processes for which stochastic integrals are defined, and he considers an important special case of what is now known as Itô's formula as an example for the unusual properties of stochastic integrals.

1947 The Ukrainian mathematician Iosif I. Gikhman (1918–1985) presents a very abstract and general approach to stochastic differential equations (Dokl. Akad. Nauk. SSSR 58, 961–964 [in Russian]). Although developed independently, Itô's and Gikhman's approaches are closely related. Gikhman shows the mean-square convergence of an approximation scheme. For a suitable subclass of his stochastic differential equations he finds the Markov property of the solutions and he obtains the infinitesimal generator (a differential operator with first and second order derivatives characterizing drift and diffusion).

1948 The US-American theoretical chemist John Gamble Kirkwood (1907–1959) begins his work on the intrinsic viscosities and diffusion constants of flexible macromolecules in solution within the framework of his statistical mechanical theory of irreversible processes. In the first paper, Kirkwood and his graduate student Jacob Riseman introduce the idea of hydrodynamic interaction into polymer kinetic theory (J. Chem. Phys. 16, 565–573). The entire series of papers on this subject written during the next decade is collected in the volume *Macromolecules* of Kirkwood's collected works (Gordon and Breach, New York, 1967).

1953 Prince E. Rouse Jr. formulates the freely draining bead-spring chain model with Hookean springs (J. Chem. Phys. 21, 1272–1280). By decoupling the equations of motion in small-amplitude oscillatory shear flow, he obtains the relaxation spectrum and the frequency-dependent material functions in the linear viscoelastic regime.

1955 In studying the relations between different approaches to a class of continuous Markov processes, one of which is based on stochastic difference equations, the Japanese mathematician Gisirô Maruyama proves the mean-square convergence of the Euler scheme for the numerical integration of stochastic differential equations (Rend. Circ. Matem. Palermo 4, 48-90).

1956 Bruno H. Zimm formulates and solves the bead-spring chain model with Hookean springs and equilibrium-averaged hydrodynamic interactions (J. Chem. Phys. 24, 269–278).

1969 Robert Zwanzig points out the formal equivalence between Fokker-Planck equations and stochastic differential equations in connection with the description of polymer dynamics in dilute solution in the presence of hydrodynamic interactions (Adv. Chem. Phys. 15, 325–331). He uses the

Stratonovich approach to stochastic differential equations with multiplicative noise, and he emphasizes the importance of stochastic differential equations for performing computer experiments.

1974 Grigorij N. Mil'shtein suggests a strong approximation scheme for stochastic differential equations that leads to a higher order of convergence than the Euler scheme (Theor. Prob. Appl. 19, 557–562). While the order of strong convergence of his general scheme for systems of equations is unity, he obtains a scheme of even higher order for one-dimensional equations. In 1978, Mil'shtein becomes a pioneer also in the development of higher-order weak integration schemes (Theor. Prob. Appl. 23, 396–401). Both for strong and for weak schemes, Mil'shtein discusses Runge-Kutta-type ideas for eliminating derivatives of coefficient functions.

1977 The foundations for the entire construction of fluid mechanics are propounded in the two-volume textbook *Dynamics of Polymeric Liquids* by Robert Byron Bird, Robert C. Armstrong, Ole Hassager, and Charles F. Curtiss (Wiley, New York; a considerably modified second edition appeared in 1987). The vast literature in the field of continuum mechanics (Vol. 1) and kinetic theory (Vol. 2) is presented in a systematic and unified manner. These volumes constitute both an excellent introduction for beginners and an invaluable source of information for researchers working in the field, and they hence had and have a great impact on the unification of the continuum mechanics and kinetic theory approaches to polymeric liquids.

1978 Donald L. Ermak and J. Andrew McCammon publish the first Brownian dynamics simulation of short chains with hydrodynamic interactions at equilibrium (J. Chem. Phys. 69, 1352–1360). Although neither the formulation of the correct stochastic differential equation (1969) nor the validity of the Euler integration scheme (1955) are new, this paper is frequently cited in the literature on Brownian dynamics simulations, probably because of the very clear algorithmic description of the simple explicit Euler scheme and of the Cholesky decomposition of the diffusion matrix.

1978 At almost the same time, Marshall Fixman independently develops and applies the mathematical background for performing Brownian dynamics simulations of polymer chains in flow (J. Chem. Phys. 69, 1527–1545). He establishes all the required relationships between diffusion equations, stochastic differential equations, and numerical integration schemes for models with constraints and hydrodynamic interactions. In the absence of hydrodynamic interactions, Fixman shows how metric correction forces due to constraints can be eliminated by means of a predictor-corrector scheme. In general, Fixman points out the usefulness of more sophisticated integration schemes for stochastic differential equations, and that "it is not at all obvious, however, that the usual integration schemes are either useful or applicable."

1978 In a series of three papers, Masao Doi and Sam F. Edwards discuss the dynamics of polymers in melts and concentrated solutions with the aim of constructing a rheological constitutive equation based on a molecular model (J. Chem. Soc. Faraday Trans. II 74, 1789–1832). Several variants on the very useful idea of reptation in a tube are employed to describe the motion of a typical chain in a macroscopically deformed concentrated polymer system, where the tube describes the constraints imposed by entanglements with other chains in the system.

1992 The progress in the development of numerical integration schemes for stochastic differential equations is reviewed by the mathematicians Peter E. Kloeden and Eckhard Platen. Their comprehensive book *Numerical Solution of Stochastic Differential Equations* (Springer, Berlin) is a rich source of rigorous information and is very valuable in designing efficient computer simulations.

Solutions to Exercises

2.2 For given $A_1 \dots A_n \in \mathcal{A}$, choose $A_{n+1} = A_{n+2} = \dots = \emptyset$ for the union or $A_{n+1} = A_{n+2} = \dots = \Omega$ for the intersection, and then apply the corresponding results for infinite sequences of events.

2.4 $\Omega = \{0, 1, 2, \dots, 36\}$.
$\mathcal{A}_1 = \mathcal{P}(\Omega)$.
$\mathcal{A}_2 = \{\Omega, \emptyset, \{7\}, \{7\}^C\}$ (lucky number "7").
$\mathcal{A}_3 = \{\Omega, \emptyset, \{1, 3, 5, \dots, 35\}, \{0, 2, 4, \dots, 36\}\}$ ("odd" or "not odd").

2.6 In what follows, $\mathcal{E}_j$ is a generating system for the σ-algebra $\mathcal{A}_j$ introduced in the solution of Exercise 2.4:
$\mathcal{E}_1 = \{\{0\}, \{1\}, \{2\}, \dots, \{36\}\}$.
$\mathcal{E}_2 = \{\{7\}\}$.
$\mathcal{E}_3 = \{\{1, 3, 5, \dots, 35\}\}$.

2.9 How could the author know?
A set $A \subset \mathbb{R}$ not contained in $\mathcal{B}$ can be constructed as follows: Say that two real numbers are equivalent if their difference is rational, and select one representative $a \in [0, 1[$ from each equivalence class. Then the set A of selected representatives a is not contained in $\mathcal{B}$ (for a proof, see Theorem 1.8.5 of [1]).

2.11 (i) Apply (2.7) to $A_j = \emptyset$ for all j in order to see that $P(\emptyset) > 0$ is impossible.
(ii) In order to see this for n sets, one can choose $A_{n+1} = A_{n+2} = \dots = \emptyset$.
(iii) This follows from the axiom (2.6) and (ii) for $n = 2$, $A_1 = A$, $A_2 = A^C$.
(iv) This follows from the axiom (2.5) and (ii) for $n = 2$, $A_1 = A$, $A_2 = B \cap A^C$.

2.17 $\boldsymbol{\alpha} = \begin{pmatrix} \frac{1}{2} \\ \frac{1}{2} \end{pmatrix}$.

$\boldsymbol{\sigma}^2 = \begin{pmatrix} \frac{1}{12} & 0 \\ 0 & \frac{1}{12} \end{pmatrix}$.

$\tilde{p}(\boldsymbol{q}) = \exp\{i\boldsymbol{q} \cdot \boldsymbol{\alpha}\} \frac{2\sin(q_1/2)}{q_1} \frac{2\sin(q_2/2)}{q_2}$.

2.18 The properties (2.5) and (2.6) follow immediately from the definition (2.15). In order to show (2.7) one needs the distributive law $(A_1 \cup A_2 \cup \dots) \cap B = (A_1 \cap B) \cup (A_2 \cap B) \cup \dots$.

2.20 No! Consider a game of dice with $P(\{j\}) = 1/6$ for $j = 1, 2, \dots, 6$. For $A_1 = \{1, 2, 3, 4\}$, $A_2 = \{4, 5, 6\}$, $A_3 = \{3, 4, 5\}$ one has $P(A_1 \cap A_2 \cap A_3) = 1/6 = P(A_1)\,P(A_2)\,P(A_3)$ and $P(A_1 \cap A_2) = 1/6 \neq 1/3 = P(A_1)\,P(A_2)$.

2.22 For $d = 2$, $\int p_{0\delta}(\boldsymbol{x})\, \mathrm{d}^2x = \int_0^\infty \exp\{-r^2/2\}\, r\, \mathrm{d}r = 1$.
Since the d-dimensional normalization integral is the product of d one-

dimensional integrals, this implies the normalization factor $(2\pi)^{-d/2}$ for arbitrary d.

2.23 Let $n_1 < n_2 \ldots < n_n$ be n arbitrary elements of $\{1, 2, \ldots, d\}$, where $n \leq d$ is also arbitrary. The assertion follows from the fact that, when all $d - n$ integrations over the full space $\mathbb{R}$ are performed, the integral

$$\int_{B_{n_1}} \mathrm{d}x_{n_1} \cdots \int_{B_{n_n}} \mathrm{d}x_{n_n} \exp\{-(x_{n_1}^2 + \ldots + x_{n_n}^2)/2\}$$

for arbitrary $B_{n_1}, \ldots, B_{n_n} \in \mathcal{B}$ is the product of n one-dimensional integrals.

2.24 A transformation of variables yields for arbitrary functions $g(\boldsymbol{y})$

$$\int g(\boldsymbol{y})\, p_{\alpha\Theta}(\boldsymbol{y})\, \mathrm{d}^d y = \int g(\boldsymbol{\alpha} + \boldsymbol{\sigma} \cdot \boldsymbol{x})\, p_{0\delta}(\boldsymbol{x})\, \mathrm{d}^d x \,.$$

The desired results are obtained from the properties of $p_{0\delta}$ by choosing $g(\boldsymbol{y}) = 1$, $g(\boldsymbol{y}) = y_j$ and $g(\boldsymbol{y}) = (y_j - \alpha_j)(y_k - \alpha_k)$.

2.25 For the assumed Gaussian distribution, the probability for a negative IQ (that is, a deviation of more than $100/16 = 6.25$ standard deviations) is $\int_{-\infty}^{-6.25} p_{0\delta}(x)\, \mathrm{d}x \approx 2 \cdot 10^{-10}$. For a world population of $5.66 \cdot 10^9$ (1994), the number of people with a negative IQ is expected to be of order unity.
In our oversimplified model, 0.62% of the population would have an IQ larger than 140 (a standard table for IQs says that the probability for an IQ larger than 140 is 1.5%).

2.30 Consider $\Omega = \{0, 1, 2, \ldots, 36\}$ and $\mathcal{A}_2 = \{\Omega, \emptyset, \{7\}, \{7\}^C\}$ which was previously introduced in the solution to Exercise 2.4. Define a "reasonable" probability measure by $P(\Omega) = 1$, $P(\emptyset) = 0$, $P(\{7\}) = 1/37$, $P(\{7\}^C) = 36/37$. The function $X : \Omega \to \Omega'$ defined by $X(7) =$ "wow!" and $X(j) =$ "damn!" for $j \in \{7\}^C$ is a random variable. The rigorous formulation of the statement that "the probability for an enthusiastic emotion is 1/37" is $P^X(\{\text{"wow!"}\}) = P(\{7\}) = 1/37$.

2.32 $\mathcal{A}^X = \{\Omega, \emptyset, \{6\}, \{1, 2, 3, 4, 5\}\}$.
$\mathcal{A}^Y = \{\Omega, \emptyset, \{1, 2, 3, 4\}, \{5\}, \{6\}, \{1, 2, 3, 4, 5\}, \{1, 2, 3, 4, 6\}, \{5, 6\}\}$.
$g(0) = g(1) =$"damn!", $g(2) =$"wow!".

2.36 We have $p^X(\boldsymbol{x}) = 1/\pi$ on Ω'. The range of $\boldsymbol{g}$ is $\Omega'' = \mathbb{R}^2 - \{(0, 0)\}$. After evaluating $\det \dfrac{\partial \boldsymbol{g}(\boldsymbol{x})}{\partial \boldsymbol{x}} = \dfrac{-2}{x_1^2 + x_2^2}$, we obtain from (2.29)

$$p^Y(\boldsymbol{y}) = \frac{1}{2\pi}\,(x_1^2 + x_2^2) = \frac{1}{2\pi} \exp\left\{-\frac{1}{2}\left(y_1^2 + y_2^2\right)\right\}.$$

2.37 We have $p^X(\boldsymbol{x}) = 2/\pi$ on Ω'. The range of $\boldsymbol{g}$ is $\Omega'' =]-1, 1[\times]-\pi/2, \pi/2[$. After evaluating $\det \dfrac{\partial \boldsymbol{g}(\boldsymbol{x})}{\partial \boldsymbol{x}} = 4$, we obtain from (2.29) $p^Y(\boldsymbol{y}) = 1/(2\pi)$.

2.42 $\langle g \rangle = \int_0^1 x\, \mathrm{d}x = \frac{1}{2}$.
Consider the sequence of simple random variables $Y_n = \sum_{j=1}^n \frac{j-1}{n} \chi_{]\frac{j-1}{n}, \frac{j}{n}]}$ with

$Y_n(x) \leq g(x)$ and $\langle Y_n \rangle = \frac{1}{n^2} \sum_{j=1}^n (j-1) = \frac{1}{2}\left(1 - \frac{1}{n}\right)$.
This implies $\langle g \rangle \geq \frac{1}{2}$. For any simple function $Y \leq g$ one has $Y \leq Y_n + \frac{1}{n}$, $\langle Y \rangle \leq \frac{1}{2}\left(1 + \frac{1}{n}\right)$ for all n, which implies $\langle g \rangle = \frac{1}{2}$.

$\langle \tilde{g} \rangle = \int_0^1 x^2 \, dx = \frac{1}{3}$.
Consider the sequence of simple random variables $Y_n = \sum_{j=1}^n \left(\frac{j-1}{n}\right)^2 \chi_{]\frac{j-1}{n},\frac{j}{n}]}$ with $Y_n(x) \leq \tilde{g}(x)$ and $\langle Y_n \rangle = \frac{1}{n^3} \sum_{j=1}^n (j-1)^2 = \frac{1}{3} - \frac{1}{2n} + \frac{1}{6n^2}$.
This implies $\langle \tilde{g} \rangle \geq \frac{1}{3}$. For any simple function $Y \leq \tilde{g}$ one has $Y \leq Y_n + \frac{2}{n}$, $\langle Y \rangle \leq \frac{1}{3} + \frac{3}{2n} + \frac{1}{6n^2}$ for all n, which implies $\langle \tilde{g} \rangle = \frac{1}{3}$.

2.44 According to Example 2.41, the expectations on both sides of (2.38) can be rewritten as integrals. The transformation between the densities p and p^X introduces the determinant of the Jacobian appearing in the usual transformation formula for integrals (cf. the transformation law (2.29) for probability densities in Example 2.35).

2.46 The joint distribution is a probability measure on the discrete space $\{$"wow!", "damn!"$\} \times \{0, 1, 2\}$. This product space consists of six elements, where only the following three have nonzero probability: ("wow!", 2), ("damn!", 0), ("damn!", 1). The corresponding probabilities are 1/6, 2/3, and 1/6, respectively.

2.52 $P(\{\omega \mid \boldsymbol{X}(\omega) \neq \boldsymbol{X}'(\omega)\}) = P(\{\omega \mid \boldsymbol{X}'(\omega) = (0,0)\}) = P^{X'}(\{(0,0)\}) = 0$, because $P^{X'}$ is a continuous probability measure so that the probability of a single point must be equal to zero.

2.53 The random variable $E(Y|\mathcal{A}(\{\{6\}\}))$ assumes the value 1/5 on $\{1, 2, 3, 4, 5\}$ and the value 2 on $\{6\}$. Here, the natural assumption $P(\{5\}) = (1/5)\, P(\{1, 2, 3, 4, 5\})$ has been made.

2.54 According to the definition of conditional expectations, one has to show only that $\langle E(X|\mathcal{A}')\chi_{A''}\rangle = \langle E(X|\mathcal{A}'')\chi_{A''}\rangle$ for all $A'' \in \mathcal{A}''$. Since A'' is also contained in $\mathcal{A}'$, the definition of conditional expectations implies that both expressions are equal to $\langle X\chi_{A''}\rangle$.

2.57 Since the covariance matrix $\boldsymbol{\Theta}$ is symmetric, there exists an orthogonal matrix $\mathbf{M}$ such that $\mathbf{M} \cdot \boldsymbol{\Theta} \cdot \mathbf{M}^T$ is diagonal (the rows of $\mathbf{M}$ are the eigenvectors of $\boldsymbol{\Theta}$). Equation (2.63) implies that the covariance matrix of $\boldsymbol{Y} = \mathbf{M} \cdot \boldsymbol{X}$ is diagonal so that the components of $\boldsymbol{Y}$ are independent.

2.59 With (2.56) and (2.64) we obtain $E(X|Y = y) = \int x\, p(X = x|Y = y)\, dx = y/2$, and thus $E(X|\mathcal{A}^Y) = Y/2$. The conditional expectation $E(Y|\mathcal{A}^X)$ can be obtained by analogous arguments. Alternatively, we can directly use the rules for manipulating conditional expectations in order to find $E(Y|\mathcal{A}^X) = X + E(Y - X|\mathcal{A}^X) = X + \langle Y - X\rangle = X$.

2.60 We assume $\boldsymbol{X} \in \mathbb{R}^d$, $\boldsymbol{Y} \in \mathbb{R}^{d'}$. From (2.56) and the subsequent arguments we obtain, after applying the results of Example 2.56,

$E(\boldsymbol{X}|\mathcal{A}^Y) = \int \boldsymbol{x}\, p^{(X,Y)}(\boldsymbol{x}, \boldsymbol{Y})\, \mathrm{d}^d x / p^Y(\boldsymbol{Y})$.
By expressing the joint probability density of $\boldsymbol{X}$ and $\boldsymbol{Y}$ in terms of the characteristic function we obtain after integrating
$\int (\boldsymbol{x} - \langle \boldsymbol{X} \rangle)\, p^{(X,Y)}(\boldsymbol{x}, \boldsymbol{Y})\, \mathrm{d}^d x =$

$$(2\pi)^{-d'}\, \boldsymbol{\Theta}' \cdot \int i\boldsymbol{q} \exp\left\{-i\boldsymbol{q}\cdot\boldsymbol{Y} + i\boldsymbol{q}\cdot\langle\boldsymbol{Y}\rangle - \tfrac{1}{2}\boldsymbol{q}\cdot\boldsymbol{\Theta}\cdot\boldsymbol{q}\right\} \mathrm{d}^{d'} q.$$

By combining these two equations we find
$E(\boldsymbol{X}|\mathcal{A}^Y) = \langle \boldsymbol{X} \rangle - \boldsymbol{\Theta}' \cdot \frac{\partial}{\partial \boldsymbol{y}} \ln p^Y(\boldsymbol{Y})$,
where the derivative with respect to the argument $\boldsymbol{y}$ of p^Y is taken before $\boldsymbol{Y}$ is inserted. The desired result is obtained after differentiating the quadratic exponent of the Gaussian distribution of $\boldsymbol{Y}$.
The second identity can be derived by the same procedure.
The results of Exercise 2.59 are obtained from $\langle X \rangle = \langle Y \rangle = 0$, $\langle X^2 \rangle = \langle XY \rangle = \sigma^2$, $\langle Y^2 \rangle = 2\sigma^2$.

2.64 Only for $\$\$ \in \{\text{as,st}\}$ is $\$\$\text{-}\lim_{n\to\infty} X_n = 0$. The sequence $\langle X_n^2 \rangle = 1$ does not converge to zero. According to Theorem 2.62, almost sure convergence implies stochastic convergence. Theorem 2.63 cannot be applied.

2.65 Only for $\$\$ \in \{\text{ms,st}\}$ is $\$\$\text{-}\lim_{n\to\infty} X_n = 0$. For any $\omega \in \Omega$, there exist arbitrarily large values of n for which $X_n(\omega) = 1$. In this exercise, Theorem 2.63 can be applied (with $Y = \chi_\Omega$); mean-square and stochastic convergence are hence equivalent.

2.68 The convergence follows by applying the strong law of large numbers to the sequences X_n and X_n^2.
The nth term can be written as
$\frac{1}{n-1} \sum_{j=1}^n \left(X_j - \frac{1}{n} \sum_{k=1}^n X_k \right)^2 \geq 0$.
Finally,

$$\frac{1}{n-1} \sum_{j=1}^{n} \left\langle X_j^2 \right\rangle - \frac{1}{n(n-1)} \sum_{j,k=1}^{n} \langle X_j X_k \rangle =$$
$$\frac{n}{n-1}(\Theta + \alpha^2) - \frac{1}{n(n-1)} \left[n(\Theta + \alpha^2) + (n^2 - n)\alpha^2 \right] = \Theta.$$

2.74 The fact that the compatibility condition (2.73) is satisfied follows immediately from the normalization of p_{t_k}.
All the X_t, $t \in \mathbb{T}$ are independent random variables.

2.75 X_t describes the deterministic motion of a particle with velocity c, starting at the position 0 at time $t = 0$.
The measure $\overline{P}_{t_1 \ldots t_n}$ on $\mathbb{R}^n$ is concentrated at the single point $(ct_1, \ldots, ct_n)$.

2.80 $a = 7/2$, $\sigma = \sqrt{35/12} \approx 1.7$.
Apply the central limit theorem to $X_t^{(j)} = \sqrt{t}\,(1/\sqrt{jt}) \sum_{n \leq jt} (X_n - a)/\sigma$, where the discrepancy between jt and the largest integer less or equal to jt does not matter for large j.

For $t_1 \leq t_2 \leq t_3 \leq t_4$, $X^{(j)}_{t_2} - X^{(j)}_{t_1}$ depends only on the X_n with $jt_1 < n \leq jt_2$, whereas $X^{(j)}_{t_4} - X^{(j)}_{t_3}$ depends on the X_n with $jt_2 \leq jt_3 < n \leq jt_4$.

2.81 The proof of (2.77) can be based on the Fourier transform of $\mathrm{e}^{-|x|}$,
$\frac{1}{\pi}\int \frac{1}{1+q^2}\,\mathrm{e}^{iqx}\,\mathrm{d}q = \mathrm{e}^{-|x|}$,
which gives the representation
$\Theta_{tt'} = \frac{1}{\pi}\int \frac{1}{1+q^2}\,(\mathrm{e}^{iqt} - \mathrm{e}^{-t})(\mathrm{e}^{-iqt'} - \mathrm{e}^{-t'})\,\mathrm{d}q$.
We hence conclude
$\sum_{j,k=1}^{n} \Theta_{t_j t_k}\, x_j x_k = \frac{1}{\pi}\int \frac{1}{1+q^2}\left|\sum_{j=1}^{n} x_j\,(\mathrm{e}^{iqt_j} - \mathrm{e}^{-t_j})\right|^2 \mathrm{d}q \geq 0$.
Therefore, we know that the functions $\alpha_t = 0$ and $\Theta_{tt'} = \mathrm{e}^{-|t-t'|} - \mathrm{e}^{-(t+t')}$ define a Gaussian process (which is a special case of the Ornstein-Uhlenbeck process discussed in Sect. 3.1 and Example 3.25).

We next estimate the moments
$\langle (X_t - X_{t'})^2\rangle = 2(1-\mathrm{e}^{-|t-t'|}) - (\mathrm{e}^{-t} - \mathrm{e}^{-t'})^2 \leq 2(1-\mathrm{e}^{-|t-t'|}) \leq 2\,|t-t'|$,
and
$\langle (X_t - X_{t'})^4\rangle = 3\,\langle (X_t - X_{t'})^2\rangle^2 \leq 12\,|t-t'|^2$,
so that the Kolmogorov-Prokhorov Theorem can be applied to establish the existence of a version of that process with continuous trajectories.

2.83 The time-evolution equation for $\overline{p}_t(x)$ follows from (2.92) and the representation $\overline{p}_t(x) = \int p_{t0}(x|x')\,p(x')\,\mathrm{d}x'$.

2.85 For any Gaussian process, the conditional probability density $p(X_t = x|X_{t'} = x')$ for $t' < t$ can be given explicitly (it is a Gaussian probability density with mean $\alpha_t + (x' - \alpha_{t'})\Theta_{tt'}/\Theta_{t't'}$ and variance $(\Theta_{t't'}\Theta_{tt} - \Theta_{tt'}^2)/\Theta_{t't'}$; see Exercise 2.60). Then, the Chapman-Kolmogorov equation is found to be equivalent to the condition $\Theta_{tt'}\Theta_{t't''} = \Theta_{tt''}\Theta_{t't'}$.

By taking the time derivative of the expression for $p(X_t = x|X_{t'} = x')$ and comparing to Kolmogorov's forward equation (2.92) we obtain
$\mathcal{L}_t = \left[\frac{\mathrm{d}\alpha_t}{\mathrm{d}t} + \frac{1}{\Theta_{tt'}}\frac{\partial\Theta_{tt'}}{\partial t}(x - \alpha_t)\right]\frac{\partial}{\partial x} + \left(\frac{1}{2}\frac{\mathrm{d}\Theta_{tt}}{\mathrm{d}t} - \frac{\Theta_{tt}}{\Theta_{tt'}}\frac{\partial\Theta_{tt'}}{\partial t}\right)\frac{\partial^2}{\partial x^2}$.
The condition proved in the first part of this exercise implies that all coefficient functions are independent of $t' < t$.

2.86 The symmetry and positivity conditions have been verified before (in establishing the fact that Gaussian processes are defined).
For the Wiener process, $\Theta_{tt'}\Theta_{t't''} = t't'' = \Theta_{tt''}\Theta_{t't'}$.
For the Gaussian process defined in Exercise 2.81,
$\Theta_{tt'}\Theta_{t't''} = \mathrm{e}^{-(t-t'')} - \mathrm{e}^{-(t+t'')} - \mathrm{e}^{-(t+2t'-t'')} + \mathrm{e}^{-(t+2t'+t'')} = \Theta_{tt''}\Theta_{t't'}$.
The results of Exercise 2.85 imply
$\mathcal{L}_t = \mathcal{L} = -x\frac{\partial}{\partial x} + \frac{\partial^2}{\partial x^2}$,
so that the process is homogeneous.

2.89 No, because the process does not possess uncorrelated increments ($\langle (X_t - X_{t'})X_{t'}\rangle = \Theta_{tt'} - \Theta_{t't'} \neq 0$).

2.90 Yes. (i) $W_t^2 - t$ is measurable with respect to the σ-algebra induced by W_t. (ii) $\langle W_t^2 - t\rangle = 0$.
(iii) $E(W_t^2 - t|\mathcal{A}_{t'}) = E(W_{t'}^2 + 2W_{t'}[W_t - W_{t'}] + [W_t - W_{t'}]^2 - t|\mathcal{A}_{t'})$
$= W_{t'}^2 + t - t' - t = W_{t'}^2 - t'$,
where (2.53) and several properties discussed in the paragraph before (2.53) have been used.

2.91 Yes. The function $\exp\{W_t - (t/2)\}$ is measurable with respect to the σ-algebra induced by W_t.
$E(\exp\{W_t - (t/2)\}|\mathcal{A}_{t'}) = \langle \exp\{W_t - W_{t'}\}\rangle \exp\{W_{t'} - (t/2)\} = \exp\{W_{t'} - (t'/2)\}$,
where the characteristic function of the Gaussian increment $W_t - W_{t'}$ for $q = -i$ has been used.
For $t' = 0$, we obtain $\langle \exp\{W_t - (t/2)\}\rangle = 1 < \infty$.

3.1 $\langle V_t V_{t'}\rangle = \dfrac{2k_BT\zeta}{M^2}\displaystyle\int_0^{\min(t,t')} e^{-\zeta(t+t'-2t'')/M}\, dt''$
$= \dfrac{k_BT}{M}\left(e^{-\zeta|t-t'|/M} - e^{-\zeta(t+t')/M}\right)$.

3.4 The condition $\int_0^{t_{max}} \langle X_t^2\rangle dt < \infty$ implies $\left\langle \int_0^{t_{max}} X_t^2 dt\right\rangle < \infty$, and hence $\int_0^{t_{max}} X_t(\omega)^2\, dt = \infty$ can hold only on a null set.

3.7 Following the ideas and notation of Example 3.5 for $X_t := W_t^2 - t$, we introduce the discretizations
$X_{t'}^{(n)}(\omega) = \sum_{j=1}^n \widetilde{X}_{j-1}^{(n)}(\omega)\,\chi_{[t_{j-1}^{(n)},t_j^{(n)}[}(t')$ with $\widetilde{X}_j^{(n)} = \widetilde{W}_j^{(n)\,2} - t_j^{(n)}$.
A straightforward calculation gives
$\int_0^t \left\langle \left(X_{t'} - X_{t'}^{(n)}\right)^2\right\rangle dt' = \frac{t^3}{n^2}\left(n - \frac{1}{3}\right)$,
and, since these integrals vanish for $n \to \infty$, $X^{(n)}$ indeed constitutes an approximating sequence for X. From the definition of the stochastic integral for simple processes we obtain after repeatedly using Wick's theorem
$\left\langle \left(\int_0^t X_{t'}^{(n)}\, dW_{t'} - \frac{1}{3}W_t^3 + tW_t\right)^2\right\rangle = \frac{2}{3}t^3 - 2\sum_{j=1}^n t_{j-1}^{(n)\,2}(t_j^{(n)} - t_{j-1}^{(n)}) = \frac{t^3}{n^2}\left(n - \frac{1}{3}\right) \to$
0.

3.10 One immediately obtains
$\langle V_t V_{t'}\rangle = \dfrac{2k_BT\zeta}{M^2}\displaystyle\int_{[0,t]\cap[0,t']} e^{-\zeta(t+t'-2t'')/M}\, dt''$.

3.14 See the solution to Exercise 2.90.

3.19 $d\left(\frac{1}{3}W_t^3 - tW_t\right) = \left(-W_t + \frac{1}{2}2W_t\right)dt + (W_t^2 - t)\, dW_t = (W_t^2 - t)\, dW_t$.

3.20 $d\boldsymbol{W}_t^2 = 2\boldsymbol{W}_t \cdot d\boldsymbol{W}_t + d'\, dt$.

3.23 We observe that the fundamental matrix $\boldsymbol{\Phi}_t$ is a solution of the deterministic homogeneous equation associated with (3.54). By first solving this equation in the time interval from 0 to t' and then proceeding with new initial conditions until t is reached, we obtain

$\boldsymbol{\Phi}_t = \mathcal{T}\text{-}\exp\left\{\int_{t'}^t \mathbf{A}(t'')\,\mathrm{d}t''\right\} \cdot \boldsymbol{\Phi}_{t'}$.

3.24 In order to prove Theorem 3.22 one needs to evaluate only the stochastic differential of the process $\boldsymbol{X}$ defined in (3.55). Since $\boldsymbol{\Phi}_t$ has the deterministic differential $\mathbf{A}(t) \cdot \boldsymbol{\Phi}_t\,dt$, the Itô formula (3.43) implies that the familiar product rule can be employed in evaluating $\mathrm{d}\boldsymbol{X}_t$ from (3.55); the stochastic differential of the process in square brackets follows directly from Definition 3.15. The peculiarities of stochastic calculus do not show up in the evaluation of the stochastic differential of $\boldsymbol{X}_t$, and this is the deeper reason why narrow-sense linear stochastic differential equations may be solved like deterministic differential equations.

3.26 $\mathrm{d}Y_t = -\frac{\zeta}{M} Y_t\,\mathrm{d}t + 2\,V_t^2\,\mathrm{d}t + 2\,\frac{\sqrt{2k_\mathrm{B}T\zeta}}{M}\,X_t\,\mathrm{d}W_t$.
Since the expectation of the last term vanishes we immediately obtain the desired result for $\langle Y_t \rangle$.

3.29 For the homogeneous solution Φ_t with $\Phi_0 = 1$, we obtain

$$\Phi_t = \exp\left\{ \int_0^t \left[A(t') - \tfrac{1}{2}B_1(t')^2\right] \mathrm{d}t' + \int_0^t B_1(t')\mathrm{d}W_{t'} \right\}.$$

The solution of the inhomogeneous problem (3.72) is then given by

$$X_t = \Phi_t \left\{X_0 + \int_0^t \Phi_{t'}^{-1} \left[a(t') - B(t')B_1(t')\right] \mathrm{d}t' + \int_0^t \Phi_{t'}^{-1} B(t')\,\mathrm{d}W_{t'}\right\}.$$

The process Y_t considered in Example 3.18 is the homogeneous solution Φ_t for the particularly simple linear stochastic differential equation with $a(t) = A(t) = B(t) = 0$ and $B_1(t) = 1$.

3.31 According to Example 3.25, the process introduced in Exercise 2.81 satisfies the stochastic differential equation $\mathrm{d}X_t = -X_t\,\mathrm{d}t + \sqrt{2}\,\mathrm{d}W_t$, that is $A(t,x) = -x$, $B(t,x) = \sqrt{2}$ in (3.75). This infinitesimal generator has previously been found in Exercise 2.86.

3.32 $\mathbf{B} = \begin{pmatrix} 9 & 0 & 0 \\ 3 & 6 & 0 \\ 2 & -1 & 7 \end{pmatrix}$.
There are eight different solutions because the sign of each column may be changed independently.

$$\mathbf{B} = \begin{pmatrix} 6 & 6 & 3 \\ 6 & 0 & -3 \\ 3 & -3 & 6 \end{pmatrix}.$$

Eight symmetric matrices $\mathbf{B}$ can be determined in a straightforward but tedious manner through diagonalization of $\mathbf{D}$ and taking square roots of the diagonal elements. The eigenvalues of $\mathbf{D}$ are
$6\left[10 + \sqrt{52}\cos\left(\frac{\phi+2\pi j}{3}\right)\right]$, where $\cos\phi = \frac{103}{26\sqrt{52}}$ and $j = 1, 2, 3$.
The symmetric $\mathbf{B}$ given above and its negative are the only symmetric solutions with integer entries.

3.34 The stationary probability density is determined by a vanishing probability current,
$-p(x) - \frac{1}{2}p'(x) = 0$ for $x \geq 0$.
The normalized solution of this equation on $[0, \infty[$ is
$p(x) = 2\mathrm{e}^{-2x}$.
We thus obtain $\langle X_t \rangle = 1/2$ for large t.

3.36 In the absence of boundary conditions we would find the solution
$$\frac{1}{2\pi}\int \mathrm{e}^{iq(x+t)}\mathrm{e}^{-q^2t/2}\mathrm{e}^{-iqx_0}\,\mathrm{d}q = \frac{1}{\sqrt{2\pi t}}\exp\left\{-\frac{1}{2}\frac{(x+t-x_0)^2}{t}\right\}.$$
The boundary condition $p(t,0) = 0$ can be fulfilled by superimposing another solution which starts at $-x_0$, that is outside $[0, \infty[$,
$$p(t,x) = \frac{1}{\sqrt{2\pi t}}\left(\exp\left\{-\frac{1}{2}\frac{(x+t-x_0)^2}{t}\right\} - \exp\left\{2x_0 - \frac{1}{2}\frac{(x+t+x_0)^2}{t}\right\}\right).$$
We then obtain
$$p^{T^{\mathrm{FP}}}(t) = -\frac{\mathrm{d}}{\mathrm{d}t}\int_0^\infty p(t,x)\,\mathrm{d}x = \frac{1}{2}\frac{\partial}{\partial x}p(t,x)\bigg|_{x=0} = \frac{x_0}{\sqrt{2\pi t^3}}\exp\left\{-\frac{1}{2}\frac{(x_0-t)^2}{t}\right\}.$$
Since the integral of $p^{T^{\mathrm{FP}}}$ from 0 to ∞ is equal to unity, all trajectories are absorbed in a finite time. The expectation of the first passage time is x_0.

3.40 By using (3.40) and (3.97), we obtain
$\int_0^t W_{t'}^2 \circ \mathrm{d}W_{t'} = \int_0^t W_{t'}^2\,\mathrm{d}W_{t'} + \int_0^t W_{t'}\,\mathrm{d}t' = \frac{1}{3}W_t^3$.
The last step follows from Itô's formula,
$\frac{1}{3}\,\mathrm{d}W_t^3 = W_t^2\,\mathrm{d}W_t + W_t\,\mathrm{d}t$.
Alternatively, we can use the result of Exercise 3.7 to obtain
$\int_0^t W_{t'}^2 \circ \mathrm{d}W_{t'} = W_t^3 - tW_t + \int_0^t (t'\,\mathrm{d}W_{t'} + W_{t'}\,\mathrm{d}t')$.
With the product rule $\mathrm{d}(t'W_{t'}) = t'\,\mathrm{d}W_{t'} + W_{t'}\,\mathrm{d}t'$ we can confirm our first result, which is the result expected according to deterministic calculus.

3.42 By applying the rules of deterministic calculus to $\mathrm{d}X_t = -\frac{1}{2}X_t\mathrm{d}t + X_t \circ \mathrm{d}W_t$ we obtain $X_t = \exp\{-\frac{1}{2}t + W_t\}$ as the desired solution.

3.44 According to (3.101), we obtain
$$\mathrm{d}\boldsymbol{X}_t = \frac{1}{2}\begin{pmatrix} 0 & 1 \\ -1 & 0 \end{pmatrix} \cdot \boldsymbol{\nabla} g(\boldsymbol{X}_t)\,\mathrm{d}t + \mathbf{B}(\boldsymbol{X}_t) \odot \mathrm{d}\boldsymbol{W}_t.$$

3.46 The range of the random variable (3.125) is $[-b, b]$, where $b = \sqrt{\Delta t_j}\,(\frac{1}{8}c_1' + \frac{1}{2}c_2')$. The probability density of X on $[-b, b]$ can be parametrized as $p^X(x) = \left\{\sqrt{\Delta t_j}\,[3c_1'(y-\frac{1}{2})^2 + c_2']\right\}^{-1}$, $x = \sqrt{\Delta t_j}\,[c_1'(y-\frac{1}{2})^3 + c_2'(y-\frac{1}{2})]$ for $0 \leq y \leq 1$.

3.47 A simple proof is based on a comparison between the suggested predictor-corrector scheme and the second-order scheme (3.121). After inserting the expressions
$\boldsymbol{A}(t_j, \overline{\boldsymbol{Y}}_{j+1}) = \boldsymbol{A}(t_j, \widetilde{\boldsymbol{Y}}_j) + \left(\frac{\partial}{\partial t} + \mathcal{L}_t\right)\boldsymbol{A}(t_j, \widetilde{\boldsymbol{Y}}_j)\,\Delta t_j + \Delta\boldsymbol{W}_j \cdot \mathbf{B}^T(t_j) \cdot \frac{\partial}{\partial \boldsymbol{x}}\boldsymbol{A}(t_j, \widetilde{\boldsymbol{Y}}_j)$
and
$\mathbf{B}(t_{j+1}) = \mathbf{B}(t_j) + \frac{\partial}{\partial t}\mathbf{B}(t_j)\,\Delta t_j$,

the equivalence of the schemes can be verified term by term. The crucial point is that the terms in (3.121) that stem from the integrals $\mathbf{I}^{(2)}_{t_j t_{j+1}}$ disappear for additive noise.

4.1 For all $j, k = 1, 2, \ldots N-1$:
$\sum_n A^{\mathrm{R}}_{jn}\, \Omega^{\mathrm{R}}_{nk} = 2\Omega^{\mathrm{R}}_{jk} - \Omega^{\mathrm{R}}_{j-1\,k} - \Omega^{\mathrm{R}}_{j+1\,k} = 2\,[1-\cos(k\pi/N)]\, \Omega^{\mathrm{R}}_{jk} = a^{\mathrm{R}}_k\, \Omega^{\mathrm{R}}_{jk}$,
where 1.314.1 and 1.321.1 of [2] have been used; for $j=1$ and $j=N-1$, the terms $\sin(0k\pi/N)=0$ and $\sin(Nk\pi/N)=0$, respectively, have been subtracted in order to keep the same structure as for intermediate j. The column vectors of Ω^{R} are hence eigenvectors of the symmetric Rouse matrix. Since all eigenvalues are different, these eigenvectors have to be orthogonal. The proper normalization follows from
$\sum_k \Omega^{\mathrm{R}}_{kj}\, \Omega^{\mathrm{R}}_{kj} = (2/N)\, \sum_k \sin^2(jk\pi/N) = 1$ for all j,
where 1.351.1 of [2] has been used.

4.2 Since the processes $\boldsymbol{W}'_\mu(t)$ for $\mu = 1, 2, \ldots, N$ are obtained by linear transformations of Wiener processes they must be Gaussian. The definitions (4.10) and (4.11) immediately imply $\left\langle \boldsymbol{W}'_\mu(t)\right\rangle = \boldsymbol{0}$. For the second moments we distinguish three different cases:
(i) For $j, k = 1, 2, \ldots, N-1$:

$$\begin{aligned}\left\langle \boldsymbol{W}'_j(t)\boldsymbol{W}'_k(t')\right\rangle &= \frac{1}{\sqrt{a^{\mathrm{R}}_j\, a^{\mathrm{R}}_k}} \sum_{l,n} \Omega^{\mathrm{R}}_{lj}\, \Omega^{\mathrm{R}}_{nk} \left[\langle \boldsymbol{W}_{l+1}(t)\, \boldsymbol{W}_{n+1}(t')\rangle \right.\\ &\quad \left. + \langle \boldsymbol{W}_l(t)\, \boldsymbol{W}_n(t')\rangle - \langle \boldsymbol{W}_{l+1}(t)\, \boldsymbol{W}_n(t')\rangle - \langle \boldsymbol{W}_l(t)\, \boldsymbol{W}_{n+1}(t')\rangle\right]\\ &= \frac{1}{\sqrt{a^{\mathrm{R}}_j\, a^{\mathrm{R}}_k}} \sum_{l,n} \Omega^{\mathrm{R}}_{lj}\, A^{\mathrm{R}}_{ln}\, \Omega^{\mathrm{R}}_{nk}\, \boldsymbol{\delta}\, \min(t,t') = \delta_{jk}\, \boldsymbol{\delta}\, \min(t,t')\,.\end{aligned}$$

(ii) $\quad \langle \boldsymbol{W}'_N(t)\boldsymbol{W}'_N(t')\rangle = \frac{1}{N}\, \sum_{\mu,\nu} \langle \boldsymbol{W}_\mu(t)\, \boldsymbol{W}_\nu(t')\rangle = \boldsymbol{\delta}\, \min(t,t')$.

(iii) For $k = 1, 2, \ldots, N-1$:

$$\begin{aligned}\langle \boldsymbol{W}'_k(t)\boldsymbol{W}'_N(t')\rangle &= \frac{1}{\sqrt{a^{\mathrm{R}}_k N}} \sum_{j,\mu} \Omega^{\mathrm{R}}_{jk}\, \langle [\boldsymbol{W}_{j+1}(t) - \boldsymbol{W}_j(t)]\, \boldsymbol{W}_\mu(t')\rangle\\ &= \frac{1}{\sqrt{a^{\mathrm{R}}_k N}} \sum_{j,\mu} \Omega^{\mathrm{R}}_{jk}\, (\delta_{j+1\,\mu} - \delta_{j\mu})\, \boldsymbol{\delta}\, \min(t,t') = \boldsymbol{0}\,.\end{aligned}$$

4.3 The first equation follows from the definition of the normal modes and the orthogonality of Ω^{R}, and the third equation follows immediately from the definition of the connector vectors. The second equation can be verified as follows,
$\boldsymbol{r}_{\mathrm{c}} - \sum_{j=1}^{N-1} \left[1 - \sum_{\mu=1}^{j}(1/N)\right] \boldsymbol{Q}_j = \boldsymbol{r}_{\mathrm{c}} + \boldsymbol{r}_1 - \boldsymbol{r}_N + (1/N)\, \sum_{\mu=1}^{N} \sum_{j=\mu}^{N-1} \boldsymbol{Q}_j$
$= \boldsymbol{r}_{\mathrm{c}} + \boldsymbol{r}_1 - \boldsymbol{r}_N + (1/N)\, \sum_{\mu=1}^{N} (\boldsymbol{r}_N - \boldsymbol{r}_\mu) = \boldsymbol{r}_1$.

4.4 For shear flows, where $\gamma = \gamma(t,t') = \int_{t'}^{t} \dot\gamma(t'')\, dt''$:

$$\mathbf{E} = \begin{pmatrix} 1 & \gamma & 0\\ 0 & 1 & 0\\ 0 & 0 & 1\end{pmatrix},\ \mathbf{C}^{-1} = \begin{pmatrix} 1+\gamma^2 & \gamma & 0\\ \gamma & 1 & 0\\ 0 & 0 & 1\end{pmatrix},\ \mathbf{C} = \begin{pmatrix} 1 & -\gamma & 0\\ -\gamma & 1+\gamma^2 & 0\\ 0 & 0 & 1\end{pmatrix}.$$

For general extensional flows, where $\epsilon_1 = \epsilon_1(t,t') = \int_{t'}^{t} \dot\epsilon(t'')\,\mathrm{d}t''$, $\epsilon_2 = \epsilon_2(t,t') = \int_{t'}^{t} m(t'')\,\dot\epsilon(t'')\,\mathrm{d}t''$, $\epsilon_3 = \epsilon_3(t,t') = -\int_{t'}^{t}[1+m(t'')]\,\dot\epsilon(t'')\,\mathrm{d}t''$:

$$\mathbf{E} = \begin{pmatrix} \mathrm{e}^{\epsilon_1} & 0 & 0 \\ 0 & \mathrm{e}^{\epsilon_2} & 0 \\ 0 & 0 & \mathrm{e}^{\epsilon_3} \end{pmatrix},$$

$$\mathbf{C}^{-1} = \begin{pmatrix} \mathrm{e}^{2\epsilon_1} & 0 & 0 \\ 0 & \mathrm{e}^{2\epsilon_2} & 0 \\ 0 & 0 & \mathrm{e}^{2\epsilon_3} \end{pmatrix}, \quad \mathbf{C} = \begin{pmatrix} \mathrm{e}^{-2\epsilon_1} & 0 & 0 \\ 0 & \mathrm{e}^{-2\epsilon_2} & 0 \\ 0 & 0 & \mathrm{e}^{-2\epsilon_3} \end{pmatrix}.$$

4.5 Inserting the expressions for the Rouse and Kramers matrices yields

$$\sum_{l=1}^{N-1} C^{\mathrm{R}}_{jl} A^{\mathrm{R}}_{lk} = \sum_{l=1}^{j-1} l\,\frac{N-j}{N}\,(2\delta_{lk} - \delta_{lk+1} - \delta_{lk-1}) + \sum_{l=j}^{N-1} j\,\frac{N-l}{N}\,(2\delta_{lk} - \delta_{lk+1} - \delta_{lk-1})\,.$$

For $j < k$, all terms in the first sum vanish, and the second sum is

$$j\left(2\frac{N-k}{N} - \frac{N-k-1}{N} - \frac{N-k+1}{N}\right) = 0\,.$$

For $j > k$, we can include the term with $l = j$ in the first rather than in the second sum, and we similarly find that both sums are zero.
For $j = k$, we obtain

$$\sum_{l=1}^{N-1} C^{\mathrm{R}}_{jl} A^{\mathrm{R}}_{lj} = -\frac{(j-1)(N-j)}{N} + 2j\,\frac{N-j}{N} - j\,\frac{N-j-1}{N} = 1\,.$$

Steady-state properties can be obtained from (4.13) for $t \to \infty$,

$$\left\langle \boldsymbol{Q}'_j \boldsymbol{Q}'_k \right\rangle = \delta_{jk}\,\frac{2k_{\mathrm{B}}T a^{\mathrm{R}}_j}{\zeta} \lim_{t\to\infty} \int_0^t \mathrm{e}^{-2H\,a^{\mathrm{R}}_j(t-t')/\zeta}\,\mathbf{E}(t,t')\cdot\mathbf{E}^T(t,t')\,\mathrm{d}t'\,,$$

where (3.31) has been used. With the results of Exercise 4.4 for the Finger tensor in steady shear flow we obtain after the substitution $t - t' \to t'$,

$$\begin{aligned} \left\langle \boldsymbol{Q}'_j \boldsymbol{Q}'_k \right\rangle &= \delta_{jk}\,\frac{2k_{\mathrm{B}}T a^{\mathrm{R}}_j}{\zeta} \int_0^\infty \mathrm{e}^{-2H\,a^{\mathrm{R}}_j t'/\zeta} \left[\boldsymbol{\delta} + t'\,(\boldsymbol{\varkappa} + \boldsymbol{\varkappa}^T) + t'^2\,\boldsymbol{\varkappa}\cdot\boldsymbol{\varkappa}^T\right] \mathrm{d}t' \\ &= \delta_{jk}\,\frac{k_{\mathrm{B}}T}{H}\left[\boldsymbol{\delta} + 2\lambda_H\,\frac{1}{a^{\mathrm{R}}_j}(\boldsymbol{\varkappa} + \boldsymbol{\varkappa}^T) + 8\lambda_H^2\,\frac{1}{(a^{\mathrm{R}}_j)^2}\,\boldsymbol{\varkappa}\cdot\boldsymbol{\varkappa}^T\right]. \end{aligned}$$

In this equation, the inverse of the Rouse matrix occurs in the diagonal representation (as a consequence of the time integration). By transforming back from normal modes to connector vectors we obtain (4.25).

4.6 After inserting (4.25) into the Kramers expression for the stress tensor (4.18) we obtain from the definitions (1.6)–(1.8),

$$\begin{aligned}\eta &= \eta_s + 2n_p\, k_B T\, \lambda_H \sum_j C^R_{jj}\,,\\ \Psi_1 &= 8n_p\, k_B T\, \lambda_H^2 \sum_{j,k} C^R_{jk} C^R_{kj}\,,\\ \Psi_2 &= 0\,.\end{aligned}$$

Due to the piecewise linear dependence of C^R_{jk} on j and k, the summations can be performed. The results are (see §11.6 of [3]),

$$\eta = \eta_s + n_p\, k_B T\, \lambda_H\, (N^2-1)/3\,, \quad \Psi_1 = 2n_p\, k_B T\, \lambda_H^2\, (N^2-1)(2N^2+7)/45\,.$$

For long chains we use $N^2\lambda_H = (\pi^2/2)\lambda_1$ in order to obtain

$$\eta = \eta_s + n_p\, k_B T\, \lambda_1\, \pi^2/6\,, \quad \Psi_1 = n_p\, k_B T\, \lambda_1^2\, \pi^4/45\,.$$

These results show that it is the combination $\lambda_1\dot\gamma$ that determines the polymer contribution to the shear stress and the first normal stress difference.

4.7 By using the transformation (4.16) we obtain
$\frac{1}{N}\sum_\mu \langle(\boldsymbol{r}_\mu - \boldsymbol{r}_c)(\boldsymbol{r}_\mu - \boldsymbol{r}_c)\rangle = \frac{1}{N}\sum_{jk} C^R_{jk} \left\langle \boldsymbol{Q}_j \boldsymbol{Q}_k \right\rangle$.
By means of (4.25) we obtain the expression
$\frac{1}{N}\sum_\mu \langle(\boldsymbol{r}_\mu - \boldsymbol{r}_c)(\boldsymbol{r}_\mu - \boldsymbol{r}_c)\rangle =$
$\frac{1}{N}\frac{k_B T}{H}\left[\mathrm{Tr}\, C^R\, \boldsymbol{\delta} + 2\lambda_H \mathrm{Tr}\,(C^R)^2(\boldsymbol{\kappa}+\boldsymbol{\kappa}^T) + 8\lambda_H^2 \mathrm{Tr}\,(C^R)^3\, \boldsymbol{\kappa}\cdot\boldsymbol{\kappa}^T\right]$,
in which the traces can be evaluated in a closed form (see p. 24 of [3]),
$\mathrm{Tr}\, C^R = \frac{N^2-1}{6}$, $\mathrm{Tr}\,(C^R)^2 = \frac{(N^2-1)(2N^2+7)}{180}$, $\mathrm{Tr}\,(C^R)^3 = \frac{(N^2-1)(8N^4+29N^2+71)}{7560}$.
For long chains we find
$\frac{1}{N}\sum_\mu \langle(\boldsymbol{r}_\mu - \boldsymbol{r}_c)(\boldsymbol{r}_\mu - \boldsymbol{r}_c)\rangle = N\frac{k_B T}{H}\left[\frac{1}{6}\boldsymbol{\delta} + \frac{\pi^2}{90}\lambda_1(\boldsymbol{\kappa}+\boldsymbol{\kappa}^T) + \frac{2\pi^4}{945}\lambda_1^2\, \boldsymbol{\kappa}\cdot\boldsymbol{\kappa}^T\right]$.

4.8 The random number generator `RANULS` developed here ("U" stands for "uniform", "LS" for "long sequence") is based on the routine `ran2` of [4]. From the sequence of integers obtained from a linear congruential generator with $k = 40014$, $l = 0$, and $n = 2147483563$ the sequence of another generator with $k = 40692$, $l = 0$, and $n = 2147483399$ is subtracted in order to obtain a much larger period ($\approx 2.3 \times 10^{18}$). In order to avoid overflow in applying the recurrence relation (4.28), the moduli n are expressed as $40014 \times 53668 + 12211$ and $40692 \times 52774 + 3791$, respectively. In addition, random shuffling in a table of length 32 is used in order to break up sequential correlations.
The random number generator `RANULS` is initialized by calling the subroutine `RANILS(ISEED)`, where `ISEED` is an integer with $0 \le$ `ISEED` $\le 2 \times 10^9$. The function `RANULS()` is called without any arguments.
`RANULS` was tested on several machines. Within the limits of floating-point precision, identical sequences of random numbers were obtained. On several different workstations, the generation of a single random number took between 2.5μs and 4.3μs. This random number generator is not suited for vector machines (for example, the generation of a single random number on a CRAY Y-MP computer took more than 3μs).

```
      SUBROUTINE RANILS(ISEED)
C     Choice of ISEED: 0 <= ISEED <= 2000000000 (2E+9);
      PARAMETER (IN=2147483563,IK=40014,IQ=53668,IR=12211,NTAB=32)
      INTEGER IV(NTAB)
      COMMON /RANBLS/ IDUM,IDUM2,IY,IV
C     Initial seeds for two random number generators
      IDUM=ISEED+123456789
      IDUM2=IDUM
C     Load the shuffle table (after 8 warm-ups)
      DO 10 J=NTAB+8,1,-1
         K=IDUM/IQ
         IDUM=IK*(IDUM-K*IQ)-K*IR
         IF(IDUM.LT.0) IDUM=IDUM+IN
         IF(J.LE.NTAB) IV(J)=IDUM
10    CONTINUE
      IY=IV(1)
      RETURN
      END

      FUNCTION RANULS()
      PARAMETER (IN1=2147483563,IK1=40014,IQ1=53668,IR1=12211,
     &           IN2=2147483399,IK2=40692,IQ2=52774,IR2=3791,
     &           NTAB=32,AN=1./IN1,INM1=IN1-1,NDIV=1+INM1/NTAB)
      INTEGER IV(NTAB)
      COMMON /RANBLS/ IDUM,IDUM2,IY,IV
C     Linear congruential generator 1
      K=IDUM/IQ1
      IDUM=IK1*(IDUM-K*IQ1)-K*IR1
      IF(IDUM.LT.0) IDUM=IDUM+IN1
C     Linear congruential generator 2
      K=IDUM2/IQ2
      IDUM2=IK2*(IDUM2-K*IQ2)-K*IR2
      IF(IDUM2.LT.0) IDUM2=IDUM2+IN2
C     Shuffling and subtracting
      J=1+IY/NDIV
      IY=IV(J)-IDUM2
      IV(J)=IDUM
      IF(IY.LT.1) IY=IY+INM1
      RANULS=AN*IY
      RETURN
      END
```

4.9 The Gaussian random number generator RANGLS developed here ("G" stands for "Gaussian", "LS" for "long sequence") uses uniform random numbers from RANULS, and it must be initialized by calling the subroutine RANILS (see solution to Exercise 4.8).

RANGLS is based on the results of Exercise 2.36. Random numbers with a uniform distribution in the unit circle can be obtained from random numbers with

a uniform distribution in $[-1,1] \times [-1,1]$ by rejecting pairs outside the unit circle (the rejection rate is $1 - \frac{\pi}{4} \approx 21\%$).
RANGLS was tested on several machines. Within the limits of floating-point precision, identical sequences of random numbers were obtained. Both on several workstations and on a CRAY Y-MP, the ratio of CPU times for generating Gaussian and uniform random numbers, respectively, was found to be only 1.9 (this is an indication of how expensive the generation of uniform random numbers with RANULS is).

```
      FUNCTION RANGLS()
      SAVE IFLAG,GAUSS2
      DATA IFLAG/0/
      IF(IFLAG.EQ.0) THEN
10       CONTINUE
C        Pair of uniform random numbers in [-1,1]x[-1,1]
         X1=2.*RANULS()-1.
         X2=2.*RANULS()-1.
C        If not in the unit circle, try again
         XSQ=X1*X1+X2*X2
         IF(XSQ.GE.1..OR.XSQ.EQ.0.) GOTO 10
C        Pair of Gaussian random numbers; return one and
C        save the other for next time
         AUX=SQRT(-2.*LOG(XSQ)/XSQ)
         RANGLS=X1*AUX
         GAUSS2=X2*AUX
         IFLAG=1
      ELSE
         RANGLS=GAUSS2
         IFLAG=0
      ENDIF
      RETURN
      END
```

4.10 The subroutine RANUVE for generating random unit vectors uses uniform random numbers from RANULS, and it must be initialized by calling the subroutine RANILS (see solution to Exercise 4.8).
RANUVE is based on the results of Exercise 2.37. The random variable Y_1 is interpreted as the 3-component of the random unit vector, and Y_2 as the polar angle. The results for the semicircle with $x_1 > 0$ have been combined with those for the equivalent semicircle with $x_1 < 0$ (boundary lines are irrelevant because they are hit with zero probability). Random numbers with a uniform distribution in the unit circle can be obtained from random numbers with a uniform distribution in $[-1,1] \times [-1,1]$ by rejecting pairs outside the unit circle (the rejection rate is $1 - \frac{\pi}{4} \approx 21\%$).
The ratio of CPU times for generating random unit vectors and uniform random numbers, respectively, is 3.7.

```
      SUBROUTINE RANUVE(U1,U2,U3)
10    CONTINUE
C     Pair of uniform random numbers in [-1,1]x[-1,1]
      X1=2.*RANULS()-1.
      X2=2.*RANULS()-1.
C     If not in the unit circle, try again
      XSQ=X1*X1+X2*X2
      IF(XSQ.GT.1.) GOTO 10
C     Random unit vector
      SQXSQ=2.*SQRT(1.-XSQ)
      U1=X1*SQXSQ
      U2=X2*SQXSQ
      U3=1.-2.*XSQ
      RETURN
      END
```

4.11 All the relevant input data are given in the header of the subsequent Brownian dynamics simulation program HOOKE1 rather than being read in. This procedure has the advantage that the reader should be able to reproduce all the steps. For reasons of clarity, this and all the subsequent simulation programs have been written in a form which is not apt for vectorization. Good vectorization can be achieved by simulating a large number of trajectories in the innermost loops of the simulation programs (that is, all these trajectories need to be propagated in each time step).

We use seven different time steps. The output consists of five columns: time-step width, polymer contribution to the viscosity, statistical error bar for the polymer contribution to the viscosity, first normal stress coefficient, statistical error bar for the first normal stress coefficient. The statistical error bars are taken as $(\Theta/n)^{1/2}$, where $n =$ NTRA and the variance Θ for a single trajectory is estimated according to Exercise 2.68. The appropriate units for the viscosity and the first normal stress coefficient are $n_p\, k_B T\, \lambda_H$ and $n_p\, k_B T\, \lambda_H^2$, respectively. According to the Boltzmann distribution, the equilibrium initial conditions for our choice of units are given by standard Gaussian random variables.

For the input data given in the header, the program uses some 2×10^{10} random numbers. The generation of the random numbers makes the dominating contribution to the total CPU time. The total time required for this example on a workstation is roughly one day. The results are extremely precise; considerably fewer trajectories would lead to satisfactory results for most purposes.

```
      PROGRAM HOOKE1
      REAL*8 AETA,VETA,APSI,VPSI
      INTEGER NTIARR(7)
C     Array: different numbers of time steps
      DATA NTIARR/2,3,4,6,8,12,25/
C     Total simulation time
      TMAX=1.
C     Shear rate
      SR=1.
```

```
C     Number of trajectories
      NTRA=100000000
C     Initial seed for random number generator
      ISEED= 290092405
      CALL RANILS(ISEED)
C
C     Loop for different time-step widths
      DO 1000 IDT=1,7
C        Auxiliary parameters
         NTIME=NTIARR(IDT)
         DELTAT=TMAX/NTIME
         OMDTH=1.-0.5*DELTAT
         SQ12DT=SQRT(12.*DELTAT)
         SRDT=SR*DELTAT
C
         AETA=0.
         VETA=0.
         APSI=0.
         VPSI=0.
C        Different trajectories
         DO 100 ITRA=1,NTRA
C           Equilibrium initial conditions
            Q1=RANGLS()
            Q2=RANGLS()
            Q3=RANGLS()
C           Time integration: Euler
            DO 10 ITIME=1,NTIME
               Q1=OMDTH*Q1+SRDT*Q2+SQ12DT*(RANULS()-0.5)
               Q2=OMDTH*Q2+SQ12DT*(RANULS()-0.5)
               Q3=OMDTH*Q3+SQ12DT*(RANULS()-0.5)
10          CONTINUE
C           "Measurement"
            Q12=Q1*Q2/SR
            Q1122=(Q1*Q1-Q2*Q2)/(SR*SR)
            AETA=AETA+Q12
            VETA=VETA+Q12*Q12
            APSI=APSI+Q1122
            VPSI=VPSI+Q1122*Q1122
100      CONTINUE
C
C        Averages, statistical errors
         AETA=AETA/NTRA
         VETA=VETA/NTRA
         VETA=SQRT((VETA-AETA*AETA)/(NTRA-1))
         APSI=APSI/NTRA
         VPSI=VPSI/NTRA
         VPSI=SQRT((VPSI-APSI*APSI)/(NTRA-1))
C        Output of results
```

```
      WRITE(6,1) DELTAT,AETA,VETA,APSI,VPSI
1     FORMAT(F11.8,2(4X,F9.6,1X,F9.6))
1000 CONTINUE
     STOP
     END
```

The output for the time step, the polymer contribution to the viscosity (with error estimate), and the first normal stress coefficient (with error estimate) produced by the above program looks as follows:

```
0.50000000   0.609218 0.000151   0.687142 0.000256
0.33333334   0.623196 0.000146   0.621646 0.000244
0.25000000   0.627294 0.000144   0.594572 0.000239
0.16666667   0.629847 0.000142   0.570055 0.000234
0.12500000   0.630938 0.000141   0.558981 0.000231
0.08333334   0.631617 0.000141   0.548361 0.000229
0.04000000   0.631973 0.000140   0.537737 0.000227
```

The order of weak convergence for the Euler scheme used in this exercise is $\upsilon = 1$. While the results for the first normal stress coefficient exhibit the expected linear dependence on Δt, the linear term for the polymer contribution to the viscosity happens to vanish.

In order to analyze the simulation results for different time steps in more detail and to extrapolate to zero time-step width we have developed the subroutine TEXTRA. Although TEXTRA is a rather simple-minded program it should be very helpful for all time-step extrapolations. The data points and the corresponding statistical error bars are stored in the arrays XARR, YARR, SIGARR. The idea is to fit polynomials of various degrees in Δt through these simulation results by a least squares method. The number of coefficients to be fitted must be less than the number of data points NDAT such that the assumption of a polynomial of a certain degree can be tested. The zeroth coefficient of the fitted polynomial is the desired extrapolated value, and its variance can be estimated from the least squares method. For any given degree, more and more data points for the largest Δt are discarded so that the fit to a polynomial of the given degree is increasingly better justified. The consistency of the considered data points with a polynomial of the given degree is tested via a χ^2 test (see Chaps. 14 and 15 of [4]); if too large values of χ^2 occur (in the top-most 25%) a polynomial of the given degree is regarded as inconsistent with the simulation results. Among the various consistent fits for different degrees of polynomials and ranges of Δt values, the one leading to the smallest statistical error bar for the zeroth coefficient is taken as the final result of the extrapolation.

For a second-order scheme it is helpful to suppress the coefficient of the linear term of the polynomial in Δt in the fitting procedure. This can be achieved by choosing IFLAG as zero in calling TEXTRA.

The subroutine TEXTRA analyses simulation results taken for given time steps. By successively reducing the time steps and analyzing the results available for some previously simulated time steps it might be possible to decide whether use of more expensive, smaller time steps would be helpful or not to achieve a

certain accuracy with a minimum amount of CPU time.
The subroutine TEXTRA uses various subroutines from [4]. The routine `lfit` is a program for general linear least squares fits, and it uses `fpoly` for the special case of fitting polynomials. For solving a linear equation and inverting a matrix, `lfit` calls the subroutine `gaussj` and, furthermore, an auxiliary routine `covsrt` is called by `lfit`. For the χ^2 test the function `gammq` is required which calls `gser`, `gcf` and `gammln`.
In the main program, the data are read in from a file before TEXTRA is called.

```
      PARAMETER (NDATM=50)
      REAL XARR(NDATM),YARR(NDATM),SIGARR(NDATM)
      CHARACTER*50 DATAFILE
C     Input of data
      WRITE(6,*) 'Data file for input?'
      READ(5,1) DATAFILE
1     FORMAT(A40)
      OPEN(8,FILE=DATAFILE,STATUS='OLD')
      NDAT=0
10    CONTINUE
         READ(8,*,END=20) Z1,Z2,Z3
         NDAT=NDAT+1
         XARR(NDAT)=Z1
         YARR(NDAT)=Z2
         SIGARR(NDAT)=Z3
      GOTO 10
20    CONTINUE
      CLOSE(8)
      WRITE(6,*) 'Number of data points: ',NDAT
      CALL TEXTRA(XARR,YARR,SIGARR,NDAT,NDATM,0)
      STOP
      END

      SUBROUTINE TEXTRA(XARR,YARR,SIGARR,NDAT,NDATM,IFLAG)
      PARAMETER (NDATP=50)
      REAL XARR(NDATM),YARR(NDATM),SIGARR(NDATM)
      REAL A(NDATP),COVAR(NDATP,NDATP)
      INTEGER LFLAG(NDATP)
      EXTERNAL fpoly
      DATA ERRLEV/0.25/
C
C     Sorting the data such that XARR(1).LE.XARR(2).LE.XARR(3) ...
C     (by straight insertion)
      DO 10 J=2,NDAT
         X=XARR(J)
         Y=YARR(J)
         SIG=SIGARR(J)
         DO 20 I=J-1,1,-1
            IF(XARR(I).LE.X) GOTO 30
```

```
            XARR(I+1)=XARR(I)
            YARR(I+1)=YARR(I)
            SIGARR(I+1)=SIGARR(I)
20       CONTINUE
         I=0
30       XARR(I+1)=X
         YARR(I+1)=Y
         SIGARR(I+1)=SIG
10    CONTINUE
C
      NOPT=-1
      SIGOPT=6.022E23
C     Fitting polynomials of various degrees N.LE.NMAX
      NMAX=NDAT-2
      IF(IFLAG.EQ.0) NMAX=NMAX+1
      DO 1000 N=0,NMAX
         NDATMI=N+2
         IF(IFLAG.EQ.0.AND.N.GE.1) NDATMI=NDATMI-1
C        Discarding data with large XARR
         DO 500 NDATU=NDAT,NDATMI,-1
            NDF=NDATU-NDATMI+1
C           Least squares fit
            DO 40 I=1,N+1
               A(I)=0.
               LFLAG(I)=1
40          CONTINUE
            DO 50 I=N+2,NDATP
               A(I)=0.
               LFLAG(I)=0
50          CONTINUE
            IF(N.GT.0) LFLAG(2)=IFLAG
            CALL lfit(XARR,YARR,SIGARR,NDATU,A,LFLAG,NDATP,
     &                               COVAR,NDATP,TEST,fpoly)
C           Chi-squared test; smaller statistical error bars?
            IF(gammq(0.5*NDF,0.5*TEST).GT.ERRLEV) THEN
               IF(SQRT(COVAR(1,1)).LT.SIGOPT) THEN
                  YOPT=A(1)
                  SIGOPT=SQRT(COVAR(1,1))
                  NOPT=N
                  NDUOPT=NDATU
                  XOPT=XARR(NDATU)
                  ALIOPT=A(2)
                  VLIOPT=SQRT(COVAR(2,2))
               ENDIF
            ENDIF
500      CONTINUE
1000  CONTINUE
      IF(NOPT.NE.-1) THEN
```

```
         WRITE(6,*) 'Extrapolated result:     ',YOPT
         WRITE(6,*) 'Statistical error bar: ',SIGOPT
         WRITE(6,*) 'Degree of polynomial:  ',NOPT
         WRITE(6,*) 'Used data points:       ',NDUOPT,' of',NDAT
         WRITE(6,*) 'Maximum value of x:    ',XOPT
         IF(IFLAG.NE.0)
     &   WRITE(6,*) 'Linear coefficient:    ',ALIOPT,' +-',VLIOPT
      ELSE
         WRITE(6,*) 'No successful fit!'
      ENDIF
      RETURN
      END
```

If this extrapolation program is applied to the above data set (with IFLAG=0) then we obtain $\eta = 0.63215(8)$ where numbers in parentheses always indicate the statistical errors in the last figure(s). The exact result is 0.63212. The output of TEXTRA actually looks as follows:

```
 Extrapolated result:      0.6321513
 Statistical error bar:    8.0670150E-05
 Degree of polynomial:               2
 Used data points:                   6 of           7
 Maximum value of x:       0.3333333
```

For the first normal stress coefficient we similarly obtain $\Psi_1 = 0.52884(33)$, which should be compared to the exact result 0.52848.

4.12 Essentially, only the DO loop 10 for the time integration needs to be changed in the program HOOKE1 of Exercise 4.11. In particular, more sophisticated random numbers must in general be used in a second-order scheme (we here use (3.125)). It is also convenient to define additional auxiliary parameters. Furthermore, for the second-order simulation scheme we omit the two smallest time steps (which are, of course, the most expensive ones). The computer time thus saved is used to simulate more trajectories (still, marginally less computer time than in Exercise 4.11 is required).

```
      PROGRAM HOOKE2
      REAL*8 AETA,VETA,APSI,VPSI
      INTEGER NTIARR(5)
C     Array: different numbers of time steps
      DATA NTIARR/2,3,4,6,8/
C     Total simulation time
      TMAX=1.
C     Shear rate
      SR=1.
C     Number of trajectories
      NTRA=200000000
```

```
C     Initial seed for random number generator
      ISEED= 816050894
      CALL RANILS(ISEED)
C
C     Loop for different time-step widths
      DO 1000 IDT=1,5
C        Auxiliary parameters
         NTIME=NTIARR(IDT)
         DELTAT=TMAX/NTIME
         OMDTH=1.-0.5*DELTAT
         DTQ=0.25*DELTAT
         SQDT=SQRT(DELTAT)
C        C1P, C2P: for generating suitable random numbers
         C1P=14.14855378*SQDT
         C2P= 1.21569221*SQDT
         SRDT=SR*DELTAT
         SRDTH=.5*SRDT
C
         AETA=0.
         VETA=0.
         APSI=0.
         VPSI=0.
C        Different trajectories
         DO 100 ITRA=1,NTRA
C           Equilibrium initial conditions
            Q1=RANGLS()
            Q2=RANGLS()
            Q3=RANGLS()
C           Time integration: predictor-corrector scheme
            DO 10 ITIME=1,NTIME
               W1=RANULS()-.5
               W2=RANULS()-.5
               W3=RANULS()-.5
               W1=W1*(C1P*W1*W1+C2P)
               W2=W2*(C1P*W2*W2+C2P)
               W3=W3*(C1P*W3*W3+C2P)
               QAUX1=OMDTH*Q1+SRDT*Q2+W1
               QAUX2=OMDTH*Q2+W2
               QAUX3=OMDTH*Q3+W3
               Q1=QAUX1-DTQ*(QAUX1-Q1)+SRDTH*(QAUX2-Q2)
               Q2=QAUX2-DTQ*(QAUX2-Q2)
               Q3=QAUX3-DTQ*(QAUX3-Q3)
10          CONTINUE
C           "Measurement"
            Q12=Q1*Q2/SR
            Q1122=(Q1*Q1-Q2*Q2)/(SR*SR)
            AETA=AETA+Q12
            VETA=VETA+Q12*Q12
```

```
            APSI=APSI+Q1122
            VPSI=VPSI+Q1122*Q1122
100       CONTINUE
C
C         Averages, statistical errors
          AETA=AETA/NTRA
          VETA=VETA/NTRA
          VETA=SQRT((VETA-AETA*AETA)/(NTRA-1))
          APSI=APSI/NTRA
          VPSI=VPSI/NTRA
          VPSI=SQRT((VPSI-APSI*APSI)/(NTRA-1))
C         Output of results
          WRITE(6,1) DELTAT,AETA,VETA,APSI,VPSI
1         FORMAT(F11.8,2(4X,F9.6,1X,F9.6))
1000  CONTINUE
      STOP
      END
```

If the time-step extrapolation is done with **TEXTRA** (see solution to Exercise 4.11), the results are $\eta_p = 0.63216(9)$ and $\Psi_1 = 0.52839(11)$. For Ψ_1, the second-order simulation algorithm is considerably more efficient; the CPU time required to achieve a specified accuracy is reduced by a factor of 9. In order to obtain a successful fit for η_p, the parameter **ERRLEV** in **TEXTRA** must be reduced from 0.25 to 0.15 in analyzing the results of the above run.

4.14 $\langle Y \rangle = \int f_c(x) x^4 \, p_{\rm red}(x) \, dx = \int x^4 \, p_{0\delta}(x) \, dx$.

By going from three- to one-dimensional Gaussians, we obtain for the probability density of the length of $\boldsymbol{Q}$ the relationship $4\pi \boldsymbol{Q}^2 \, p_{0\delta}(\boldsymbol{Q}) = 2x^2 \, p_{0\delta}(x)$ for $x = |\boldsymbol{Q}| \geq 0$. Since we are interested only in even powers of x, we may assume that x takes even and odd values with equal probability, so that $p_{\rm red}(x) = x^2 \, p_{0\delta}(x)$ on $\mathbb{R}$. From (4.36), we indeed obtain $Y = (1/X^2) \, X^4 = X^2 = \boldsymbol{Q}^2$.

According to (4.36), the choice $p_{\rm red}(x) \sim x^4 \, p_{0\delta}(x)$ makes Y deterministic and hence reduces the variance to zero. The proper normalization of $p_{\rm red}(x)$, that follows from the comments after (4.35), requires the desired result for $\langle Y \rangle$. In general, perfect variance reduction is possible only when exact results are available.

4.15 The stochastic differential equation for Hookean dumbbells is a special case of the narrow-sense linear equations discussed in Sect. 3.3.2. More precisely, in dimensionless units, we have according to Sect. 4.1.1 $\boldsymbol{a}(t) = \boldsymbol{0}$, $\mathbf{B}(t) = \boldsymbol{\delta}$, $\boldsymbol{\Phi}_t = e^{-t/2} \, \mathbf{E}(t,0) = e^{-t/2} \, (\boldsymbol{\delta} + t\boldsymbol{\varkappa})$ and hence $\boldsymbol{\Phi}_t \cdot \boldsymbol{\Phi}_{t''}^{-1} = e^{-(t-t'')/2} \, [\boldsymbol{\delta} + (t-t'')\boldsymbol{\varkappa}]$, where the results of Exercise 4.4 for shear flow have been used. The solutions of narrow-sense linear stochastic differential equations are Gaussian, and their first and second moments can be obtained from (3.57) and (3.58).
For $\boldsymbol{X}_0 = \boldsymbol{Q}_0$, we obtain (4.46) with
$\boldsymbol{\Theta} = \int_0^t e^{-(t-t')} \, [\boldsymbol{\delta} + (t-t') \, (\boldsymbol{\varkappa} + \boldsymbol{\varkappa}^T) + (t-t')^2 \, \boldsymbol{\varkappa} \cdot \boldsymbol{\varkappa}^T] dt'$.

For $\langle \boldsymbol{X}_0 \rangle = \boldsymbol{0}$ and $\langle \boldsymbol{X}_0 \boldsymbol{X}_0 \rangle = \boldsymbol{\delta}$, we obtain (4.48) with
$\boldsymbol{\Theta}_\mathrm{f} = \boldsymbol{\Theta} + \mathrm{e}^{-t}\,[\boldsymbol{\delta} + t\,(\boldsymbol{\varkappa} + \boldsymbol{\varkappa}^T) + t^2\,\boldsymbol{\varkappa} \cdot \boldsymbol{\varkappa}^T]$,
where $t = t_\mathrm{max}$.
Finally, (4.50) and (4.51) follow from Exercise 2.60 where $\boldsymbol{X} = \boldsymbol{Q}(0)$, $\boldsymbol{Y} = \boldsymbol{Q}(t_\mathrm{max})$ takes on the value $\boldsymbol{Q}_n$, $\langle \boldsymbol{X}\boldsymbol{X} \rangle - \langle \boldsymbol{X} \rangle \langle \boldsymbol{X} \rangle = \boldsymbol{\delta}$, $\langle \boldsymbol{Y}\boldsymbol{Y} \rangle - \langle \boldsymbol{Y} \rangle \langle \boldsymbol{Y} \rangle = \boldsymbol{\Theta}_\mathrm{f}$, and $\langle \boldsymbol{X}\boldsymbol{Y} \rangle - \langle \boldsymbol{X} \rangle \langle \boldsymbol{Y} \rangle = \boldsymbol{\Phi}_t^T$ [from (3.58)]. In order to verify (4.51), the identity $\boldsymbol{\Theta} = \boldsymbol{\Theta}_\mathrm{f} - \mathrm{e}^{-t}\,[\boldsymbol{\delta} + t\,(\boldsymbol{\varkappa} + \boldsymbol{\varkappa}^T) + t^2\,\boldsymbol{\varkappa} \cdot \boldsymbol{\varkappa}^T]$ can be used.

4.16 In the notation of the solution to Exercise 4.15, $\boldsymbol{\Phi}_\mathrm{app}(t_{j+1}, t_n) = \boldsymbol{\Phi}^{-1}_{t_n - t_{j+1}}$. Since $\boldsymbol{\Theta}(t_{j+1}, t_n)$ corresponds to $\boldsymbol{\Theta}$ in (4.47) with t replaced by $t_n - t_{j+1}$, (4.55) follows after evaluating the products.

4.18 We use the notation $\boldsymbol{Q} = \boldsymbol{Q}_1$, $\boldsymbol{F}^\mathrm{c} = \boldsymbol{F}_1 = -\boldsymbol{F}_2$. We then obtain by taking the difference and the arithmetic mean of the equations for the two bead positions,

$$\mathrm{d}\boldsymbol{Q}(t) = \left\{ \boldsymbol{\varkappa}(t) \cdot \boldsymbol{Q}(t) - \frac{2}{\zeta}[\boldsymbol{\delta} - \zeta \boldsymbol{\Omega}(\boldsymbol{Q}(t))] \cdot \boldsymbol{F}^\mathrm{c}(t) \right\} \mathrm{d}t + \sqrt{\frac{2k_\mathrm{B}T}{\zeta}} \left\{ [\mathbf{B}_{21}(t) - \mathbf{B}_{11}(t)] \cdot \mathrm{d}\boldsymbol{W}_1(t) + [\mathbf{B}_{22}(t) - \mathbf{B}_{12}(t)] \cdot \mathrm{d}\boldsymbol{W}_2(t) \right\},$$

$$\mathrm{d}\boldsymbol{r}_\mathrm{c}(t) = [\boldsymbol{v}_0(t) + \boldsymbol{\varkappa}(t) \cdot \boldsymbol{r}_\mathrm{c}]\,\mathrm{d}t + \sqrt{\frac{k_\mathrm{B}T}{2\zeta}} \{ [\mathbf{B}_{21}(t) + \mathbf{B}_{11}(t)] \cdot \mathrm{d}\boldsymbol{W}_1(t) + [\mathbf{B}_{22}(t) + \mathbf{B}_{12}(t)] \cdot \mathrm{d}\boldsymbol{W}_2(t) \}.$$

The 6×6 diffusion matrix corresponding to the noise terms in these equations can by means of the properties of $\mathbf{B}_{\mu\nu}(t)$ be shown to have the block structure
$\begin{pmatrix} (4k_\mathrm{B}T/\zeta)\,[\boldsymbol{\delta} - \zeta\boldsymbol{\Omega}(\boldsymbol{Q})] & \boldsymbol{0} \\ \boldsymbol{0} & (k_\mathrm{B}T/\zeta)\,[\boldsymbol{\delta} + \zeta\boldsymbol{\Omega}(\boldsymbol{Q})] \end{pmatrix}$.

An alternative decomposition of this diffusion matrix leads to the more compact, decoupled stochastic differential equations,

$$\mathrm{d}\boldsymbol{Q}(t) = \left\{ \boldsymbol{\varkappa}(t) \cdot \boldsymbol{Q}(t) - \frac{2}{\zeta}[\boldsymbol{\delta} - \zeta \boldsymbol{\Omega}(\boldsymbol{Q}(t))] \cdot \boldsymbol{F}^\mathrm{c}(t) \right\} \mathrm{d}t + \sqrt{\frac{4k_\mathrm{B}T}{\zeta}}\, \mathbf{B}_1(\boldsymbol{Q}(t)) \cdot \mathrm{d}\boldsymbol{W}'_1(t),$$

$$\mathrm{d}\boldsymbol{r}_\mathrm{c}(t) = [\boldsymbol{v}_0(t) + \boldsymbol{\varkappa}(t) \cdot \boldsymbol{r}_\mathrm{c}]\,\mathrm{d}t + \sqrt{\frac{k_\mathrm{B}T}{\zeta}}\, \mathbf{B}_2(\boldsymbol{Q}(t)) \cdot \mathrm{d}\boldsymbol{W}'_2(t),$$

where
$\mathbf{B}_1(\boldsymbol{Q}) \cdot \mathbf{B}_1^T(\boldsymbol{Q}) = \boldsymbol{\delta} - \zeta\boldsymbol{\Omega}(\boldsymbol{Q})$, $\mathbf{B}_2(\boldsymbol{Q}) \cdot \mathbf{B}_2^T(\boldsymbol{Q}) = \boldsymbol{\delta} + \zeta\boldsymbol{\Omega}(\boldsymbol{Q})$,
and $\boldsymbol{W}'_1(t)$, $\boldsymbol{W}'_2(t)$ are independent Wiener processes (the increments of which are related to those of $\boldsymbol{W}_1(t)$, $\boldsymbol{W}_2(t)$ by a configuration-dependent orthogonal linear transformation).

The Fokker-Planck equation for $p = p(t, \boldsymbol{Q})$ is

$$\frac{\partial p}{\partial t} = -\frac{\partial}{\partial \boldsymbol{Q}} \cdot \left\{ \left[\boldsymbol{\kappa} \cdot \boldsymbol{Q} - \frac{2}{\zeta} [\boldsymbol{\delta} - \zeta \boldsymbol{\Omega}(\boldsymbol{Q})] \cdot \boldsymbol{F}^{\mathrm{c}} \right] p \right\}$$
$$+ \frac{2 k_{\mathrm{B}} T}{\zeta} \frac{\partial}{\partial \boldsymbol{Q}} \cdot [\boldsymbol{\delta} - \zeta \boldsymbol{\Omega}(\boldsymbol{Q})] \cdot \frac{\partial}{\partial \boldsymbol{Q}} p .$$

We find that $\boldsymbol{Q} \cdot [\boldsymbol{\delta} - \zeta \boldsymbol{\Omega}(\boldsymbol{Q})] \cdot \boldsymbol{Q} = Q^2 - [\zeta Q / (4\pi\eta_{\mathrm{s}})]$ is negative for $Q < \zeta/(4\pi\eta_{\mathrm{s}})$. The diffusion tensor is hence not positive semidefinite, and the decomposition $\mathbf{B}_1(\boldsymbol{Q}) \cdot \mathbf{B}_1^T(\boldsymbol{Q}) = \boldsymbol{\delta} - \zeta \boldsymbol{\Omega}(\boldsymbol{Q})$ is impossible for small Q.

4.19 From (4.25) it follows that the covariance matrix of the Gaussian distribution of $\boldsymbol{r}_\mu - \boldsymbol{r}_\nu$ at equilibrium is $\boldsymbol{\sigma} = c\,\boldsymbol{\delta}$ with $c = |\mu - \nu|\, k_{\mathrm{B}} T / H$. Hence $\langle \boldsymbol{\Omega}(\boldsymbol{r}_\mu(t) - \boldsymbol{r}_\nu(t)) \rangle_{\mathrm{eq}} = \int \boldsymbol{\Omega}(\boldsymbol{r})\, p_{0\sigma}(\boldsymbol{r})\, \mathrm{d}^3 r$.
Since the result must be of the isotropic form $c'\,\boldsymbol{\delta}$, we obtain by taking the trace

$$3c' = \int_0^\infty \frac{1}{8\pi\eta_{\mathrm{s}} r} \frac{4}{\sqrt{2\pi c}^{\,3}} \exp\{-r^2/(2c)\}\, 4\pi r^2 \, \mathrm{d}r = \frac{\sqrt{2}}{2\pi\eta_{\mathrm{s}} \sqrt{\pi c}} .$$

4.20 Since the processes $\boldsymbol{W}'_\mu(t)$ for $\mu = 1, 2, \ldots, N$ are obtained by linear transformations of Wiener processes they must be Gaussian. The definitions (4.77) and (4.81) immediately imply $\left\langle \boldsymbol{W}'_\mu(t) \right\rangle = \boldsymbol{0}$. For the second moments we distinguish three different cases:
(i) For $j, k = 1, 2, \ldots, N-1$:

$$\left\langle \boldsymbol{W}'_j(t) \boldsymbol{W}'_k(t') \right\rangle = \frac{1}{\sqrt{a_j^{\mathrm{Z}}\, a_k^{\mathrm{Z}}}} \sum_{l,n} \Omega_{lj}^{\mathrm{Z}}\, \Omega_{nk}^{\mathrm{Z}} \Big(H_{l+1\,n+1} + H_{ln}$$
$$- H_{l+1\,n} - H_{l\,n+1} \Big)\, \boldsymbol{\delta} \, \min(t, t')$$
$$= \frac{1}{\sqrt{a_j^{\mathrm{Z}}\, a_k^{\mathrm{Z}}}} \sum_{l,n} \Omega_{lj}^{\mathrm{Z}}\, A_{ln}^{\mathrm{Z}}\, \Omega_{nk}^{\mathrm{Z}}\, \boldsymbol{\delta} \, \min(t, t') = \delta_{jk}\, \boldsymbol{\delta} \, \min(t, t') .$$

(ii)

$$\langle \boldsymbol{W}'_N(t) \boldsymbol{W}'_N(t') \rangle = \Big(\sum_{\mu,\nu} H_{\mu\nu}^{-1} \Big) \Big(\sum_{\mu,\nu} l_\mu\, H_{\mu\nu}\, l_\nu \Big)\, \boldsymbol{\delta} \, \min(t, t') = \boldsymbol{\delta} \, \min(t, t') .$$

(iii) For $k = 1, 2, \ldots, N-1$:

$$\langle \boldsymbol{W}'_k(t) \boldsymbol{W}'_N(t') \rangle \sim \sum_{j,\mu} \Omega_{jk}^{\mathrm{Z}}\, (H_{j+1\,\mu} - H_{j\mu})\, l_\mu = 0 .$$

4.21 From (4.76), we obtain the following expression in exactly the same manner as in the solution to Exercise 4.5,

$$\left\langle \boldsymbol{Q}'_j \boldsymbol{Q}'_k \right\rangle = \delta_{jk} \frac{k_{\mathrm{B}} T}{H} \left[\boldsymbol{\delta} + 2\lambda_H \frac{1}{a_j^{\mathrm{Z}}} (\boldsymbol{\kappa} + \boldsymbol{\kappa}^T) + 8\lambda_H^2 \frac{1}{(a_j^{\mathrm{Z}})^2} \boldsymbol{\kappa} \cdot \boldsymbol{\kappa}^T \right] .$$

In this equation, the inverse of the Zimm matrix occurs in the diagonal representation. By transforming back from normal modes to connector vectors we obtain the desired formula.

4.22 The first equation follows from the definition of the normal modes and the orthogonality of Ω^Z, and the third equation follows immediately from the definition of the connector vectors. The second equation can be verified as follows,
$\boldsymbol{r}_\mathrm{h} - \sum_{j=1}^{N-1}\left(1 - \sum_{\mu=1}^{j} l_\mu\right) \boldsymbol{Q}_j = \boldsymbol{r}_\mathrm{h} + \boldsymbol{r}_1 - \boldsymbol{r}_N + \sum_{\mu=1}^{N} l_\mu \sum_{j=\mu}^{N-1} \boldsymbol{Q}_j$
$= \boldsymbol{r}_\mathrm{h} + \boldsymbol{r}_1 - \boldsymbol{r}_N + \sum_{\mu=1}^{N} l_\mu(\boldsymbol{r}_N - \boldsymbol{r}_\mu) = \boldsymbol{r}_1.$

4.24 A nonzero velocity perturbation can arise only when $\langle \boldsymbol{F}_\mu(\boldsymbol{r}_\mu - \boldsymbol{r}_\mathrm{h})\rangle$ is anisotropic. Therefore, one has $\Delta\boldsymbol{v}(\boldsymbol{r}_\mathrm{h} + \boldsymbol{r}) = \boldsymbol{0}$ for $\boldsymbol{\varkappa} = \boldsymbol{0}$ and, in obtaining the first-order contribution in $\boldsymbol{\varkappa}$, we may replace $\langle(\boldsymbol{r}_\mu - \boldsymbol{r}_\mathrm{h})(\boldsymbol{r}_\mu - \boldsymbol{r}_\mathrm{h})\rangle$ in the exponential by its equilibrium average. The limitation to first order in $\boldsymbol{\varkappa}$ seems to be appropriate in view of the equilibrium averaging of hydrodynamic interaction in the Zimm model. By using the results of Exercise 4.21 and Example 4.23, we obtain after a series of elementary manipulations,

$$\Delta\boldsymbol{v}(\boldsymbol{r}_\mathrm{h} + \boldsymbol{r}) = -\frac{3h^*}{32\pi^2}\sqrt{\pi\frac{k_\mathrm{B}T}{H}}\sum_{\mu\nu} H^{-1}_{\mu\nu}\left(c_\mu + c_\nu - |\mu-\nu|\right)\frac{1}{c_\mu}$$
$$\int \exp\left\{-i\boldsymbol{q}\cdot\frac{\boldsymbol{r}}{\sqrt{2c_\mu\, k_\mathrm{B}T/H}}\right\}\frac{1}{q^2}\left(\boldsymbol{\delta} - \frac{\boldsymbol{q}\boldsymbol{q}}{q^2}\right)\cdot(\boldsymbol{\varkappa} + \boldsymbol{\varkappa}^T)\cdot(i\boldsymbol{q})\,\mathrm{e}^{-q^2/4}\,\mathrm{d}^3q,$$

where $c_\mu := \sum_{k=1}^{\mu-1}(\sum_{\nu=1}^{k} l_\nu)^2 + \sum_{k=1}^{N-\mu}(\sum_{\nu=1}^{k} l_\nu)^2$. The coefficients c_μ correspond to $\sum_k \tilde{B}^2_{\mu k}$ in the notation of §15.4 of [3]. Very useful is the identity $2\sum_k \tilde{B}_{\mu k}\tilde{B}_{\nu k} = c_\mu + c_\nu - |\mu - \nu|$.

For a more detailed discussion of the velocity perturbation it is very helpful that the q-integrations in the above expression for $\Delta\boldsymbol{v}$ can be performed in terms of error functions. The following auxiliary equation (the simplicity of which should be noticed in spite of its length) can be derived by means of the approach described in Appendix A of [5],

$$\frac{2}{\pi^2}\int \mathrm{e}^{-i\boldsymbol{q}\cdot\boldsymbol{r}}\,\frac{1}{q^2}\left(\boldsymbol{\delta} - \frac{\boldsymbol{q}\boldsymbol{q}}{q^2}\right)\cdot\boldsymbol{\Theta}\cdot(i\boldsymbol{q})\,\mathrm{e}^{-q^2/4}\,\mathrm{d}^3q =$$
$$2\left[\frac{3}{r^4}\,\mathrm{erf}(r) - \left(\frac{2}{r} + \frac{3}{r^3}\right)\frac{2}{\sqrt{\pi}}\mathrm{e}^{-r^2}\right]\boldsymbol{\Theta}\cdot\frac{\boldsymbol{r}}{r}$$
$$+\left[\left(\frac{6}{r^2} - \frac{15}{r^4}\right)\mathrm{erf}(r) + \left(\frac{4}{r} + \frac{15}{r^3}\right)\frac{2}{\sqrt{\pi}}\mathrm{e}^{-r^2}\right]\frac{\boldsymbol{r}\cdot\boldsymbol{\Theta}\cdot\boldsymbol{r}}{r^2}\,\frac{\boldsymbol{r}}{r},$$

where $\boldsymbol{\Theta}$ is an arbitrary traceless symmetric matrix, and
$$\mathrm{erf}(r) := \frac{2}{\sqrt{\pi}}\int_0^r \mathrm{e}^{-x^2}\,\mathrm{d}x.$$
The numerical evaluation of the above expression for small r is subtle; we have used series expansions consisting of ten terms for $r \le 1$.

For small r, the right side of the above equation is equal to $16\,\boldsymbol{\Theta}\cdot\boldsymbol{r}/(5\sqrt{\pi})$. We hence find near the center of resistance $\Delta\boldsymbol{v}(\boldsymbol{r}_\mathrm{h} + \boldsymbol{r}) = -c\,(\boldsymbol{\varkappa} + \boldsymbol{\varkappa}^T)\cdot\boldsymbol{r}$ with

$$c = \frac{3h^*}{20\sqrt{2}} \sum_{\mu\nu} H^{-1}_{\mu\nu} \left(c_\mu + c_\nu - |\mu - \nu|\right) \frac{1}{\sqrt{c_\mu}^{\,3}}.$$

A coefficient $c = 1/2$ would imply a pure rotation with angular velocity $\dot{\gamma}/2$. For large N, the above expression for c becomes independent both of N and of h^*. By numerical extrapolation of the results for N up to 1000 we have obtained $c \approx 0.4931(3)$, so that the shear flow is not perfectly screened at the center of resistance. The ratio in Fig. 4.7 for $r = 0$ is equal to $1 - 2c$.

4.25 For $N = 2$: $\lambda_1 = \dfrac{\lambda_H}{1 - \sqrt{2}h^*}$.

For $N = 3$: $\lambda_1 = \dfrac{\lambda_H}{1 - \sqrt{2}h^* - [1 - h^*(2\sqrt{2} - 1)]/2} = \dfrac{2\lambda_H}{1 - h^*}$,

$$\lambda_2 = \frac{\lambda_H}{1 - \sqrt{2}h^* + [1 - h^*(2\sqrt{2} - 1)]/2} = \frac{2\lambda_H}{3 - (4\sqrt{2} - 1)h^*}.$$

The polymer contribution to the viscosity is given by $n_\mathrm{p}\, k_\mathrm{B}T\, \lambda_1$ and $n_\mathrm{p}\, k_\mathrm{B}T\, (\lambda_1 + \lambda_2)$, respectively. Analytical expressions for the eigenvalues for $N = 4$ and $N = 5$ can be found in Table III of [6].

4.26 The results of Exercise 4.6 imply that $U_{\eta\lambda} = \pi^2/6$, $U_{\Psi\eta} = 4/5$, and $U_{\Psi\Psi} = 0$. If we further use $D_\mathrm{c} = k_\mathrm{B}T/(\zeta N)$ and the solution of Exercise 4.7 then we obtain $U_\mathrm{RD} = 0$ and $U_{\eta\mathrm{R}} = \infty$; these results indicate the excessively strong molecular weight dependence of the diffusion coefficient and of the viscosity predicted by the Rouse model. The universal prefactors should hence not be taken too seriously.

When keeping the relationship $\lambda_1 = 6\lambda_\eta/\pi^2$ in mind, we see that (4.23) and the results of Exercise 4.7 imply that the normalized extensional viscosity is a universal function of the reduced strain rate $\lambda_\eta\dot{\epsilon}$, and that the normalized mean-square gyration tensor in shear flow is a universal function of the reduced shear rate $\lambda_\eta\dot{\gamma}$. For example,

$\frac{3}{N}\sum_\mu \langle (\boldsymbol{r}_\mu - \boldsymbol{r}_\mathrm{c})(\boldsymbol{r}_\mu - \boldsymbol{r}_\mathrm{c})\rangle / \overline{R}_\mathrm{g}^2 = \boldsymbol{\delta} + \frac{2}{5}\,\lambda_\eta(\boldsymbol{\kappa} + \boldsymbol{\kappa}^T) + \frac{16}{35}\,\lambda_\eta^2\,\boldsymbol{\kappa}\cdot\boldsymbol{\kappa}^T$.

4.28 The following program `HIEUL` is based on the simulation of a single trajectory. A single time step is performed by the subroutine `EULER`. For the first time-step width, we start from equilibrium initial conditions; for further time-step widths, we start from the final configuration of the preceding time step. In any case, the respective initial conditions are relaxed by a simulation over several relaxation times. The estimation of error bars is based on the idea of introducing blocks, as discussed in Sect. 4.2.5.

We use five different time steps. The output consists of three columns: time-step width, second normal stress coefficient, and statistical error bar for the second normal stress coefficient. The appropriate unit for the second normal stress coefficient is $n_\mathrm{p}\, k_\mathrm{B}T\, \lambda_H^2$. According to the Boltzmann distribution, the equilibrium initial conditions for our choice of units are given by standard Gaussian random variables.

For the input data given in the header, the total time required for this example on a workstation is roughly two days.

```
      PROGRAM HIEUL
      REAL*8 APSI2,BPSI2,VPSI2
      REAL DTARR(5)
      INTEGER NTIARR(5)
      COMMON /STEPBL/ AB,A2,A4,SR,SQ12DT,DELTAT
C     Array: different time steps
      DATA DTARR/0.5,0.4,0.3,0.2,0.1/
C     Array: different numbers of time steps per block
      DATA NTIARR/200000,250000,333333,500000,1000000/
C     Number of blocks
      NBLOCK=5000
C     HI parameter
      HSTAR=0.15
C     Shear rate
      SR=1.
C     Initial seed for random number generator
      ISEED=1510241094
      CALL RANILS(ISEED)
      PI=3.1415926
C     Bead radius
      AB=SQRT(PI)*HSTAR
      A2=AB*AB
      A4=A2*A2
C     Equilibrium initial conditions
      Q1=RANGLS()
      Q2=RANGLS()
      Q3=RANGLS()
C
C     Loop for different time step widths
      DO 1000 IDT=1,5
C        Auxiliary parameters
         DELTAT=DTARR(IDT)
         SQ12DT=SQRT(12.*DELTAT)
C        Relaxation of initial conditions
         NTIME=10./DELTAT
         DO 50 ITIME=1,NTIME
            CALL EULER(Q1,Q2,Q3)
50       CONTINUE
         APSI2=0.
         VPSI2=0.
         NTIME=NTIARR(IDT)
C        Loop for different blocks
         DO 100 IBLOCK=1,NBLOCK
            BPSI2=0.
C           Time integration: Euler
            DO 10 ITIME=1,NTIME
               CALL EULER(Q1,Q2,Q3)
C              "Measurement"
```

```
               BPSI2=BPSI2+Q2*Q2-Q3*Q3
10          CONTINUE
C           Summation of the results for the blocks
            BPSI2=BPSI2/(NTIME*SR*SR)
            APSI2=APSI2+BPSI2
            VPSI2=VPSI2+BPSI2*BPSI2
100      CONTINUE
C
C        Average, statistical error
         APSI2=APSI2/NBLOCK
         VPSI2=VPSI2/NBLOCK
         VPSI2=SQRT((VPSI2-APSI2*APSI2)/(NBLOCK-1))
C        Output of results
         WRITE(6,1) DELTAT,APSI2,VPSI2
1        FORMAT(F11.8,4X,F9.6,1X,F9.6)
1000  CONTINUE
      STOP
      END


      SUBROUTINE EULER(Q1,Q2,Q3)
C     Time step: Euler scheme for Hookean dumbbells with hydrodynamic
C     interaction (regularized Oseen tensor)
      COMMON /STEPBL/ AB,A2,A4,SR,SQ12DT,DELTAT
      DATA C43/1.3333333/, C143/4.6666667/, C83/2.6666667/
C     Auxiliary parameters
      QM2=Q1*Q1+Q2*Q2+Q3*Q3
      QM1=SQRT(QM2)
      QM3=QM2*QM1
      QM5=QM3*QM2
C     Evaluation of the hydrodynamic interaction tensor
      AUX1=QM2+C43*A2
      AUX1=0.75*AB/(AUX1*AUX1*AUX1)
      AUX2=1.-AUX1*(QM5+C143*A2*QM3+8.*A4*QM1)
      AUX3=AUX2-AUX1*(QM5+2.*A2*QM3-C83*A4*QM1)
      AUX1=1.-0.5*DELTAT*AUX3
C     Generation of the required random numbers
      W1=SQ12DT*(RANULS()-.5)
      W2=SQ12DT*(RANULS()-.5)
      W3=SQ12DT*(RANULS()-.5)
C     Time step
      AUXA=SQRT(AUX2)
      AUXB=(SQRT(AUX3)-AUXA)*(Q1*W1+Q2*W2+Q3*W3)/QM2
      Q1=AUX1*Q1+SR*DELTAT*Q2+AUXA*W1+AUXB*Q1
      Q2=AUX1*Q2             +AUXA*W2+AUXB*Q2
      Q3=AUX1*Q3             +AUXA*W3+AUXB*Q3
      RETURN
      END
```

The output for the time step and the second normal stress coefficient (with error estimate) produced by the above program looks as follows:

```
0.50000000    -0.008329  0.000137
0.40000001    -0.009336  0.000137
0.30000001    -0.010300  0.000135
0.20000000    -0.011532  0.000137
0.10000000    -0.012403  0.000136
```

If the time-step extrapolation is done with `TEXTRA` (see solution to Exercise 4.11), the result is $\Psi_2 = -0.01348(14)$.

4.29 The following Maple code generates a FORTRAN subroutine based on the second-order simulation algorithm for Hookean dumbbells with hydrodynamic interaction in steady shear flow. More precisely, the program performs a single time step, where the initial dumbbell configuration is `X1`, `X2`, `X3`, and the final configuration is `Q1`, `Q2`, `Q3`. The quantity `DELTAT` is the time step, `W1`, `W2`, `W3` are independent Gaussian random numbers with mean 0 and variance Δt (or other independent random numbers satisfying the moment conditions appropriate for a second-order simulation algorithm), `V1`, `V2`, `V3` are further independent random numbers that take the values $\pm\Delta t$ with probability 1/2, `SR` is the shear rate, and `AB` is the bead radius.
The first part of the Maple code is taken from [7]. The line
`convert([ seq(b.k.ip*Diff(F,v[k]),k=1..d)],'+');`
of the code displayed in [7] contains a typographical error. It has been changed into
`convert([ seq(b.k.ip(v)*Diff(F,v[k]),k=1..d)],'+');`
A hydrodynamic-interaction tensor of the regularized type (4.67) is assumed. Our Maple code is neither very elegant nor particularly compact nor as general as it might be desirable (the second part should be written for stochastic differential equations of arbitrary dimensionality). As a beginner in Maple, the author only wanted to save time. The following code produces 914 lines of FORTRAN code in only a few seconds.

```
# Definition of differential operators occurring in numerical
# integration schemes (from Kloeden and Scott)
L := proc(F)
  local ip,v,k,j,l;
  ip := op(procname); v := op(F);
  if ip = 0 then
    seq(a.k(v)*Diff(F,v[k]),k=1..d);
    seq(seq(seq(b.k.j(v)*b.l.j(v)*Diff(Diff(F,v[k]),v[l])
                                        ,j=1..m),l=1..d),k=1..d);
    Diff(F,v[d+1]) + convert([""],'+') + 1/2*convert(["],'+');
  else
    convert([ seq(b.k.ip(v)*Diff(F,v[k]),k=1..d)],'+');
  fi;
end:
```

```
# Second-order algorithm for general 3-dimensional SDEs
d :=3: m :=3:
v := X1,X2,X3,t:
CO := U(v) + 1/2*(L[0](U(v))*DELTAT + L[1](U(v))*W1 + L[2](U(v))*W2
    + L[3](U(v))*W3):
seq(b1.j(v)*W.j,j=1..3): a1(v)*DELTAT + convert(["],`+`):
X1N := X1 + subs(U(v)=",CO):
seq(b2.j(v)*W.j,j=1..3): a2(v)*DELTAT + convert(["],`+`):
X2N := X2 + subs(U(v)=",CO):
seq(b3.j(v)*W.j,j=1..3): a3(v)*DELTAT + convert(["],`+`):
X3N := X3 + subs(U(v)=",CO):
X1N := X1N - 1/2*DELTAT*(L[1](b11(v)) + L[2](b12(v)) + L[3](b13(v))):
X2N := X2N - 1/2*DELTAT*(L[1](b21(v)) + L[2](b22(v)) + L[3](b23(v))):
X3N := X3N - 1/2*DELTAT*(L[1](b31(v)) + L[2](b32(v)) + L[3](b33(v))):
X1N := X1N - 1/2*(L[1](cr1(v)) + L[2](cr2(v)) + L[3](cr3(v))):
X1N := subs({cr1(v)=V3*b12(v)-V2*b13(v), cr2(v)=V1*b13(v)-V3*b11(v),
            cr3(v)=V2*b11(v)-V1*b12(v)},"):
X2N := X2N - 1/2*(L[1](cr1(v)) + L[2](cr2(v)) + L[3](cr3(v))):
X2N := subs({cr1(v)=V3*b22(v)-V2*b23(v), cr2(v)=V1*b23(v)-V3*b21(v),
            cr3(v)=V2*b21(v)-V1*b22(v)},"):
X3N := X3N - 1/2*(L[1](cr1(v)) + L[2](cr2(v)) + L[3](cr3(v))):
X3N := subs({cr1(v)=V3*b32(v)-V2*b33(v), cr2(v)=V1*b33(v)-V3*b31(v),
            cr3(v)=V2*b31(v)-V1*b32(v)},"):

# Evaluation and insertion of coefficient functions,
# generation of FORTRAN code
AA2:=AB*AB: AA4:=AA2*AA2:
XM2:=X1*X1+X2*X2+X3*X3: XM1:=sqrt(XM2): XM3:=XM2*XM1: XM5:=XM3*XM2:
denom:=XM2+4/3*AA2: denom:=3/4*AB/(denom*denom*denom):
auxf:=1-denom*(XM5+14/3*AA2*XM3+8*AA4*XM1):
auxfpg:=auxf-denom*(XM5+2*AA2*XM3-8/3*AA4*XM1):
sauxf:=sqrt(auxf): dsaux:=sqrt(auxfpg)-sauxf:
slist:={a1(v)=SR*X2-auxfpg*X1/2, a2(v)=-auxfpg*X2/2,
    a3(v)=-auxfpg*X3/2, b11(v)=sauxf+dsaux*X1*X1/XM2,
    b22(v)=sauxf+dsaux*X2*X2/XM2, b33(v)=sauxf+dsaux*X3*X3/XM2,
    b12(v)=dsaux*X1*X2/XM2, b13(v)=dsaux*X1*X3/XM2,
    b23(v)=dsaux*X2*X3/XM2, b21(v)=dsaux*X1*X2/XM2,
    b31(v)=dsaux*X1*X3/XM2, b32(v)=dsaux*X2*X3/XM2,
    Diff=diff}:
X1N:=eval(subs(slist,X1N)):
X2N:=eval(subs(slist,X2N)):
X3N:=eval(subs(slist,X3N)):
readlib(fortran);
fortran([Q1=X1N, Q2=X2N, Q3=X3N],filename=`secord.f`,optimized);
```

4.30 The potential corresponding to (4.117) is $(1-\nu_{\rm ev})\,H\,|\boldsymbol{Q}|^{1/(1-\nu_{\rm ev})}$, so that we obtain the following result by averaging with the Boltzmann distribution:

$$\langle Q^2 \rangle = \frac{\int Q^2 \exp\left\{-(1-\nu_{\rm ev})\, H\, Q^{1/(1-\nu_{\rm ev})}/(k_{\rm B}T)\right\} {\rm d}^3Q}{\int \exp\left\{-(1-\nu_{\rm ev})\, H\, Q^{1/(1-\nu_{\rm ev})}/(k_{\rm B}T)\right\} {\rm d}^3Q}$$

$$= \left[\frac{k_{\rm B}T}{H(1-\nu_{\rm ev})}\right]^{2(1-\nu_{\rm ev})} \frac{\int_0^\infty x^{4-5\nu_{\rm ev}} {\rm e}^{-x}\, {\rm d}x}{\int_0^\infty x^{2-3\nu_{\rm ev}} {\rm e}^{-x}\, {\rm d}x} = \frac{\Gamma(5(1-\nu_{\rm ev}))}{\Gamma(3(1-\nu_{\rm ev}))} \left[\frac{k_{\rm B}T}{H(1-\nu_{\rm ev})}\right]^{2(1-\nu_{\rm ev})}.$$

4.31 One can estimate b by comparing the length of a fully stretched chain segment with its equilibrium size. If b is a rather large number (this assumption is consistent with the result of our estimation), we obtain the equilibrium mean-square size of the segment approximately from the Hookean result, $3k_{\rm B}T/H$ (see Example 4.13). The ratio of the square of the maximum possible spring extension and the equilibrium mean-square size is hence equal to $b/3$. If L is the bond length, the definition of the steric factor for polymers with pure carbon backbone implies that the mean-square size at equilibrium is $\sigma_{\rm sf}^2\, 2(N_{\rm C}/N)L^2$. For the square of the size of a fully stretched planar zig-zag chain segment we have $(2/3)(N_{\rm C}/N)^2L^2$. The ratio gives the desired estimate for b.

4.32 $$\frac{1}{2}\frac{\sqrt{\bar b(\Delta t)}\,\Delta t}{1-\frac{\bar b(\Delta t)}{b}} = \sqrt{b} - \sqrt{\bar b(\Delta t)} \approx \frac{b-\bar b(\Delta t)}{2\sqrt{b}} \quad\Longrightarrow\quad \bar b(\Delta t) \approx b(1-\sqrt{\Delta t}).$$

4.33 If the lengths of $\boldsymbol{Q}(t_{j+1})$ and of the vector on the right-hand side of (4.122) are x and L, respectively, then the cubic equation for x is
$g(x) := x^3 - L\,x^2 - (1+\Delta t/4)b\,x + L\,b = 0.$
For $L = 0$, we obtain $x = 0$. For $L > 0$, we find $g(0) > 0$ and $g(\sqrt{b}) < 0$, so that there is a solution in $[0, \sqrt{b}]$. The solution must be unique because there are two further solutions outside $[0, \sqrt{b}]$.

4.34 The following program `FENESI` produces three of the data points shown in Fig. 4.16. It is based on the simulation of a single trajectory (cf. Exercise 4.28). A single time step is performed by the subroutine `SEMIMP`. For the precision necessary for our purposes, the exact solution of the cubic equation occurring in the implicit step (see p. 179 of [4]) is more efficiently determined than a numerical solution (based on the combination of bisection and Newton-Raphson realized in the program `rtsafe` of [4]). The output of `FENESI` consists of five columns: time-step width, polymer contribution to the viscosity, statistical error bar for the polymer contribution to the viscosity, first normal stress coefficient, statistical error bar for the first normal stress coefficient. The appropriate units for the viscosity and the first normal stress coefficient are $n_{\rm p}\, k_{\rm B}T\, \lambda_H$ and $n_{\rm p}\, k_{\rm B}T\, \lambda_H^2$, respectively. The total time required for this example on a workstation is roughly four hours.

```
      PROGRAM FENESI
      REAL*8 AETA,BETA,VETA,APSI,BPSI,VPSI
      REAL DTARR(3)
      INTEGER NTIARR(3)
      COMMON /STEPBL/ B,DTQ,BAUXQ,BAUXR,SRDT,SRDTH,C1P,C2P
C     Array: different time steps
      DATA DTARR/0.5,0.4,0.3/
```

```
C     Array: different numbers of time steps per block
      DATA NTIARR/20000,25000,33333/
C     Number of blocks
      NBLOCK=11000
C     FENE parameter
      B=50.
C     Shear rate
      SR=5.
C     Initial seed for random number generator
      ISEED= 321041094
      CALL RANILS(ISEED)
C     Deterministic initial conditions
      Q1=0.
      Q2=0.
      Q3=0.
C
C     Loop for different time-step widths
      DO 1000 IDT=1,3
C        Auxiliary parameters
         DELTAT=DTARR(IDT)
         DTQ=0.25*DELTAT
         SQDT=SQRT(DELTAT)
         C1P=14.14855378*SQDT
         C2P= 1.21569221*SQDT
         BAUXQ=3.*B*(1.+0.25*DELTAT)
         BAUXR=9.*B*(1.-0.125*DELTAT)
         SRDT=SR*DELTAT
         SRDTH=.5*SRDT
C        Relaxation of initial conditions
         NTIME=10./DELTAT
         DO 50 ITIME=1,NTIME
            CALL SEMIMP(Q1,Q2,Q3)
50       CONTINUE
C
         AETA=0.
         VETA=0.
         APSI=0.
         VPSI=0.
         NTIME=NTIARR(IDT)
C        Loop for different blocks
         DO 100 IBLOCK=1,NBLOCK
            BETA=0.
            BPSI=0.
C           Time integration: semi-implicit scheme
            DO 10 ITIME=1,NTIME
               CALL SEMIMP(Q1,Q2,Q3)
C              "Measurement"
               QL=Q1*Q1+Q2*Q2+Q3*Q3
```

```
            AMPL=1./(1.-QL/B)
            BETA=BETA+AMPL*Q1*Q2
            BPSI=BPSI+AMPL*(Q1*Q1-Q2*Q2)
10        CONTINUE
C         Summation of the results for the blocks
          BETA=BETA/(NTIME*SR)
          BPSI=BPSI/(NTIME*SR*SR)
          AETA=AETA+BETA
          VETA=VETA+BETA*BETA
          APSI=APSI+BPSI
          VPSI=VPSI+BPSI*BPSI
100     CONTINUE
C
C       Averages, statistical errors
        AETA=AETA/NBLOCK
        VETA=VETA/NBLOCK
        VETA=SQRT((VETA-AETA*AETA)/(NBLOCK-1))
        APSI=APSI/NBLOCK
        VPSI=VPSI/NBLOCK
        VPSI=SQRT((VPSI-APSI*APSI)/(NBLOCK-1))
C       Output of results
        WRITE(6,1) DELTAT,AETA,VETA,APSI,VPSI
1       FORMAT(F11.8,2(4X,F9.6,1X,F9.6))
1000  CONTINUE
      STOP
      END


      SUBROUTINE SEMIMP(Q1,Q2,Q3)
C     Time step: semi-implicit scheme for FENE model
      PARAMETER (FOURPI=12.566371)
      COMMON /STEPBL/ B,DTQ,BAUXQ,BAUXR,SRDT,SRDTH,C1P,C2P
C     Auxiliary parameters
      QL=Q1*Q1+Q2*Q2+Q3*Q3
      DTQM=DTQ/(1.-QL/B)
      OMDTHM=1.-2.*DTQM
C     Construction of suitable random numbers
      W1=RANULS()-.5
      W2=RANULS()-.5
      W3=RANULS()-.5
      W1=W1*(C1P*W1*W1+C2P)
      W2=W2*(C1P*W2*W2+C2P)
      W3=W3*(C1P*W3*W3+C2P)
C
C     Predictor step
      QAUX1=OMDTHM*Q1+SRDT*Q2+W1
      QAUX2=OMDTHM*Q2+W2
      QAUX3=OMDTHM*Q3+W3
C     Corrector step for the flow term
```

```
      Q1=QAUX1+DTQM*Q1+SRDTH*(QAUX2-Q2)
      Q2=QAUX2+DTQM*Q2
      Q3=QAUX3+DTQM*Q3
C     Exact solution of the implicit equation for the
C        length of the connector vector
      QL=Q1*Q1+Q2*Q2+Q3*Q3
      SQQL=SQRT(QL)
      AUXQ=(BAUXQ+QL)/9.
      AUXR=SQQL*(BAUXR-QL)/27.
      SQAUXQ=SQRT(AUXQ)
      AUX=ACOS(AUXR/(AUXQ*SQAUXQ))
      XL=-2.*SQAUXQ*COS((AUX+FOURPI)/3.)+SQQL/3.
C     Rescaling to obtain the proper length
      RED=XL/SQQL
      Q1=RED*Q1
      Q2=RED*Q2
      Q3=RED*Q3
      RETURN
      END
```

The output for the time step, the polymer contribution to the viscosity (with error estimate), and the first normal stress coefficient (with error estimate) produced by the above program looks as follows:

```
0.50000000     0.629354  0.000110     0.640420  0.000089
0.40000001     0.621531  0.000108     0.641418  0.000088
0.30000001     0.615380  0.000108     0.642575  0.000089
```

5.1 After rearranging terms by means of the product rule, (3.107) gives the following diffusion equation corresponding to $\mathrm{d}\boldsymbol{X}_t = \boldsymbol{A}\,\mathrm{d}t + \mathbf{B}(\boldsymbol{X}_t) \odot \mathrm{d}\boldsymbol{W}_t$: $\frac{\partial}{\partial t} p(t,\boldsymbol{x}) = \frac{1}{2}\frac{\partial}{\partial \boldsymbol{x}} \cdot \frac{\partial}{\partial \boldsymbol{x}} p(t,\boldsymbol{x})$.

After introducing the constraints and simplifying the terms with second-order derivatives, the difference between the right sides of the respective Fokker-Planck equations is $-\frac{\partial}{\partial \boldsymbol{x}} \cdot [\mathbf{P}(\boldsymbol{x}) \cdot \boldsymbol{A}\, p(t,\boldsymbol{x})]$, which corresponds to a surplus rotational term $\mathbf{P}(\boldsymbol{X}_t) \cdot \boldsymbol{A}\,\mathrm{d}t$.

5.2 Equation (5.10) follows from $(\delta_{\mu\nu}\boldsymbol{\zeta}_\mu^{-1} + \boldsymbol{\Omega}_{\mu\nu})^T = \delta_{\nu\mu}\boldsymbol{\zeta}_\nu^{-1} + \boldsymbol{\Omega}_{\nu\mu}$, and (5.10) immediately implies $\mathbf{Z}^T = \mathbf{Z}$ and then $\tilde{\boldsymbol{\zeta}}_{\mu\nu}^T = \tilde{\boldsymbol{\zeta}}_{\nu\mu}$.

Equation (5.13) follows from the definition of $\mathbf{Z}$, and this implies $\sum_\mu \tilde{\boldsymbol{\zeta}}_{\mu\nu} = \mathbf{Z} \cdot \boldsymbol{\lambda}_\nu - \mathbf{Z} \cdot \boldsymbol{\lambda}_\nu = \mathbf{0}$ (the second part of the identity follows by transposition).

By summing (5.3) over μ we obtain (5.15) after multiplying by $\mathbf{Z}^{-1}$. Equation (5.16) is a combination of (5.3) and (5.15).

The symmetry $\tilde{\mathrm{g}}_{jk} = \tilde{\mathrm{g}}_{kj}$ (and hence the symmetry of the inverse) follows from (5.11).

5.3 Equation (5.18) follows after inserting $\boldsymbol{\Omega}_{\mu\mu'} = \boldsymbol{\Omega}(\boldsymbol{r}_\mu - \boldsymbol{r}_{\mu'})$.

Equation (5.19) follows by differentiating (5.3), multiplying with $\boldsymbol{\zeta}_{\mu'\nu'}$ from the right, and summing over μ, μ', ν'; (5.20) is obtained from the fact that the derivative of $\mathbf{Z} \cdot \mathbf{Z}^{-1}$ vanishes.

Equation (5.21) follows by differentiating (5.15), multiplying with $\boldsymbol{\zeta}_{\nu\nu'}$ from the right, and summing over ν.

Equation (5.22) follows by differentiating (5.16), multiplying with $\boldsymbol{\zeta}_{\mu'\nu'}$ from the right, and summing over μ'.

5.4 $\boldsymbol{R}_2 = -\boldsymbol{R}_1 = \dfrac{L}{2}\begin{pmatrix} \sqrt{1-z^2}\cos\phi \\ \sqrt{1-z^2}\sin\phi \\ z \end{pmatrix}$.

$$\frac{\partial \boldsymbol{R}_2}{\partial z} = \frac{L}{2\sqrt{1-z^2}}\begin{pmatrix} -z\cos\phi \\ -z\sin\phi \\ \sqrt{1-z^2} \end{pmatrix}, \qquad \frac{\partial \boldsymbol{R}_2}{\partial \phi} = \frac{L\sqrt{1-z^2}}{2}\begin{pmatrix} -\sin\phi \\ \cos\phi \\ 0 \end{pmatrix}.$$

$$\boldsymbol{\zeta}_{\mu\nu} = \delta_{\mu\nu}\,\zeta\,\boldsymbol{\delta}, \qquad \boldsymbol{\lambda}_1 = \boldsymbol{\lambda}_2 = \tfrac{1}{2}\,\boldsymbol{\delta}, \qquad \tilde{\boldsymbol{\zeta}}_{\mu\nu} = (-1)^{\mu+\nu}\,\tfrac{\zeta}{2}\,\boldsymbol{\delta}.$$

$$\tilde{\mathsf{g}}_{jk} = 2\zeta\,\frac{\partial \boldsymbol{R}_2}{\partial Q_j}\cdot\frac{\partial \boldsymbol{R}_2}{\partial Q_k}, \qquad (\tilde{\mathsf{g}}_{jk}) = \frac{\zeta L^2}{2}\begin{pmatrix} \dfrac{1}{1-z^2} & 0 \\ 0 & 1-z^2 \end{pmatrix}.$$

$$(\tilde{\mathsf{G}}_{jk}) = \frac{2}{\zeta L^2}\begin{pmatrix} 1-z^2 & 0 \\ 0 & \dfrac{1}{1-z^2} \end{pmatrix}, \qquad \mathsf{g}_{jk} = 2M\,\frac{\partial \boldsymbol{R}_2}{\partial Q_j}\cdot\frac{\partial \boldsymbol{R}_2}{\partial Q_k}, \qquad \mathsf{g} = \frac{M^2L^4}{4}.$$

$$\begin{aligned}\frac{\partial p}{\partial t} = &-\frac{\partial}{\partial z}\left\{-\left[\frac{1}{2}\dot{\gamma}\,z(1-z^2)\,\sin 2\phi + \frac{1}{6\lambda}\left((1-z^2)\frac{\partial V^{(\mathrm{e})}}{\partial z} + 2z\right)\right]p\right\} \\ &-\frac{\partial}{\partial \phi}\left\{-\left[\dot{\gamma}\sin^2\phi + \frac{1}{6\lambda(1-z^2)}\,\frac{\partial V^{(\mathrm{e})}}{\partial \phi}\right]p\right\} \\ &+\frac{1}{6\lambda}\left\{\frac{\partial^2}{\partial z^2}[(1-z^2)p] + \frac{\partial^2}{\partial \phi^2}\left(\frac{p}{1-z^2}\right)\right\}.\end{aligned}$$

The stochastic differential equations (5.27) and (5.28) immediately follow from the above diffusion equation and the results of Sect. 3.3.3.

5.5 The identity (5.34) is obtained by multiplying the definition (5.29) with $\sum_{\mu'}(\delta_{\mu\mu'}\boldsymbol{\zeta}_\mu^{-1} + \boldsymbol{\Omega}_{\mu\mu'}) \cdot \partial Q_k/\partial \boldsymbol{R}_{\mu'}$, and by summing over μ [(5.16), (5.30), and (5.31) need to be used].

5.6 $\Delta \ln\det(\mathsf{M}_{jk}) = \ln\det(\mathsf{M}_{jk} + \Delta\mathsf{M}_{jk}) - \ln\det(\mathsf{M}_{jk})$
$= \ln\det(\delta_{jk} + \sum_{l=1}^{d}\mathsf{M}_{jl}^{-1}\Delta\mathsf{M}_{lk}) \approx \ln(1 + \sum_{j,l=1}^{d}\mathsf{M}_{jl}^{-1}\Delta\mathsf{M}_{lj}) \approx \sum_{j,l=1}^{d}\mathsf{M}_{jl}^{-1}\Delta\mathsf{M}_{lj}$.

5.7 Both equations follow after inserting the definition (5.37) and using (5.7) and (5.8).

5.8 $\sum_\nu \mathbf{P}_{\mu\nu} = \boldsymbol{\delta}$ follows from (5.41), (5.31), and (5.13).

$\sum_\mu \boldsymbol{\lambda}_\mu \cdot \mathbf{P}_{\mu\nu} = \boldsymbol{\lambda}_\nu$ follows from (5.41) and (5.13).

By multiplying (5.44) from the left with $\tilde{\boldsymbol{\zeta}}_{\nu'\mu}$ and using (5.14) one obtains $\sum_{\mu'} \tilde{\boldsymbol{\zeta}}_{\mu\mu'} \cdot \mathbf{P}_{\mu'\nu} = \sum_{\mu'\nu'} \tilde{\boldsymbol{\zeta}}_{\mu\mu'} \cdot \mathbf{H}_{\mu'\nu'} \cdot \tilde{\boldsymbol{\zeta}}_{\nu'\nu}$, which can be transposed in order to obtain the next identity (where (5.11) and an immediate consequence of (5.37), $\mathbf{H}^T_{\mu\nu} = \mathbf{H}_{\nu\mu}$, need to be used).

The alternative expression for $\sum_\mu (M_\mu/M_{\rm p}) \mathbf{P}_{\mu\nu}$ follows from the definition (5.41) when (5.2) and (5.43) are used. The next identity follows from (5.15) and (5.16), and the equation for $\sum_{\mu'} \mathbf{P}_{\mu\mu'} \cdot (\delta_{\mu'\nu} \boldsymbol{\zeta}^{-1}_{\mu'} + \boldsymbol{\Omega}_{\mu'\nu})$ is then obtained from (5.44) and (5.16).

$\sum_{\mu'} \mathbf{P}_{\mu\mu'} \cdot (\delta_{\mu'\nu} \boldsymbol{\zeta}^{-1}_{\mu'} + \boldsymbol{\Omega}_{\mu'\nu}) = \sum_{\mu'} (\delta_{\mu\mu'} \boldsymbol{\zeta}^{-1}_{\mu} + \boldsymbol{\Omega}_{\mu\mu'}) \cdot \mathbf{P}^T_{\nu\mu'}$ follows from the explicit equation for $\sum_{\mu'} \mathbf{P}_{\mu\mu'} \cdot (\delta_{\mu'\nu} \boldsymbol{\zeta}^{-1}_{\mu'} + \boldsymbol{\Omega}_{\mu'\nu})$ and the symmetry properties of all the tensors involved. The validity of the final, explicitly symmetric formulation then follows by means of the projection operator property (5.42).

5.9 We use the results of Exercise 5.4 in order to obtain

$$\mathbf{H}_{\mu\nu} = \frac{(-1)^{\mu+\nu}}{2\zeta} (\boldsymbol{\delta} - \boldsymbol{U}\boldsymbol{U}), \qquad \sum_{\mu'} \mathbf{H}_{\mu\mu'} \cdot \tilde{\boldsymbol{\zeta}}_{\mu'\nu} = \tfrac{1}{2}(-1)^{\mu+\nu} (\boldsymbol{\delta} - \boldsymbol{U}\boldsymbol{U}).$$

$$\mathbf{P}_{\mu\nu} = \delta_{\mu\nu}\, \boldsymbol{\delta} - \tfrac{1}{2}(-1)^{\mu+\nu} \boldsymbol{U}\boldsymbol{U}, \qquad \mathbf{P}_{2\nu} - \mathbf{P}_{1\nu} = (-1)^\nu (\boldsymbol{\delta} - \boldsymbol{U}\boldsymbol{U}).$$

$$\mathbf{B}_{\mu\nu} = \delta_{\mu\nu}\, \boldsymbol{\delta} \frac{1}{\sqrt{\zeta}}.$$

$$\begin{aligned} \mathrm{d}\boldsymbol{U} &= \frac{1}{L} \sum_\nu (\mathbf{P}_{2\nu} - \mathbf{P}_{1\nu}) \cdot \left[(\boldsymbol{\kappa} \cdot \boldsymbol{r}_\nu + \frac{1}{\zeta} \boldsymbol{F}^{(\mathrm{e})}_\nu)\, \mathrm{d}t + \sqrt{\frac{2k_{\mathrm{B}}T}{\zeta}}\, \mathrm{d}\boldsymbol{W}_\nu \right] \\ &\quad + \frac{k_{\mathrm{B}}T}{L} \sum_{j,k=1}^{d} \left[\tilde{G}_{jk} \frac{\partial^2 (\boldsymbol{R}_2 - \boldsymbol{R}_1)}{\partial Q_j \partial Q_k} + \frac{1}{\sqrt{g}} \frac{\partial (\boldsymbol{R}_2 - \boldsymbol{R}_1)}{\partial Q_j} \frac{\partial}{\partial Q_k} \left(\sqrt{g}\, \tilde{G}_{jk} \right) \right] \mathrm{d}t \\ &= (\boldsymbol{\delta} - \boldsymbol{U}\boldsymbol{U}) \cdot \left\{ \left[\boldsymbol{\kappa} \cdot \boldsymbol{U} + \frac{1}{\zeta L} \left(\boldsymbol{F}^{(\mathrm{e})}_2 - \boldsymbol{F}^{(\mathrm{e})}_1 \right) \right] \mathrm{d}t + \frac{1}{\sqrt{3\lambda}} \mathrm{d}\boldsymbol{W} \right\} - \frac{1}{3\lambda} \boldsymbol{U}\, \mathrm{d}t, \end{aligned}$$

where $\boldsymbol{W}(t) = [\boldsymbol{W}_2(t) - \boldsymbol{W}_1(t)]/\sqrt{2}$.

By expressing $(\boldsymbol{\delta} - \boldsymbol{U}\boldsymbol{U})$ in terms of $\partial \boldsymbol{R}_2/\partial z$ and $\partial \boldsymbol{R}_2/\partial \phi$ for the external potential of Exercise 5.4 one obtains the explicit representation

$$(\boldsymbol{\delta} - \boldsymbol{U}\boldsymbol{U}) \cdot \frac{1}{\zeta L} \left(\boldsymbol{F}^{(\mathrm{e})}_2 - \boldsymbol{F}^{(\mathrm{e})}_1 \right) =$$

$$-\frac{1}{6\lambda} \left[\begin{pmatrix} -z \cos\phi \\ -z \sin\phi \\ \sqrt{1-z^2} \end{pmatrix} \sqrt{1-z^2} \frac{\partial V^{(\mathrm{e})}}{\partial z} + \begin{pmatrix} -\sin\phi \\ \cos\phi \\ 0 \end{pmatrix} \frac{1}{\sqrt{1-z^2}} \frac{\partial V^{(\mathrm{e})}}{\partial \phi} \right].$$

Alternative derivation:

$$\mathrm{d}\boldsymbol{U} = \frac{\partial \boldsymbol{U}}{\partial z} \mathrm{d}z + \frac{\partial \boldsymbol{U}}{\partial \phi} \mathrm{d}\phi + \frac{1}{6\lambda} \left[(1 - z^2) \frac{\partial^2 \boldsymbol{U}}{\partial z^2} + \frac{1}{1 - z^2} \frac{\partial^2 \boldsymbol{U}}{\partial \phi^2} \right] \mathrm{d}t, \text{ where}$$

$$\boldsymbol{U} = \begin{pmatrix} \sqrt{1-z^2} \cos\phi \\ \sqrt{1-z^2} \sin\phi \\ z \end{pmatrix},$$

$$\frac{\partial \boldsymbol{U}}{\partial z} = \frac{1}{\sqrt{1-z^2}} \begin{pmatrix} -z\cos\phi \\ -z\sin\phi \\ \sqrt{1-z^2} \end{pmatrix}, \quad \frac{\partial \boldsymbol{U}}{\partial \phi} = \sqrt{1-z^2} \begin{pmatrix} -\sin\phi \\ \cos\phi \\ 0 \end{pmatrix},$$

$$\frac{\partial^2 \boldsymbol{U}}{\partial z^2} = -\frac{1}{\sqrt{1-z^2}^3} \begin{pmatrix} \cos\phi \\ \sin\phi \\ 0 \end{pmatrix}, \quad \frac{\partial^2 \boldsymbol{U}}{\partial \phi^2} = -\sqrt{1-z^2} \begin{pmatrix} \cos\phi \\ \sin\phi \\ 0 \end{pmatrix}.$$

The replacement of $\begin{pmatrix} -z\cos\phi \\ -z\sin\phi \\ \sqrt{1-z^2} \end{pmatrix} \mathrm{d}W^z + \begin{pmatrix} -\sin\phi \\ \cos\phi \\ 0 \end{pmatrix} \mathrm{d}W^\phi$ with $(\boldsymbol{\delta} - \boldsymbol{U}\boldsymbol{U}) \cdot \mathrm{d}\boldsymbol{W}$ corresponds to a change in the Cholesky decomposition of the diffusion matrix.

5.10 In a first step, we need to generalize the diffusion equation (5.23) by incorporating the center of mass position. This requires reworking §18.2 of [3] and, in particular, generalization of (18.2-44). We only outline the procedure. The starting point is (17.5-6) for only one species of polymer molecules from which the center of mass velocity is eliminated by means of (18.2-17) (where $\overline{N} = 1$ for dilute solutions). Expressions for the additional force terms occurring in the resulting equation are given in (18.2-27) and (18.3-6). After these steps one obtains the desired diffusion equation from polymer kinetic theory.
We still have to see whether the same diffusion equation is obtained from the coupled stochastic differential equations derived from (5.40) for the generalized coordinates Q_j (that is (5.32)) and the center of mass position (that is, the weighted sum of (5.40) with weights M_μ/M_p). In verifying the equality of the diffusion equations the results of Exercises 5.7 and 5.8 are very useful.

5.11 $\displaystyle\sum_\mu \frac{\partial g_j}{\partial \boldsymbol{r}_\mu} \cdot \frac{\partial \boldsymbol{R}_\mu}{\partial g_k} = \sum_{l=1}^{d'} \bar{\mathsf{g}}_{kl} \sum_{\mu\nu} \frac{\partial g_j}{\partial \boldsymbol{r}_\mu} \cdot (\delta_{\mu\nu}\boldsymbol{\zeta}_\mu^{-1} + \boldsymbol{\Omega}_{\mu\nu}) \cdot \frac{\partial g_l}{\partial \boldsymbol{r}_\nu} = \sum_{l=1}^{d'} \bar{\mathsf{g}}_{kl}\, \overline{\mathsf{G}}_{lj} = \delta_{jk}.$

$\displaystyle\sum_\mu \frac{\partial g_j}{\partial \boldsymbol{r}_\mu} \cdot \frac{\partial \boldsymbol{R}_\mu}{\partial Q_k} = \sum_\mu \frac{\partial g_j}{\partial Q_k} = 0$ because $g_j = 0$ in the manifold of constrained configurations.

$$\sum_\mu \frac{\partial Q_j}{\partial \boldsymbol{R}_\mu} \cdot \frac{\partial \boldsymbol{R}_\mu}{\partial g_k} = \sum_{\mu\nu\nu'} \sum_{l=1}^{d} \sum_{n=1}^{d'} \tilde{\mathsf{G}}_{jl}\, \bar{\mathsf{g}}_{kn} \frac{\partial \boldsymbol{R}_\nu}{\partial Q_l} \cdot \tilde{\boldsymbol{\zeta}}_{\nu\mu} \cdot (\delta_{\mu\nu'}\boldsymbol{\zeta}_\mu^{-1} + \boldsymbol{\Omega}_{\mu\nu'}) \cdot \frac{\partial g_n}{\partial \boldsymbol{r}_{\nu'}}$$

$$= \sum_{\nu\nu'} \sum_{l=1}^{d} \sum_{n=1}^{d'} \tilde{\mathsf{G}}_{jl}\, \bar{\mathsf{g}}_{kn} \frac{\partial \boldsymbol{R}_\nu}{\partial Q_l} \cdot (\delta_{\nu\nu'}\boldsymbol{\delta} - \boldsymbol{\lambda}_\nu^T) \cdot \frac{\partial g_n}{\partial \boldsymbol{r}_{\nu'}} = 0$$

because of the already proved first part of (5.52) and (5.46).

$\displaystyle\sum_\mu \boldsymbol{\lambda}_\mu \cdot \frac{\partial \boldsymbol{R}_\mu}{\partial g_k} = \boldsymbol{0}$ follows after inserting the definition (5.50) and using (5.15) and (5.46).

5.12 Let $Q_1, Q_2, \ldots, Q_d$ be an arbitrary set of generalized coordinates, and let their values at the given point be $Q_1^0, Q_2^0, \ldots, Q_d^0$. We first select a $d \times d$-matrix with elements b_{jk} such that $\mathsf{g}_{jk} = \sum_{l=1}^{d} \mathsf{b}_{lj}\mathsf{b}_{lk}$, and the elements of the inverse matrix are denoted by B_{jk}. We furthermore introduce the coefficients

$c_{jkl} = \sum_\mu \sum_{n=1}^{d} M_\mu \frac{\partial^2 \boldsymbol{R}_\mu}{\partial Q_k \partial Q_l} \cdot \frac{\partial \boldsymbol{R}_\mu}{\partial Q_n} \mathsf{B}_{nj}.$

All these coefficients are evaluated at the given point with coordinates $Q_1^0, Q_2^0, \ldots, Q_d^0$. We then introduce a new set of generalized coordinates, $Q_1', Q_2', \ldots, Q_d'$, by

$Q_j' = \sum_{k=1}^{d} \mathsf{b}_{jk}(Q_k - Q_k^0) + (1/2) \sum_{k,l=1}^{d} c_{jkl}(Q_k - Q_k^0)(Q_l - Q_l^0).$

Since, at the given point, we have $\partial Q_j / \partial Q_k' = \mathsf{B}_{jk}$, we obtain

$\sum_\mu M_\mu \frac{\partial \boldsymbol{R}_\mu}{\partial Q_j'} \cdot \frac{\partial \boldsymbol{R}_\mu}{\partial Q_k'} = \sum_{l,n=1}^{d} \mathsf{B}_{lj} \mathsf{g}_{ln} \mathsf{B}_{nk} = \delta_{jk}.$

By evaluating $\partial^2 \boldsymbol{R}_\mu / (\partial Q_j \partial Q_k)$ at the given point we obtain

$\sum_{l,n=1}^{d} \mathsf{b}_{lj} \mathsf{b}_{nk} \frac{\partial^2 \boldsymbol{R}_\mu}{\partial Q_l' \partial Q_n'} = \frac{\partial^2 \boldsymbol{R}_\mu}{\partial Q_j \partial Q_k} - \sum_{l=1}^{d} c_{ljk} \frac{\partial \boldsymbol{R}_\mu}{\partial Q_l'}.$

Equation (5.57) follows after multiplying this identity with $M_\mu \partial \boldsymbol{R}_\mu / \partial Q_{n'}'$ and summing over μ.

5.13 The d conditions (5.57) for $l = 1, 2, \ldots, d$ imply that, for any j and k, $\partial^2 \boldsymbol{R}_\mu / (\partial Q_j \partial Q_k)$ is from a $3N - d = d' + 3$ dimensional linear subspace of the space of $(3N)$-dimensional column vectors. Therefore, $\partial^2 \boldsymbol{R}_\mu / (\partial Q_j \partial Q_k)$ can be represented as a linear combination of the d' $(3N)$-dimensional column vectors $(1/M_\mu)\, \partial g_l / \partial \boldsymbol{r}_\mu$ and of the three $(3N)$-dimensional vectors obtained by setting all the components for different μ equal to three independent three-dimensional vectors. Due to the condition $\sum_\mu M_\mu \partial^2 \boldsymbol{R}_\mu / (\partial Q_j \partial Q_k) = 0$, the representation must be of the form

$$\frac{\partial^2 \boldsymbol{R}_\mu}{\partial Q_j \partial Q_k} = \sum_{l=1}^{d'} c_l^{jk} \frac{1}{M_\mu} \frac{\partial g_l}{\partial \boldsymbol{r}_\mu}.$$

The coefficients c_l^{jk} follow by inserting this result into (5.58).

5.14 If we introduce the expression (5.54) for the projection operators into the given compact expression, five terms result. Two terms can be shown to vanish by combining the last result of Exercise 5.8 and the identity $\sum_\mu \mathbf{P}_{\mu\nu}^T \cdot \frac{\partial g_k}{\partial \boldsymbol{r}_\mu} = \boldsymbol{0}$. The remaining three terms reproduce the last three lines of (5.61).

5.15 By evaluating (5.62) for the situation of this exercise we obtain

$$F_{\rm rel}^{(\rm m)} = 2k_{\rm B}T \left(\frac{\partial g}{\partial \boldsymbol{Q}} \cdot \frac{\partial g}{\partial \boldsymbol{Q}} \right)^{-3/2} \times \left[\frac{\partial g}{\partial \boldsymbol{Q}} \cdot \frac{\partial^2 g}{\partial \boldsymbol{Q} \partial \boldsymbol{Q}} \cdot \left(\frac{\partial g}{\partial \boldsymbol{Q}} \cdot \frac{\partial g}{\partial \boldsymbol{Q}} \boldsymbol{\delta} - \frac{\partial g}{\partial \boldsymbol{Q}} \frac{\partial g}{\partial \boldsymbol{Q}} \right) \cdot \frac{\partial^2 g}{\partial \boldsymbol{Q} \partial \boldsymbol{Q}} \cdot \frac{\partial g}{\partial \boldsymbol{Q}} \right]^{1/2},$$

where $\boldsymbol{Q} := \boldsymbol{r}_2 - \boldsymbol{r}_1$. For $g(\boldsymbol{Q}) = \boldsymbol{Q}^2 - L^2$ we find $F_{\rm rel}^{(\rm m)} = 0$.

5.16 There is only one constraint, $g_1(\boldsymbol{r}_1, \boldsymbol{r}_2) = (\boldsymbol{r}_2 - \boldsymbol{r}_1)^2 - L^2 = 0$.

$\frac{\partial g_1}{\partial \boldsymbol{r}_\mu} = 2(-1)^\mu (\boldsymbol{r}_2 - \boldsymbol{r}_1), \qquad \frac{\partial^2 g_1}{\partial \boldsymbol{r}_\mu \partial \boldsymbol{r}_\nu} = 2(-1)^{\mu+\nu} \boldsymbol{\delta}.$

$\overline{\mathsf{G}}_{11} = \frac{8}{\zeta} (\boldsymbol{r}_2 - \boldsymbol{r}_1)^2, \qquad \hat{\mathsf{G}} = \hat{\mathsf{G}}_{11} = \frac{8}{M} (\boldsymbol{r}_2 - \boldsymbol{r}_1)^2.$

$$\frac{\partial \boldsymbol{R}_\mu}{\partial g_1} = \frac{1}{4}(-1)^\mu \frac{\boldsymbol{r}_2 - \boldsymbol{r}_1}{(\boldsymbol{r}_2 - \boldsymbol{r}_1)^2}, \qquad \boldsymbol{F}_\mu^{(\mathrm{m})} = 4\, k_\mathrm{B} T \frac{\partial \boldsymbol{R}_\mu}{\partial g_1}.$$

$$\mathbf{P}_{\mu\nu} = \delta_{\mu\nu}\, \boldsymbol{\delta} - \tfrac{1}{2}(-1)^{\mu+\nu}\boldsymbol{U}\boldsymbol{U}, \qquad \mathbf{P}_{2\nu} - \mathbf{P}_{1\nu} = (-1)^\nu\,(\boldsymbol{\delta} - \boldsymbol{U}\boldsymbol{U}).$$

$$\begin{aligned}
\mathrm{d}\boldsymbol{U} &= \frac{1}{L}\sum_\nu(\mathbf{P}_{2\nu} - \mathbf{P}_{1\nu})\cdot\left[(\boldsymbol{\kappa}\cdot\boldsymbol{r}_\nu + \frac{1}{\zeta}\boldsymbol{F}_\nu^{(\mathrm{e})} + \frac{1}{\zeta}\boldsymbol{F}_\nu^{(\mathrm{m})})\,\mathrm{d}t + \sqrt{\frac{2k_\mathrm{B}T}{\zeta}}\,\mathrm{d}\boldsymbol{W}_\nu\right] \\
&\quad -2\frac{k_\mathrm{B}T}{\zeta L}\sum_{\nu\nu'}\left(\mathbf{P}_{\nu\nu'} : \frac{\partial^2 g_1}{\partial \boldsymbol{r}_{\nu'}\partial \boldsymbol{r}_\nu}\right)\frac{\partial \boldsymbol{R}_2}{\partial g_1}\,\mathrm{d}t \\
&\quad -2c\frac{k_\mathrm{B}T}{\zeta L}\sum_{\nu\nu'}(\mathbf{P}_{2\nu} - \mathbf{P}_{1\nu})\cdot\left(\frac{\partial g_1}{\partial \boldsymbol{r}_{\nu'}}\cdot\frac{\partial}{\partial \boldsymbol{r}_{\nu'}}\right)\frac{\partial \boldsymbol{R}_\nu}{\partial g_1}\,\mathrm{d}t \\
&= (\boldsymbol{\delta} - \boldsymbol{U}\boldsymbol{U})\cdot\left\{\left[\boldsymbol{\kappa}\cdot\boldsymbol{U} + \frac{1}{\zeta L}\left(\boldsymbol{F}_2^{(\mathrm{e})} - \boldsymbol{F}_1^{(\mathrm{e})}\right)\right]\mathrm{d}t + \frac{1}{\sqrt{3\lambda}}\,\mathrm{d}\boldsymbol{W}\right\} - \frac{1}{3\lambda}\boldsymbol{U}\,\mathrm{d}t,
\end{aligned}$$

where $\boldsymbol{W}(t) = [\boldsymbol{W}_2(t) - \boldsymbol{W}_1(t)]/\sqrt{2}$.

5.17 On the manifold of constrained configurations we find

$$\frac{\partial g}{\partial \boldsymbol{Q}} = \tilde{g}\,\mathbf{G}\cdot\boldsymbol{Q}, \qquad \frac{\partial^2 g}{\partial \boldsymbol{Q}\partial \boldsymbol{Q}} = \mathbf{G}\cdot\boldsymbol{Q}\frac{\partial g}{\partial \boldsymbol{Q}} + \frac{\partial g}{\partial \boldsymbol{Q}}\boldsymbol{Q}\cdot\mathbf{G} + \tilde{g}\,\mathbf{G},$$

where $\mathbf{G}$ is a diagonal matrix with diagonal elements $2/a^2$, $2/b^2$, and 0.

For $\tilde{g} = 1$, we obtain by means of the result of Exercise 5.15,

$$\begin{aligned}
F_\mathrm{rel}^{(\mathrm{m})} &= 2k_\mathrm{B}T(\boldsymbol{Q}\cdot\mathbf{G}^2\cdot\boldsymbol{Q})^{-3/2}\left[(\boldsymbol{Q}\cdot\mathbf{G}^4\cdot\boldsymbol{Q})(\boldsymbol{Q}\cdot\mathbf{G}^2\cdot\boldsymbol{Q}) - (\boldsymbol{Q}\cdot\mathbf{G}^3\cdot\boldsymbol{Q})^2\right]^{1/2} \\
&= 2k_\mathrm{B}T\left[(Q_1^2/a^4) + (Q_2^2/b^4)\right]^{-3/2}|(a^2 - b^2)Q_1Q_2|/(a^4b^4).
\end{aligned}$$

For $\tilde{g} = \left[(Q_1^2/a^4) + (Q_2^2/b^4)\right]^{-1/2} = 2(\boldsymbol{Q}\cdot\mathbf{G}^2\cdot\boldsymbol{Q})^{-1/2}$, we find that $\dfrac{\partial^2 g}{\partial \boldsymbol{Q}\partial \boldsymbol{Q}}\cdot\dfrac{\partial g}{\partial \boldsymbol{Q}}$ is parallel to $\mathbf{G}\cdot\boldsymbol{Q}$ and hence to $\dfrac{\partial g}{\partial \boldsymbol{Q}}$. The result of Exercise 5.15 then gives $F_\mathrm{rel}^{(\mathrm{m})} = 0$, which is certainly different from the corresponding result for $\tilde{g} = 1$.

5.19 For Hookean dumbbells we have $\mathbf{K} = \frac{1}{2}\zeta\,\langle\boldsymbol{Q}\boldsymbol{Q}\rangle = \frac{1}{2}\zeta\left(\frac{k_\mathrm{B}T}{H}\boldsymbol{\delta} - \frac{1}{n_\mathrm{p}H}\boldsymbol{\tau}^p\right)$.
From (5.72), we hence obtain
$\mathrm{d}\boldsymbol{\tau}^p/\mathrm{d}t - \boldsymbol{\kappa}\cdot\boldsymbol{\tau}^p - \boldsymbol{\tau}^p\cdot\boldsymbol{\kappa}^T + \boldsymbol{\tau}^p/\lambda_H = -n_\mathrm{p}\,k_\mathrm{B}T\,(\boldsymbol{\kappa} + \boldsymbol{\kappa}^T)$.
This constitutive equation coincides with (4.20), as can be seen after taking the time derivative of the latter equation and using the property
$\mathrm{d}\mathbf{C}^{-1}(t,t')/\mathrm{d}t = \boldsymbol{\kappa}(t)\cdot\mathbf{C}^{-1}(t,t') + \mathbf{C}^{-1}(t,t')\cdot\boldsymbol{\kappa}^T(t)$.

5.20 From the equation of motion derived in Exercise 5.9 we obtain:
$\boldsymbol{A}_\nu - \boldsymbol{\kappa}\cdot\boldsymbol{R}_\nu = (-1)^\nu\,\frac{1}{2\zeta}\,(\boldsymbol{\delta} - \boldsymbol{U}\boldsymbol{U})\cdot\left(\boldsymbol{F}_2^{(\mathrm{e})} - \boldsymbol{F}_1^{(\mathrm{e})}\right) - \boldsymbol{U}\boldsymbol{U}\cdot\boldsymbol{\kappa}\cdot\boldsymbol{R}_\nu - \frac{1}{3\lambda}\,\boldsymbol{R}_\nu$.
We then obtain the desired result from (5.70).

5.21 By using the result $\tilde{\boldsymbol{\zeta}}_{\mu\nu} = (-1)^{\mu+\nu}\,\frac{\zeta}{2}\,\boldsymbol{\delta}$ of Exercise 5.4 we obtain for the tensor $\mathbf{K}$ introduced in Example 5.18:
$\mathbf{K} = \frac{\zeta}{2}\,\langle(\boldsymbol{R}_2 - \boldsymbol{R}_1)(\boldsymbol{R}_2 - \boldsymbol{R}_1)\rangle = \frac{\zeta}{2}\,L^2\,\langle\boldsymbol{U}\boldsymbol{U}\rangle$.
From the definition of λ in Exercise 5.4, we have $\zeta L^2 = 12k_\mathrm{B}T\lambda$, so that (5.72) gives the desired expression for the polymer contribution to the stress tensor.

5.22 The following program RIGID2 is based on the simulation of a single trajectory (cf. Exercise 4.28). A single time step is performed by the subroutine SECRES. The output of RIGID2 consists of five columns: time-step width, polymer contribution to the viscosity, statistical error bar for the polymer contribution to the viscosity, first normal stress coefficient, statistical error bar for the first normal stress coefficient. For each time step, we first give the viscometric functions obtained from the Giesekus expression for the stress tensor (see Exercise 5.21) and then those from the expression (5.74). The appropriate units for the viscosity and the first normal stress coefficient are $n_p\, k_B T\, \lambda$ and $n_p\, k_B T\, \lambda^2$, respectively. The total time required for this example on a workstation is roughly two days.

```
      PROGRAM RIGID2
      REAL*8 AETA,BETA,VETA,APSI,BPSI,VPSI
      REAL*8 AETAP,BETAP,VETAP,APSIP,BPSIP,VPSIP
      REAL DTARR(5)
      INTEGER NTIARR(5)
      COMMON /STEPBL/ SR,PRE1,PRE2,PRE3,PRE4,PRE5,PRE6,PRE7,PRE8,PRE9
C     Array: different time steps
      DATA DTARR/0.5,0.4,0.3,0.2,0.1/
C     Array: different numbers of time steps per block
      DATA NTIARR/200000,250000,333333,500000,1000000/
C     Number of blocks
      NBLOCK=2500
C     Shear rate
      SR=1.
C     Initial seed for random number generator
      ISEED= 914081194
      CALL RANILS(ISEED)
C     Deterministic initial conditions
      U1=0.
      U2=1.
      U3=0.
C
C     Loop for different time-step widths
      DO 1000 IDT=1,5
C        Auxiliary parameters
         DELTAT=DTARR(IDT)
         PRE1=SQRT(DELTAT/3.)
         PRE2=SR*DELTAT
         PRE3=DELTAT/6.
         PRE4=.5*DELTAT*SQRT(DELTAT/3.)
         PRE5=2./3.
         PRE6=4./3.
         PRE7=DELTAT*DELTAT
         PRE8=1./18.
         PRE9=7./6.
C        Relaxation of initial conditions
```

```
      NTIME=10./DELTAT
      DO 50 ITIME=1,NTIME
         CALL SECRES(U1,U2,U3)
50    CONTINUE
C
      AETA=0.
      VETA=0.
      APSI=0.
      VPSI=0.
      AETAP=0.
      VETAP=0.
      APSIP=0.
      VPSIP=0.
      NTIME=NTIARR(IDT)
C     Loop for different blocks
      DO 100 IBLOCK=1,NBLOCK
         BETA=0.
         BPSI=0.
         BETAP=0.
         BPSIP=0.
C        Time integration: explicit second-order scheme
C                          followed by rescaling
         DO 10 ITIME=1,NTIME
            CALL SECRES(U1,U2,U3)
C           "Measurement"
            BETA=BETA+U2*U2
            BPSI=BPSI+U1*U2
            AUX=3./SR+6.*U1*U2
            BETAP=BETAP+AUX*U1*U2
            BPSIP=BPSIP+AUX*(U1*U1-U2*U2)
10       CONTINUE
C        Summation of the results for the blocks
         BETA=3.*BETA/NTIME
         AETA=AETA+BETA
         VETA=VETA+BETA*BETA
         BPSI=6.*BPSI/(NTIME*SR)
         APSI=APSI+BPSI
         VPSI=VPSI+BPSI*BPSI
         BETAP=BETAP/NTIME
         AETAP=AETAP+BETAP
         VETAP=VETAP+BETAP*BETAP
         BPSIP=BPSIP/(NTIME*SR)
         APSIP=APSIP+BPSIP
         VPSIP=VPSIP+BPSIP*BPSIP
100   CONTINUE
C
C     Averages, statistical errors
      AETA=AETA/NBLOCK
```

```
        VETA=VETA/NBLOCK
        VETA=SQRT((VETA-AETA*AETA)/(NBLOCK-1))
        APSI=APSI/NBLOCK
        VPSI=VPSI/NBLOCK
        VPSI=SQRT((VPSI-APSI*APSI)/(NBLOCK-1))
        AETAP=AETAP/NBLOCK
        VETAP=VETAP/NBLOCK
        VETAP=SQRT((VETAP-AETAP*AETAP)/(NBLOCK-1))
        APSIP=APSIP/NBLOCK
        VPSIP=VPSIP/NBLOCK
        VPSIP=SQRT((VPSIP-APSIP*APSIP)/(NBLOCK-1))
C       Output of results
        WRITE(6,1) DELTAT,AETA,VETA,APSI,VPSI
        WRITE(6,1) DELTAT,AETAP,VETAP,APSIP,VPSIP
1       FORMAT(F11.8,2(3X,F9.6,1X,F9.6))
1000  CONTINUE
      STOP
      END


      SUBROUTINE SECRES(U1,U2,U3)
      COMMON /STEPBL/ SR,PRE1,PRE2,PRE3,PRE4,PRE5,PRE6,PRE7,PRE8,PRE9
      PARAMETER (C1P=14.14855378, C2P=1.21569221)
C     Generation of the required random numbers
      W1=RANULS()-.5
      W2=RANULS()-.5
      W3=RANULS()-.5
      W1=W1*(C1P*W1*W1+C2P)
      W2=W2*(C1P*W2*W2+C2P)
      W3=W3*(C1P*W3*W3+C2P)
      V1=1.
      V2=1.
      V3=1.
      IF(RANULS().LT..5) V1=-V1
      IF(RANULS().LT..5) V2=-V2
      IF(RANULS().LT..5) V3=-V3
C
C     Explicit second-order algorithm
C     (terms grouped in blocks of increasing order in DELTAT)
      UW=U1*W1+U2*W2+U3*W3
      WW=W1*W1+W2*W2+W3*W3-2.*UW*UW
      U1U2=U1*U2
      SRU1U2=SR*U1U2
      U11=PRE1*(W1-UW*U1)
      U12=PRE1*(W2-UW*U2)
      U13=PRE1*(W3-UW*U3)
C
      U21=PRE2*(U2-U1U2*U1)-PRE3*(UW*W1+WW*U1)
      U22=-PRE2*U1U2*U2-PRE3*(UW*W2+WW*U2)
```

```
      U23=-PRE2*U1U2*U3-PRE3*(UW*W3+WW*U3)
      U21=U21+PRE3*( V1*U2+V2*U3)
      U22=U22+PRE3*(-V1*U1+V3*U3)
      U23=U23+PRE3*(-V2*U1-V3*U2)
C
      AUX1=-PRE4*(PRE5+SRU1U2)
      AUX2=PRE4*(PRE6*UW-SR*U1*W2-2.*SR*U2*W1+5.*SRU1U2*UW)
      U31=AUX1*W1+AUX2*U1+PRE4*SR*(W2-2.*UW*U2)
      U32=AUX1*W2+AUX2*U2
      U33=AUX1*W3+AUX2*U3
C
      AUX=PRE7*(PRE8+PRE9*SRU1U2+1.5*SRU1U2*SRU1U2-.5*SR*SR*U2*U2)
      U41=AUX*U1-PRE7*(.5+SRU1U2)*SR*U2
      U42=AUX*U2-PRE7*SR*U1/6.
      U43=AUX*U3
      U1=U1+U11+U21+U31+U41
      U2=U2+U12+U22+U32+U42
      U3=U3+U13+U23+U33+U43
C
C     Rescaling
      AUX=1./SQRT(U1*U1+U2*U2+U3*U3)
      U1=AUX*U1
      U2=AUX*U2
      U3=AUX*U3
      RETURN
      END
```

The extrapolated results for $\Delta t = 0$ are $\eta_p = 0.76748(7)$, $\Psi_1 = 0.64648(14)$ from the Giesekus expression for the stress tensor, and $\eta_p = 0.76745(8)$, $\Psi_1 = 0.64696(20)$ from the expression (5.74). For lower shear rates, the Giesekus expression gives much better results than (5.74) because one needs to divide by a smaller power of shear rate.

6.1 If we define $\gamma(t,t') = \int_{t'}^{t} \dot\gamma(t'')\,dt''$, then we have

$$\hat{\boldsymbol{u}}(\boldsymbol{u},t,t') = \begin{pmatrix} u_1 + \gamma(t,t')\,u_2 \\ u_2 \\ u_3 \end{pmatrix} \Bigg/ \left| \begin{pmatrix} u_1 + \gamma(t,t')\,u_2 \\ u_2 \\ u_3 \end{pmatrix} \right| .$$

This expression can be verified by differentiating with respect to t.

6.2 Let $(\boldsymbol{U}_t, S_t)$ and T_t^{LR} play the roles of X and Y in (2.60). Then,

$$\langle g(\boldsymbol{U}_t, S_t)\rangle = \int_{-\infty}^{t} dt' \int_0^1 ds \int d^3\hat{u}\; g(\hat{\boldsymbol{u}}, s)\, p(\boldsymbol{U}_t = \hat{\boldsymbol{u}}, S_t = s | T_t^{\mathrm{LR}} = t')\, p^{T_t^{\mathrm{LR}}}(t').$$

Note that, under the condition $T_t^{\mathrm{LR}} = t'$, $\boldsymbol{U}_t$ is the composite of $\hat{\boldsymbol{u}}(\boldsymbol{u},t,t')$ and $\boldsymbol{U}_{t'}$, where $\boldsymbol{U}_{t'}$ possesses a uniform distribution over the unit sphere. From the independent time evolution of $\boldsymbol{U}_t$ and S_t after the last reflection before t, and from the transformation formula (2.29), we obtain

$$\mathrm{d}^3\hat{u}\, p(\boldsymbol{U}_t = \hat{\boldsymbol{u}}, S_t = s | T_t^{\mathrm{LR}} = t') = \mathrm{d}^3 u\, \frac{\delta(|\boldsymbol{u}|-1)}{4\pi}\, p(S_t = s | T_t^{\mathrm{LR}} = t').$$

After using (3.95) and (3.96) we obtain (6.7).

6.3 Equation (6.6) implies

$$\langle \boldsymbol{U}_t \boldsymbol{U}_t \rangle = \int_{-\infty}^{t} \mathrm{d}t'\, \overline{\mu}(t-t') \int \mathrm{d}^3 u\, \frac{\delta(|\boldsymbol{u}|-1)}{4\pi}\, \hat{\boldsymbol{u}}\hat{\boldsymbol{u}}.$$

If the results of Example 3.37 are introduced into (6.7) then we obtain

$$\langle g(\boldsymbol{U}_t, S_t) \rangle = \frac{1}{\lambda} \sum_{\substack{n=1 \\ n \text{ odd}}}^{\infty} n \int_{-\infty}^{t} \mathrm{d}t'\, \mathrm{e}^{-n^2\pi^2(t-t')/\lambda} \int_0^1 \mathrm{d}s \int \mathrm{d}^3 u\, \delta(|\boldsymbol{u}|-1)\, \sin(n\pi s)\, g(\hat{\boldsymbol{u}}(\boldsymbol{u}, t, t'), s).$$

By using the formula $\int_0^1 \mathrm{d}s\, \sin(n\pi s)\, s(1-s) = 4/(n\pi)^3$ for odd n we find

$$\langle S_t(1-S_t)\, \boldsymbol{U}_t\boldsymbol{U}_t\boldsymbol{U}_t\boldsymbol{U}_t \rangle = \int_{-\infty}^{t} \mathrm{d}t' \left[\frac{16}{\pi^2\lambda} \sum_{\substack{n=1 \\ n \text{ odd}}}^{\infty} \frac{1}{n^2}\, \mathrm{e}^{-n^2\pi^2(t-t')/\lambda} \right] \int \mathrm{d}^3 u\, \frac{\delta(|\boldsymbol{u}|-1)}{4\pi}\, \hat{\boldsymbol{u}}\hat{\boldsymbol{u}}\hat{\boldsymbol{u}}\hat{\boldsymbol{u}}.$$

If these results are introduced into (6.5) we immediately obtain the stress tensor in the desired form.

6.4 The following program REPT is based on the simulation of a single trajectory (cf. Exercise 4.28). A single time step is performed by the subroutine REPTNA. The output of REPT consists of five columns: time step-width, Doi-Edwards viscosity η_Ω, statistical error bar for η_Ω, additional Curtiss-Bird contribution to the viscosity η_Υ, statistical error bar for η_Υ. The unit for the viscosity is $N n_{\mathrm{p}}\, k_{\mathrm{B}} T\, \lambda$. The normal stress coefficients can be obtained in an analogous manner. The total time required for this example on a workstation is some 45 minutes.

```
      PROGRAM REPT
      REAL*8 AETA,BETA,VETA,AETAP,BETAP,VETAP
      REAL DTARR(2)
      INTEGER NTIARR(2)
      COMMON /STEPBL/ DELTAT,SRDT,SQ2DT
C     Array: different time steps
      DATA DTARR/0.04,0.02/
C     Array: different numbers of time steps per block
      DATA NTIARR/100000,200000/
C     Number of blocks
      NBLOCK=1000
C     Shear rate
      SR=1.
C     Initial seed for random number generator
      ISEED= 415080295
      CALL RANILS(ISEED)
```

```
C     Equilibrium initial conditions
      CALL RANUVE(U1,U2,U3)
      S=RANULS()
C
C     Loop for different time-step widths
      DO 1000 IDT=1,2
C        Auxiliary parameters
         DELTAT=DTARR(IDT)
         SQ2DT=SQRT(2.*DELTAT)
         SRDT=SR*DELTAT
C        Relaxation of initial conditions
         NTIME=2./DELTAT
         DO 50 ITIME=1,NTIME
            CALL REPTNA(U1,U2,U3,S)
50       CONTINUE
C
         AETA=0.
         VETA=0.
         AETAP=0.
         VETAP=0.
         NTIME=NTIARR(IDT)
C        Loop for different blocks
         DO 100 IBLOCK=1,NBLOCK
            BETA=0.
            BETAP=0.
C           Time integration
            DO 10 ITIME=1,NTIME
               CALL REPTNA(U1,U2,U3,S)
C              "Measurement"
               U12=U1*U2
               BETA=BETA+U12
               BETAP=BETAP+U12*U12*S*(1.-S)
10          CONTINUE
C           Summation of the results for the blocks
            BETA=BETA/(NTIME*SR)
            AETA=AETA+BETA
            VETA=VETA+BETA*BETA
            BETAP=BETAP/NTIME
            AETAP=AETAP+BETAP
            VETAP=VETAP+BETAP*BETAP
100      CONTINUE
C
C        Averages, statistical errors
         AETA=AETA/NBLOCK
         VETA=VETA/NBLOCK
         VETA=SQRT((VETA-AETA*AETA)/(NBLOCK-1))
         AETAP=AETAP/NBLOCK
         VETAP=VETAP/NBLOCK
```

```
      VETAP=SQRT((VETAP-AETAP*AETAP)/(NBLOCK-1))
C     Output of results
      WRITE(6,1) DELTAT,AETA,VETA,AETAP,VETAP
1     FORMAT(F10.7,2(3X,F9.6,1X,F9.6))
1000  CONTINUE
      STOP
      END
```

The subroutine REPTNA for performing a single time step for the reptation process has the following, remarkably simple form:

```
      SUBROUTINE REPTNA(U1,U2,U3,S)
C     Time integration: naive algorithm
      COMMON /STEPBL/ DELTAT,SRDT,SQ2DT
C     Stochastic time evolution (pointer process)
      S=S+SQ2DT*RANGLS()
C     Boundary conditions
      IF(S.GT.1) THEN
C     Reflection at the right boundary
         S=2-S
         CALL RANUVE(U1,U2,U3)
      ELSE
         IF (S.LT.0.) THEN
C        Reflection at the left boundary
            S=-S
            CALL RANUVE(U1,U2,U3)
         ELSE
C        No reflection
C           Deterministic time evolution (link process)
            U1=U1+SRDT*U2
            AUX=1./SQRT(U1*U1+U2*U2+U3*U3)
            U1=AUX*U1
            U2=AUX*U2
            U3=AUX*U3
         ENDIF
      ENDIF
      RETURN
      END
```

6.5 Without loss of generality we look only at the boundary $b = 0$. Let $B_1, B_2 \subset [0, 1]$ be open intervals, $B_2 =]a, c[$ and $B_2' =]-c, -a[$, and let A be the event that at least one reflection happened in the time interval $[j\Delta t, (j+1)\Delta t]$. We then have from the properties of elementary conditional probabilities:

$$P(A|S_j \in B_1, S_{j+1} \in B_2) = P(S_{j+1} \in B_2|S_j \in B_1, A)\,\frac{P(S_j \in B_1, A)}{P(S_j \in B_1, S_{j+1} \in B_2)}.$$

The reflection principle implies that, if we do not reflect the trajectories of S

at 0, then $P(S_{j+1} \in B_2|S_j \in B_1, A) = P(S_{j+1} \in B_2'|S_j \in B_1, A)$. By inserting this and again using the properties of elementary conditional probabilities we obtain

$$P(A|S_j \in B_1, S_{j+1} \in B_2) = \frac{P(S_{j+1} \in B_2'|S_j \in B_1)}{P(S_{j+1} \in B_2|S_j \in B_1)}.$$

By letting B_1, B_2 become small intervals around a_1, a_2, respectively, we obtain after introducing Gaussian probability densities

$$P_{\mathrm{u}} = \frac{p_{0\,2\Delta t/\lambda}(-a_2 - a_1)}{p_{0\,2\Delta t/\lambda}(a_2 - a_1)} = \exp\left\{-\frac{\lambda}{\Delta t} a_1\, a_2\right\}.$$

6.6 In the program REPT listed in the solution to Exercise 6.4, only the subroutine REPTNA needs to be replaced by the following subroutine REPTIM:

```
      SUBROUTINE REPTIM(U1,U2,U3,S)
C     Time integration: improved algorithm (unobserved reflections)
      COMMON /STEPBL/ DELTAT,SRDT,SQ2DT
C     Stochastic time evolution (pointer process)
      SOLD=S
      S=S+SQ2DT*RANGLS()
C     Boundary conditions
      IF(S.GT.1) THEN
C     Reflection at the right boundary
         S=2-S
         CALL RANUVE(U1,U2,U3)
      ELSE
         IF (S.LT.0.) THEN
C        Reflection at the left boundary
            S=-S
            CALL RANUVE(U1,U2,U3)
         ELSE
C        No observed reflection
            BOUND=0.
            IF(S+SOLD.GT.1.) BOUND=1.
            CRIT=EXP((S-BOUND)*(BOUND-SOLD)/DELTAT)
            IF(RANULS().LE.CRIT) THEN
C           Unobserved reflection
               CALL RANUVE(U1,U2,U3)
            ELSE
C           No observed or unobserved reflection
C              Deterministic time evolution (link process)
               U1=U1+SRDT*U2
               AUX=1./SQRT(U1*U1+U2*U2+U3*U3)
               U1=AUX*U1
               U2=AUX*U2
               U3=AUX*U3
            ENDIF
         ENDIF
```

```
ENDIF
RETURN
END
```

6.7 We first assume $\boldsymbol{\kappa} = \mathbf{0}$. Consider a segment of length $\Delta s\, L_C$ with velocity v_C along the tube. Since the frictional force has to be balanced by the tension in the rope, we have $\zeta \Delta s\, v_C = \Delta s\, \partial\sigma_F/\partial s$, where ζ is the friction coefficient for the full chain in the tube. By taking the difference of this expression for two segments separated along the contour by l_C, where l_C is so small that the orientation of the rope can be assumed to be constant, we obtain $\zeta\, \Delta l_C/\Delta t = (l_C/L_C)\, \partial^2\sigma_F/\partial s^2$, so that the reason for the occurrence of the rate of change of the relative stretching of the rope becomes clear. The flow field can be taken into account by introducing the difference between the relative stretching of rope and tube, $(1/\kappa)\partial\sigma_F/\partial t - g = [1/(\zeta L_C)]\, \partial^2\sigma_F/\partial s^2$. With the diffusion coefficient $k_B T/\zeta = L_C^2/\lambda$, we obtain the desired equation.

6.9 According to (1.16) we have $\eta(0) = \int_0^\infty G(t)\, dt$. With the convention $\int_0^\infty \delta(t)\, dt = 1/2$, we immediately obtain the terms proportional to ϵ and $\bar{\bar{\epsilon}}$. The remaining integral is:

$$\int_0^\infty \overline{\nu}(t) e^{-\bar{\epsilon}^2\pi^2 t/\lambda}\, dt = \frac{16}{\pi^4} \sum_{\substack{n=1 \\ n \text{ odd}}}^{\infty} \frac{1}{n^2(n^2+\bar{\epsilon}^2)}.$$

The sum can be performed by applying 1.421.2 of [2] in order to evaluate

$$\lim_{x\to 0} \frac{\tanh(\pi x/2)}{\pi x/2} - \frac{\tanh(\pi\bar{\epsilon}/2)}{\pi\bar{\epsilon}/2}.$$

References

1. Bauer H (1981) *Probability Theory and Elements of Measure Theory*, 2nd Edn. Academic, London New York Toronto Sydney San Francisco
2. Gradshteyn IS, Ryzhik IM (1980) *Table of Integrals, Series, and Products*, corrected and enlarged edition. Academic, San Diego New York Berkeley Boston London Sydney Tokyo Toronto
3. Bird RB, Curtiss CF, Armstrong RC, Hassager O (1987) *Dynamics of Polymeric Liquids, Vol 2, Kinetic Theory*, 2nd Edn. Wiley-Interscience, New York Chichester Brisbane Toronto Singapore
4. Press WH, Teukolsky SA, Vetterling WT, Flannery BP (1992) *Numerical Recipes in FORTRAN. The Art of Scientific Computing*, 2nd Edn. Cambridge University Press, Cambridge
5. Zylka W, Öttinger HC (1991) Macromolecules 24: 484
6. Sammler RL, Schrag JL (1988) Macromolecules 21: 1132
7. Kloeden PE, Scott WD (1993) Maple Technical Newsletter Issue 10: 60

Author Index

Subject Index

Springer-Verlag and the Environment

We at Springer-Verlag firmly believe that an international science publisher has a special obligation to the environment, and our corporate policies consistently reflect this conviction.

We also expect our business partners – paper mills, printers, packaging manufacturers, etc. – to commit themselves to using environmentally friendly materials and production processes.

The paper in this book is made from low- or no-chlorine pulp and is acid free, in conformance with international standards for paper permanency.

GPSR Compliance

The European Union's (EU) General Product Safety Regulation (GPSR) is a set of rules that requires consumer products to be safe and our obligations to ensure this.

If you have any concerns about our products, you can contact us on ProductSafety@springernature.com

In case Publisher is established outside the EU, the EU authorized representative is:

Springer Nature Customer Service Center GmbH
Europaplatz 3
69115 Heidelberg, Germany

Batch number: 10682323

Printed by Printforce, the Netherlands